T0203037

MATHEMATICS TEXTBOOKS FOR SCIENCE AND ENGINEERING

VOLUME 1

SERIES EDITOR: **C.K. CHUI**

Mathematics Textbooks for Science and Engineering

Series Editor:

C.K. Chui

Stanford University, USA

(ISSN: 2213-0853)

Aims and scope of the series

Textbooks in the series 'Mathematics Textbooks for Science and Engineering' will be aimed at the broad mathematics, science and engineering undergraduate and graduate levels, covering all areas of applied and applicable mathematics, interpreted in the broadest sense.

For more information on this series and our other book series, please visit our website at:

www.atlantis-press.com/publications/books

ATLANTIS
PRESS

AMSTERDAM – PARIS – BEIJING

© **ATLANTIS PRESS**

Mathematics of Approximation

Johan de Villiers

Department of Mathematical Sciences
Stellenbosch University
South Africa

ATLANTIS
PRESS

AMSTERDAM – PARIS – BEIJING

Atlantis Press

8, square des Bouleaux
75019 Paris, France

For information on all Atlantis Press publications, visit our website at: *www.atlantis-press.com*

Copyright

This book, or any parts thereof, may not be reproduced for commercial purposes in any form or by any means, electronic or mechanical, including photocopying, recording or any information storage and retrieval system known or to be invented, without prior permission from the Publisher.

ISBNs
Print: 978-94-6239-046-1
E-Book: 978-94-91216-50-3
ISSN: 2213-0853

© 2012 ATLANTIS PRESS
Softcover re-print of the Hardcover 1st edition 2013

To my mother, and
in remembrance of my father

MTSE : Editorial for Volume I

Recent years have witnessed an extraordinarily rapid advance in the direction of information technology within both the scientific and engineering disciplines. In addition, the current profound technological advances of data acquisition devices and transmission systems contribute enormously to the continuing exponential growth of data information that requires much better data processing tools. To meet such urgent demands, innovative mathematical theory, methods, and algorithms must be developed, with emphasis on such application areas as complex data organization, contaminated noise removal, corrupted data repair, lost data recovery, reduction of data volume, data dimensionality reduction, data compression, data understanding and visualization, as well as data security and encryption.

The revolution of the data information explosion as mentioned above demands early mathematical training and emphasis on data manipulation at the college level and beyond. This new Atlantis/Springer book series, "Mathematics Textbooks for Science and Engineering (MTSE)" is founded to meet the needs of such mathematics textbooks that can be used for both classroom teaching and self-study. For the benefit of students and readers from the interdisciplinary areas of mathematics, computer science, physical and biological sciences, and various engineering specialities, contributing authors are requested to keep in mind that the writings for the MTSE book series should be elementary and relatively easy to read, with sufficient examples and exercises. We welcome submission of such book manuscripts from all who agree with us on this point of view.

This first volume is intended to be an elementary textbook for "Mathematics of Approximation", with emphasis on constructive methods and derivation of error formulas. This book is elementary, self-contained, and friendly towards teacher and reader. It is a suitable textbook for teaching in a variety of courses both at the undergraduate and beginning graduate levels. The author, Professor Johan de Villiers, is congratulated for contributing this very nice textbook.

Foreword

Mathematics of approximation plays a key role in bridging the gap between abstract mathematical theory and numerical implementation. With the recent exponential increase of available data that are easily accessible and relevant to our daily lives, understanding of such data often requires sophisticated mathematical manipulation by taking advantage of the continuing rapid technological advancement of computational capability. However, without assurance of accuracy in mathematical manipulation, results from simply number crunching by the powerful computer might be meaningless. In addition, for those data governed by certain physical phenomena or biological models, approximate solutions of the associated complex systems in terms of commonly used basis functions must guarantee accurate representation within a given tolerance.

This book by a leading expert in the field is intended to meet the need of such mathematical contents for classroom teaching, particularly for the undergraduate college level. Approximation and interpolation by algebraic and trigonometric polynomials for global representation, and spline functions for local analysis, are discussed from first principles with examples and full sets of carefully prepared exercises for each chapter. Convergence results are derived to assure meaningful mathematical manipulation, and error estimates are developed for data representation, with a guarantee of accuracy within a given tolerance. On the other hand, in contrast with the vast literature on Approximation Theory and Computational Mathematics, Professor De Villiers has taken great care to avoid using powerful advanced mathematics from Real Analysis and Functional Analysis, in that only elementary theory and methods from Linear Algebra and basic Advanced Calculus are applied in the derivations and discussions throughout the entire presentation. As a result, the writing is elementary and self-contained.

Furthermore, by including all necessary computational details, the writing of the book is friendly for classroom teaching as well as for self-reading. For adoption as a textbook, a variety of courses are mentioned with teaching guides outlined in the Preface. The author has taught from the preliminary notes that constitute the contents of this most welcomed textbook from a span of some twenty-five years to undergraduate students.

Charles K. Chui

Menlo Park, California

Preface

The approximation of functions by algebraic polynomials, trigonometric polynomials, and splines, is not only an important topic of mathematical studies, but also provides powerful mathematical tools to such application areas as data representation, signal processing, non-parametric time-series analysis, computer-aided geometric design, numerical analysis, and solutions of differential equations. This book is an introduction to the mathematical analysis of such approximation, with a strong emphasis on explicit approximation formulations, corresponding error bounds and convergence results, as well as applications in quadrature. A mathematically rigorous approach is adopted throughout, and, apart from an assumed prerequisite knowledge of advanced calculus and linear algebra, the presentation of the material is self-contained.

The book is suitable for use as textbook for courses at both upper undergraduate and graduate level. Each of the ten chapters is concluded by a set of exercises, with a total of altogether 220 exercises, some of which are routine, whereas others are concerned with further development of the material.

The book evolved from lecture notes compiled by the author during approximately 25 years of teaching courses in Computational Mathematics and Approximation Theory at Department of Mathematical Science, Stellenbosch University. Several standard textbooks were consulted and used, some more extensively than others, during the preparation of the lecture notes and the resulting book, and include the following, as listed alphabetically according to author:

- N.I. Achieser, *Theory of Approximation* (translated by C.J. Hyman), Frederick Ungar Publishing Co., New York, 1956.
- H. Brass, *Quadraturverfahren*, Vandenhoek & Rupert, Göttingen, 1977.
- E.W. Cheney, *Introduction to Approximation Theory*, McGraw-Hill, New York, 1966.

- Ronald A. DeVore and George G. Lorentz, *Constructive Approximation*, Springer, Berlin, 1993.
- Walter Gautschi, *Numerical Analysis: An Introduction*, Birkhäuser, Boston, 1997.
- Eugene Isaacson and Herbert Bishop Keller, *Analysis of Numerical Methods*, John Wiley & Sons, Inc., New York, 1966.
- Gunther Nürnberger, *Approximation by Spline Functions*, Springer, Berlin, 1989.
- M.J.D. Powell, *Approximation Theory and Methods,* Cambridge University Press, Cambridge, 1981.
- T.J. Rivlin, *An Introduction to the Approximation of Functions*, Blaisdell Publishing Co., Waltham, Mass, 1969.
- L.L. Schumaker, *Spline Functions: Basic Theory*, John Wiley & Sons, Inc., New York, 1981.

It should be pointed out that, whereas in some of the above-listed books concepts from Functional Analysis, like metric spaces, normed linear spaces and inner product spaces, as well as operators on those spaces, are assumed as prerequisite knowledge, the approach followed in this book is to develop any such concepts from first principles.

The contents of the respective chapters of the book can be summarized as follows.

- **Chapter 1: Polynomial Interpolation Formulas**

  The specific approximation method of polynomial interpolation is introduced, for which an existence and uniqueness theorem is then established by means of a Vandermonde matrix. Next, both the Lagrange and Newton interpolation formulas, as based on, respectively, the Lagrange fundamental polynomials and divided differences, are derived. Finally, an existence and uniqueness result, as well as a recursive computational method, are developed for Hermite interpolation, where also function derivatives are interpolated.

- **Chapter 2: Error Analysis for Polynomial Interpolation**

  A formulation in terms of a divided difference is established for the polynomial interpolation error, and a corresponding error bound is derived. Chebyshev polynomials are introduced, and shown to possess real zeros which yield interpolation points which minimize the error bound for polynomial interpolation.

- **Chapter 3: Polynomial Uniform Convergence**

  The concept of uniform convergence of a sequence of polynomial approximations to a given continuous function on a bounded interval is defined, and shown to apply in selected polynomial interpolation examples. It is then proved rigorously that, in contrast, divergence to infinity occurs in the Runge example. Next, the Bernstein polynomials are introduced and their properties analyzed, by means of which it is then proved that the sequence of Bernstein polynomial approximations to any given continuous function on a bounded interval uniformly converges to that function, and thereby immediately yielding also the Weierstrass theorem, according to which a continuous function on a bounded interval can be uniformly approximated with arbitrary "closeness" by a polynomial. It is furthermore shown that a Bernstein polynomial approximation has remarkable shape-preservation properties, and, in addition, a convergence rate result is established for the case where the approximated function is continuously differentiable.

- **Chapter 4: Best Approximation**

In the general setting of normed linear spaces, it is proved that, if the approximation set is a finite-dimensional subspace, then the existence of a best approximation is guaranteed. In addition, a uniqueness result for best approximation is established for those cases where the norm is generated by an inner product. The examples of best uniform polynomial approximation and best weighted L^2 polynomial approximation are highlighted as applications of the theoretical results.

- **Chapter 5: Approximation Operators**

The notion of an approximation operator from a normed linear space to an approximation set is introduced, and examples from previous chapters are provided. The properties of linearity, exactness and boundedness with respect to approximation operators are discussed and analyzed, with particular attention devoted to operator norms, or Lebesgue constants. In addition, the Lebesgue inequality for bounding an approximation error with respect to the best approximation error is derived. Applications of the theory to particularly the polynomial interpolation operator are provided.

- **Chapter 6: Best Uniform Polynomial Approximation**

It is proved that the equi-oscillation property of a polynomial approximation error is a necessary and sufficient condition for best uniform polynomial approximation, by means of which the uniqueness of a best uniform polynomial approximation is then established. The resulting best uniform polynomial approximation operator is shown in particular to satisfy a convergence rate which increases with the number of continuous derivatives of the approximated function, and by means of which, together with the Lebesgue inequality from Chapter 5, an interpolation error bound is obtained, from which it is then immediately seen that uniform convergence is obtained for the Runge example of Chapter 3 if the uniformly distributed interpolation points are replaced by the Chebyshev points of Chapter 2. Finally, examples are provided of cases where explicit calculation of the best uniform polynomial approximation can be performed in a straightforward manner.

- **Chapter 7: Orthogonality**

In the general setting of inner product spaces, it is proved that best approximation with respect to the norm generated by an inner product is achieved if and only if the approximation error is orthogonal to the approximation subspace, and properties of the corresponding best approximation operator are established. For those cases where the approximation subspace is finite-dimensional, a construction procedure for the best approximation, as based on the inversion of the corresponding Gram matrix, is derived, and it is moreover shown that the availability of an orthogonal basis yields an explicitly formulated best approximation. The Gram-Schmidt procedure for the construction of an orthogonal basis from any given basis is then derived by means of best approximation principles. For polynomials, an efficient three-term recursion formula for orthogonal polynomials with respect to any weighted inner product is obtained, and specialized to the important cases of Legendre polynomials and Chebyshev polynomials.

- **Chapter 8: Interpolatory Quadrature**

After defining an interpolatory quadrature rule for the numerical approximation of a weighted integral as the (weighted) integral of a polynomial interpolant of the integrand, and introducing the concept of the polynomial exactness degree of a quadrature rule, it is shown that non-negative quadrature weights guarantee quadrature convergence for continuous integrands, with convergence rate increasing with the number of continuous derivatives of the integrand. Next, by choosing the underlying interpolation points as the zeros of a certain orthogonal polynomial, it is shown that optimal polynomial exactness is achieved, and thereby yielding the Gauss quadrature rules, for which the weights are then proved to be positive. The related Clenshaw-Curtis quadrature rule, as based on interpolation points obtained from extremal properties of Chebyshev polynomials, is shown to possess positive weights with explicit formulations. The rest of the chapter is devoted to Newton-Cotes quadrature, as based on uniformly distributed interpolation points. By employing results on polynomial interpolation from Chapters 1 and 2, explicit formulations in terms of Laplace coefficients of Newton-Cotes weights and error expressions are derived. In addition, composite Newton-Cotes quadrature rules, including the trapezoidal, midpoint and Simpson rules as special cases, are explicitly constructed and subjected to error and convergence analysis.

- **Chapter 9: Approximation of Periodic Functions**

  The trigonometric polynomials are introduced and analyzed, with the view to using them to approximate any given continuous periodic (with period $= 2\pi$) function. Next, as a Weierstrass-type theorem, it is proved that any continuous periodic function can be approximated with arbitrary "closeness" in the maximum norm by a trigonometric polynomial. The Fourier series operator is defined as the best L^2 approximation operator into the finite-dimensional space of trigonometric polynomials of a given degree, and explicitly calculated in terms of Fourier coefficients expressed as integrals. For those instances where these integrals are to be numerically approximated, the Euler-Maclaurin formula is derived, and shown to imply the remarkable efficiency of the trapezoidal rule when applied to the integral of a smooth periodic function over its full period, as in the Fourier coefficient case, and thereby yielding the discrete Fourier series operator. In the limit, the Fourier series operator yields the (infinite) Fourier series of a continuous periodic function, and for which L^2 convergence is immediately deduced. In order to investigate the uniform convergence of Fourier series, upper and lower bounds for the Lebesgue constant (with respect to the maximum norm) of the Fourier series operator are first derived, thereby showing that this Lebesgue constant grows to infinity at a logarithmic rate. After furthermore establishing two Jackson theorems on best approximation convergence rates, the Lebesgue inequality of Chapter 5 is then applied to prove the Dini-Lipschitz theorem, according to which Lipschitz continuity of a periodic function is a sufficient condition for the uniform convergence of its Fourier series, that is, the Fourier series converges pointwise to the function itself.

- **Chapter 10: Spline Approximation**

  Splines are introduced as piecewise polynomials of a given degree, and with breakpoints, or knots, at a finite number of specified points, together with the maximal smoothness requirement providing a proper extension of the corresponding polynomial space. Preliminary properties of splines are established, and it is shown that truncated powers provide a basis for spline spaces. Next, the compactly supported, and hence more efficient, B-spline basis is constructed and analyzed, with particular attention devoted to the case where the knots are placed at the integers, yielding the cardinal B-splines. The Schoenberg-Whitney theorem, which is an existence and uniqueness result for (non-local) spline interpolation, is then proved. In order to obtain a local spline approximation method, an explicit construction method is developed for spline quasi-interpolation, which combines the approximation properties of linearity, optimal polynomial exactness, and locality. In order to achieve, in addition, the property of interpolation, it is shown that such a local spline interpolation operator can be constructed explicitly by choosing the interpolation points as a specified subset of the spline knots. The Schoenberg operator is introduced, and shown to be a local spline approximation operator, which, for any fixed spline degree, and for the maximum spacing between spline knots tending to zero, possesses the uniform convergence property with respect to any continuous function on a bounded interval, and thereby establishing a Weierstrass-type theorem for splines. Similarly, sufficient conditions on the spline knot placement for the uniform convergence of the above-mentioned local spline interpolation operator are derived, as well as a corresponding convergence rate result by means of the Peano kernel theorem. Finally, the interpolatory quadrature rule obtained from the uniformly distributed knots case of this local spline interpolation operator is analyzed, and shown to yield a class of trapezoidal rules with endpoints corrections, which are precisely the classical Gregory rules of even order, after which results from Chapter 8 are employed to explicitly evaluate, in terms of Laplace coefficients, the weights and error expressions for these Gregory rules, and with particular attention devoted to the special case of the Lacroix rule.

Examples of courses that could be taught from this book are as follows:

- A one-semester mathematics course of "Mathematics of Approximation" can be taught from Chapters 1-8. If the course is oriented towards numerical methods and numerical analysis, then Chapters 9-10 can be adopted to replace Chapters 6-7.
- A one-year course of "Mathematics of Approximation" can be taught with first semester based on Chapters 1-7, and second semester on Chapters 8-10.
- A one-semester course "Polynomial and Spline Approximation" can be taught from Chapters 1-7 and Chapter 10 up to Section 10.6.
- A one-semester course of "Polynomial Approximation and Quadrature" can be taught from Chapters 1-8.
- A one-semester course of "Polynomial Approximation and Fourier series" can be taught from Chapters 1-7 and Chapter 9.

Material that could be regarded as optional in the courses listed above are Sections 6.5, 8.3 and 10.7; Section 9.3 from Theorem 9.3.3 onwards; as well as the proof of (3.1.8) in Example 3.1.3 of Section 3.1.

The author wishes to express his sincere gratitude to:

- Lauretta Adams, who expertly LATEX-ed the entire manuscript, and whose friendly and kind attitude throughout is much appreciated;
- Maryke van der Walt, who displayed excellent skills in proofreading the whole text and preparing the index, and whose many meaningful suggestions enhanced the presentation of the material;
- Charles Chui, whose academic inspiration and generous editorial leadership contributed substantially to this book project;
- Carl Rohwer, for productive research collaboration yielding the local spline interpolation operator of Section 10.5 onwards;
- Nick Trefethen, whose plenary lecture in March 2011 at the SANUM conference in Stellenbosch was the inspiration for the inclusion of Sections 6.5 and 8.3;
- Mike Neamtu, Dirk Laurie, Sizwe Mabizela, André Weideman, David Kubayi and Ben Herbst, for many stimulating discussions which contributed to improving the author's insight into the book's material;
- Department of Mathematical Sciences, Stellenbosch University, and in particular its Head, Ingrid Rewitzky, for providing the author with a friendly and conducive environment for the writing of this book;
- The students in the author's courses in Computational Mathematics and Approximation Theory since 1986 at Stellenbosch University, for their enthusiastic participation, and whose consistent feedback contributed significantly to the gradual improvement of the lecture notes on which this book is based;
- Keith Jones of Atlantis Press, for his encouragement and patience;
- My wife, Louwina, who sacrificed much as a result of the many hours I spent on the preparation of the manuscript, and without whose unconditional love and devotion this book would not have been possible.

Johan de Villiers
Stellenbosch, South Africa

Contents

Chapter 1

Polynomial Interpolation Formulas

When an (algebraic) polynomial P is used to approximate a certain function f, the polynomial P is said to interpolate the function f on a given finite sample set of distinct points in the domain of f, if P is obtained to satisfy the condition that P agrees with f on the sample set. The objective of this chapter is to establish a fundamental existence and uniqueness result for polynomial interpolation as well as to derive explicit formulations of the interpolation polynomial P.

1.1 Existence based on the Vandermonde matrix

For any non-negative integer k, we write π_k for the set of algebraic polynomials of degree $\leqslant k$, and with real coefficients. Then π_k is a finite-dimensional linear space with dimension

$$\dim(\pi_k) = k+1, \tag{1.1.1}$$

and where the standard basis of π_k is given by $\{1, x, \ldots, x^k\}$.

Let f denote a real-valued function with domain $D_f \subset \mathbb{R}$, with $\mathbb{R}$ denoting the set of real numbers, and, for any given non-negative integer n, suppose

$$\triangle_n := \{x_0, \ldots, x_n\} \tag{1.1.2}$$

is a sequence of $n+1$ distinct points in D_f. We investigate the existence and construction of a polynomial P satisfying the interpolatory conditions

$$P(x_j) = f(x_j), \quad j = 0, \ldots, n, \tag{1.1.3}$$

in which case we say that the polynomial P interpolates the function f at the interpolation points $\triangle_n$. Based on (1.1.1), together with the fact that there are precisely $n+1$ interpolation conditions in (1.1.3), we consider the polynomial construction

$$P(x) := \sum_{j=0}^{n} c_j x^j, \tag{1.1.4}$$

1

with $\{c_0, \ldots, c_n\}$ denoting a real coefficient sequence, that is, $P \in \pi_n$. By substituting (1.1.4) into (1.1.3), we deduce that there exists a polynomial $P \in \pi_n$ satisfying the interpolation conditions (1.1.3) if and only if there exists a (column vector) solution

$$\mathbf{c} = [c_0, c_1, \ldots, c_n]^T \in \mathbb{R}^{n+1} \tag{1.1.5}$$

of the $(n+1) \times (n+1)$ linear system

$$V_n \mathbf{c} = \mathbf{f}, \tag{1.1.6}$$

where

$$V_0 := 1 \quad ; \quad V_n := \begin{bmatrix} 1 & x_0 & x_0^2 & \cdots & x_0^n \\ 1 & x_1 & x_1^2 & \cdots & x_1^n \\ \vdots & \vdots & \vdots & & \vdots \\ 1 & x_n & x_n^2 & \cdots & x_n^n \end{bmatrix} \quad \text{if } n \geqslant 1, \tag{1.1.7}$$

a socalled $(n+1) \times (n+1)$ Vandermonde matrix, and where the column vector $\mathbf{f} \in \mathbb{R}^{n+1}$ is defined by

$$\mathbf{f} := [f(x_0), \ldots, f(x_n)]^T. \tag{1.1.8}$$

Here, as throughout the book, we adopt, for any $m \in \mathbb{N}$, the notation $\mathbb{R}^m$ to denote the m-dimensional Euclidean space, according to which $\mathbb{R}^1 = \mathbb{R}$.

In order to investigate the invertibility of the Vandermonde matrix V_n in (1.1.7), we proceed to prove the following determinant formula, where, as in the rest of this book, we use the convention $\prod_{j=\sigma}^{\tau} a_j := 1$ if $\tau < \sigma$.

Theorem 1.1.1. *For a non-negative integer n, and any sequence $\{x_0, \ldots, x_n\}$ of $n+1$ (not necessarily distinct) points in $\mathbb{R}$, the Vandermonde matrix V_n in (1.1.7) has determinant*

$$\det(V_n) = \prod_{j=0}^{n-1} \prod_{k=j+1}^{n} (x_k - x_j). \tag{1.1.9}$$

Proof. After first noting from (1.1.7) that (1.1.9) trivially holds for $n = 0$, we suppose next that $n \geqslant 1$, and consider the sequence of functions defined, for $x \in \mathbb{R}$, by

$$D_0(x) := 1; \tag{1.1.10}$$

$$D_r(x) := \begin{vmatrix} 1 & x_0 & x_0^2 & \cdots & x_0^r \\ 1 & x_1 & x_1^2 & \cdots & x_1^r \\ \vdots & \vdots & \vdots & & \vdots \\ 1 & x_{r-1} & x_{r-1}^2 & \cdots & x_{r-1}^r \\ 1 & x & x^2 & \cdots & x^r \end{vmatrix}, \quad r = 1, \ldots, n. \tag{1.1.11}$$

Let $r \in \{1,\ldots,n\}$ be fixed. Observe from (1.1.11) that

$$D_r(x_j) = 0, \quad j = 0,\ldots,r-1. \tag{1.1.12}$$

Moreover, from the co-factor expansion with respect to the last row of the determinant in (1.1.11), we deduce that

$$D_r(x) = D_{r-1}(x_{r-1})x^r + E(x), \tag{1.1.13}$$

where $E \in \pi_{r-1}$, after having used also (1.1.10) for the case $r = 1$.

Next, we use (1.1.12), together with the fact that, from (1.1.13), the polynomial $D_r(x)$ has degree $\leqslant r$, to deduce that

$$D_r(x) = K_r \prod_{j=0}^{r-1}(x - x_j),$$

for some constant K_r, and thus

$$D_r(x) = K_r x^r + F(x), \tag{1.1.14}$$

where $F \in \pi_{r-1}$. It follows from (1.1.13) and (1.1.14) that $K_r = D_{r-1}(x_{r-1})$ and $E = F$, yielding the formula

$$D_r(x) = D_{r-1}(x_{r-1}) \prod_{j=0}^{r-1}(x - x_j). \tag{1.1.15}$$

Now use (1.1.7) and (1.1.11), before repeatedly applying (1.1.15), and eventually (1.1.10), to deduce that

$$\det(V_n) = D_n(x_n) = D_{n-1}(x_{n-1}) \prod_{j=0}^{n-1}(x_n - x_j)$$

$$= \left[D_{n-2}(x_{n-2}) \prod_{j=0}^{n-2}(x_{n-1} - x_j) \right] \prod_{j=0}^{n-1}(x_n - x_j)$$

$$= \cdots$$

$$= [D_0(x_0)(x_1 - x_0)] \left[\prod_{j=0}^{1}(x_2 - x_j) \right] \cdots \left[\prod_{j=0}^{n-1}(x_n - x_j) \right]$$

$$= \prod_{k=0}^{n-1} \prod_{j=0}^{k}(x_{k+1} - x_j) = \prod_{j=0}^{n-1} \prod_{k=j}^{n-1}(x_{k+1} - x_j),$$

which is equivalent to the desired formula (1.1.9). ∎

According to a standard result in linear algebra, a (square) matrix is invertible if and only if its determinant is non-zero. Now observe from (1.1.9) in Theorem 1.1.1 that $\det(V_n) \neq 0$ if and only if the points $\{x_0,\ldots,x_n\}$ are distinct. Since also the interpolation conditions

(1.1.3) are satisfied by a polynomial P of the form (1.1.4) if and only if there exists a solution $\mathbf{c} \in \mathbb{R}^{n+1}$ of the $(n+1) \times (n+1)$ linear system (1.1.6), we therefore have the following fundamental existence and uniqueness result with respect to polynomial interpolation.

Theorem 1.1.2. *For any non-negative integer n, let $\{x_0,\ldots,x_n\}$ denote any sequence of $n+1$ distinct points in $\mathbb{R}$. Then the Vandermonde matrix V_n in (1.1.7) is invertible, and the polynomial*

$$P(x) = P_n^I(x) := \sum_{j=0}^{n} (V_n^{-1}\mathbf{f})_j x^j, \tag{1.1.16}$$

with the column vector $\mathbf{f} \in \mathbb{R}^{n+1}$ defined by (1.1.8), is the unique polynomial in π_n satisfying the interpolation conditions (1.1.3).

Observe from the uniqueness statement in Theorem 1.1.2 that P_n^I is the polynomial of least degree such that $P = P_n^I$ satisfies the interpolation conditions (1.1.3).

Example 1.1.1. According to Theorem 1.1.2, there exists a unique polynomial $P = P_3^I \in \pi_3$ satisfying the interpolatory conditions

$$P(-1) = 1; \quad P(0) = -1; \quad P(2) = 3; \quad P(3) = 2. \tag{1.1.17}$$

Since we have here $n = 3$, and

$$\{x_0,x_1,x_2,x_3\} = \{-1,0,2,3\}, \tag{1.1.18}$$

the corresponding Vandermonde matrix in (1.1.7) is given by

$$V_3 = \begin{bmatrix} 1 & -1 & 1 & -1 \\ 1 & 0 & 0 & 0 \\ 1 & 2 & 4 & 8 \\ 1 & 3 & 9 & 27 \end{bmatrix}, \tag{1.1.19}$$

the inverse matrix of which is calculated to be given by

$$V_3^{-1} = \frac{1}{12} \begin{bmatrix} 0 & 12 & 0 & 0 \\ -6 & 2 & 6 & -2 \\ 5 & -8 & 4 & -1 \\ -1 & 2 & -2 & 1 \end{bmatrix}. \tag{1.1.20}$$

Also, (1.1.8) is given here by

$$\mathbf{f} = [1,-1,3,2]^T,$$

and thus

$$V_3^{-1}\mathbf{f} = \frac{1}{12}[-12,6,23,-7]^T,$$

so that, by using the formula (1.1.16), we obtain

$$P_3^I(x) = \frac{1}{12}\left(-12 + 6x + 23x^2 - 7x^3\right). \tag{1.1.21}$$

∎

Although Theorem 1.1.2 is a useful existence and uniqueness result, the resulting polynomial interpolation formula (1.1.16) involves the computation of the inverse of the Vandermonde matrix V_n. We proceed in Section 1.2 to deduce a more efficient interpolation formula than (1.1.16), by expressing P_n^I in terms of a basis for π_n that is better suited to polynomial interpolation than the standard basis $\{1, x, \ldots, x^n\}$.

1.2 The Lagrange interpolation formula

As in Section 1.1, for any non-negative integer n, let $\{x_0, \ldots, x_n\}$ denote a sequence of $n + 1$ distinct points in $\mathbb{R}$. The polynomials

$$L_{0,0}(x) := 1; \quad L_{n,j}(x) := \prod_{\substack{j \neq k = 0}}^{n} \frac{x - x_k}{x_j - x_k}, \quad j = 0, \ldots, n, \quad (\text{if } n \geqslant 1), \tag{1.2.1}$$

are called the Lagrange fundamental polynomials with respect to the sequence $\{x_0, \ldots, x_n\}$, and satisfy the following properties, in which we adopt the Kronecker delta notation

$$\delta_j := \begin{cases} 1, & j = 0; \\ 0, & j \in \mathbb{Z} \setminus \{0\}, \end{cases} \tag{1.2.2}$$

with $\mathbb{Z}$ denoting the set of integers.

Theorem 1.2.1. *The Lagrange fundamental polynomials, as given by (1.2.1), satisfy:*

(a)

$$L_{n,j}(x_\ell) = \delta_{j-\ell}, \quad j, \ell = 0, \ldots, n; \tag{1.2.3}$$

(b) *the sequence* $\{L_{n,j} : j = 0, \ldots, n\}$ *is a basis for* π_n.

Proof. (a) The interpolatory property (1.2.3) is an immediate consequence of (1.2.1).

(b) Since (1.1.1) gives $\dim(\pi_n) = n + 1$, and since the sequence $\{L_{n,j} : j = 0, \ldots, n\}$ contains precisely $n + 1$ functions, with, from (1.2.1), $L_{n,j} \in \pi_n$ for $j = 0, \ldots, n$, it will suffice to prove, according to a standard result in linear algebra, that $\{L_{n,j} : j = 0, \ldots, n\}$ is a linearly independent set. Suppose therefore that $\{c_0, \ldots, c_n\}$ are real coefficients such that

$$\sum_{j=0}^{n} c_j L_{n,j}(x) = 0, \quad x \in \mathbb{R}. \tag{1.2.4}$$

For any $\ell \in \{0,\ldots,n\}$, it follows from (1.2.4), (1.2.3) and (1.2.2) that

$$0 = \sum_{j=0}^{n} c_j L_{n,j}(x_\ell) = \sum_{j=0}^{n} c_j \delta_{j-\ell} = c_\ell,$$

that is, $c_j = 0, j = 0,\ldots,n$, which proves the desired linear independence. ∎

The following alternative to the polynomial interpolation formula (1.1.16) then holds.

Theorem 1.2.2 (Lagrange interpolation formula). *The interpolation polynomial P_n^I of Theorem* 1.1.2 *satisfies the formulation*

$$P_n^I(x) = \sum_{j=0}^{n} f(x_j) L_{n,j}(x), \tag{1.2.5}$$

where the Lagrange fundamental polynomials $\{L_{n,j} : j = 0,\ldots n\}$ are defined by (1.2.1).

Proof. Since $P_n^I \in \pi_n$, it follows from Theorem 1.2.1(b) that there exists a (unique) real coefficient sequence $\{\alpha_0,\ldots,\alpha_n\}$ such that

$$P_n^I(x) = \sum_{j=0}^{n} \alpha_j L_{n,j}(x), \quad x \in \mathbb{R}. \tag{1.2.6}$$

For any $\ell \in \{0,\ldots,n\}$, we may now choose $x = x_\ell$ in (1.2.6), and apply (1.1.3), as well as (1.2.3) in Theorem 1.2.1(a), to obtain

$$f(x_\ell) = P_n^I(x_\ell) = \sum_{j=0}^{n} \alpha_j L_{n,j}(x_\ell) = \sum_{j=0}^{n} \alpha_j \delta_{j-\ell} = \alpha_\ell,$$

which, together with (1.2.6), proves the formula (1.2.5). ∎

We proceed to prove the following polynomial identity.

Theorem 1.2.3. *The Lagrange fundamental polynomials $\{L_{n,j} : j = 0,\ldots,n\}$, as defined by* (1.2.1), *satisfy the identity*

$$\sum_{j=0}^{n} P(x_j) L_{n,j}(x) = P(x), \quad x \in \mathbb{R}, \quad P \in \pi_n, \tag{1.2.7}$$

with, in particular,

$$\sum_{j=0}^{n} L_{n,j}(x) = 1, \quad x \in \mathbb{R}. \tag{1.2.8}$$

Proof. Let $P \in \pi_n$. According to Theorem 1.2.2, the polynomial

$$P_n^I(x) := \sum_{j=0}^{n} P(x_j) L_{n,j}(x) \tag{1.2.9}$$

then satisfies the interpolatory conditions

$$P_n^I(x_j) = P(x_j), \quad j = 0,\ldots,n.$$

Since also P trivially interpolates itself at the points $\{x_0,\dots,x_n\}$, and since P and P_n^I both belong to π_n, we deduce from the uniqueness statement in Theorem 1.1.2 that $P = P_n^I$, which, together with (1.2.9), yields the desired identity (1.2.7).

The identity (1.2.8) follows immediately by choosing in (1.2.7) the polynomial

$$P(x) = 1, \quad x \in \mathbb{R},$$

which belongs to π_n for each non-negative integer n. ∎

Example 1.2.1. Consider the interpolation conditions (1.1.17) of Example 1.1.1. Then, by using (1.2.1) and (1.1.18), we calculate the Lagrange fundamental polynomials

$$L_{3,0}(x) = \frac{(x-0)(x-2)(x-3)}{(-1-0)(-1-2)(-1-3)} = -\frac{1}{12}x^3 + \frac{5}{12}x^2 - \frac{1}{2}x;$$

$$L_{3,1}(x) = \frac{(x-(-1))(x-2)(x-3)}{(0-(-1))(0-2)(0-3)} = \frac{1}{6}x^3 - \frac{2}{3}x^2 + \frac{1}{6}x + 1;$$

$$L_{3,2}(x) = \frac{(x-(-1))(x-0)(x-3)}{(2-(-1))(2-0)(2-3)} = -\frac{1}{6}x^3 + \frac{1}{3}x^2 + \frac{1}{2}x;$$

$$L_{3,3}(x) = \frac{(x-(-1))(x-0)(x-2)}{(3-(-1))(3-0)(3-2)} = \frac{1}{12}x^3 - \frac{1}{12}x^2 - \frac{1}{6}x,$$

from which we verify that the identity (1.2.8) is indeed satisfied. By using the Lagrange interpolation formula (1.2.5), together with (1.1.17), we obtain

$$P_3^I(x) = 1\left(-\frac{1}{12}x^3 + \frac{5}{12}x^2 - \frac{1}{2}x\right) + (-1)\left(\frac{1}{6}x^3 - \frac{2}{3}x^2 + \frac{1}{6}x + 1\right)$$

$$+ 3\left(-\frac{1}{6}x^3 + \frac{1}{3}x^2 + \frac{1}{2}x\right) + 2\left(\frac{1}{12}x^3 - \frac{1}{12}x^2 - \frac{1}{6}x\right)$$

$$= \frac{1}{12}\left(-7x^3 + 23x^2 + 6x - 12\right), \tag{1.2.10}$$

which agrees with (1.1.21) in Example 1.1.1. ∎

In the notation of Theorems 1.1.2 and 1.2.2, suppose one more interpolation point x_{n+1} is added such that $\{x_0,\dots,x_n,x_{n+1}\}$ is a sequence of $n+2$ distinct points in $\mathbb{R}$. Then, if the Lagrange interpolation formula (1.2.5) is used to calculate the corresponding interpolation polynomial P_{n+1}^I, the updated Lagrange fundamental polynomials $\{L_{n+1,j} : j = 0,\dots,n+1\}$ need to be calculated from scratch, without recourse to $\{L_{n,j} : j = 0,\dots,n\}$. We proceed, in Section 1.3, to establish yet another basis for π_n, which will allow the computation of P_{n+1}^I from P_n^I by the addition of a single term.

1.3 Divided differences and the Newton interpolation formula

Let the interpolation polynomial P_n^I be as in Theorem 1.1.2. Then the Lagrange interpolation formula (1.2.5), together with (1.2.1), yields

$$P_0^I(x) = \beta_0; \quad P_n^I(x) = \beta_n x^n + Q(x), \quad n \in \mathbb{N}, \tag{1.3.1}$$

with $\mathbb{N} := \{1, 2, \ldots\}$, where $Q \in \pi_{n-1}$, and with the leading coefficient of the polynomial P_n^I given by

$$\beta_0 := f(x_0); \qquad \beta_n := \sum_{j=0}^{n} \frac{f(x_j)}{\displaystyle\prod_{\substack{j \neq k=0}}^{n} (x_j - x_k)}, \quad n \in \mathbb{N}. \tag{1.3.2}$$

Adopting the notation

$$f[x_0, \ldots, x_n] := \beta_n, \tag{1.3.3}$$

we see from (1.3.2) that, for $n = 0$ and $n = 1$,

$$f[x_0] = f(x_0); \qquad f[x_0, x_1] = \frac{f(x_1) - f(x_0)}{x_1 - x_0}. \tag{1.3.4}$$

We call $f[x_0, \ldots, x_n]$ the nth-order divided difference of f with respect to the (distinct) point sequence $\triangle_n := \{x_0, \ldots, x_n\}$.

Now introduce an additional interpolation point x_{n+1}, such that

$$\triangle_{n+1} := \triangle_n \cup \{x_{n+1}\} \tag{1.3.5}$$

is a sequence of $n+2$ distinct points in $\mathbb{R}$. The following recursive result for polynomial interpolation then holds.

Theorem 1.3.1. *The interpolatory polynomial $P_{n+1}^I \in \pi_{n+1}$, as uniquely defined by the interpolatory conditions*

$$P_{n+1}^I(x) = f(x), \quad x \in \triangle_{n+1}, \tag{1.3.6}$$

with $\triangle_{n+1}$ as in (1.3.5), satisfies

$$P_{n+1}^I(x) = P_n^I(x) + f[x_0, \ldots, x_{n+1}] \prod_{j=0}^{n} (x - x_j), \tag{1.3.7}$$

with P_n^I denoting the interpolation polynomial of Theorem 1.1.2.

Proof. Let

$$P(x) := P_n^I(x) + f[x_0, \ldots, x_{n+1}] \prod_{j=0}^{n} (x - x_j), \qquad (1.3.8)$$

according to which, since also $P_n^I \in \pi_n$,

$$P(x) = (f[x_0, \ldots, x_{n+1}]) x^{n+1} + \widetilde{P}(x), \qquad (1.3.9)$$

where $\widetilde{P} \in \pi_n$. Next, we use (1.3.1) and (1.3.3) to obtain

$$P_{n+1}^I(x) = (f[x_0, \ldots, x_{n+1}]) x^{n+1} + \widetilde{Q}(x), \qquad (1.3.10)$$

where $\widetilde{Q} \in \pi_n$. It follows from (1.3.9) and (1.3.10) that the difference polynomial $Q := P - P_{n+1}^I$ satisfies $Q \in \pi_n$. Moreover, from (1.3.8) and (1.3.6), together with the fact that $P_n^I(x) = f(x)$, $x \in \triangle_n$, we have, for $j = 0, \ldots, n$,

$$Q(x_j) = P(x_j) - P_{n+1}^I(x_j) = P_n^I(x_j) - P_{n+1}^I(x_j) = f(x_j) - f(x_j) = 0.$$

Hence Q is a polynomial in π_n with $n+1$ distinct zeros at the points $\{x_0, \ldots, x_n\}$. It follows that Q must be the zero polynomial, and thus $P = P_{n+1}^I$, which, together with (1.3.8), proves (1.3.7). ∎

Motivated by the recursive formulation (1.3.7) in Theorem 1.3.1, we now define, for any distinct point sequence $\{x_j : j = 0, 1, \ldots\}$ in $\mathbb{R}$, the polynomial sequence

$$Q_0(x) := 1; \qquad Q_j(x) := \prod_{k=0}^{j-1} (x - x_k), \quad j = 1, 2, \ldots, \qquad (1.3.11)$$

which satisfies the following properties:

Theorem 1.3.2. *The polynomial sequence* $\{Q_j : j = 0, 1, \ldots\}$, *as defined by* (1.3.11), *satisfies:*

(a)

$$Q_j \in \pi_j, \quad j = 0, 1, \ldots; \qquad (1.3.12)$$

(b) *for any non-negative integer n, the sequence* $\{Q_j : j = 0, \ldots, n\}$ *is a basis for* π_n.

Proof (a) The property (1.3.12) is an immediate consequence of the definition (1.3.11).

(b) Since the case $n = 0$ is trivial, we suppose next $n \in \mathbb{N}$. As in the proof of Theorem 1.2.1(b), it will suffice to prove that $\{Q_j : j = 0, \ldots, n\}$ is a linearly independent set. To this end, we let the real sequence $\{c_0, \ldots, c_n\}$ be such that

$$\sum_{j=0}^{n} c_j Q_j(x) = 0, \quad x \in \mathbb{R}. \qquad (1.3.13)$$

By setting $x = x_0$ in (1.3.13), and using (1.3.11), we obtain $c_0 = 0$, which, together with (1.3.13), implies

$$\sum_{j=1}^{n} c_j Q_j(x) = 0, \quad x \in \mathbb{R}. \tag{1.3.14}$$

According to (1.3.11), we may now divide the identity (1.3.14) by $x - x_0$, and set $x = x_1$ in the resulting identity, to obtain $c_1 = 0$. Repeated further applications of this procedure yield $c_2 = \cdots = c_n = 0$, and thereby completing our linear independence proof. ∎

It follows from Theorem 1.3.2(b) that there exists a unique real coefficient sequence $\{\beta_0, \ldots, \beta_n\}$ such that the interpolation polynomial P_n^I of Theorem 1.1.2 is given by

$$P_n^I(x) = \sum_{j=0}^{n} \beta_j Q_j(x). \tag{1.3.15}$$

By applying the formula (1.3.7) of Theorem 1.3.1 recursively, we obtain the coefficient sequence

$$\beta_j = f[x_0, \ldots, x_j], \quad j = 0, \ldots, n, \tag{1.3.16}$$

which is consistent with (1.3.1)–(1.3.4). The following interpolation formula is now an immediate consequence of (1.3.15) and (1.3.16).

Theorem 1.3.3 (Newton interpolation formula). *The interpolation polynomial P_n^I of Theorem 1.1.2 satisfies the formulation*

$$P_n^I(x) = \sum_{j=0}^{n} f[x_0, \ldots, x_j] Q_j(x), \tag{1.3.17}$$

with the divided differences $\{f[x_0, \ldots, x_j] : j = 0, \ldots, n\}$ defined by (1.3.3), (1.3.2), and with the polynomial sequence $\{Q_j : j = 0, \ldots, n\}$ given as in (1.3.11).

Observe from (1.3.4) that the second order divided difference is given by

$$f[x_0, x_1] = \frac{f[x_1] - f[x_0]}{x_1 - x_0}. \tag{1.3.18}$$

We proceed to prove that the divided difference property of the right hand side of (1.3.18) extends to higher order divided differences. To this end, for integers m and n, with $n \geqslant 0$, let

$$\triangle_{m,n} := \{x_m, \ldots, x_{m+n}\} \tag{1.3.19}$$

denote a sequence of $n+1$ distinct points in $\mathbb{R}$. Then Theorem 1.1.2 implies the existence of a (unique) polynomial $P_{m,n}^I$ in π_n such that the interpolation conditions

$$P_{m,n}^I(x) = f(x), \quad x \in \triangle_{m,n}, \tag{1.3.20}$$

are satisfied, where the function f is such that $\triangle_{m,n} \subset D_f$. The n-th order divided difference $f[x_m,\ldots,x_{m+n}]$ of f with respect to the point sequence $\triangle_{m,n}$ is then defined by means of the leading coefficient property

$$P^I_{m,n}(x) = (f[x_m,\ldots,x_{m+n}])x^n + Q(x),$$

where $Q \in \pi_{n-1}$. Observe that the case $m=0$ corresponds precisely with (1.3.1)–(1.3.4). Our following result extends (1.3.18) recursively.

Theorem 1.3.4. *For integers j and k, with $k \geqslant 0$, let $\{x_j,\ldots,x_{j+k+1}\}$ denote a sequence of $k+2$ distinct points in $\mathbb{R}$. Then*

$$f[x_j,\ldots,x_{j+k+1}] = \frac{f[x_{j+1},\ldots,x_{j+k+1}] - f[x_j,\ldots,x_{j+k}]}{x_{j+k+1} - x_j}. \qquad (1.3.21)$$

Proof. Define the polynomial

$$Q(x) := \frac{(x-x_{j+k})P^I_{j+1,j+k+1}(x) + (x_{j+k+1}-x)P^I_{j,j+k}(x)}{x_{j+k+1} - x_j}, \qquad (1.3.22)$$

with $P^I_{j+1,j+k+1}$ and $P^I_{j,j+k}$ denoting the interpolation polynomials in π_k as uniquely determined by means of (1.3.20), (1.3.19). It follows from (1.3.22) that $Q \in \pi_{k+1}$, with, moreover,

$$Q(x_\ell) = f(x_\ell), \quad \ell = j,\ldots,j+k+1, \qquad (1.3.23)$$

and thus, from Theorem 1.1.2, we have $Q = P^I_{j,j+k+1}$, which, together with (1.3.22), yields

$$P^I_{j,j+k+1}(x) = \frac{(x-x_{j+k})P^I_{j+1,j+k+1}(x) + (x_{j+k+1}-x)P^I_{j,j+k}(x)}{x_{j+k+1} - x_j}. \qquad (1.3.24)$$

The desired result (1.3.21) is then obtained by equating the leading coefficients of the two sides of equation (1.3.24), and using the definition of a divided difference. ∎

We see from (1.3.21) that, for example,

$$f[x_0,x_1,x_2] = \frac{f[x_1,x_2] - f[x_0,x_1]}{x_2 - x_0},$$

which does indeed extend the divided difference pattern of (1.3.18).

The recursive computation of divided differences by means of (1.3.21) is considerably more efficient than using instead the explicit formulation (1.3.3), (1.3.2). The resulting iterative scheme for calculating the interpolation polynomial P^I_n by means of the Newton interpolation formula (1.3.17), together with (1.3.21), is illustrated in Figure 1.3.1.

x_0 $f(x_0)$

$\qquad\qquad f[x_0,x_1]$

x_1 $f(x_1)$ $f[x_0,x_1,x_2]$

$\qquad\qquad\qquad\qquad\qquad\qquad\ddots$

$\qquad\qquad f[x_1,x_2]$ $f[x_0,\ldots,x_{n-1}]$

x_2 $f(x_2)$ $f[x_0,\ldots,x_n]$

$\vdots$ $\vdots$ $f[x_1,\ldots,x_{n-1}]$

x_{n-1} $f(x_{n-1})$

$\qquad\qquad f[x_{n-1},x_n]$

x_n $f(x_n)$

Figure 1.3.1 *Divided differences by means of the recursive formulation* (1.3.21)

Example 1.3.1. Consider, as was also done in Example 1.2.1, the interpolation conditions (1.1.17) of Example 1.1.1. As in Figure 1.3.1, we calculate the relevant divided differences in (1.3.17) by means of (1.3.21), as follows:

x_j	$f(x_j)$	order 1	order 2	order 3
-1	1			
		-2		
0	-1		$\frac{4}{3}$	
		2		$-\frac{7}{12}$
2	3		-1	
		-1		
3	2			

The above table of values are obtained by the calculations

$$f[-1,0] = \frac{f(0)-f(-1)}{0-(-1)} = \frac{-1-1}{1} = -2;$$

$$f[0,2] = \frac{f(2)-f(0)}{2-0} = \frac{3-(-1)}{2} = 2;$$

$$f[2,3] = \frac{f(3)-f(2)}{3-2} = \frac{2-3}{1} = -1;$$

$$f[-1,0,2] = \frac{f[0,2]-f[-1,0]}{2-(-1)} = \frac{2-(-2)}{3} = \frac{4}{3};$$

$$f[0,2,3] = \frac{f[2,3]-f[0,2]}{3-0} = \frac{-1-2}{3} = -1;$$

$$f[-1,0,2,3] = \frac{f[0,2,3]-f[-1,0,2]}{3-(-1)} = \frac{-1-\frac{4}{3}}{4} = -\frac{7}{12}.$$

The Newton interpolation formula (1.3.17), together with (1.3.11), then yields the interpolation polynomial

$$P_3^I(x) = 1 + (-2)(x-(-1)) + \frac{4}{3}(x-(-1))(x-0) - \frac{7}{12}(x-(-1))(x-0)(x-2)$$

$$= 1 - 2(x+1) + \frac{4}{3}x(x+1) - \frac{7}{12}x(x+1)(x-2)$$

$$= \frac{1}{12}(-12+6x+23x^2-7x^3),$$

which agrees with (1.1.21) and (1.2.10) in, respectively, Examples 1.1.1 and 1.1.2. ■

We proceed to state the following identity, the proof of which, as based on Theorems 1.3.3 and 1.1.2, is analogous to the proof of Theorem 1.2.3.

Theorem 1.3.5. *The polynomials $\{Q_j : j = 0,1,\ldots\}$, as given by (1.3.11), satisfy, for any non-negative integer n, the identity*

$$\sum_{j=0}^{n} P[x_0,\ldots,x_j]Q_j(x) = P(x), \quad x \in \mathbb{R}, \quad P \in \pi_n, \tag{1.3.25}$$

with the divided differences $\{P[x_0,\ldots,x_j] : j = 0,\ldots,n\}$ obtained from (1.3.2), (1.3.3).

Divided differences satisfy the following symmetry condition, which is an immediate consequence of the definition (1.3.2), (1.3.3), and which, along with Theorem 1.3.5, we shall rely on in Section 1.4.

Theorem 1.3.6. *For any positive integer n and (distinct) point sequence $\{x_0,\ldots,x_n\}$ in $\mathbb{R}$, the divided difference $f[x_0,\ldots,x_n]$, as defined by (1.3.2), (1.3.3), is a symmetric function of its arguments, that is, for any permutation $\{j_0,\ldots,j_n\}$ of the index set $\{0,\ldots,n\}$, we have*

$$f[x_{j_0},\ldots,x_{j_n}] = f[x_0,\ldots,x_n]. \tag{1.3.26}$$

Observe in particular that, throughout Sections 1.1, 1.2 and 1.3, the (distinct) interpolation points $\{x_0,\ldots,x_n\}$ have not been required to satisfy any ordering condition like, for example, $x_0 < x_1 < \cdots < x_n$.

1.4   Hermite interpolation

As in the preceding sections, for any non-negative integer n, let $\triangle_n := \{x_0,\ldots,x_n\}$ denote a sequence of $n+1$ distinct points in $\mathbb{R}$. We proceed to investigate the existence and construction of a polynomial P satisfying the interpolatory conditions

$$P^{(k)}(x_j) = f^{(k)}(x_j), \quad k=0,\ldots,r_j; \quad j=0,\ldots,n, \qquad (1.4.1)$$

with $\{r_0,\ldots,r_n\}$ denoting a given sequence of $n+1$ non-negative integers, and where f is a real-valued function such that $\triangle_n \subset D_f$, and f has all the derivatives implied by the right hand side of (1.4.1). The conditions (1.4.1) are called Hermite polynomial interpolation conditions, and a polynomial P satisfying (1.4.1) is called a Hermite interpolation polynomial of f with respect to the points $\triangle_n$ and the sequence $\mathbf{r} = \{r_0,\ldots,r_n\}$.

Observe that the precise number of Hermite interpolation conditions in (1.4.1) is given by

$$\sum_{j=0}^{n} (r_j+1) = (n+1) + \sum_{j=0}^{n} r_j.$$

Based on (1.1.1), we therefore consider the polynomial construction

$$P(x) := \sum_{j=0}^{\nu} c_j x^j, \qquad (1.4.2)$$

where

$$\nu := n + \sum_{j=0}^{n} r_j, \qquad (1.4.3)$$

and with $\{c_0,\ldots,c_\nu\}$ denoting a real coefficient sequence, that is, $P \in \pi_\nu$. For any $k \in \{1,\ldots,\nu\}$, we now differentiate (1.4.2) k times to obtain the formula

$$P^{(k)}(x_j) = \sum_{\ell=k}^{\nu} [k!\binom{\ell}{k}x_j^{\ell-k}]c_\ell, \quad j=0,\ldots,n. \qquad (1.4.4)$$

By substituting (1.4.4) into (1.4.1), we deduce that there exists a polynomial $P \in \pi_\nu$ satisfying the Hermite interpolation conditions (1.4.1) if and only if there exists a (column vector) solution

$$\mathbf{c} = [c_0,c_1,\ldots,c_\nu]^T \in \mathbb{R}^{\nu+1} \qquad (1.4.5)$$

of the $(\nu+1) \times (\nu+1)$ linear system

$$W_{n,\mathbf{r}}\mathbf{c} = \mathbf{f}, \qquad (1.4.6)$$

where the successive rows of the (square) matrix $W_{n,\mathbf{r}}$ are defined by setting, in (1.4.4), $k=0,\ldots,r_j$ for $j=0,1,\ldots,n$, and where the column vector $\mathbf{f} \in \mathbb{R}^{\nu+1}$ is defined by

$$\mathbf{f} := [f(x_0),f'(x_0),\ldots,f^{(r_0)}(x_0),f(x_1),\ldots,f^{(r_n)}(x_n)]^T. \qquad (1.4.7)$$

The following result then holds.

Theorem 1.4.1. *The $(\nu+1) \times (\nu+1)$ matrix $W_{n,\mathbf{r}}$ in (1.4.6) is invertible.*

Proof. Let $\mathbf{c} = \{c_0, \ldots, c_\nu\}^T \in \mathbb{R}^{\nu+1}$ be such that

$$W_{n,\mathbf{r}}\mathbf{c} = \mathbf{0}. \tag{1.4.8}$$

We proceed to prove that $\mathbf{c} = \mathbf{0}$, the zero sequence, from which, according to a standard result in linear algebra, it will then follow that $W_{n,\mathbf{r}}$ is indeed an invertible matrix.

To this end, we first use (1.4.4), (1.4.6), (1.4.7) and (1.4.8) to deduce that the polynomial $P \in \pi_\nu$ defined by

$$P(x) := \sum_{j=0}^{\nu} c_j x^j \tag{1.4.9}$$

satisfies

$$P^{(k)}(x_j) = 0, \quad k = 0, \ldots, r_j, \quad j = 0, \ldots, n,$$

and thus

$$P(x) = K \prod_{j=0}^{n} \prod_{k=1}^{r_j+1} (x - x_j)^k = Kx^{\nu+1} + \widetilde{P}(x), \tag{1.4.10}$$

for some constant K, where $\widetilde{P} \in \pi_\nu$, and with ν defined by (1.4.3). It follows from (1.4.9) and (1.4.10) that $K = 0$, which, together with the first equation in (1.4.10), shows that P is the zero polynomial, and thus, from (1.4.9), $c_0 = c_1 = \cdots = c_\nu = 0$, as required. ∎

As an immediate consequence of Theorem 1.4.1, together with (1.4.1)–(1.4.7), we have the following existence and uniqueness result for Hermite polynomial interpolation.

Theorem 1.4.2. *For any non-negative integer n, let $\{x_0, \ldots, x_n\}$ denote a sequence of $n+1$ distinct points in $\mathbb{R}$, and let $\mathbf{r} = \{r_0, \ldots, r_n\}$ be a given sequence of non-negative integers. Then there exists a unique Hermite interpolation polynomial $P = P_{n,\mathbf{r}}^I$ in π_ν, with ν defined by (1.4.3), such that the conditions (1.4.1) are satisfied.*

Note from Theorems 1.1.2 and 1.4.2 that

$$P_{n,\mathbf{0}}^I = P_n^I.$$

We proceed to establish an explicit construction method for the Hermite interpolation polynomial $P_{n,\mathbf{r}}^I$ of Theorem 1.4.2. To this end, we first observe from (1.3.18) that, for a function f that is differentiable in the smallest interval containing x_0 and x_1, we have

$$f[x_0, x_1] = \int_0^1 f'(t(x_1 - x_0) + x_0) dt. \tag{1.4.11}$$

In general, for sufficiently differentiable functions f, divided differences of order $m \geqslant 2$ can be expressed as follows in terms of an iterated integral. We shall denote by $f^{(n)}$ the nth derivative of a function f, with the convention also that $f^{(0)} := f$.

Theorem 1.4.3. *For any integer $n \in \mathbb{N}$, let $\{x_0,\ldots,x_n\}$ denote a sequence of $n+1$ distinct points in $\mathbb{R}$, and suppose the real-valued function f has n continuous derivatives in the smallest interval containing the points $\{x_0,\ldots,x_n\}$. Then the divided difference $f[x_0,\ldots,x_n]$, as obtained from (1.3.2), (1.3.3), satisfies the iterated integral formulation*

$$f[x_0,\ldots,x_n] = \int_0^{t_0} \cdots \int_0^{t_{n-1}} f^{(n)}\left(t_n(x_n - x_{n-1}) + \cdots + t_1(x_1 - x_0) + x_0\right)dt_n \ldots dt_1, \quad (1.4.12)$$

where

$$t_0 := 1. \qquad\qquad (1.4.13)$$

Remark. For any integer $n \in \mathbb{N}$, observe that the number

$$\gamma_n := t_n(x_n - x_{n-1}) + \cdots + t_1(x_1 - x_0) + x_0$$

satisfies

$$\gamma_n = (1 - t_1)x_0 + (t_1 - t_2)x_1 + \cdots + (t_{n-1} - t_n)x_{n-1} + t_n x_n,$$

and thus, with the notation

$$\alpha_n := \min\{x_0,\ldots,x_n\} \quad ; \quad \beta_n := \max\{x_0,\ldots,x_n\},$$

we have, for $0 \leqslant t_n \leqslant t_{n-1},\ldots, 0 \leqslant t_2 \leqslant t_1, 0 \leqslant t_1 \leqslant t_0 = 1$, the inequalities

$$\alpha_n[(1 - t_1) + (t_1 - t_2) + \cdots + (t_{n-1} - t_n) + t_n] \leqslant \gamma_n \leqslant \beta_n[(1 - t_1) + (t_1 - t_2) + \cdots + (t_{n-1} - t_n) + t_n],$$

which yields $\alpha_n \leqslant \gamma_n \leqslant \beta_n$. Hence the argument $(= \gamma_n)$ of the integrand $f^{(n)}$ in (1.4.12) is indeed in the smallest interval containing the points $\{x_0,\ldots,x_n\}$, and thereby guaranteeing the existence of the iterated integral in (1.4.12).

Proof of Theorem 1.4.3. Our proof is by induction on the integer n. After noting from (1.4.11) that (1.4.12), (1.4.13) are satisfied for $n = 1$, we suppose next that (1.4.12), (1.4.13) hold for any fixed integer $n \in \mathbb{N}$. But then, by using also (1.3.21) in Theorem 1.3.4, as well as Theorem 1.3.6, we obtain, by integrating with respect to t_{n+1},

$$\int_0^{t_0} \cdots \int_0^{t_n} f^{(n+1)}\left(t_{n+1}(x_{n+1} - x_n) + \cdots + t_1(x_1 - x_0) + x_0\right)dt_{n+1}dt_n \ldots dt_1$$

$$= \frac{1}{x_{n+1} - x_n}\left[\int_0^{t_0} \cdots \int_0^{t_{n-1}} f^{(n)}\left(t_n(x_{n+1} - x_{n-1}) + t_{n-1}(x_{n-1} - x_{n-2}) + \cdots\right.\right.$$

$$\left. + t_1(x_1 - x_0) + x_0\right)dt_{n-1} \ldots dt_1$$

$$\left. - \int_0^{t_0} \cdots \int_0^{t_{n-1}} f^{(n)}\left(t_n(x_n - x_{n-1}) + \cdots + t_1(x_1 - x_0) + x_0\right)dt_{n-1} \ldots dt_1\right]$$

$$= \frac{f[x_0,\ldots,x_{n-1},x_{n+1}] - f[x_0,\ldots,x_{n-1},x_n]}{x_{n+1} - x_n}$$

$$= \frac{f[x_0,\ldots,x_{n-1},x_{n+1}] - f[x_n,x_0,\ldots,x_{n-1}]}{x_{n+1} - x_n}$$

$$= f[x_n,x_0,\ldots,x_{n-1},x_{n+1}] = f[x_0,x_1,\ldots,x_n,x_{n+1}],$$

and thus (1.4.12), (1.4.13) are satisfied with n replaced by $n+1$, which completes our inductive proof. ∎

The definition (1.3.2), (1.3.3) of the divided difference $f[x_0,\ldots,x_n]$ holds for any distinct point sequence $\{x_0,\ldots,x_n\}$. Based on (1.4.12), (1.4.13) in Theorem 1.4.3, we now define, for any non-negative integer ν and (not necessarily distinct) point sequence $\{\xi_0,\ldots,\xi_\nu\}$ in $\mathbb{R}$, the νth order divided difference of a function f, with ν continuous derivatives in the smallest interval containing the points $\{\xi_0,\ldots,\xi_\nu\}$, by

$$\left.\begin{array}{l} f[\xi_0] := f(\xi_0); \\[2em] f[\xi_0,\ldots,\xi_\nu] := \int_0^{t_0} \cdots \int_0^{t_{\nu-1}} f^{(\nu)}(t_\nu(\xi_\nu - \xi_{\nu-1}) + \cdots + t_1(\xi_1 - \xi_0) + \xi_0) dt_\nu \ldots dt_1, \text{ if } \nu \geqslant 1, \\[1em] \text{where} \\[1em] \qquad\qquad t_0 := 1. \end{array}\right\}$$

$$(1.4.14)$$

Observe from Theorem 1.4.3 that the definition (1.4.14) yields (1.3.2), (1.3.3) if $\{\xi_0,\ldots,\xi_\nu\}$ is a distinct point sequence in $\mathbb{R}$.

As an immediate consequence of the definition (1.4.14), we have the following continuity result.

Theorem 1.4.4. *For any non-negative integer ν, let $\{\xi_0,\ldots,\xi_\nu\}$ denote a (not necessarily distinct) point sequence in $\mathbb{R}$, and suppose the real-valued function f has ν continuous derivatives in the smallest interval containing the points $\{\xi_0,\ldots,\xi_\nu\}$. Then the divided difference $f[\xi_0,\ldots,\xi_\nu]$, as defined by (1.4.14), is a continuous function on $\mathbb{R}^{\nu+1}$.*

Henceforth in this section we shall assume that the sequence $\{x_0,\ldots,x_n\}$ of Theorem 1.4.2 is strictly increasing, that is,

$$x_0 < x_1 < \cdots < x_n. \qquad\qquad (1.4.15)$$

In the notation of Theorem 1.4.2, according to which the integer v is defined by (1.4.3), we now define the sequence $\{\xi_0, \ldots, \xi_v\}$ of $v+1$ points in $\mathbb{R}$ by

$$
\left.
\begin{aligned}
\xi_0 &= \cdots = \xi_{r_0} & &:= x_0; \\
\xi_{r_0+1} &= \cdots = \xi_{r_0+r_1+1} & &:= x_1; \\
&\;\;\vdots & &\;\;\vdots \\
\xi_{v-r_n} &= \cdots = \xi_v & &:= x_n,
\end{aligned}
\right\}
\tag{1.4.16}
$$

and, for $k \in \mathbb{N}$, we denote by $\{\xi_{0,k}, \ldots, \xi_{v,k}\}$ any sequence of $v+1$ distinct points in $\mathbb{R}$, such that

$$
\xi_{0,k} < \xi_{1,k} < \cdots < \xi_{v,k},
\tag{1.4.17}
$$

with, moreover,

$$
\lim_{k \to \infty} \xi_{j,k} = \xi_j, \quad j = 0, \ldots, v.
\tag{1.4.18}
$$

Following (1.3.11), the polynomial sequences $\{\widetilde{Q}_j : j = 0, \ldots, v\}$ and $\{\widetilde{Q}_{j,k} : j = 0, \ldots, v\}$ are then defined accordingly by

$$
\widetilde{Q}_0(x) := 1; \quad \widetilde{Q}_j(x) := \prod_{\ell=0}^{j-1}(x - \xi_\ell), \quad j = 1, \ldots, v,
\tag{1.4.19}
$$

and

$$
\widetilde{Q}_{0,k}(x) := 1; \quad \widetilde{Q}_{j,k}(x) := \prod_{\ell=0}^{j-1}(x - \xi_{\ell,k}), \quad j = 1, \ldots, v,
\tag{1.4.20}
$$

for $k = 1, 2, \ldots$.

Since, according to (1.4.17), $\{\xi_{0,k}, \ldots, \xi_{v,k}\}$ is, for any fixed $k \in \mathbb{N}$, a sequence of $v+1$ distinct points in $\mathbb{R}$, and since the Hermite interpolation polynomial $P_{n,\mathbf{r}}^I$ of Theorem 1.4.2 satisfies $P_{n,\mathbf{r}}^I \in \pi_v$, we may apply (1.3.25) in Theorem 1.3.5 to obtain the identity

$$
P_{n,\mathbf{r}}^I(x) = \sum_{j=0}^{v} P_{n,\mathbf{r}}^I[\xi_{0,k}, \ldots, \xi_{j,k}]\widetilde{Q}_{j,k}(x), \quad x \in \mathbb{R},
\tag{1.4.21}
$$

for $k = 1, 2, \ldots$.

By applying Theorem 1.4.4, we may now deduce from (1.4.18) that

$$
\lim_{k \to \infty} P_{n,\mathbf{r}}^I[\xi_{0,k}, \ldots, \xi_{j,k}] = P_{n,\mathbf{r}}^I[\xi_0, \ldots, \xi_j], \quad j = 0, \ldots, v,
\tag{1.4.22}
$$

whereas (1.4.20), (1.4.18) and (1.4.19) imply

$$
\lim_{k \to \infty} \widetilde{Q}_{j,k}(x) = \widetilde{Q}_j(x), \quad j = 0, \ldots, v,
\tag{1.4.23}
$$

and for any $x \in \mathbb{R}$. Since, moreover, the left hand side of (1.4.21) is independent of k, we may now combine (1.4.21), (1.4.22) and (1.4.23) to obtain the identity

$$P_{n,\mathbf{r}}^I(x) = \sum_{j=0}^{\nu} P_{n,\mathbf{r}}^I[\xi_0, \dots, \xi_j]\widetilde{Q}_j(x), \quad x \in \mathbb{R}. \tag{1.4.24}$$

The recursive computation of the divided differences in the right hand side of (1.4.24) will be based on the following polynomial extension of Theorem 1.3.4.

Theorem 1.4.5. *Let P be any polynomial, and, for integers j and k, with $k \geqslant 0$, let $\{\xi_j, \dots, \xi_{j+k+1}\}$ be a sequence of $k+1$ points in $\mathbb{R}$ such that*

$$\xi_j \leqslant \xi_{j+1} \leqslant \cdots \leqslant \xi_{j+k+1}. \tag{1.4.25}$$

Then the divided difference $P[\xi_j, \dots, \xi_{j+k+1}]$, as obtained from (1.4.12), (1.4.13), satisfies

$$P[\xi_j, \dots, \xi_{j+k+1}] = \begin{cases} \dfrac{P[\xi_{j+1}, \dots, \xi_{j+k+1}] - P[\xi_j, \dots, \xi_{j+k}]}{\xi_{j+k+1} - \xi_j}, & \text{if } \xi_j \neq \xi_{j+k+1}; \\[2mm] \dfrac{P^{(k+1)}(\xi)}{(k+1)!}, & \text{if } \xi_j = \xi_{j+1} = \cdots = \xi_{j+k+1} =: \xi. \end{cases} \tag{1.4.26}$$

Proof. Suppose first $\xi_j \neq \xi_{j+k+1}$. For $\ell \in \mathbb{N}$, analogously to (1.4.17), (1.4.18), let $\{\xi_{r,\ell} : r = j, \dots, j+k+1\}$ denote a sequence of $k+2$ distinct points in $\mathbb{R}$ such that

$$\xi_{j,\ell} < \xi_{j+1,\ell} < \cdots < \xi_{j+k+1,\ell}, \tag{1.4.27}$$

with, moreover,

$$\lim_{\ell \to \infty} \xi_{r,\ell} = \xi_r, \quad r = j, \dots, j+k+1. \tag{1.4.28}$$

Based on (1.4.27) and (1.4.28), we may now apply (1.3.21) in Theorem 1.3.4, as well as Theorem 1.4.4, to deduce that, for $\ell \in \mathbb{N}$,

$$\left| P[\xi_j, \dots, \xi_{j+k+1}] - \frac{P[\xi_{j+1}, \dots, \xi_{j+k+1}] - P[\xi_j, \dots, \xi_{j+k}]}{\xi_{j+k+1} - \xi_j} \right|$$

$$\leqslant |P[\xi_j, \dots, \xi_{j+k+1}] - P[\xi_{j,\ell}, \dots, \xi_{j+k+1,\ell}]|$$

$$+ \left| \frac{P[\xi_{j+1,\ell}, \dots, \xi_{j+k+1,\ell}] - P[\xi_{j,\ell}, \dots, \xi_{j+k,\ell}]}{\xi_{j+k+1,\ell} - \xi_{j,\ell}} - \frac{P[\xi_{j+1}, \dots, \xi_{j+k+1}] - P[\xi_j, \dots, \xi_{j+k}]}{\xi_{j+k+1} - \xi_j} \right|$$

$$\to 0 + 0 = 0, \quad \ell \to \infty,$$

and thereby proving the first line of (1.4.26).

Next, if $\xi_j = \xi_{j+k+1}$, and thus, from (1.4.25), $\xi_j = \xi_{j+1} = \cdots = \xi_{j+k+1} =: \xi$, it follows from (1.4.12), (1.4.13) that

$$P[\xi_j,\ldots,\xi_{j+k+1}] = P^{(k+1)}(\xi) \int_0^{t_0} \int_0^{t_1} \cdots \int_0^{t_k} dt_{k+1}dt_k \ldots dt_1$$

$$= P^{(k+1)}(\xi) \int_0^{t_0} \cdots \int_0^{t_{k-1}} t_k dt_k \ldots dt_1$$

$$= \frac{P^{(k+1)}(\xi)}{2} \int_0^{t_0} \cdots \int_0^{t_{k-2}} (t_{k-1})^2 dt_{k-1} \ldots dt_1$$

$$= \cdots$$

$$= \frac{P^{(k+1)}(\xi)}{k!} \int_0^{t_0} (t_1)^k dt_1 = \frac{P^{(k+1)}(\xi)}{k!} \int_0^1 (t_1)^k dt_1 = \frac{P^{(k+1)}(\xi)}{(k+1)!},$$

which gives the second line of (1.4.26).   ∎

It follows from (1.4.26) in Theorem 1.4.5 that the divided differences $\{P_{n,\mathbf{r}}^I[\xi_0,\ldots,\xi_j] : j = 0,\ldots,\nu\}$ in the right hand side of (1.4.24) are determined uniquely by the values

$$\{(P_{n,\mathbf{r}}^I)^{(k)}(x_j) : k = 0,\ldots,r_j; \quad j = 0,\ldots,n\}.$$

Since also the Hermite interpolation conditions (1.4.1) are satisfied, we deduce from (1.4.24), together with Theorem 1.4.5, the following Hermite interpolation formula.

Theorem 1.4.6. *The Hermite interpolation polynomial $P_{n,\mathbf{r}}^I$ of Theorem 1.4.2 satisfies the formulation*

$$P_{n,\mathbf{r}}^I(x) = \sum_{j=0}^{\nu} f[\xi_0,\ldots,\xi_j]\widetilde{Q}_j(x), \tag{1.4.29}$$

with the polynomial sequence $\{\widetilde{Q}_j : j = 0,\ldots,\nu\}$ given by (1.4.19) in terms of the sequence $\{\xi_0,\ldots,\xi_\nu\}$ defined in (1.4.16), and with the divided differences $\{f[\xi_0,\ldots,\xi_j] : j = 0,\ldots,\nu\}$ defined for $0 \leqslant j < j+k+1 \leqslant \nu$ by

$$f[\xi_j,\ldots,\xi_{j+k+1}] := \begin{cases} \dfrac{f[\xi_{j+1},\ldots,\xi_{j+k+1}] - f[\xi_j,\ldots,\xi_{j+k}]}{\xi_{j+k+1} - \xi_j}, & \text{if } \xi_j \neq \xi_{j+k+1}; \\ \dfrac{f^{(k+1)}(\xi)}{(k+1)!}, & \text{if } \xi_j = \xi_{j+k+1} =: \xi. \end{cases}$$
$$\tag{1.4.30}$$

Example 1.4.1. According to Theorem 1.4.2, there exists a unique polynomial $P = P_{2,\mathbf{r}}^I \in \pi_5$, where $\mathbf{r} = \{r_0,r_1,r_2\} := \{1,0,2\}$, such that the Hermite interpolation conditions

$$P(0) = 2; \quad P'(0) = -2; \quad P(2) = 3; \quad P(3) = -1; \quad P'(3) = 1; \quad P''(3) = 2$$

are satisfied. Observe that here $n = 2$, $v = 5$, $\{x_0, x_1, x_2\} = \{0, 2, 3\}$ and, from (1.4.16), $\{\xi_0, \xi_1, \xi_2, \xi_3, \xi_4, \xi_5\} = \{0, 0, 2, 3, 3, 3\}$. The divided differences in the formula (1.4.29) are then computed by means of (1.4.30), as follows:

ξ_j	$f(\xi_j)$	order 1	order 2	order 3	order 4	order 5
0	2					
		-2				
0	2		$\frac{5}{4}$			
		$\frac{1}{2}$		$-\frac{11}{12}$		
2	3		$-\frac{3}{2}$		$\frac{37}{36}$	
		-4		$\frac{13}{6}$		$-\frac{37}{36}$
3	-1		5		$-\frac{37}{18}$	
		1		-4		
3	-1		1			
		1				
3	-1					

The above table of values are obtained by the calculations

$$f[0,0] = f'(0) = -2;$$

$$f[0,2] = \frac{f(2) - f(0)}{2 - 0} = \frac{3 - 2}{2} = \frac{1}{2};$$

$$f[2,3] = \frac{f(3) - f(2)}{3 - 2} = \frac{-1 - 3}{3 - 2} = -4;$$

$$f[3,3] = f'(3) = 1;$$

$$f[0,0,2] = \frac{f[0,2] - f[0,0]}{2 - 0} = \frac{\frac{1}{2} - (-2)}{2} = \frac{5}{4};$$

$$f[0,2,3] = \frac{f[2,3] - f[0,2]}{3 - 0} = \frac{-4 - \frac{1}{2}}{3} = -\frac{3}{2};$$

$$f[2,3,3] \quad = \frac{f[3,3]-f[2,3]}{3-2} = \frac{1-(-4)}{1} = 5;$$

$$f[3,3,3] \quad = \frac{f''(3)}{2} = \frac{2}{2} = 1;$$

$$f[0,0,2,3] \quad = \frac{f[0,2,3]-f[0,0,2]}{3-0} = \frac{-\frac{3}{2}-\frac{5}{4}}{3} = -\frac{11}{12};$$

$$f[0,2,3,3] \quad = \frac{f[2,3,3]-f[0,2,3]}{3-0} = \frac{5-(-\frac{3}{2})}{3} = \frac{13}{6};$$

$$f[2,3,3,3] \quad = \frac{f[3,3,3]-f[2,3,3]}{3-2} = \frac{1-5}{1} = -4;$$

$$f[0,0,2,3,3] \quad = \frac{f[0,2,3,3]-f[0,0,2,3]}{3-0} = \frac{\frac{13}{6}-(-\frac{11}{12})}{3} = \frac{37}{36};$$

$$f[0,2,3,3,3] \quad = \frac{f[2,3,3,3]-f[0,2,3,3]}{3-0} = \frac{-4-\frac{13}{6}}{3} = -\frac{37}{18};$$

$$f[0,0,2,3,3,3] = \frac{f[0,2,3,3,3]-f[0,0,2,3,3]}{3-0} = \frac{-\frac{37}{18}-\frac{37}{36}}{3} = -\frac{37}{36}.$$

The Hermite interpolation polynomial $P^I_{2,\mathbf{r}}$ is then given, according to (1.4.29) and (1.4.30), by

$$P^I_{2,\mathbf{r}}(x) = 2 - 2(x-0) + \frac{5}{4}(x-0)^2 - \frac{11}{12}(x-0)^2(x-2)$$

$$+\frac{37}{36}(x-0)^2(x-2)(x-3) - \frac{37}{36}(x-0)^2(x-2)(x-3)^2$$

$$= \frac{1}{36}(72 - 72x + 999x^2 - 995x^3 + 333x^4 - 37x^5).$$

∎

1.5   Exercises

Exercise 1.1 For the Vandermonde matrix V_3 corresponding to the point sequence

$$x_j = j^2, \quad j = 0,1,2,3,$$

calculate the inverse matrix V_3^{-1}, and then apply the interpolation formula (1.1.16) in Theorem 1.1.2 to obtain the polynomial P of least degree such that the interpolation conditions

$$P(x_j) = \sqrt{x_j}, \quad j = 0,1,2,3,$$

are satisfied.

Exercise 1.2 Verify the polynomial P calculated in Exercise 1.1 by means of (a) the Lagrange interpolation formula; (b) the Newton interpolation formula. In (a), it should also be verified that the corresponding Lagrange fundamental polynomials $\{L_{3,j} : j = 0,\ldots,3\}$ satisfy the identity (1.2.8) in Theorem 1.2.3.

Exercise 1.3 Let $\{P_j^I : j = 0,\ldots,4\}$ denote the sequence of interpolation polynomials, with $P_j^I \in \pi_j, j = 0,\ldots,4$, such that

$$P_j^I(x_k) = f(x_k), \quad k = 0,\ldots,j; \quad j = 0,\ldots,4,$$

with

$$\{x_0,\ldots,x_4\} = \{0,\ldots,4\},$$

and where

$$f(x_0) = 1; \quad f(x_1) = -1; \quad f(x_2) = 0; \quad f(x_3) = 3; \quad f(x_4) = -2.$$

Calculate the sequence $\{P_j^I : j = 0,\ldots,4\}$ by means of (a) the Lagrange interpolation formula; (b) the Newton interpolation formula, and compare the efficiency of the two methods.

Exercise 1.4 For the function

$$f(x) = e^x,$$

calculate the divided difference $f[0,1,1,1]$ (a) directly by means of an iterated integral as in (1.4.14); (b) recursively as in Example 1.4.1.

Exercise 1.5 For the function

$$f(x) = \frac{16}{x},$$

use the recursive method, as applied in Example 1.4.1, to find the polynomial P of least degree such that P satisfies the Hermite interpolation conditions

$$P(1) = f(1); \quad P'(1) = f'(1); \quad P''(1) = f''(1);$$

$$P(2) = f(2);$$

$$P(4) = f(4) \quad ; \quad P'(4) = f'(4).$$

Exercise 1.6 As a continuation of Exercise 1.5, write down the corresponding vector $\mathbf{r}$ and matrix $W_{2,\mathbf{r}}$, as appearing in equation (1.4.6).

Exercise 1.7 As a further continuation of Exercise 1.5, write down the corresponding point sequences $\{x_0, x_1, x_2\}$ and $\{\xi_0,\ldots,\xi_5\}$, as appearing in (1.4.16), and give an explicit formulation of a sequence $\{\xi_{j,k} : k = 0,1,\ldots; j = 0,\ldots,5\}$ satisfying (1.4.17) and (1.4.18).

Exercise 1.8 Show that the set

$$S := \{1, x, x(x-1), x(x-1)^2, x(x-1)^3, x(x-1)^3(x-2)\}$$

is a basis for the polynomial space π_5.

Exercise 1.9 As a continuation of Exercise 1.8, formulate and solve, by means of the recursive method as in Example 1.4.1, the Hermite interpolation problem for the function

$$f(x) = \frac{1}{x+1},$$

and in which S is the appropriate basis for π_5.

Exercise 1.10 Verify Theorem 1.3.6 for the case $n = 3$; $\{x_0, x_1, x_2, x_3\} = \{0, 1, 2, 3\}$; $\{j_0, j_1, j_2, j_3\} = \{2, 0, 3, 1\}$, and

$$f(x_0) = 3; \quad f(x_1) = -2; \quad f(x_2) = 1; \quad f(x_3) = -1,$$

by calculating both sides of equation (1.3.26) by means of the recursive formulation (1.3.21) in Theorem 1.3.4.

Chapter 2

Error Analysis For Polynomial Interpolation

As a continuation of Chapter 1, the notion of divided difference is applied to deduce the uniform error bound for polynomial interpolation for any given finite sample point set. In addition, an optimal sample point set, on which the minimum uniform error bound is achieved among all sample point sets with the same cardinality, is derived.

2.1 General error estimate

Let $[a,b]$ denote a bounded interval in $\mathbb{R}$, and suppose $f \in C[a,b]$, with $C[a,b]$ denoting the linear space of continuous functions $f : [a,b] \to \mathbb{R}$. For any non-negative integer n, let $\triangle_n := \{x_0, \ldots, x_n\}$ be a sequence of $n+1$ distinct points such that

$$\triangle_n \subset [a,b], \tag{2.1.1}$$

and, as in Theorem 1.1.2, denote by P_n^I the unique polynomial in π_n satisfying the interpolation conditions

$$P_n^I(x) = f(x), \quad x \in \triangle_n. \tag{2.1.2}$$

The corresponding polynomial interpolation error function is then defined by

$$E_n^I := f - P_n^I. \tag{2.1.3}$$

Hence $E_n^I \in C[a,b]$, with

$$E_n^I(x) = 0, \quad x \in \triangle_n. \tag{2.1.4}$$

The function E_n^I has the following explicit formulation in terms of a divided difference.

Theorem 2.1.1. *The error function E_n^I, as defined by (2.1.3), satisfies, for any non-negative integer n,*

$$E_n^I(x) = \begin{cases} 0 & , x \in \triangle_n; \\ f[x, x_0, \ldots, x_n] Q_{n+1}(x) & , x \in [a,b] \setminus \triangle_n, \end{cases} \tag{2.1.5}$$

with $Q_{n+1} \in \pi_{n+1}$ defined as in (1.3.11), that is,

$$Q_{n+1}(x) := \prod_{j=0}^{n}(x - x_j). \tag{2.1.6}$$

Proof. The first line of (2.1.5) has already been noted in (2.1.4).

Let $x \in [a,b] \setminus \triangle_n$ be fixed, and denote by P the unique interpolation polynomial in π_{n+1} such that

$$P(t) = f(t), \quad t \in \triangle_n \cup \{x\}. \tag{2.1.7}$$

It follows from (2.1.1) and (1.3.7) in Theorem 1.3.1 that

$$P(t) = P_n^I(t) + f[x_0, \ldots, x_n, x]Q_{n+1}(t), \tag{2.1.8}$$

with the polynomial Q_{n+1}^I defined as in (2.1.6). By setting $t = x$ in (2.1.8), and using (2.1.7), we obtain

$$f(x) = P_n^I(x) + f[x_0, \ldots, x_n, x]Q_{n+1}(x). \tag{2.1.9}$$

The second line of (2.1.5) is now a consequence of (2.1.8), (2.1.3), as well as the symmetry result of Theorem 1.3.6.   ∎

In order to obtain a useful estimate for the error function E_n^I, we first prove the following property of divided differences.

Theorem 2.1.2. *For any non-negative integer n, let $\triangle_n := \{x_0, \ldots, x_n\}$ denote a sequence of $n+1$ distinct points in $\mathbb{R}$, and suppose f has n continuous derivatives in the smallest interval containing the points $\{x_0, \ldots, x_n\}$. Then the divided difference $f[x_0, \ldots, x_n]$, as defined by (1.3.2), (1.3.3), satisfies*

$$f[x_0, \ldots, x_n] = \frac{f^{(n)}(\xi)}{n!}, \tag{2.1.10}$$

for some point ξ in the smallest interval containing the points $\{x_0, \ldots, x_n\}$.

Proof. By applying (1.4.12), (1.4.13) in Theorem 1.4.3, and recalling the remark following the statement of Theorem 1.4.3, we deduce by means of the mean value theorem for integrals, together with (2.1.1), that there is a point ξ in the smallest interval containing the points $\{x_0, \ldots, x_n\}$ such that

$$f[x_0, \ldots, x_n] = f^{(n)}(\xi) \int_0^{t_0} \cdots \int_0^{t_{n-1}} dt_n dt_{n-1} \ldots dt_1$$

$$= f^{(n)}(\xi) \int_0^{t_0} \cdots \int_0^{t_{n-2}} t_{n-1} dt_{n-1} \ldots dt_1$$

$$= \cdots = \frac{f^{(n)}(\xi)}{n!},$$

analogously to the final argument in the proof of Theorem 1.4.5. ∎

We now combine Theorems 2.1.1 and 2.1.2, and use the fact that (2.1.6) implies

$$Q_{n+1}(x) = 0, \quad x \in \Delta_n, \tag{2.1.11}$$

to immediately deduce the following result, in which, as throughout the book, we adopt, for any non-negative integer m, the notation $C^m[a,b]$ to denote the linear space of functions $f : [a,b] \to \mathbb{R}$ such that $f^{(k)} \in C[a,b], k = 0, \ldots, m$, according to which $C^0[a,b] = C[a,b]$.

Theorem 2.1.3. *For a non-negative integer n, suppose $f \in C^{n+1}[a,b]$. Then, for any $x \in [a,b]$, there is a point $\xi \in (a,b)$ such that the error function E_n^I in (2.1.3) satisfies*

$$E_n^I(x) = \frac{f^{(n+1)}(\xi)}{(n+1)!}Q_{n+1}(x), \tag{2.1.12}$$

with the polynomial $Q_{n+1} \in \pi_{n+1}$ given by (2.1.6).

Next, for any function $g \in C[a,b]$, we introduce the notation

$$\|g\|_\infty := \max_{a \leqslant x \leqslant b} |g(x)|, \tag{2.1.13}$$

in terms of which the following interpolation error estimate holds.

Theorem 2.1.4. *The interpolation error function E_n^I in Theorem 2.1.3 satisfies the estimate*

$$\|E_n^I\|_\infty \leqslant \frac{\|f^{(n+1)}\|_\infty}{(n+1)!}\|Q_{n+1}\|_\infty. \tag{2.1.14}$$

Proof. Let $x \in [a,b]$ be fixed. It follows from (2.1.12) in Theorem 2.1.3, together with (2.1.13), that

$$|E_n^I(x)| \leqslant \frac{\|f^{(n+1)}\|_\infty}{(n+1)!}\|Q_{n+1}\|_\infty,$$

from which the desired estimate (2.1.14) then immediately follows. ∎

Example 2.1.1. Consider the case $f(x) = \cos x$, and $[a,b] = [0, \frac{\pi}{2}]$.

(a) For $n = 2$, let

$$\Delta_2 = \{x_0, x_1, x_2\} := \{0, \frac{\pi}{4}, \frac{\pi}{2}\}. \tag{2.1.15}$$

Then, by using either of the interpolation formulas (1.2.5) or (1.3.17), we obtain

$$P_2^I(x) = \frac{8}{\pi^2}(1 - \sqrt{2})x^2 + \frac{2}{\pi}(2\sqrt{2} - 3)x + 1.$$

Moreover, the error estimate (2.1.14) yields

$$\max_{0 \leqslant x \leqslant \frac{\pi}{2}} \left| \cos x - P_2^I(x) \right| \leqslant \frac{1}{3!} \left[\max_{0 \leqslant x \leqslant \frac{\pi}{2}} |\sin x| \right] \max_{0 \leqslant x \leqslant \frac{\pi}{2}} \left| x \left(x - \frac{\pi}{4} \right) \left(x - \frac{\pi}{2} \right) \right|$$

$$= \frac{1}{6} \frac{\sqrt{3}\pi^3}{288} = \frac{\sqrt{3}\pi^3}{1728} \approx 0.031. \tag{2.1.16}$$

(b) For $n = 9$, let $\triangle_9 = \{x_0, \ldots, x_9\}$ denote any sequence of 10 distinct points in $[0, \frac{\pi}{2}]$. Then the corresponding interpolation polynomial P_9^I can be calculated by means of either (1.2.5) or (1.3.17), and the error estimate (2.1.14) gives

$$\max_{0 \leqslant x \leqslant 2} \left| \cos x - P_9^I(x) \right| \leqslant \frac{1}{10!} \left[\max_{0 \leqslant x \leqslant \frac{\pi}{2}} |\cos x| \right] \max_{0 \leqslant x \leqslant \frac{\pi}{2}} \prod_{j=0}^{9} |x - x_j|$$

$$\leqslant \frac{1}{10!} \left(\frac{\pi}{2} \right)^{10} \approx 2.52 \times 10^{-5}. \tag{2.1.17}$$

■

Observe that the upper bound on $||E_n^I||_\infty$, as given by the right hand side of (2.1.14), depends on f, n and $\triangle_n := \{x_0, \ldots, x_n\}$, with the dependence on $\triangle_n$ entirely restricted to the factor $||Q_{n+1}||_\infty$. Moreover, $||Q_{n+1}||_\infty$ is independent of f. We shall proceed in Section 2.2 to investigate the existence of a sequence $\triangle_n$ which minimizes $||Q_{n+1}||_\infty$.

2.2   The Chebyshev interpolation points

The Chebyshev polynomials $\{T_j : j = 0, 1, \ldots\}$ are defined recursively by

$$\left. \begin{array}{l} T_0(x) := 1 \quad ; \quad T_1(x) := x \quad ; \\[2mm] T_{j+1}(x) := 2xT_j(x) - T_{j-1}(x), \quad j = 1, 2, \ldots. \end{array} \right\} \tag{2.2.1}$$

By using (2.2.1), we obtain

$$\left. \begin{array}{l} T_2(x) = 2x^2 - 1 \quad ; \quad T_3(x) = 4x^3 - 3x \quad ; \quad T_4(x) = 8x^4 - 8x^2 + 1; \\[2mm] T_5(x) = 16x^5 - 20x^3 + 5x \quad ; \quad T_6(x) = 32x^6 - 48x^4 + 18x^2 - 1. \end{array} \right\} \tag{2.2.2}$$

The following properties are satisfied by the Chebyshev polynomials.

Theorem 2.2.1. *For $j \in \mathbb{N}$, the Chebyshev polynomial T_j, as defined in (2.2.1), satisfies:*

(a) *T_j is a polynomial of degree j such that the leading coefficient in*

$$T_j(x) = \sum_{k=0}^{j} c_{j,k} x^k \tag{2.2.3}$$

is given by

$$c_{j,j} = 2^{j-1}; \tag{2.2.4}$$

(b)
$$T_j(x) = \cos(j \arccos x), \quad x \in [-1,1]; \tag{2.2.5}$$

(c)
$$|T_j(x)| \leqslant 1, \quad x \in [-1,1]; \tag{2.2.6}$$

(d)
$$T_j\left(\cos\left(\frac{j-k}{j}\pi\right)\right) = (-1)^{j-k}, \quad k = 0, \ldots, j; \tag{2.2.7}$$

(e)
$$T_j\left(\cos\left(\frac{2j-1-2k}{2j}\pi\right)\right) = 0, \quad k = 0, \ldots, j-1; \tag{2.2.8}$$

(f)
$$T_j(x) = 2^{j-1} \prod_{k=0}^{j-1}\left[x - \cos\left(\frac{2j-1-2k}{2j}\pi\right)\right], \quad x \in \mathbb{R}. \tag{2.2.9}$$

Proof. (a) The properties (2.2.3) and (2.2.4) follow inductively from the definition (2.2.1).
(b) Let the function sequence $\{g_j : j = 0, 1, \ldots\}$ be defined by

$$g_j(x) := \cos(j \arccos x), \quad x \in [-1,1], \quad j = 0, 1, \ldots, \tag{2.2.10}$$

and introduce the one-to-one mapping between the intervals $[0, \pi]$ and $[-1, 1]$ as given by

$$x = \cos\theta, \quad \theta \in [0, \pi], \tag{2.2.11}$$

or equivalently,

$$\theta = \arccos x, \quad x \in [-1,1], \tag{2.2.12}$$

in terms of which (2.2.10) may be written as

$$g_j(x) = \cos(j\theta), \quad \theta \in [0, \pi], \quad j = 0, 1, \ldots. \tag{2.2.13}$$

The trigonometric identity

$$\cos[(j+1)\theta] + \cos[(j-1)\theta] = 2(\cos\theta)\cos(j\theta),$$

together with (2.2.13) and (2.2.11), yields the identity

$$g_{j+1}(x) + g_{j-1}(x) = 2xg_j(x), \quad x \in [-1,1], \quad j = 1, 2, \ldots,$$

and thus, by using also (2.2.13) for $j = 0$ and $j = 1$, as well as (2.2.11), we obtain

$$\left.\begin{array}{l} g_0(x) = 1; \quad g_1(x) = x; \\ g_{j+1}(x) = 2xg_j(x) - g_{j-1}(x), \quad j = 1, 2, \ldots, \end{array}\right\} x \in [-1,1]. \tag{2.2.14}$$

It follows from (2.2.1) and (2.2.14) that $g_j(x) = T_j(x), x \in [-1,1],\ j = 0,1,\ldots,$ which, together with (2.2.10), proves the formula (2.2.5).

(c) The property (2.2.6) is an immediate consequence of (2.2.5).

(d) For $j \in \mathbb{N}$ and $k = 0,\ldots,j$, we have, from (2.2.5),

$$T_j\left(\cos\left(\frac{j-k}{j}\pi\right)\right) = \cos\left(j\arccos\left(\cos\left(\frac{j-k}{j}\pi\right)\right)\right)$$

$$= \cos((j-k)\pi) = (-1)^{j-k},$$

which proves (2.2.7).

(e) Similarly, for $j \in \mathbb{N}$ and $k = 0,\ldots,j-1$, we deduce from (2.2.5) that

$$T_j\left(\cos\left(\frac{2j-1-2k}{2j}\pi\right)\right) = \cos\left(j\arccos\left(\cos\left(\frac{2j-2k-1}{2j}\pi\right)\right)\right)$$

$$= \cos\left(\left(j-k-\frac{1}{2}\right)\pi\right) = 0,$$

and thereby proving (2.2.8).

(f) The explicit formulation (2.2.9) is an immediate consequence of (2.2.3), (2.2.4) and (2.2.8).   ∎

Observe from Theorem 2.2.1(f) that, for $j \in \mathbb{N}$, the Chebyshev polynomial T_j of degree j has precisely j distinct zeros in $(-1,1)$, with, more precisely,

$$T_j(t_{j,k}) = 0, \quad k = 0,\ldots,j-1, \tag{2.2.15}$$

where

$$t_{j,k} := \cos\left(\frac{2j-1-2k}{2j}\pi\right), \quad k = 0,\ldots,j-1, \tag{2.2.16}$$

and thus

$$-1 < t_{j,0} < t_{j,1} < \cdots < t_{j,j-1} < 1. \tag{2.2.17}$$

Moreover, according to Theorem 2.2.1(d), the Chebyshev polynomial T_j attains, for $j \in \mathbb{N}$, its maximum $(= 1)$ and minimum $(= -1)$ on $[-1,1]$ alternately, in the sense that

$$T_j(\xi_{j,k}) = (-1)^{j-k}, \quad k = 0,\ldots,j, \tag{2.2.18}$$

where

$$\xi_{j,k} := \cos\left(\frac{j-k}{j}\pi\right), \quad k = 0,\ldots,j, \tag{2.2.19}$$

and thus

$$-1 = \xi_{j,0} < \xi_{j,1} < \cdots < \xi_{j,j} = 1. \tag{2.2.20}$$

For any non-negative integer k, if $P(x) = \sum_{j=1}^{k} c_j x^j$, with leading coefficient $c_k = 1$, we say that P is a monic polynomial. The set of all monic polynomials in π_k will be denoted by the symbol $\widetilde{\pi}_k$. Observe from Theorem 2.2.1(a) that the normalized Chebyshev polynomials

$$\widetilde{T}_j := 2^{1-j} T_j, \quad j = 1, 2, \ldots, \tag{2.2.21}$$

are monic polynomials, that is,

$$\widetilde{T}_j \in \widetilde{\pi}_j, \quad j \in \mathbb{N}. \tag{2.2.22}$$

We shall rely on the following minimization property of $\widetilde{T}_j$.

Theorem 2.2.2. *For any $j \in \mathbb{N}$,*

$$\min_{P \in \widetilde{\pi}_j} \max_{-1 \leqslant x \leqslant 1} |P(x)| = \max_{-1 \leqslant x \leqslant 1} |\widetilde{T}_j(x)| = 2^{1-j}, \tag{2.2.23}$$

where $\widetilde{T}_j$ is the normalized Chebyshev polynomial defined by (2.2.21).

Proof. Let $j \in \mathbb{N}$. First, observe that (2.2.21), (2.2.6) and (2.2.7) imply the second equation in (2.2.23).

We use a proof by contradiction to prove the first equation in (2.2.23). Suppose therefore that there exists a polynomial $Q \in \widetilde{\pi}_j$ such that

$$\max_{-1 \leqslant x \leqslant 1} |Q(x)| < 2^{1-j}, \tag{2.2.24}$$

according to which $Q \neq \widetilde{T}_j$, and define the polynomial

$$R := (-1)^j (\widetilde{T}_j - Q), \tag{2.2.25}$$

for which it then follows that R is not the zero polynomial. Since $\widetilde{T}_j$ and Q are both monic polynomials in $\widetilde{\pi}_j$, it follows from (2.2.25) that

$$R \in \pi_{j-1}. \tag{2.2.26}$$

Now observe from (2.2.21) and (2.2.18) that

$$\widetilde{T}_j(\xi_{j,k}) = (-1)^{j-k} 2^{1-j}, \quad k = 0, \ldots, j, \tag{2.2.27}$$

where the sequence $\{\xi_{j,k} : k = 0, \ldots, j\}$ is given by (2.2.19), and satisfies (2.2.20). By using (2.2.25), (2.2.27) and (2.2.24), we deduce that

$$R(\xi_{j,0}) = 2^{1-j} - (-1)^j Q(\xi_{j,0}) > 0;$$

$$R(\xi_{j,1}) = -2^{1-j} - (-1)^j Q(\xi_{j,1}) < 0,$$

and it follows from the intermediate value theorem that there is a point $\eta_1 \in (\xi_{j,0}, \xi_{j,1})$ such that $R(\eta_1) = 0$. Similarly it can be shown by means of (2.2.25), (2.2.27) and (2.2.24) that $R(\xi_{j,k})$ alternates in sign for $k = 1, \ldots, j$, and that there consequently exist points $\eta_k \in (\xi_{j,k-1}, \xi_{j,k})$, $k = 2, \ldots, j$, such that $R(\eta_k) = 0$, $k = 2, \ldots, j$. Hence R has j distinct real zeros at $\{\eta_1, \ldots, \eta_j\}$. Since also (2.2.26) holds, it follows that R must be the zero polynomial, which is a contradiction, and thereby concluding our proof of the first equation in (2.2.23).   ∎

We proceed to show how Theorem 2.2.2 can be used to minimize the factor

$$\|Q_{n+1}\|_\infty := \max_{a \leqslant x \leqslant b} |Q_{n+1}(x)| \tag{2.2.28}$$

in (2.1.14) with respect to the choice of the interpolation point sequence $\triangle_n := \{x_0, \ldots, x_n\}$. To this end, we introduce the one-to-one mapping between the intervals $[-1, 1]$ and $[a, b]$ as given by

$$x = \frac{1}{2}(b - a)t + \frac{1}{2}(a + b), \quad t \in [-1, 1], \tag{2.2.29}$$

or equivalently,

$$t = \frac{2}{b - a}\left[x - \frac{1}{2}(a + b)\right], \quad x \in [a, b]. \tag{2.2.30}$$

Based on (2.2.15), (2.2.16) and (2.2.17), for $n \in \mathbb{N}$ and $j = n + 1$, we now define the Chebyshev interpolation points

$$x_{n,j}^C := \frac{1}{2}(b - a)\cos\left(\frac{2n + 1 - 2j}{2n + 2}\pi\right) + \frac{1}{2}(a + b), \quad j = 0, \ldots, n, \tag{2.2.31}$$

which then satisfy

$$a < x_{n,0}^C < x_{n,1}^C < \cdots < x_{n,n}^C < b. \tag{2.2.32}$$

Observe from (2.2.31) that the Chebyshev interpolation points are concentrated more densely towards the endpoints of the interval $[a, b]$. The following minimization property can now be proved by means of Theorem 2.2.2.

Theorem 2.2.3. *The factor* $\|Q_{n+1}\|_\infty$ *in the polynomial interpolation error estimate* (2.1.14) *of Theorem* 2.1.4 *is minimized by*

$$\min_{x_0, \ldots, x_n \in [a,b]} \|Q_{n+1}\|_\infty = \max_{a \leqslant x \leqslant b}\left|\prod_{j=0}^{n}(x - x_{n,j}^C)\right| = 2^{-n}\left(\frac{b - a}{2}\right)^{n+1}, \tag{2.2.33}$$

with $\{x_{n,j}^C : j = 0, \ldots, n\}$ *denoting the Chebyshev interpolation points, as defined in* (2.2.31).

Proof. First, we use the one-to-one mapping (2.2.29), (2.2.30) between the intervals $[a,b]$ and $[-1,1]$ to deduce that

$$\min_{x_0,\ldots,x_n \in [a,b]} \max_{a \leqslant x \leqslant b} \left| \prod_{j=0}^{n} (x-x_j) \right|$$

$$= \min_{x_0,\ldots,x_n \in [a,b]} \max_{-1 \leqslant t \leqslant 1} \left| \prod_{j=0}^{n} \frac{b-a}{2} \left[t - \frac{2}{b-a} \left(x_j - \frac{1}{2}(a+b) \right) \right] \right|$$

$$= \left(\frac{b-a}{2} \right)^{n+1} \min_{t_0,\ldots,t_n \in [-1,1]} \max_{-1 \leqslant t \leqslant 1} \left| \prod_{j=0}^{n} (t-t_j) \right|. \tag{2.2.34}$$

For the sequence $\{t_{n+1,j} : j = 0,\ldots,n\}$ as defined by means of (2.2.16), it follows from Theorem 2.2.2, together with (2.2.21) and (2.2.9), that

$$2^{-n} = \max_{-1 \leqslant t \leqslant 1} \left| \widetilde{T}_{n+1}(t) \right| = \max_{-1 \leqslant t \leqslant 1} \left| \prod_{j=0}^{n} (t-t_{n+1,j}) \right| \geqslant \min_{t_0,\ldots,t_n \in [-1,1]} \max_{-1 \leqslant t \leqslant 1} \left| \prod_{j=0}^{n} (t-t_j) \right|$$

$$\geqslant \min_{P \in \widetilde{\pi}_{n+1}} \max_{-1 \leqslant t \leqslant 1} |P(t)| = 2^{-n},$$

and thus

$$\min_{t_0,\ldots,t_n \in [-1,1]} \max_{-1 \leqslant t \leqslant 1} \left| \prod_{j=0}^{n} (t-t_j) \right| = \max_{-1 \leqslant t \leqslant 1} \left| \prod_{j=0}^{n} (t-t_{n+1,j}) \right| = 2^{-n},$$

which, together with (2.2.34), and (2.1.6), yields the desired result (2.2.33). ∎

By combining Theorems 2.1.4 and 2.2.3, we immediately derive the following optimal polynomial interpolation error estimate.

Theorem 2.2.4. *In Theorem* 2.1.3, *for any positive integer n, let the interpolation points be chosen as the Chebyshev interpolation points, that is,*

$$x_j = x_{n,j}^C, \quad j = 0,\ldots,n, \tag{2.2.35}$$

as defined by (2.2.31). *Then the error estimate*

$$||E_n^I||_\infty \leqslant \frac{1}{2^n (n+1)!} \left(\frac{b-a}{2} \right)^{n+1} ||f^{(n+1)}||_\infty \tag{2.2.36}$$

is satisfied.

Example 2.2.1. As in Example 2.1.1, we consider the case $f(x) = \cos x$, and $[a,b] = [0, \frac{\pi}{2}]$, in which case, for any $n \in \mathbb{N}$, the Chebyshev interpolation points are given, according to (2.2.31), by

$$x_{n,j}^C = \frac{\pi}{4} \left[\cos \left(\frac{2n+1-2j}{2n+2} \pi \right) + 1 \right], \quad j = 0,\ldots,n, \tag{2.2.37}$$

and the corresponding error estimate (2.2.36) is

$$\max_{0 \leqslant x \leqslant \frac{\pi}{2}} \left| \cos x - P_n^I(x) \right| \leqslant \frac{1}{2^n (n+1)!} \left(\frac{\pi}{4} \right)^{n+1}. \tag{2.2.38}$$

(a) For $n = 2$, it follows from (2.2.37) that

$$\{x_{2,0}^C, x_{2,1}^C, x_{2,2}^C\} = \left\{ \frac{2 - \sqrt{3}}{8} \pi, \frac{\pi}{4}, \frac{2 + \sqrt{3}}{8} \pi \right\},$$

and (2.2.38) gives the estimate

$$\max_{0 \leqslant x \leqslant \frac{\pi}{2}} \left| \cos x - P_2^I(x) \right| \leqslant \frac{1}{24} \left(\frac{\pi}{4} \right)^3 \approx 0.02,$$

which improves on the error estimate (2.1.16) in Example 2.1.1(a).

(b) For $n = 9$, the formula (2.2.37) yields the Chebyshev interpolation points

$$x_{9,j}^C = \frac{\pi}{4} \left[\cos \left(\frac{19 - 2j}{20} \pi \right) + 1 \right], \quad j = 0, \ldots, 9,$$

and (2.2.38) gives the estimate

$$\max_{0 \leqslant x \leqslant \frac{\pi}{2}} \left| \cos x - P_9^I(x) \right| \leqslant \frac{1}{2^9 10!} \left(\frac{\pi}{4} \right)^{10} \approx 4.81 \times 10^{-11},$$

which is a considerable improvement on the error estimate (2.1.17) in Example 2.1.1(b).

∎

2.3 Exercises

Exercise 2.1 For the function

$$f(x) = \frac{1}{\sqrt{x}},$$

find a point $\xi \in [\frac{1}{9}, 1]$, as guaranteed by Theorem 2.1.2, for which it holds that

$$f[\tfrac{1}{9}, \tfrac{1}{4}, 1] = \tfrac{1}{2} f''(\xi).$$

Exercise 2.2 Let

$$f(x) = \ln(x+2), \quad x \in [0, 2],$$

and, for $n \in \{1, 2\}$, denote by P_n^I the interpolation polynomial in π_n such that

$$P_n^I(x) = f(x), \quad x \in \Delta_n,$$

where

$$\Delta_1 := \{\tfrac{1}{2}, \tfrac{3}{2}\} \quad ; \quad \Delta_2 := \{\tfrac{1}{2}, 1, \tfrac{3}{2}\}.$$

For $n = 1$ and $n = 2$, calculate the polynomial P_n^I, as well as the interpolation error estimate (2.1.14) in Theorem 2.1.4, with $[a,b] = [0,2]$. Also, for $n = 1$ and $n = 2$, investigate the sharpness of these estimates by calculating the exact value of $||E_n^I||_\infty$.

Exercise 2.3 As a continuation of Exercise 2.2, let n be any positive integer, and suppose

$$\triangle_n := \{x_0, \ldots, x_n\} \subset [0,2]$$

is an arbitrary point sequence in $[0,2]$. Apply the interpolation error estimate (2.1.14) in Theorem 2.1.4 to show that

$$\max_{0 \leqslant x \leqslant 2} |\ln(x+2) - P_n^I(x)| \leqslant \frac{1}{n+1}, \qquad (*)$$

with P_n^I denoting the interpolation polynomial in π_n with respect to the interpolation point sequence $\triangle_n$.

Exercise 2.4 Calculate the Chebyshev polynomials T_7 and T_8, thereby extending the formulas in (2.2.2).

Exercise 2.5 Calculate, for $n = 1$ and $n = 2$, the sequences $\triangle_n^C$ defined by

$$\triangle_n^C := \{x_{n,0}^C, \ldots, x_{n,n}^C\}, \quad n \in \mathbb{N},$$

with $\{x_{n,0}^C, \ldots, x_{n,n}^C\}$ denoting the Chebyshev interpolation points, as given in (2.2.31), for the interval $[0,2]$.

Exercise 2.6 As a continuation of Exercise 2.5, repeat Exercises 2.2 and 2.3 with $\triangle_n$ replaced by $\triangle_n^C$, and with the interpolation error estimate (2.1.14) replaced by (2.2.36) in Theorem 2.2.4. In particular, obtain the analogue of the estimate $(*)$ in Exercise 2.3.

Exercise 2.7 As a continuation of Exercise 2.6, find, according to the error estimate obtained there, the smallest possible value of n for which it holds that

$$\max_{0 \leqslant x \leqslant 2} |\ln(x+2) - P_n^I(x)| < \frac{1}{100}.$$

Exercise 2.8 Apply Theorem 2.2.2 to obtain the minimum value

$$\min_{a,b,c \in \mathbb{R}} \max_{-1 \leqslant x \leqslant 1} |x^3 + ax^2 + bx + c|,$$

as well as the corresponding optimal values of the coefficients a, b and c.

Exercise 2.9 Prove that, for any fixed $j \in \mathbb{N}$, the sum of the coefficients of the Chebyshev polynomial T_j is equal to one.

[*Hint:* Use Theorem 2.2.1(b).]

Exercise 2.10 Prove that the Chebyshev polynomials $\{T_0, T_1, \ldots\}$ satisfy the condition

$$\int_{-1}^{1} \frac{1}{\sqrt{1-x^2}} T_j(x) T_k(x) dx = 0, \quad \text{if } j \neq k.$$

[*Hint:* Apply the transformation (2.2.11), (2.2.12).]

Chapter 3

Polynomial Uniform Convergence

For a sequence of polynomials that approximate a function $f \in C[a,b]$, an important issue for investigation is if the corresponding sequence of approximation errors converges uniformly to zero on $[a,b]$. In this chapter, we first show that such convergence is not guaranteed in the case that the polynomials are constructed to interpolate $f \in C[a,b]$ on some sequence of equally-spaced sample point sets with cardinalities increasing to infinity. We then proceed to give an explicit construction of polynomials that approximate any given function $f \in C[a,b]$, for which the sequence of approximation errors does indeed converge uniformly to zero, as the degrees of the polynomials tend to infinity.

3.1 General definition and examples

Let $f \in C[a,b]$. If a function sequence $\{f_n : n = 1,2,\ldots\} \subset C[a,b]$ is such that

$$||f - f_n||_\infty := \max_{a \leqslant x \leqslant b} |f(x) - f_n(x)| \to 0, \quad n \to \infty, \tag{3.1.1}$$

we say that the sequence $\{f_n\}$ converges uniformly on $[a,b]$ to f.

We proceed to provide two examples from polynomial interpolation.

Example 3.1.1. As in Examples 2.1.1 and 2.2.1, let $f(x) = \cos x$, and $[a,b] = [0, \frac{\pi}{2}]$, and denote by P_n^I the interpolation polynomial with respect to the Chebyshev interpolation points (2.2.37). Then, since

$$\frac{1}{2^n(n+1)!} \left(\frac{\pi}{4}\right)^{n+1} \to 0, \quad n \to \infty,$$

the corresponding polynomial interpolation error estimate (2.2.38) yields

$$\max_{0 \leqslant x \leqslant \frac{\pi}{2}} |\cos x - P_n^I(x)| \to 0, \quad n \to \infty,$$

that is, the sequence $\{P_n^I : n = 1,2,\ldots\}$ converges uniformly on $[0, \frac{\pi}{2}]$ to f. ∎

Example 3.1.2. Let $f(x) = \ln x$, and $[a,b] = [\frac{1}{2}, \frac{5}{2}]$. Then, for $n = 1, 2, \ldots$, the Chebyshev interpolation points are given, according to (2.2.31), by

$$x_{n,j}^C = \cos\left(\frac{2n+1-2j}{2n+2}\pi\right) + \frac{3}{2}, \quad j = 0, \ldots, n,$$

and the corresponding interpolation polynomial sequence $\{P_n^I : n = 1, 2, \ldots\}$ can be computed by means of, for example, the Newton interpolation formula (1.3.17). Moreover, we may apply (2.2.36) in Theorem 2.2.4 to obtain the error estimate

$$\max_{\frac{1}{2} \leqslant x \leqslant \frac{5}{2}} \left|\ln x - P_n^I(x)\right| \leqslant \frac{1}{2^n(n+1)!} \max_{\frac{1}{2} \leqslant x \leqslant \frac{5}{2}} \left|\left(\frac{d}{dx}\right)^{n+1}(\ln x)\right|, \tag{3.1.2}$$

for $n = 1, 2, \ldots$. But

$$\left(\frac{d}{dx}\right)^{n+1}(\ln x) = (-1)^n \frac{n!}{x^{n+1}},$$

and thus

$$\max_{\frac{1}{2} \leqslant x \leqslant \frac{5}{2}} \left|\left(\frac{d}{dx}\right)^{n+1}(\ln x)\right| = n! \max_{\frac{1}{2} \leqslant x \leqslant \frac{5}{2}} \frac{1}{x^{n+1}} = \frac{n!}{(\frac{1}{2})^{n+1}} = n!2^{n+1},$$

which, together with (3.1.2), gives

$$\max_{\frac{1}{2} \leqslant x \leqslant \frac{5}{2}} \left|\ln x - P_n^I(x)\right| \leqslant \frac{2}{n+1}. \tag{3.1.3}$$

Since

$$\frac{2}{n+1} \to 0, \quad n \to \infty,$$

we deduce from (3.1.3) that

$$\max_{\frac{1}{2} \leqslant x \leqslant \frac{5}{2}} \left|\ln x - P_n^I(x)\right| \to 0, \quad n \to \infty,$$

that is, the interpolation polynomial sequence $\{P_n^I : n = 1, 2, \ldots\}$ converges uniformly on $[\frac{1}{2}, \frac{5}{2}]$ to f. ■

The uniform convergence result

$$\|f - P_n^I\|_\infty := \max_{a \leqslant x \leqslant b} \left|f(x) - P_n^I(x)\right| \to 0, \quad n \to \infty, \tag{3.1.4}$$

is not obtained for all choices of $f \in C[a,b]$, $\triangle_n := \{x_0, \ldots, x_n\}$ and $[a,b]$, as illustrated by the following example.

Example 3.1.3. (Runge example) For $[a,b] = [-5,5]$, let

$$f(x) = \frac{1}{1+x^2}, \tag{3.1.5}$$

and choose, for $n = 1, 2, \ldots$, the interpolation points

$$x_j = x_{n,j} := -5 + \frac{10j}{n}, \quad j = 0, \ldots, n, \tag{3.1.6}$$

that is, $\{x_{n,j} : j = 0, \ldots, n\}$ are the uniformly distributed partition points of the interval $[-5, 5]$, with

$$-5 = x_{n,0} < x_{n,1} < \cdots < x_{n,n} = 5. \tag{3.1.7}$$

We proceed to prove the divergence result

$$\max_{-5 \leqslant x \leqslant 5} |E_n^I(x)| := \max_{-5 \leqslant x \leqslant 5} \left| \frac{1}{1 + x^2} - P_n^I(x) \right| \to \infty, \quad n \to \infty. \tag{3.1.8}$$

To this end, we let $\{\widetilde{x}_n : n \in \mathbb{N}\}$ denote the midpoints of the intervals $\{[x_{n,n-1}, x_{n,n}] : n \in \mathbb{N}\}$, that is,

$$\widetilde{x}_n := \frac{1}{2}(x_{n,n-1} + x_{n,n}) = 5 - \frac{5}{n}, \quad n \in \mathbb{N}, \tag{3.1.9}$$

from (3.1.6). We shall show that

$$|E_n(\widetilde{x}_n)| \to \infty, \quad n \to \infty, \tag{3.1.10}$$

which will then imply the desired divergence result (3.1.8).

To prove (3.1.10), we first apply the second line of (2.1.5) in Theorem 2.1.1, together with the definition (2.1.6), to obtain

$$E_n^I(\widetilde{x}_n) = f[\widetilde{x}_n, x_{n,0}, \ldots, x_{n,n}] \prod_{j=0}^{n} (\widetilde{x}_n - x_{n,j}), \quad n \in \mathbb{N}. \tag{3.1.11}$$

Our next step is to explicitly calculate the divided difference in (3.1.11). To this end, for any integer $m \in \mathbb{N}$, let $\{t_0, \ldots, t_m\}$ denote a point sequence in $\mathbb{R}$ satisfying

$$t_0 < t_1 < \cdots < t_m, \tag{3.1.12}$$

as well as symmetry with respect to the origin, in the sense that

$$t_{m-j} = -t_j, \quad j = 0, \ldots, m. \tag{3.1.13}$$

Observe from (3.1.13) that then

$$t_{m/2} = 0, \quad \text{if} \quad m \text{ is even.} \tag{3.1.14}$$

Also, let $x \in \mathbb{R} \setminus \{t_0, \ldots, t_m\}$.

For the case $m = 1$ in (3.1.12), (3.1.13), according to which $t_1 = -t_0$, we now apply the recursive formulation (1.3.21) in Theorem 1.3.4, together with the definition (3.1.5), to obtain, for $x \in \mathbb{R} \setminus \{t_0, t_1\}$,

$$f[x, t_0, t_1] = \frac{1}{t_1 - x} \left[\frac{\frac{1}{1+t_1^2} - \frac{1}{1+t_0^2}}{t_1 - t_0} - \frac{\frac{1}{1+t_0^2} - \frac{1}{1+x^2}}{t_0 - x} \right] = \frac{1}{t_0 + x} \left[\frac{x^2 - t_0^2}{(t_0 - x)(1 + t_0^2)(1 + x^2)} \right],$$

and thus

$$f[x,t_0,t_1] = -\frac{f(x)}{1+t_0^2}. \tag{3.1.15}$$

Next, for the case $m = 2$ in (3.1.12),(3.1.13), for which, by using also (3.1.14), we have $\{t_0,t_1,t_2\} = \{t_0,0,-t_0\}$, it follows from (1.3.21) and (3.1.5) that

$$f[x,t_0,t_1] = f[x,t_0,0] = \frac{1}{0-x}\left[\frac{1-\frac{1}{1+t_0^2}}{0-t_0} - \frac{\frac{1}{1+t_0^2}-\frac{1}{1+x^2}}{t_0-x}\right] = \frac{t_0x-1}{(1+t_0^2)(1+x^2)}, \tag{3.1.16}$$

whereas

$$f[t_0,t_1,t_2] = f[t_0,0,t_2] = \frac{1}{t_2-t_0}\left[\frac{\frac{1}{1+t_2^2}-1}{t_2-0} - \frac{1-\frac{1}{1+t_0^2}}{0-t_0}\right] = \frac{1}{2t_0}\left[\frac{-t_0}{1+t_0^2} + \frac{-t_0}{1+t_0^2}\right],$$

that is,

$$f[t_0,t_1,t_2] = -\frac{1}{1+t_0^2}. \tag{3.1.17}$$

It follows from (1.3.21), (3.1.16) and (3.1.17), as well as $t_2 = -t_0$, that

$$f[x,t_0,t_1,t_2] = \frac{1}{t_2-x}\left[-\frac{1}{1+t_0^2} - \frac{t_0x-1}{(1+t_0^2)(1+x^2)}\right] = \frac{1}{x+t_0}\left[\frac{x(x+t_0)}{(1+t_0^2)(1+x^2)}\right],$$

and thus

$$f[x,t_0,t_1,t_2] = \frac{xf(x)}{1+t_0^2}. \tag{3.1.18}$$

Now observe from (3.1.15) and (3.1.18) that the statement

$$f[x,t_0,\ldots,t_m] = \begin{cases} (-1)^{(m+1)/2}\dfrac{f(x)}{\displaystyle\prod_{j=0}^{(m-1)/2}(1+t_j^2)}, & \text{if } m \text{ is odd;} \\[2em] (-1)^{(m/2)-1}\dfrac{xf(x)}{\displaystyle\prod_{j=0}^{(m/2)-1}(1+t_j^2)}, & \text{if } m \text{ is even,} \end{cases} \tag{3.1.19}$$

is true for $m = 1$ and $m = 2$.

Proceeding inductively, suppose that (3.1.19) holds for a fixed $m \in \mathbb{N}$, with $\{t_0,\ldots,t_m\}$ denoting any sequence in $\mathbb{R}$ such that (3.1.12) and (3.1.13) are satisfied. Let $\{t_0,\ldots,t_{m+2}\}$ denote any sequence in $\mathbb{R}$ satisfying

$$t_0 < t_1 < \cdots < t_{m+2}; \tag{3.1.20}$$

$$t_{m+2-j} = -t_j, \quad j = 0,\ldots,m+2, \tag{3.1.21}$$

and define the function

$$g(x) := f[x, t_1, \ldots, t_{m+1}], \quad x \in \mathbb{R} \setminus \{t_0, \ldots, t_{m+2}\}. \qquad (3.1.22)$$

By applying the recursive formulation (1.3.21) in Theorem 1.3.4, as well as the symmetry result (1.3.26) of Theorem 1.3.6, it follows from (3.1.22) that, for any $x \in \mathbb{R} \setminus \{t_0, \ldots, t_{m+2}\}$,

$$g[x, t_0, t_{m+2}] = \frac{1}{t_{m+2} - x} \left[\frac{g(t_{m+2}) - g(t_0)}{t_{m+2} - t_0} - \frac{g(t_0) - g(x)}{t_0 - x} \right]$$

$$= \frac{1}{t_{m+2} - x} \left[\frac{f[t_1, \ldots, t_{m+2}] - f[t_0, \ldots, t_{m+1}]}{t_{m+2} - t_0} \right.$$

$$\left. - \frac{f[t_1, \ldots, t_{m+1}, t_0] - f[x, t_1, \ldots, t_{m+1}]}{t_0 - x} \right]$$

$$= \frac{f[t_0, \ldots, t_{m+2}] - f[x, t_1, \ldots, t_{m+1}, t_0]}{t_{m+2} - x}$$

$$= \frac{f[t_0, \ldots, t_{m+2}] - f[x, t_0, \ldots, t_{m+1}]}{t_{m+2} - x} = f[x, t_0, \ldots, t_{m+2}],$$

that is,

$$g[x, t_0, t_{m+2}] = f[x, t_0, \ldots, t_{m+2}], \quad x \in \mathbb{R} \setminus \{t_0, \ldots, t_{m+2}\}. \qquad (3.1.23)$$

With the definition

$$\tau_j := t_{j+1}, \quad j = 0, \ldots, m, \qquad (3.1.24)$$

it follows from (3.1.20), (3.1.21) that

$$\tau_0 < \tau_1 < \cdots < \tau_m; \qquad (3.1.25)$$

$$\tau_{m-j} = t_{m+2-(j+1)} = -t_{j+1} = -\tau_j, \quad j = 0, \ldots, m. \qquad (3.1.26)$$

Hence we may apply the inductive hypothesis (3.1.19) to deduce from (3.1.22) that

$$g(x) = \begin{cases} (-1)^{(m+1)/2} \dfrac{f(x)}{\prod\limits_{j=0}^{(m-1)/2} (1+\tau_j^2)}, & \text{if } m \text{ is odd;} \\[4ex] (-1)^{(m/2)-1} \dfrac{x f(x)}{\prod\limits_{j=0}^{(m/2)-1} (1+\tau_j^2)}, & \text{if } m \text{ is even,} \end{cases}$$

and thus, from (3.1.24),

$$g(x) = \begin{cases} (-1)^{(m+1)/2} \dfrac{f(x)}{\prod\limits_{j=1}^{(m+1)/2} (1+t_j^2)}, & \text{if } m \text{ is odd;} \\[4ex] (-1)^{(m/2)-1} \dfrac{x f(x)}{\prod\limits_{j=1}^{m/2} (1+t_j^2)}, & \text{if } m \text{ is even.} \end{cases} \qquad (3.1.27)$$

Suppose m is odd. It then follows from the first line of (3.1.27), together with (1.3.3), (1.3.2), that

$$g[x,t_0,t_{m+2}] = \frac{(-1)^{(m+1)/2}}{\prod\limits_{j=1}^{(m+1)/2}(1+t_j^2)} f[x,t_0,t_{m+2}]. \tag{3.1.28}$$

With the definition

$$\tilde{\tau}_0 := t_0; \quad \tilde{\tau}_1 := t_{m+2}, \tag{3.1.29}$$

we observe from (3.1.20), (3.1.21) that

$$\tilde{\tau}_0 < \tilde{\tau}_1; \quad \tilde{\tau}_1 = -\tilde{\tau}_0.$$

Hence we may apply (3.1.15) and (3.1.29) to obtain

$$f[x,t_0,t_{m+2}] = f[x,\tilde{\tau}_0,\tilde{\tau}_1] = -\frac{f(x)}{1+\tilde{\tau}_0^2} = -\frac{f(x)}{1+t_0^2},$$

which can now be substituted into (3.1.28) to obtain

$$g[x,t_0,t_{m+2}] = \frac{(-1)^{(m+3)/2}}{\prod\limits_{j=0}^{(m+1)/2}(1+t_j^2)} f(x). \tag{3.1.30}$$

It follows from (3.1.30) and (3.1.23) that the first line of (3.1.19) also holds with m replaced by $m+2$, and thereby completing our inductive proof of the first line of (3.1.19).

Next, suppose m is even. Analogously to the derivation (3.1.28), we deduce from the second line of (3.1.27) that

$$g[x,t_0,t_{m+2}] = \frac{(-1)^{(m/2)-1}}{\prod\limits_{j=1}^{m/2}(1+t_j^2)} h[x,t_0,t_{m+2}], \tag{3.1.31}$$

where

$$h(x) := xf(x) = \frac{x}{1+x^2}, \tag{3.1.32}$$

from (3.1.5). By applying (1.3.21), (3.1.32), and $t_{m+2} = -t_0$, as follows from (3.1.21), we obtain

$$h[x,t_0,t_{m+2}] = h[x,t_0,-t_0] = \frac{1}{-t_0-x}\left[\frac{\frac{(-t_0)}{1+t_0^2}-\frac{t_0}{1+t_0^2}}{-2t_0} - \frac{\frac{t_0}{1+t_0^2}-\frac{x}{1+x^2}}{t_0-x}\right]$$

$$= -\frac{1}{x+t_0}\left[\frac{1}{1+t_0^2} + \frac{t_0x^2-(1+t_0^2)x+t_0}{(x-t_0)(1+t_0^2)(1+x^2)}\right]$$

$$= -\frac{1}{x+t_0}\left[\frac{1}{1+t_0^2} + \frac{t_0 x - 1}{(1+t_0^2)(1+x^2)}\right]$$

$$= -\frac{x}{(1+t_0^2)(1+x^2)},$$

which can now be substituted into (3.1.31) to obtain

$$g[x, t_0, t_{m+2}] = \frac{(-1)^{((m+2)/2)-1}}{\displaystyle\prod_{j=0}^{m/2}(1+t_j^2)}[xf(x)]. \tag{3.1.33}$$

It follows from (3.1.33) and (3.1.23) that the second line of (3.1.19) also holds with m replaced by $m+2$, and thereby completing our inductive proof of the second line of (3.1.19). Observing from (3.1.6) that, for any $n \in \mathbb{N}$, it holds that

$$x_{n,n-j} = -5 + \frac{10(n-j)}{n} = -\left(-5 + \frac{10j}{n}\right) = -x_{n,j}, \quad j = 0, \ldots, n, \tag{3.1.34}$$

and recalling also (3.1.7), it follows that the sequence

$$\{t_0, \ldots, t_n\} = \{x_{n,0}, \ldots, x_{n,n}\} \tag{3.1.35}$$

satisfies the conditions (3.1.12),(3.1.13), with $m = n$. Also, the definition (3.1.9), together with (3.1.6), shows that $\tilde{x}_n \in \mathbb{R} \setminus \{x_{n,0}, \ldots, x_{n,n}\}$. Hence we may apply (3.1.19) with $m = n$, and with the sequence $\{t_0, \ldots, t_n\}$ given by (3.1.35), to obtain the formulas

$$f[\tilde{x}_n, x_{n,0}, \ldots, x_{n,n}] = \begin{cases} (-1)^{(n+1)/2}\dfrac{f(\tilde{x}_n)}{\displaystyle\prod_{j=0}^{(n-1)/2}(1+x_{n,j}^2)}, & \text{if } n \text{ is odd;} \\[4ex] (-1)^{(n/2)-1}\dfrac{\tilde{x}_n f(\tilde{x}_n)}{\displaystyle\prod_{j=0}^{(n/2)-1}(1+x_{n,j}^2)}, & \text{if } n \text{ is even.} \end{cases} \tag{3.1.36}$$

Next, to evaluate the product in (3.1.11), suppose first n is odd, and denote by ν the non-negative integer such that $n = 2\nu + 1$. It then follows from (3.1.34), with $n = 2\nu + 1$, that

$$\prod_{j=0}^{n}(\tilde{x}_n - x_{n,j}) = \prod_{j=0}^{2\nu+1}(\tilde{x}_{2\nu+1} - x_{2\nu+1,j})$$

$$= \prod_{j=0}^{\nu}(\tilde{x}_{2\nu+1} - x_{2\nu+1,j}) \prod_{j=\nu+1}^{2\nu+1}(\tilde{x}_{2\nu+1} - x_{2\nu+1,j})$$

$$= \prod_{j=0}^{\nu}(\tilde{x}_{2\nu+1} - x_{2\nu+1,j}) \prod_{j=0}^{\nu}(\tilde{x}_{2\nu+1} - x_{2\nu+1,2\nu+1-j})$$

$$= \prod_{j=0}^{v}(\tilde{x}_{2v+1}-x_{2v+1,j})(\tilde{x}_{2v+1}+x_{2v+1,j}) = \prod_{j=0}^{v}(\tilde{x}_{2v+1}^2-x_{2v+1,j}^2),$$

and thus, since $v = (n-1)/2$,

$$\prod_{j=0}^{n}(\tilde{x}_n - x_{n,j}) = \prod_{j=0}^{(n-1)/2}(\tilde{x}_n^2 - x_{n,j}^2), \text{ if } n \text{ is odd.} \qquad (3.1.37)$$

If n is even, and v denotes the positive integer such that $n = 2v$, we apply (3.1.34), together with the fact that

$$x_{n,n/2} = x_{2v,v} = 0, \qquad (3.1.38)$$

as follows from (3.1.34), to obtain

$$\prod_{j=0}^{n}(\tilde{x}_n - x_{n,j}) = \prod_{j=0}^{2v}(\tilde{x}_{2v} - x_{2v,j})$$

$$= \left[\prod_{j=0}^{v-1}(\tilde{x}_{2v} - x_{2v,j})\right]\tilde{x}_{2v}\left[\prod_{j=v+1}^{2v}(\tilde{x}_{2v} - x_{2v,j})\right]$$

$$= \tilde{x}_{2v}\prod_{j=0}^{v-1}(\tilde{x}_{2v} - x_{2v,j})\prod_{j=0}^{v-1}(\tilde{x}_{2v} - x_{2v,2v-j})$$

$$= \tilde{x}_{2v}\prod_{j=0}^{v-1}(\tilde{x}_{2v} - x_{2v,j})(\tilde{x}_{2v} + x_{2v,j})$$

$$= \tilde{x}_{2v}\prod_{j=0}^{v-1}(\tilde{x}_{2v}^2 - x_{2v,j}^2) = \frac{1}{\tilde{x}_{2v}}\prod_{j=0}^{v}(\tilde{x}_{2v}^2 - x_{2v,j}^2),$$

and thus, since $v = n/2$,

$$\prod_{j=0}^{n}(\tilde{x}_n - x_{n,j}) = \frac{1}{\tilde{x}_n}\prod_{j=0}^{n/2}(\tilde{x}_n^2 - x_{n,j}^2), \text{ if } n \text{ is even.} \qquad (3.1.39)$$

By substituting (3.1.36), (3.1.37) and (3.1.39) into (3.1.11), and by using also the fact that (3.1.38) implies

$$\prod_{j=0}^{(n/2)-1}(1+x_{n,j}^2) = \prod_{j=0}^{n/2}(1+x_{n,j}^2), \text{ if } n \text{ is even,}$$

we obtain the interpolation error expression

$$E_n^I(\tilde{x}_n) = \begin{cases} (-1)^{(n+1)/2}f(\tilde{x}_n)\displaystyle\prod_{j=0}^{(n-1)/2}\frac{\tilde{x}_n^2 - x_{n,j}^2}{1+x_{n,j}^2}, & \text{if } n \text{ is odd;} \\[4mm] (-1)^{(n/2)-1}f(\tilde{x}_n)\displaystyle\prod_{j=0}^{n/2}\frac{\tilde{x}_n^2 - x_{n,j}^2}{1+x_{n,j}^2}, & \text{if } n \text{ is even.} \end{cases} \qquad (3.1.40)$$

With the standard notation $\lfloor x \rfloor$ for the largest integer $\leqslant x$, it follows from (3.1.40) that

$$|E_n^I(\tilde{x}_n)| = |f(\tilde{x}_n)| \prod_{j=0}^{\lfloor n/2 \rfloor} \frac{|\tilde{x}_n^2 - x_{n,j}^2|}{1 + x_{n,j}^2}, \quad n \in \mathbb{N}. \tag{3.1.41}$$

Since (3.1.9) gives $\tilde{x}_n \in [0,5), n \in \mathbb{N}$, whereas (3.1.5) yields $f(x) \geqslant \frac{1}{26}, x \in [-5,5]$, it follows that

$$|f(\tilde{x}_n)| \geqslant \frac{1}{26}, \quad n \in \mathbb{N},$$

which, together with (3.1.41), yields

$$|E_n^I(\tilde{x}_n)| \geqslant \frac{1}{26} \prod_{j=0}^{\lfloor n/2 \rfloor} \frac{|\tilde{x}_n^2 - x_{n,j}^2|}{1 + x_{n,j}^2}, \quad n \in \mathbb{N}. \tag{3.1.42}$$

By noting from (3.1.9) and (3.1.6) that, for any $n \in \mathbb{N}$,

$$\tilde{x}_n^2 - x_{n,j}^2 = \begin{cases} \left(5 - \frac{5}{n}\right)^2 - 25 = 25\left[\left(1 - \frac{1}{n}\right)^2 - 1\right] < 0, j = 0; \\ \left(5 - \frac{5}{n}\right)^2 - \left(5 - \frac{10j}{n}\right)^2 > 0, \quad j = 1,2,\ldots,\lfloor n/2 \rfloor, \end{cases}$$

we deduce that, with the convention that $\prod_{j=j_1}^{j_2} a_j := 1$ if $j_2 < j_1$,

$$\prod_{j=0}^{\lfloor n/2 \rfloor} \frac{|\tilde{x}_n^2 - x_{n,j}^2|}{1 + x_{n,j}^2} = \frac{25}{26}\left(\frac{2n-1}{n^2}\right) \prod_{j=1}^{\lfloor n/2 \rfloor} \frac{\tilde{x}_n^2 - x_{n,j}^2}{1 + x_{n,j}^2}, \quad n \in \mathbb{N}. \tag{3.1.43}$$

By using (3.1.43) in (3.1.42), we obtain the lower bound

$$|E_n^I(\tilde{x}_n)| \geqslant \frac{25}{(26)^2}\alpha_n, \quad n \in \mathbb{N}, \tag{3.1.44}$$

where

$$\alpha_n := \frac{2n-1}{n^2} \prod_{j=1}^{\lfloor n/2 \rfloor} \frac{\tilde{x}_n^2 - x_{n,j}^2}{1 + x_{n,j}^2}, \quad n \in \mathbb{N}. \tag{3.1.45}$$

We proceed to prove that

$$\alpha_n \to \infty, \quad n \to \infty, \tag{3.1.46}$$

which, together with (3.1.45) and (3.1.44), will then yield the desired divergence result (3.1.10).

To this end, we note from (3.1.45) that

$$\alpha_n = e^{\beta_n}, \quad n \in \mathbb{N}, \tag{3.1.47}$$

where, with the convention that $\sum_{j=j_0}^{j_1} a_j := 0$ if $j_1 < j_0$,

$$\beta_n := \ln(2n-1) - 2\ln n + \sum_{j=1}^{\lfloor n/2\rfloor} \ln(\tilde{x}_n + x_{n,j}) + \sum_{j=1}^{\lfloor n/2\rfloor} \ln(\tilde{x}_n - x_{n,j}) - \sum_{j=1}^{\lfloor n/2\rfloor} \ln(1 + x_{n,j}^2), \quad n \in \mathbb{N}.$$
(3.1.48)

We shall show that

$$\beta_n \to \infty, \quad n \to \infty,$$
(3.1.49)

which, together with (3.1.47), will then prove the desired divergence result (3.1.46).

To prove (3.1.49), we fix $n \in \mathbb{N}$, and first note from (3.1.9) and (3.1.6) that

$$\sum_{j=1}^{\lfloor n/2\rfloor} \ln(\tilde{x}_n + x_{n,j}) + \sum_{j=1}^{\lfloor n/2\rfloor} \ln(\tilde{x}_n - x_{n,j})$$

$$= 2(\ln 5)\lfloor n/2\rfloor - 2(\ln n)\lfloor n/2\rfloor + \sum_{j=1}^{\lfloor n/2\rfloor} \ln(2j-1) + \sum_{j=1}^{\lfloor n/2\rfloor} \ln(2n-1-2j)$$

$$\geqslant (\ln 5)(n-1) - n\ln n + \sum_{j=1}^{\lfloor n/2\rfloor-1} \ln(2j+1) + \sum_{j=1}^{\lfloor n/2\rfloor} \ln(2n-1-2j).$$
(3.1.50)

Let the piecewise constant functions u_n and v_n be defined by

$$u_n(x) := \ln(2j+1), \ x \in [j, j+1), \ j = 1, \ldots, \lfloor n/2\rfloor - 1 \ (\text{if } n \geqslant 4);$$
(3.1.51)

$$v_n(x) := \ln(2n-1-2j), \ x \in [j, j+1), \ j = 1, \ldots, \lfloor n/2\rfloor \ (\text{if } n \geqslant 2).$$
(3.1.52)

It then follows from (3.1.51), (3.1.52) that

$$u_n(x) \geqslant \ln(2x-1), \ x \in [1, \lfloor n/2\rfloor] \quad (\text{if } n \geqslant 4);$$
(3.1.53)

$$v_n(x) \geqslant \ln(2n-1-2x), \ x \in [1, \lfloor n/2\rfloor + 1] \quad (\text{if } n \geqslant 2).$$
(3.1.54)

By using (3.1.51) and (3.1.53), together with the fact that $\lfloor n/2\rfloor \geqslant (n-1)/2$, as well as integration by parts, we deduce that, for $n \geqslant 4$,

$$\sum_{j=1}^{\lfloor n/2\rfloor-1} \ln(2j+1) = \int_1^{\lfloor n/2\rfloor} u_n(x)dx$$

$$\geqslant \int_1^{\lfloor n/2\rfloor} \ln(2x-1)dx$$

$$\geqslant \int_1^{(n-1)/2} \ln(2x-1)dx$$

$$= \frac{1}{2} \int_1^{(n-1)/2} \ln(2x-1) \frac{d}{dx}(2x-1)dx$$

$$= \frac{1}{2} \left\{ \left[(2x-1)\ln(2x-1) \right]_1^{(n-1)/2} - 2 \left[\frac{n-1}{2} - 1 \right] \right\}$$

$$= \frac{1}{2}(n-2)\ln(n-2) - \frac{1}{2}n + \frac{3}{2}, \tag{3.1.55}$$

and similarly, from (3.1.52) and (3.1.54), for $n \geqslant 2$,

$$\sum_{j=1}^{\lfloor n/2 \rfloor} \ln(2n-1-2j) = \int_1^{\lfloor n/2 \rfloor + 1} v_n(x)dx$$

$$\geqslant \int_1^{\lfloor n/2 \rfloor + 1} \ln(2n-1-2x)dx$$

$$\geqslant \int_1^{(n+1)/2} \ln(2n-1-2x)dx$$

$$= -\frac{1}{2} \int_1^{(n+1)/2} \ln(2n-1-2x) \frac{d}{dx}(2n-1-2x)dx$$

$$= -\frac{1}{2} \left\{ \left[(2n-1-2x)\ln(2n-1-2x) \right]_1^{(n+1)/2} + 2 \left[\frac{n+1}{2} - 1 \right] \right\}$$

$$= -\frac{1}{2}(n-2)\ln(n-2) + \frac{1}{2}(2n-3)\ln(2n-3) - \frac{1}{2}n + \frac{1}{2}. \tag{3.1.56}$$

It follows from (3.1.50), (3.1.55) and (3.1.56) that, for $n \geqslant 4$, we have

$$\sum_{j=1}^{\lfloor n/2 \rfloor} \ln(\tilde{x}_n + x_{n,j}) + \sum_{j=1}^{\lfloor n/2 \rfloor} \ln(\tilde{x}_n - x_{n,j})$$

$$\geqslant (\ln 5)(n-1) - n\ln n + \left(n - \frac{3}{2}\right)\ln(2n-3) - n + 2$$

$$= n \left[\ln \frac{5(2n-3)}{n} - 1 \right] - \frac{3}{2}\ln(2n-3) - \ln 5 + 2. \tag{3.1.57}$$

Next, to bound the third sum in (3.1.48), we define, for $n \geqslant 2$, the piecewise constant function w_n by

$$w_n(x) := \ln \left(1 + 25 \left(\frac{2j}{n} - 1 \right)^2 \right), \quad x \in [j, j+1), \quad j = 1, \dots, \lfloor n/2 \rfloor. \tag{3.1.58}$$

It follows from (3.1.58) that

$$w_n(x) \leqslant \ln \left(1 + 25 \left(\frac{2x}{n} - 1 \right)^2 \right), \quad x \in [1, \lfloor n/2 \rfloor + 1] \quad \text{(if } n \geqslant 2\text{)}. \tag{3.1.59}$$

By using (3.1.6), (3.1.58) and (3.1.59), together with the fact that $\lfloor n/2 \rfloor \leqslant n/2$, as well as integration by parts, we deduce that, for $n \geqslant 2$,

$$- \sum_{j=1}^{\lfloor n/2 \rfloor} \ln(1 + x_{n,j}^2)$$

$$= - \sum_{j=1}^{\lfloor n/2 \rfloor} \ln \left(1 + 25 \left(\frac{2j}{n} - 1 \right)^2 \right)$$

$$= - \int_1^{\lfloor n/2 \rfloor + 1} w_n(x) dx$$

$$\geqslant - \int_1^{\lfloor n/2 \rfloor + 1} \ln \left(1 + 25 \left(\frac{2x}{n} - 1 \right)^2 \right) dx$$

$$\geqslant - \int_1^{(n+3)/2} \ln \left(1 + 25 \left(\frac{2x}{n} - 1 \right)^2 \right) dx$$

$$= - \frac{n}{10} \int_{-5(1-\frac{2}{n})}^{15/n} \ln(1 + \xi^2) d\xi$$

$$= - \frac{n}{10} \left\{ \left[\xi \ln(1 + \xi^2) \right]_{-5(1-\frac{2}{n})}^{15/n} - 2 \int_{-5(1-\frac{2}{n})}^{15/n} \left(1 - \frac{1}{1+\xi^2} \right) d\xi \right\}$$

$$= - \frac{n}{10} \left\{ \frac{15 \ln(1 + \frac{225}{n^2})}{n} + 5 \left(1 - \frac{2}{n} \right) \ln \left(1 + 25 \left(1 - \frac{2}{n} \right)^2 \right) \right.$$

$$\left. - 10 \left(1 + \frac{1}{n} \right) + 2 \left(\arctan \left(\frac{15}{n} \right) \right) + \arctan \left(5 \left(1 - \frac{2}{n} \right) \right) \right\}$$

$$= n \left[- \frac{1}{2} \ln \left(1 + 25 \left(1 - \frac{2}{n} \right)^2 \right) + 1 - \frac{1}{5} \left\{ \arctan \left(\frac{15}{n} \right) + \arctan \left(5 \left(1 - \frac{2}{n} \right) \right) \right\} \right]$$

$$- \frac{3}{2} \ln \left(1 + \frac{225}{n^2} \right) + \ln \left(1 + 25 \left(1 - \frac{2}{n} \right)^2 \right) + 1. \tag{3.1.60}$$

It now follows from (3.1.48), (3.1.57) and (3.1.60) that, for $n \geqslant 4$,

$$\beta_n \geqslant \ln(2n - 1) - 2 \ln n + \left\{ n \left[\ln \frac{5(2n-3)}{n} - 1 \right] - \frac{3}{2} \ln(2n - 3) - \ln 5 + 2 \right\}$$

$$+ \left\{ n \left[- \frac{1}{2} \ln \left(1 + 25 \left(1 - \frac{2}{n} \right)^2 \right) + 1 - \frac{\arctan \left(\frac{15}{n} \right) + \arctan(5(1 - \frac{2}{n}))}{5} \right] \right.$$

$$-\frac{3}{2}\ln\left(1+\frac{225}{n^2}\right)+\ln\left(1+25\left(1-\frac{2}{n}\right)^2\right)+1\right\}$$

$$= a_n n + b_n,\qquad(3.1.61)$$

where

$$a_n := \ln\left(\frac{5\left(2-\frac{3}{n}\right)}{\sqrt{1+25\left(1-\frac{2}{n}\right)^2}}\right)-\frac{\arctan\left(\frac{15}{n}\right)+\arctan\left(5\left(1-\frac{2}{n}\right)\right)}{5}$$

$$-\frac{\ln n}{n}-\frac{3}{2}\frac{\ln(2n+3)}{n};\qquad(3.1.62)$$

$$b_n := \ln\left(2-\frac{1}{n}\right)-\frac{3}{2}\ln\left(1+\frac{225}{n^2}\right)+\ln\left(1+25\left(1-\frac{2}{n}\right)^2\right)-\ln 5+3.\qquad(3.1.63)$$

Since applications of L'Hospital's rule yield

$$\lim_{x\to\infty}\frac{\ln x}{x}=\lim_{x\to\infty}\frac{1/x}{1}=0;\quad \lim_{x\to\infty}\frac{\ln(2x-3)}{x}=\lim_{x\to\infty}\frac{2/(2x-3)}{1}=0,$$

according to which

$$\lim_{n\to\infty}\frac{\ln n}{n}=0;\quad \lim_{n\to\infty}\frac{\ln(2n-3)}{n}=0,$$

we deduce from (3.1.62) that

$$\gamma := \lim_{n\to\infty} a_n = \ln\frac{10}{\sqrt{26}}-\frac{\arctan 5}{5}\approx 0.399,\qquad(3.1.64)$$

and thus

$$\gamma > 0,\qquad(3.1.65)$$

whereas (3.1.63) gives

$$\delta := \lim_{n\to\infty} b_n = 3+\ln\frac{52}{5}\approx 5.342,\qquad(3.1.66)$$

so that also

$$\delta > 0.\qquad(3.1.67)$$

It follows from (3.1.64) - (3.1.67) that there exists a positive integer $\widetilde{N}$ such that

$$\left.\begin{array}{l}-\dfrac{\gamma}{2}<a_n-\gamma<\dfrac{\gamma}{2},\\[2mm]-\dfrac{\delta}{2}<b_n-\delta<\dfrac{\delta}{2},\end{array}\right\}\quad n\geqslant\widetilde{N},$$

and thus

$$\left.\begin{aligned} a_n > \frac{\gamma}{2} > 0, \\ b_n > \frac{\delta}{2} > 0, \end{aligned}\right\} \quad n \geqslant \widetilde{N}. \tag{3.1.68}$$

By applying (3.1.68) in (3.1.61), we deduce that

$$\beta_n \geqslant (\gamma/2)n + (\delta/2), \quad n \geqslant \widetilde{N},$$

and thus, for any given positive number M, if we define the positive number N by

$$N := \max\{\widetilde{N}, \lceil 2(M - (\delta/2))/\gamma \rceil\},$$

with $\lceil x \rceil$ denoting the smallest integer $\geqslant x$, we have

$$\beta_n > M, \quad n > N,$$

according to which (3.1.49) holds, and thereby completing our proof of the desired divergence result (3.1.10).

If, however, we replace the uniformly distributed interpolation points (3.1.6) by the Chebyshev interpolation points, that is, from (2.2.31),

$$x_j := x_{n,j}^C = 5\cos\left(\frac{2n+1-2j}{2n+2}\pi\right), \quad j = 0,\dots,n, \tag{3.1.69}$$

thereby concentrating the interpolation points more densely towards -5 and 5, the uniform convergence result (3.1.4) is indeed satisfied, as will follow from Theorem 6.5.3 in Section 6.5 of Chapter 6. ∎

The results of Example 3.1.3 lead to the following interesting question: For any prescribed sequence

$$\triangle_n := \{x_{n,0},\dots,x_{n,n}\}, \quad n = 1,2,\dots, \tag{3.1.70}$$

of (distinct) interpolation point sequences satisfying (2.1.1), like, for example, the Chebyshev interpolation points (2.2.31), is it possibly true that the uniform convergence result (3.1.4) is obtained for each $f \in C[a,b]$? The answer is negative, due to a known result, the proof of which is beyond the scope of this book, and according to which, for any prescribed sequence $\{\triangle_n : n = 1,2,\dots\}$, there exists a function $f \in C[a,b]$ such that

$$\|f - P_n^I\|_\infty \to \infty, \quad n \to \infty. \tag{3.1.71}$$

Hence, to investigate whether the uniform convergence result (3.1.4) holds for given $f \in C[a,b]$, $\triangle_n$ and $[a,b]$, error estimates like (2.2.36) in Theorem 2.2.4 need to be applied.

In the rest of this chapter, we proceed to establish, for any given $f \in C[a,b]$, a sequence $\{P_n : n = 1, 2, \ldots\}$ of approximating polynomials such that

$$\|f - P_n\|_\infty := \max_{a \leqslant x \leqslant b} |f(x) - P_n(x)| \to 0, \quad n \to \infty. \tag{3.1.72}$$

Observe from (3.1.72) that any given function $f \in C[a,b]$ can therefore be approximated with arbitrary (uniform) "closeness" by a polynomial, and thereby providing justification to the attention given to specifically polynomial approximation.

3.2   The Bernstein polynomials

For any non-negative integer j, and $k \in \mathbb{Z}$, we adopt the standard binomial coefficient notation

$$\binom{j}{k} := \begin{cases} \dfrac{j!}{k!(j-k)!} & , k = 0, \ldots, j; \\ 0 & , k \notin \{0, \ldots, j\}, \end{cases} \tag{3.2.1}$$

and with the convention that $0! := 1$.

Let n denote any non-negative integer, and suppose $[a,b]$ is a bounded interval in $\mathbb{R}$. The polynomials

$$B_{n,j}(x) := \binom{n}{j} \left(\frac{x-a}{b-a}\right)^j \left(\frac{b-x}{b-a}\right)^{n-j}, \quad j = 0, \ldots, n, \tag{3.2.2}$$

are called the Bernstein polynomials of degree n with respect to the interval $[a,b]$.

We shall rely on the one-to-one mapping between the intervals $[a,b]$ and $[0,1]$ given by

$$t = \frac{x-a}{b-a}, \quad a \leqslant x \leqslant b, \tag{3.2.3}$$

or equivalently,

$$x = (b-a)t + a, \quad 0 \leqslant t \leqslant 1. \tag{3.2.4}$$

By using (3.2.3), (3.2.4), we observe from (3.2.2) that

$$\left. \begin{aligned} B_{n,j}(x) &= \binom{n}{j} t^j (1-t)^{n-j}, \quad j = 0, \ldots, n, \\ \text{where} \qquad t &:= \frac{x-a}{b-a}. \end{aligned} \right\} \tag{3.2.5}$$

We proceed to prove the following properties of the Bernstein polynomials.

Theorem 3.2.1. *For any non-negative integer n, and bounded interval $[a,b] \subset \mathbb{R}$, the corresponding Bernstein polynomials, as defined by (3.2.2), satisfy:*

(a)

$$\left.\begin{array}{l} B_{n,j}(a) = \delta_j, \\[2mm] B_{n,j}(b) = \delta_{n-j}, \end{array}\right\} \quad j = 0, \ldots, n; \tag{3.2.6}$$

(b)

$$B_{n,j}(x) > 0, \quad x \in (a,b); \tag{3.2.7}$$

(c)

$$\sum_{j=0}^{n} B_{n,j}(x) = 1, \quad x \in \mathbb{R}; \tag{3.2.8}$$

(d) *the polynomial sequence* $\{B_{n,j} : j = 0, \ldots, n\}$ *is a basis for* π_n.

Proof. (a), (b) These properties are immediate consequences of the definition (3.2.2).

(c) For any $x \in \mathbb{R}$, an application of (3.2.5) yields

$$\sum_{j=0}^{n} B_{n,j}(x) = \sum_{j=0}^{n} \binom{n}{j} t^j (1-t)^{n-j} = [t + (1-t)]^n = 1^n = 1,$$

which proves (3.2.8).

(d) As in the proofs of Theorem 1.2.1(b) and Theorem 1.3.2(b), it will suffice to prove that $\{B_{n,j} : j = 0, \ldots, n\}$ is a linearly independent set. After noting that such linear independence is trivial if $n = 0$, we suppose next that $n \geqslant 1$, and let the coefficient sequence $\{c_0, c_1, \ldots, c_n\} \subset \mathbb{R}$ be such that

$$\sum_{j=0}^{n} c_j B_{n,j}(x) = 0, \quad x \in \mathbb{R},$$

or equivalently, from (3.2.5),

$$\sum_{j=0}^{n} c_j \binom{n}{j} t^j (1-t)^{n-j} = 0, \quad t \in \mathbb{R}. \tag{3.2.9}$$

By setting successively $t = 0$ and $t = 1$ in (3.2.9), we obtain $c_0 = c_n = 0$, which then proves the desired linear independence result for $n = 1$. If $n \geqslant 2$, we may set $c_0 = c_n = 0$ in (3.2.9) to obtain

$$t(1-t) \sum_{j=1}^{n-1} c_j \binom{n}{j} t^{j-1} (1-t)^{n-1-j} = 0, \quad t \in \mathbb{R}, \tag{3.2.10}$$

and thus

$$\sum_{j=1}^{n-1} c_j \binom{n}{j} t^{j-1} (1-t)^{n-1-j} = 0, \quad t \in \mathbb{R}. \tag{3.2.11}$$

If $n = 2$, (3.2.11) immediately gives $c_1 = 0$, and thus $c_0 = c_1 = c_2 = 0$, which shows that linear independence is also obtained for $n = 2$. If $n \geqslant 3$, we may successively set $t = 0$ and

$t = 1$ in (3.2.11) to obtain $c_1 = c_{n-1} = 0$. By applying the same argument as in (3.2.10) and (3.2.11), sufficiently many times, we eventually prove that $c_0 = \cdots = c_n = 0$, and thereby establishing the desired linear independence result. ∎

It follows from Theorem 3.2.1(d) that, for any polynomial $P \in \pi_n$, there exists a unique coefficient sequence $\{b_{n,j} : j = 0, \ldots, n\} \subset \mathbb{R}$ such that

$$P(x) = \sum_{j=0}^{n} b_{n,j} B_{n,j}(x). \tag{3.2.12}$$

The expression (3.2.12) is called the Bernstein representation in π_n with respect to the interval $[a, b]$ of the polynomial P, and has practical applications in e.g. interactive geometric design.

We proceed in Section 3.3 to construct a polynomial approximation in π_n of a given function $f \in C[a, b]$ by means of an appropriate choice of the coefficient sequence $\{b_{n,j} : j = 0, \ldots, n\}$ in (3.2.12).

3.3 Bernstein polynomial approximation

For a given function $f \in C[a, b]$ and any integer $n \in \mathbb{N}$, we define the Bernstein polynomial approximation P_n^B in π_n of f with respect to the interval $[a, b]$ by

$$\left. \begin{aligned} P_n^B(x) &:= \sum_{j=0}^{n} f(x_{n,j}) B_{n,j}(x), \\ x_{n,j} &:= a + j \left(\frac{b-a}{n} \right), \quad j = 0, \ldots, n. \end{aligned} \right\} \tag{3.3.1}$$

where

Observe that the point sequence $\{x_{n,j} : j = 0, \ldots, n\}$ in (3.3.1) partitions the interval $[a, b]$ into n subintervals of equal length $\left(= \dfrac{b-a}{n} \right)$, and with

$$a = x_{n,0} < x_{n,1} < \cdots < x_{n,n} = b. \tag{3.3.2}$$

The following properties of Bernstein polynomial approximations can now be proved by means of Theorem 3.2.1.

Theorem 3.3.1. *Let $f \in C[a, b]$ and $n \in \mathbb{N}$, and denote by P_n^B the Bernstein polynomial approximation in π_n of f with respect to $[a, b]$, as defined in (3.3.1). Then:*

(a) The polynomial P_n^B interpolates f at a and b, that is,

$$P_n^B(a) = f(a) \quad ; \quad P_n^B(b) = f(b). \tag{3.3.3}$$

(b) *Sign-preservation on $[a,b]$ is satisfied, in the sense that, if*

$$f(x) \geqslant 0, \quad x \in [a,b], \tag{3.3.4}$$

then

$$P_n^B(x) \geqslant 0, \quad x \in [a,b]. \tag{3.3.5}$$

(c) *Linear polynomials are reproduced, that is, if*

$$f \in \pi_1, \tag{3.3.6}$$

then

$$P_n^B = f. \tag{3.3.7}$$

Proof. (a) By using (3.3.1) and Theorem 3.2.1(a), we obtain

$$P_n^B(a) = \sum_{j=0}^{n} f(x_{n,j})\delta_j = f(x_{n,0}) = f(a),$$

and

$$P_n^B(b) = \sum_{j=0}^{n} f(x_{n,j})\delta_{n-j} = f(x_{n,n}) = f(b).$$

(b) If (3.3.4) is satisfied, it follows from (3.3.1) and (3.2.7) that (3.3.5) holds.

(c) Let $f \in \pi_1$, that is,

$$f(x) = c_0 + c_1 x, \tag{3.3.8}$$

for some real coefficients c_0 and c_1. It follows from (3.3.1), (3.3.8), Theorem 3.2.1(c) and (3.2.5) that, for any $x \in \mathbb{R}$,

$$P_n^B(x) = c_0 \sum_{j=0}^{n} B_{n,j}(x) + c_1 \sum_{j=0}^{n} \left[a + j\left(\frac{b-a}{n}\right) \right] B_{n,j}(x)$$

$$= c_0 + c_1 \left[a + (b-a) \sum_{j=1}^{n} \frac{j}{n}\binom{n}{j} t^j (1-t)^{n-j} \right], \tag{3.3.9}$$

where

$$t := \frac{x-a}{b-a}. \tag{3.3.10}$$

Now use (3.2.1), and the index transformation $k = j - 1$, to deduce that

$$\sum_{j=1}^{n} \frac{j}{n}\binom{n}{j} t^j (1-t)^{n-j} = \sum_{j=1}^{n} \frac{(n-1)!}{(j-1)!(n-j)!} t^j (1-t)^{n-j}$$

$$= \sum_{k=0}^{n-1} \frac{(n-1)!}{k!(n-(k+1))!} t^{k+1} (1-t)^{n-(k+1)}$$

$$= t \sum_{k=0}^{n-1} \frac{(n-1)!}{k!(n-1-k)!} t^k (1-t)^{n-1-k}$$

$$= t \sum_{k=0}^{n-1} \binom{n-1}{k} t^k (1-t)^{n-1-k}$$

$$= t[t + (1-t)]^{n-1} = t(1^{n-1}) = t. \tag{3.3.11}$$

It follows from (3.3.9), (3.3.11), (3.3.10) and (3.3.8) that

$$P_n^B(x) = c_0 + c_1 \left[a + (b-a)\frac{x-a}{b-a} \right] = c_0 + c_1 x = f(x),$$

which completes the proof. ∎

Our next result shows that, although quadratic polynomials are not reproduced by Bernstein polynomial approximation, the uniform convergence result

$$||f - P_n^B||_\infty := \max_{a \leqslant x \leqslant b} |f(x) - P_n^B(x)| \to 0, \quad n \to \infty, \tag{3.3.12}$$

is indeed achieved for any $f \in \pi_2$.

Theorem 3.3.2. (a) *Let the function f in the definition (3.3.1) be given by*

$$f(x) = x^2, \quad x \in [a,b]. \tag{3.3.13}$$

Then, for any $n \in \mathbb{N}$, the corresponding Bernstein polynomial approximation P_n^B in π_n of f is given by

$$P_n^B(x) = x^2 + \frac{(x-a)(b-x)}{n}, \tag{3.3.14}$$

with, moreover,

$$||f - P_n^B||_\infty = \frac{(b-a)^2}{4n}. \tag{3.3.15}$$

(b) *The uniform convergence result (3.3.12) is satisfied for any $f \in \pi_2$.*

Proof. (a) First, we observe from (3.3.1), (3.3.13), Theorem 3.2.1(c), as well as (3.2.5) and (3.3.11), that, for any $x \in \mathbb{R}$,

$$P_n^B(x) = \sum_{j=0}^{n} \left[a + j\left(\frac{b-a}{n}\right) \right]^2 R_{n,j}(x)$$

$$= a^2 \sum_{j=0}^{n} B_{n,j}(x) + 2a(b-a) \sum_{j=0}^{n} \frac{j}{n} \binom{n}{j} t^j (1-t)^{n-j} + (b-a)^2 \sum_{j=0}^{n} \frac{j^2}{n^2} \binom{n}{j} t^j (1-t)^{n-j}$$

$$= a^2 + 2a(b-a)\left(\frac{x-a}{b-a}\right) + (b-a)^2 \sum_{j=1}^{n} \frac{j^2}{n^2}\binom{n}{j} t^j (1-t)^{n-j}$$

$$= 2ax - a^2 + (b-a)^2 \sum_{j=1}^{n} \frac{j^2}{n^2}\binom{n}{j} t^j (1-t)^{n-j}, \tag{3.3.16}$$

where t is given by (3.3.10). For $n \geqslant 2$, we now use (3.2.1), as well as the index transformations $k = j-1$ and $\ell = k-1$, to obtain, with the convention $\sum_{j=j_0}^{j_1} \alpha_j := 0$ if $j_1 < j_0$,

$$\sum_{j=1}^{n} \frac{j^2}{n^2}\binom{n}{j} t^j (1-t)^{n-j} = \sum_{j=1}^{n} \frac{j^2}{n^2} \frac{n!}{j!(n-j)!} t^j (1-t)^{n-j}$$

$$= \sum_{j=1}^{n} \frac{j}{n} \frac{(n-1)!}{(j-1)!(n-j)!} t^j (1-t)^{n-j}$$

$$= \sum_{k=0}^{n-1} \frac{k+1}{n} \frac{(n-1)!}{k!(n-(k+1))!} t^{k+1} (1-t)^{n-(k+1)}$$

$$= t \sum_{k=1}^{n-1} \frac{k}{n} \frac{(n-1)!}{k!(n-1-k)!} t^k (1-t)^{n-1-k}$$

$$+ \frac{t}{n} \sum_{k=0}^{n-1} \frac{(n-1)!}{k!(n-1-k)!} t^k (1-t)^{n-1-k}$$

$$= \frac{t}{n} \sum_{k=1}^{n-1} \frac{(n-1)!}{(k-1)!(n-1-k)!} t^k (1-t)^{n-1-k}$$

$$+ \frac{t}{n} \sum_{k=0}^{n-1} \binom{n-1}{k} t^k (1-t)^{n-1-k}$$

$$= \frac{t}{n} \sum_{\ell=0}^{n-2} \frac{(n-1)!}{\ell!(n-1-(\ell+1))!} t^{\ell+1} (1-t)^{n-1-(\ell+1)}$$

$$+ \frac{t}{n} [t + (1-t)]^{n-1}$$

$$= \frac{(n-1)t^2}{n} \sum_{\ell=0}^{n-2} \frac{(n-2)!}{\ell!(n-2-\ell)!} t^\ell (1-t)^{n-2-\ell} + \frac{t}{n}(1^{n-1})$$

$$= \frac{(n-1)t^2}{n} \sum_{\ell=0}^{n-2} \binom{n-2}{\ell} t^\ell (1-t)^{n-2-\ell} + \frac{t}{n}$$

$$= \frac{(n-1)t^2}{n}[t+(1-t)]^{n-2}+\frac{t}{n}$$

$$= \frac{(n-1)t^2}{n}(1^{n-2})+\frac{t}{n}$$

$$= \frac{(n-1)t^2}{n}+\frac{t}{n}=t^2+\frac{t(1-t)}{n}. \qquad (3.3.17)$$

Observe that (3.3.17) also holds for $n=1$. It follows from (3.3.16), (3.3.17) and (3.3.10) that

$$P_n^B(x) = 2ax-a^2+(b-a)^2\left[\left(\frac{x-a}{b-a}\right)^2+\frac{1}{n}\left(\frac{x-a}{b-a}\right)\left(\frac{b-x}{b-a}\right)\right]$$

$$= 2ax-a^2+(x^2-2ax+a^2)+\frac{1}{n}(x-a)(b-x),$$

which gives the formula (3.3.14).

We deduce from (3.3.14) and (3.3.13) that

$$\|f-P_n^B\|_\infty = \max_{a\leqslant x\leqslant b}\left|x^2-\left[x^2+\frac{(x-a)(b-x)}{n}\right]\right|$$

$$= \max_{a\leqslant x\leqslant b}\frac{(x-a)(b-x)}{n}$$

$$= \frac{1}{n}\left[\frac{1}{2}(a+b)-a\right]\left[b-\frac{1}{2}(a+b)\right] = \frac{(b-a)^2}{4n},$$

which proves (3.3.15).

(b) Let $f\in\pi_2$, that is,

$$f(x) = c_0+c_1x+c_2x^2$$

for some real coefficients c_0, c_1 and c_2, so that (3.3.1), Theorem 3.3.1(c), together with (3.3.14), yield, for any $x\in\mathbb{R}$,

$$P_n^B(x) = \sum_{j=0}^{n}\left\{c_0+c_1\left[a+j\left(\frac{b-a}{n}\right)\right]+c_2\left[a+j\left(\frac{b-a}{n}\right)\right]^2\right\}B_{n,j}(x)$$

$$= \sum_{j=0}^{n}\left\{c_0+c_1\left[a+j\left(\frac{b-a}{n}\right)\right]\right\}B_{n,j}(x)+c_2\sum_{j=0}^{n}\left[a+i\left(\frac{b-a}{n}\right)\right]^2R_{n,j}(x)$$

$$= (c_0+c_1x)+c_2\left(x^2+\frac{(x-a)(b-x)}{n}\right)$$

$$= f(x)+c_2\frac{(x-a)(b-x)}{n},$$

and thus

$$||f - P_n^B||_\infty = |c_2| \frac{[\frac{1}{2}(a+b) - a][b - \frac{1}{2}(a+b)]}{n} = \frac{|c_2|(b-a)^2}{4n},$$

which implies the uniform convergence result (3.3.12). ∎

Observe that (3.3.14) in Theorem 3.3.2(a) is in accordance with Theorem 3.3.1(a) and (b). Also, observe from (3.3.14) that, if f is given by (3.3.13), then $P_n^B \in \pi_2$ for each $n = 1, 2, \ldots$. According to (3.2.8), (3.2.5), (3.3.11) and (3.3.17), the three identities

$$\sum_{j=0}^n \binom{n}{j} t^j (1-t)^{n-j} = 1; \tag{3.3.18}$$

$$\sum_{j=1}^n \frac{j}{n} \binom{n}{j} t^j (1-t)^{n-j} = t; \tag{3.3.19}$$

$$\sum_{j=1}^n \frac{j^2}{n^2} \binom{n}{j} t^j (1-t)^{n-j} = t^2 + \frac{t(1-t)}{n}, \tag{3.3.20}$$

are satisfied for all $t \in \mathbb{R}$ and $n \in \mathbb{N}$. We proceed to show how these identities can be used to prove the following theorem, which extends the uniform convergence result of Theorem 3.3.2(b) from π_2 to all of $C[a,b]$.

Theorem 3.3.3. *Let $f \in C[a,b]$. Then the corresponding Bernstein polynomial approximation sequence $\{P_n^B : n = 1, 2, \ldots\}$, as defined in (3.3.1), satisfies the uniform convergence result (3.3.12).*

Proof. Let $\varepsilon > 0$. We shall prove that there exists an integer $N = N(\varepsilon) \in \mathbb{N}$ such that

$$||f - P_n^B||_\infty := \max_{a \leqslant x \leqslant b} |f(x) - P_n^B(x)| < \varepsilon, \quad n > N, \tag{3.3.21}$$

which is equivalent to (3.3.12).

Let $x \in [a,b]$, and for any $n \in \mathbb{N}$, denote by $\{x_{n,j} : j = 0, \ldots, n\}$ the uniform partition points of $[a,b]$, as given in the second line of (3.3.1). Since f is continuous on the closed and bounded interval $[a,b]$, we know from a standard result in calculus that f is uniformly continuous on $[a,b]$, according to which there exists a positive number $\delta = \delta(\varepsilon)$, which is independent of x, and such that

$$|f(x) - f(x_{n,j})| < \frac{\varepsilon}{2}, \quad j \in J, \tag{3.3.22}$$

where

$$J := \{j \in \{0, \ldots, n\} : |x - x_{n,j}| < \delta\}. \tag{3.3.23}$$

With the definition

$$\tilde{J} := \{ j \in \{0, \ldots, n\} : |x - x_{n,j}| \geqslant \delta \}, \tag{3.3.24}$$

it follows that

$$J \cup \tilde{J} = \{0, \ldots, n\}; \quad J \cap \tilde{J} = \emptyset. \tag{3.3.25}$$

By using (3.3.1), Theorem 3.2.1(c), (3.3.25), as well as Theorem 3.2.1(b), we obtain

$$|f(x) - P_n^B(x)| = \left| \sum_{j=0}^{n} [f(x) - f(x_{n,j})] B_{n,j}(x) \right|$$

$$\leqslant \sum_{j=0}^{n} |f(x) - f(x_{n,j})| B_{n,j}(x)$$

$$= \sum_{j \in J} |f(x) - f(x_{n,j})| B_{n,j}(x) + \sum_{j \in \tilde{J}} |f(x) - f(x_{n,j})| B_{n,j}(x). \tag{3.3.26}$$

Now apply (3.3.22) and Theorem 3.2.1(b) to deduce that

$$\sum_{j \in J} |f(x) - f(x_{n,j})| B_{n,j}(x) < \frac{\varepsilon}{2} \sum_{j \in J} B_{n,j}(x) \leqslant \frac{\varepsilon}{2} \sum_{j=0}^{n} B_{n,j}(x) = \frac{\varepsilon}{2},$$

from Theorem 3.2.1(c), and thus

$$\sum_{j \in J} |f(x) - f(x_{n,j})| B_{n,j}(x) < \frac{\varepsilon}{2}. \tag{3.3.27}$$

Next, with the notation

$$M := \max_{a \leqslant x \leqslant b} |f(x)|, \tag{3.3.28}$$

and observing from (3.3.24) that

$$\frac{(x - x_{n,j})^2}{\delta^2} \geqslant 1, \quad j \in \tilde{J}, \tag{3.3.29}$$

we obtain

$$\sum_{j \in \tilde{J}} |f(x) - f(x_{n,j})| B_{n,j}(x) \leqslant \sum_{j \in \tilde{J}} [|f(x)| + |f(x_{n,j})|] B_{n,j}(x)$$

$$\leqslant 2M \sum_{j \in \tilde{J}} B_{n,j}(x)$$

$$\leqslant 2M \sum_{j \in \tilde{J}} \frac{(x - x_{n,j})^2}{\delta^2} B_{n,j}(x)$$

$$\leqslant \frac{2M}{\delta^2} \sum_{j=0}^{n} (x - x_{n,j})^2 B_{n,j}(x), \tag{3.3.30}$$

after recalling also Theorem 3.2.1(b).

By applying (3.2.5) and the definition in (3.3.1) of the point sequence $\{x_{n,j} : j = 0, \ldots, n\}$, as well as the identities (3.3.18), (3.3.19) and (3.3.20), we deduce that

$$\sum_{j=0}^{n}(x - x_{n,j})^2 B_{n,j}(x) = \sum_{j=0}^{n}\left[\{a + (b-a)t\} - \left\{a + j\left(\frac{b-a}{n}\right)\right\}\right]^2 B_{n,j}(x)$$

$$= (b-a)^2 \sum_{j=0}^{n}\left(t - \frac{j}{n}\right)^2 \binom{n}{j} t^j (1-t)^{n-j}$$

$$= (b-a)^2 \left[t^2 - 2t \sum_{j=0}^{n}\frac{j}{n}\binom{n}{j}t^j(1-t)^{n-j} + \sum_{j=0}^{n}\frac{j^2}{n^2}\binom{n}{j}t^j(1-t)^{n-j}\right]$$

$$= (b-a)^2 \left[t^2 - 2t(t) + \left(t^2 + \frac{t(1-t)}{n}\right)\right]$$

$$= (b-a)^2 \left[\frac{1}{n}\left(\frac{x-a}{b-a}\right)\left(\frac{b-x}{b-a}\right)\right] = \frac{(x-a)(b-x)}{n}. \qquad (3.3.31)$$

It follows from (3.3.30) and (3.3.31) that

$$\sum_{j\in\tilde{J}}|f(x) - f(x_{n,j})|B_{n,j}(x) \leqslant \frac{2M}{\delta^2}\frac{(x-a)(b-x)}{n}$$

$$\leqslant \frac{2M[\frac{1}{2}(a+b) - a][b - \frac{1}{2}(a+b)]}{\delta^2 n} = \frac{M(b-a)^2}{2\delta^2 n}. \qquad (3.3.32)$$

Let the positive integer $N = N(\varepsilon)$ be defined by

$$N := \left\lceil\frac{M(b-a)^2}{\varepsilon\delta^2}\right\rceil, \qquad (3.3.33)$$

where we adopt the standard notation $\lceil y \rceil$ for the smallest integer $\geqslant y$. Observe from (3.3.33) that N is independent of x. It follows from (3.3.33) that the inequality $n > N$ implies

$$n > \frac{M(b-a)^2}{\varepsilon\delta^2},$$

or equivalently,

$$\frac{M(b-a)^2}{2\delta^2 n} < \frac{\varepsilon}{2},$$

which, together with (3.3.32), implies that

$$\sum_{j\in\tilde{J}}|f(x) - f(x_{n,j})|B_{n,j}(x) < \frac{\varepsilon}{2}, \quad n > N. \qquad (3.3.34)$$

Finally, we combine (3.3.26), (3.3.27) and (3.3.34) to deduce that

$$|f(x) - P_n^B(x)| < \varepsilon, \quad n > N,$$

from which, since $N = N(\varepsilon)$ is independent of x, the desired result (3.3.21) then follows.

∎

The following result can now easily be deduced from Theorem 3.3.3.

Theorem 3.3.4 (Weierstrass). *Let $f \in C[a,b]$. Then, for each $\varepsilon > 0$, there exists a polynomial P such that*

$$||f - P||_\infty := \max_{a \leqslant x \leqslant b} |f(x) - P(x)| < \varepsilon. \tag{3.3.35}$$

Proof. Let $\varepsilon > 0$. According to (3.3.21) in the proof of Theorem 3.3.3, the polynomial

$$P := P_{N+1}^B, \tag{3.3.36}$$

where $N = N(\varepsilon) \in \mathbb{N}$ is defined by (3.3.33), satisfies the inequality (3.3.35). ∎

3.4 Shape-preservation

In this section we show that, in Theorem 3.3.3, if f has a continuous k-th derivative $f^{(k)}$ on $[a,b]$, then Bernstein polynomial approximation is shape-preserving in the sense that the sequence $\{(P_n^B)^{(k)} : n = 1, 2, \ldots\}$ is uniformly convergent to $f^{(k)}$ on $[a,b]$. We shall rely on the following explicit formulation for divided differences in the special case of uniformly spaced points.

Theorem 3.4.1. *Let $\{x_j : j \in \mathbb{Z}\}$ denote a uniformly spaced point sequence in $\mathbb{R}$, that is,*

$$x_{j+1} - x_j = h, \quad j \in \mathbb{Z}, \tag{3.4.1}$$

for a constant $h > 0$. Then, for any integers μ and v, with $v - \mu \geqslant 1$, the divided difference $f[x_\mu, \ldots, x_v]$ has the explicit formulation

$$f[x_\mu, \ldots, x_v] = \frac{(-1)^{v-\mu}}{(v-\mu)! h^{v-\mu}} \sum_{j=0}^{v-\mu} (-1)^j \binom{v-\mu}{j} f(x_{\mu+j}). \tag{3.4.2}$$

Proof. Our proof is by induction on the integer $v - \mu$. If $v - \mu = 1$, the right hand side of (3.4.2) is given by

$$\frac{f(x_v) - f(x_\mu)}{h} = \frac{f(x_v) - f(x_\mu)}{x_v - x_\mu},$$

from (3.4.1), and it follows from (1.3.21) in Theorem 1.3.4 that (3.4.2) holds.

Suppose next that (3.4.2) is satisfied for any integers μ and v such that $v - \mu \geqslant 1$. By applying the inductive hypothesis (3.4.2), together with (3.4.1), as well as (1.3.21) in Theorem 1.3.4, we obtain

$$\sum_{j=0}^{v-\mu+1} (-1)^j \binom{v-\mu+1}{j} f(x_{\mu+j})$$

$$= \sum_{j=0}^{v-\mu+1} (-1)^j \left[\binom{v-\mu}{j} + \binom{v-\mu}{j-1} \right] f(x_{\mu+j})$$

$$= \sum_{j=0}^{v-\mu} (-1)^j \binom{v-\mu}{j} f(x_{\mu+j}) + \sum_{j=1}^{v-\mu+1} (-1)^j \binom{v-\mu}{j-1} f(x_{\mu+j})$$

$$= \sum_{j=0}^{v-\mu} (-1)^j \binom{v-\mu}{j} f(x_{\mu+j}) - \sum_{j=0}^{v-\mu} (-1)^j \binom{v-\mu}{j} f(x_{\mu+j+1})$$

$$= (-1)^{v-\mu}(v-\mu)! h^{v-\mu} \{ f[x_\mu, \ldots, x_v] - f[x_{\mu+1}, \ldots, x_{v+1}] \}$$

$$= (-1)^{v-\mu+1}(v-\mu)! h^{v-\mu} \{ (x_{v+1} - x_\mu) f[x_\mu, \ldots, x_{v+1}] \}$$

$$= (-1)^{v-\mu+1}(v-\mu)! h^{v-\mu} \{ (v+1-\mu) h f[x_\mu, \ldots, x_{v+1}] \}$$

$$= (-1)^{v-\mu+1}(v-\mu+1)! h^{v-\mu+1} f[x_\mu, \ldots, x_{v+1}],$$

which shows that (3.4.2) holds if the index difference $v - \mu$ is advanced to $v - \mu + 1$, and thereby completing our inductive proof. ∎

For a function $f \in C[a,b]$ with a continuous k-th order derivative $f^{(k)}$ on $[a,b]$, it follows from Theorem 3.3.3 that the polynomial sequence

where
$$\left. \begin{aligned} P_n^{B,k}(x) &:= \sum_{j=0}^{n} f^{(k)}(x_{n,j}) B_{n,j}(x), \quad n = 1, 2, \ldots, \\ x_{n,j} &:= a + j \left(\frac{b-a}{n} \right), \qquad j = 0, \ldots, n, \end{aligned} \right\} \tag{3.4.3}$$

satisfies the uniform convergence result

$$\| f^{(k)} - P_n^{B,k} \|_\infty := \max_{a \leqslant x \leqslant b} |f^{(k)}(x) - P_n^{B,k}(x)| \to 0, \quad n \to \infty. \tag{3.4.4}$$

We proceed to show how (3.4.4) can be used to prove the following shape-preserving property of Bernstein polynomial approximation.

In our proof below, we shall rely, for sufficiently differentiable functions u and v, on the Leibniz formula

$$(uv)^{(k)} = \sum_{\ell=0}^{k} \binom{k}{\ell} u^{(k-\ell)} v^{(\ell)}, \tag{3.4.5}$$

(see Exercise 3.8), and the fact that the one-to-one transformation (3.2.3), (3.2.4) implies

$$u^{(k)}(x) = \frac{1}{(b-a)^k} \left(\frac{d}{dt} \right)^k u(a+(b-a)t). \tag{3.4.6}$$

Also, we shall use the differentiation formula

$$\left(\frac{d}{dt} \right)^\ell t^j = \ell! \binom{j}{\ell} t^{j-\ell}, \tag{3.4.7}$$

for any non-negative integers j and ℓ, with the binomial coefficient $\binom{j}{\ell}$ defined as in (3.2.1).

Theorem 3.4.2. *In Theorem 3.3.3, suppose that, moreover, $f \in C^m[a,b]$ for an integer $m \in \mathbb{N}$. Then*

$$||f^{(k)} - (P_n^B)^{(k)}||_\infty := \max_{a \leqslant x \leqslant b} \left| f^{(k)}(x) - (P_n^B)^{(k)}(x) \right| \to 0, \quad n \to \infty, \tag{3.4.8}$$

for any $k \in \{1,\dots,m\}$.

Proof. Let $k \in \{1,\dots,m\}$, and observe that, for $n = 1,2,\dots$, we have

$$||f^{(k)} - (P_n^B)^{(k)}||_\infty \leqslant ||f^{(k)} - P_n^{B,k}||_\infty + ||P_n^{B,k} - (P_n^B)^{(k)}||_\infty, \tag{3.4.9}$$

with the polynomial sequence $\{P_n^{B,k} : n = 1,2,\dots\}$ defined as in (3.4.3). It follows from (3.4.9) and (3.4.4) that it will suffice to prove that

$$||P_n^{B,k} - (P_n^B)^{(k)}||_\infty \to 0, \quad n \to \infty. \tag{3.4.10}$$

We shall in fact prove that

$$||P_n^{B,k} - (P_{n+k}^B)^{(k)}||_\infty \to 0, \quad n \to \infty, \tag{3.4.11}$$

which is equivalent to (3.4.10). Hence, for each $\varepsilon > 0$, we shall prove the existence of a positive integer $N = N(\varepsilon)$ such that

$$||P_n^{B,k} - (P_{n+k}^B)^{(k)}||_\infty < \varepsilon, \quad n > N. \tag{3.4.12}$$

To this end, we fix $x \in \mathbb{R}$, and use (3.3.1), (3.2.5) and (3.4.6) to obtain

$$(P_{n+k}^B)^{(k)}(x) = \frac{1}{(b-a)^k} \sum_{j=0}^{n+k} f(x_{n+k,j}) \binom{n+k}{j} \left(\frac{d}{dt} \right)^k \left[t^j (1-t)^{n+k-j} \right]. \tag{3.4.13}$$

Now use (3.4.5) and (3.4.7) to deduce that, for any $j \in \{0,\dots,n+k\}$,

$$\left(\frac{d}{dt} \right)^k \left[t^j (1-t)^{n+k-j} \right] = \sum_{\ell=0}^{k} \binom{k}{\ell} (k-\ell)! \binom{j}{k-\ell} t^{j-k+\ell} (-1)^\ell \ell! \binom{n+k-j}{\ell} (1-t)^{n+k-j-\ell}. \tag{3.4.14}$$

By substituting (3.4.14) into (3.4.13) and interchanging the summation order (see Exercise 3.9), we get, by recalling also the definition (3.2.1),

$$(P_{n+k}^B)^{(k)}(x) = \frac{1}{(b-a)^k} \sum_{\ell=0}^{k} (-1)^\ell (k-\ell)! \ell! \binom{k}{\ell} \sum_{j=k-\ell}^{n+k-\ell} f(x_{n+k,j}) \binom{n+k}{j}$$

$$\times \binom{j}{k-\ell} \binom{n+k-j}{\ell} t^{j-k+\ell} (1-t)^{n+k-j-\ell}$$

$$= \frac{1}{(b-a)^k} \sum_{\ell=0}^{k} (-1)^\ell \binom{k}{\ell} \sum_{j=k-\ell}^{n+k-\ell} f(x_{n+k,j}) \frac{(n+k)!}{j!(n+k-j)!} \frac{j!}{(j-k+\ell)!}$$

$$\times \frac{(n+k-j)!}{(n+k-j-\ell)!} t^{j-k+\ell} (1-t)^{n+k-j-\ell}$$

$$= \frac{1}{(b-a)^k} \sum_{\ell=0}^{k} (-1)^\ell \binom{k}{\ell} \sum_{j=k-\ell}^{n+k-\ell} f(x_{n+k,j}) \frac{(n+k)!}{(j-k+\ell)!(n+k-j-\ell)!}$$

$$\times t^{j-k+\ell} (1-t)^{n+k-j-\ell}$$

$$= \frac{(n+k)!}{(b-a)^k n!} \sum_{\ell=0}^{k} (-1)^\ell \binom{k}{\ell} \sum_{j=0}^{n} f(x_{n+k,j+k-\ell}) \frac{n!}{j!(n-j)!} t^j (1-t)^{n-j}$$

$$= \frac{(n+k)!}{(b-a)^k n!} \sum_{j=0}^{n} \left[\sum_{\ell=0}^{k} (-1)^\ell \binom{k}{\ell} f(x_{n+k,j+k-\ell}) \right] B_{n,j}(x), \qquad (3.4.15)$$

from (3.2.5).

Since the second line of (3.3.1) gives

$$x_{n+k,j+1} - x_{n+k,j} = \frac{b-a}{n+k}, \quad j = 0,\dots,n+k-1,$$

we may now apply the formula (3.4.2) in Theorem 3.4.1 to deduce that

$$\sum_{\ell=0}^{k} (-1)^\ell \binom{k}{\ell} f(x_{n+k,j+k-\ell}) = \sum_{\ell=0}^{k} (-1)^\ell \binom{k}{k-\ell} f(x_{n+k,j+k-\ell})$$

$$= (-1)^k \sum_{\ell=0}^{k} (-1)^\ell \binom{k}{\ell} f(x_{n+k,j+\ell})$$

$$= k! \left(\frac{b-a}{n+k} \right)^k f[x_{n+k,j},\dots,x_{n+k,j+k}]. \qquad (3.4.16)$$

Next, we apply Theorem 2.1.2 to deduce the existence of a point

$$\xi_{n,j} \in [x_{n+k,j}, x_{n+k,j+k}] \qquad (3.4.17)$$

such that

$$f[x_{n+k,j},\ldots,x_{n+k,j+k}] = \frac{f^{(k)}(\xi_{n,j})}{k!}. \tag{3.4.18}$$

By combining (3.4.3), (3.4.15), (3.4.16) and (3.4.18), we obtain

$$P_n^{B,k}(x) - (P_{n+k}^B)^{(k)}(x) = \sum_{j=0}^{n} \left[f^{(k)}(x_{n,j}) - \frac{(n+k)!}{n!(n+k)^k} f^{(k)}(\xi_{n,j}) \right] B_{n,j}(x),$$

and thus, by using also Theorem 3.2.1(b),

$$|P_n^{B,k}(x) - (P_{n+k}^B)^{(k)}(x)| \leqslant \sum_{j=0}^{n} \left| f^{(k)}(x_{n,j}) - \frac{(n+k)!}{n!(n+k)^k} f^{(k)}(\xi_{n,j}) \right| B_{n,j}(x). \tag{3.4.19}$$

Let $\varepsilon > 0$ be given. Since $f \in C^m[a,b]$ and $k \in \{1,\ldots,m\}$, we know that $f^{(k)} \in C[a,b]$, and hence $f^{(k)}$ is uniformly continuous on $[a,b]$, thereby implying the existence of a positive number $\delta = \delta(\varepsilon) < b-a$, which is independent of n and j, such that

$$|f^{(k)}(x_{n,j}) - f^{(k)}(\xi_{n,j})| < \frac{\varepsilon}{2} \tag{3.4.20}$$

for all n and $j \in \{0,\ldots,n\}$ satisfying

$$|x_{n,j} - \xi_{n,j}| < \delta. \tag{3.4.21}$$

Now observe from the second line of (3.3.1), together with (3.4.17), that, for $n \in \mathbb{N}$ and $j = 0,\ldots,n$,

$$x_{n,j} - \xi_{n,j} \leqslant \left[a + j\left(\frac{b-a}{n}\right) \right] - \left[a + j\left(\frac{b-a}{n+k}\right) \right]$$

$$= (b-a)\left(\frac{j}{n}\right)\frac{k}{n+k} \leqslant (b-a)\frac{k}{n+k},$$

whereas

$$x_{n,j} - \xi_{n,j} \geqslant \left[a + j\left(\frac{b-a}{n}\right) \right] - \left[a + (j+k)\left(\frac{b-a}{n+k}\right) \right]$$

$$= -(b-a)\left(1 - \frac{j}{n}\right)\frac{k}{n+k} \geqslant -(b-a)\frac{k}{n+k},$$

and thus

$$|x_{n,j} - \xi_{n,j}| \leqslant (b-a)\frac{k}{n+k}. \tag{3.4.22}$$

With the definition

$$\widetilde{N} = \widetilde{N}(\varepsilon) := \left\lceil \frac{k(b-a-\delta)}{\delta} \right\rceil, \tag{3.4.23}$$

it follows that (3.4.21), and therefore also (3.4.20), are satisfied for $n > \widetilde{N}$ and $j = 0,\ldots,n$.

Next, we observe that

$$\frac{(n+k)!}{n!(n+k)^k} = \frac{(n+k)(n+k-1)\dots(n+k-(k-1))}{(n+k)^k}$$

$$= 1\left(1-\frac{1}{n+k}\right)\dots\left(1-\frac{k-1}{n+k}\right) \to 1, \quad n \to \infty,$$

according to which there is a positive integer $N^* = N^*(\varepsilon)$ such that

$$\left|\frac{(n+k)!}{n!(n+k)^k} - 1\right| < \frac{\varepsilon}{2M_k}, \quad n > N^*, \tag{3.4.24}$$

where

$$M_k := \max_{a\leqslant x\leqslant b} |f^{(k)}(x)|. \tag{3.4.25}$$

Hence, if we define

$$N := \max\{\widetilde{N}, N^*\},$$

it follows from (3.4.20), (3.4.25), and (3.4.24) that, for $n > N$ and $j = 0,\dots,n$,

$$\left|f^{(k)}(x_{n,j}) - \frac{(n+k)!}{n!(n+k)^k}f^{(k)}(\xi_{n,j})\right|$$

$$\leqslant \left|f^{(k)}(x_{n,j}) - f^{(k)}(\xi_{n,j})\right| + \left|\frac{(n+k)!}{n!(n+k)^k} - 1\right| |f^{(k)}(\xi_{n,j})|$$

$$< \frac{\varepsilon}{2} + \frac{\varepsilon}{2M_k}M_k = \varepsilon. \tag{3.4.26}$$

By inserting (3.4.26) into (3.4.19), and using Theorem 3.2.1(c), we obtain

$$\left|P_n^{B,k}(x) - (P_{n+k}^B)^{(k)}(x)\right| < \varepsilon, \quad n > N,$$

which then immediately implies the desired result (3.4.12). ∎

Observe that, if in (3.4.15) we set $k = 1$ and replace n by $n-1$, we obtain

$$(P_n^B)'(x) = \frac{n}{b-a} \sum_{j=0}^{n-1} \left[f(x_{n,j+1}) - f(x_{n,j})\right] B_{n-1,j}(x), \tag{3.4.27}$$

after having noted also that (3.4.15) holds for any $f \in C[a,b]$ and $n \in \mathbb{N}$. It follows from (3.4.27) and (3.3.2), together with (3.2.7) in Theorem 3.2.1(b), that, if f is strictly increasing on $[a,b]$, then $(P_n^B)'(x) > 0, x \in [a,b]$, whereas, if f is strictly decreasing on $[a,b]$, then $(P_n^B)'(x) < 0, x \in [a,b]$. Hence we have the following strict monotonicity-preserving result.

Theorem 3.4.3. *For $n \in \mathbb{N}$, the Bernstein polynomial approximation P_n^B in π_n to any $f \in C[a,b]$, as defined by (3.3.1), satisfies the following:*

(a) *If f is strictly increasing on $[a,b]$, then P_n^B is strictly increasing on $[a,b]$.*

(b) *If f is strictly decreasing on $[a,b]$, then P_n^B is strictly decreasing on $[a,b]$.*

3.5 Convergence rate

We proceed to investigate, for a function f in Theorem 3.3.3 that is also continuously differentiable on $[a,b]$, the rate of convergence at which $||f - P_n^B||_\infty$ tends to zero for $n \to \infty$. Specifically we shall establish an explicit bound of the form

$$||f - P_n^B||_\infty \leqslant g(n)||f'||_\infty, \quad n = 1, 2, \ldots, \tag{3.5.1}$$

where $g(n) \to 0$, $n \to \infty$.

Let the function $H : \mathbb{R} \to \mathbb{R}$ be defined by

$$H(x) := \begin{cases} 1, & x \geqslant 0; \\ 0, & x < 0. \end{cases} \tag{3.5.2}$$

Observe that, for $f \in C^1[a,b]$ and $[\alpha, \beta] \subset [a,b]$, the definition (3.5.2) yields, for any $x \in [a,b]$,

$$\int_\alpha^\beta H(x-t)f'(t)dt = \int_\alpha^x f'(t)dt = f(x) - f(\alpha),$$

and thus

$$f(x) = f(\alpha) + \int_\alpha^\beta H(x-t)f'(t)dt. \tag{3.5.3}$$

The following convergence rate result is satisfied by Bernstein polynomial approximation.

Theorem 3.5.1. *In Theorem 3.3.3, suppose that, moreover,* $f \in C^1[a,b]$. *Then*

$$||f - P_n^B||_\infty \leqslant \frac{b-a}{n}||f'||_\infty, \quad n = 1, 2, \ldots. \tag{3.5.4}$$

Proof. Let $x \in [a,b]$, and for any fixed $n \in \mathbb{N}$, denote by k the (unique) integer in $\{0, \ldots, n-1\}$ such that

$$x \in [x_{n,k}, x_{n,k+1}), \tag{3.5.5}$$

where $\{x_{n,j} : j = 0, \ldots, n\}$ is the uniform partition of $[a,b]$ as given in the second line of (3.3.1). By choosing $[\alpha, \beta] = [x_{n,k}, x_{n,k+1}]$ in (3.5.3), we obtain

$$f(x) = f(x_{n,k}) + \int_{x_{n,k}}^{x_{n,k+1}} H(x-t)f'(t)dt, \tag{3.5.6}$$

where the function H is defined by (3.5.2). It then follows from (3.3.1) and (3.5.6), together with Theorem 3.2.1(c), that

$$P_n^B(x) = f(x_{n,k}) \sum_{j=0}^n B_{n,j}(x) + \sum_{j=0}^n \left[\int_{x_{n,k}}^{x_{n,k+1}} H(x_{n,j} - t)f'(t)dt \right] B_{n,j}(x)$$

$$= f(x_{n,k}) + \int_{x_{n,k}}^{x_{n,k+1}} \left[\sum_{j=0}^{n} H(x_{n,j} - t) B_{n,j}(x) \right] f'(t) dt,$$

which, together with (3.5.6), yields

$$f(x) - P_n^B(x) = \int_{x_{n,k}}^{x_{n,k+1}} \left[H(x - t) - \sum_{j=0}^{n} H(x_{n,j} - t) B_{n,j}(x) \right] f'(t) dt,$$

and thus, by using also (3.5.2), Theorem 3.2.1(b) and (c), (3.5.5), and the second line of (3.3.1), we deduce that

$$|f(x) - P_n^B(x)| \leqslant ||f'||_\infty \int_{x_{n,k}}^{x_{n,k+1}} \left| H(x - t) - \sum_{j=0}^{n} H(x_{n,j} - t) B_{n,j}(x) \right| dt$$

$$= ||f'|| \left\{ \int_{x_{n,k}}^{x} \left| 1 - \sum_{j=k+1}^{n} B_{n,j}(x) \right| dt + \int_{x}^{x_{n,k+1}} \left| \sum_{j=k+1}^{n} B_{n,j}(x) \right| dt \right\}$$

$$= ||f'||_\infty \left\{ \left[\sum_{j=0}^{k} B_{n,j}(x) \right] (x - x_{n,k}) + \left[\sum_{j=k+1}^{n} B_{n,j}(x) \right] (x_{n,k+1} - x) \right\}$$

$$\leqslant ||f'||_\infty \left\{ \left[\sum_{j=0}^{k} B_{n,j}(x) \right] (x_{n,k+1} - x_{n,k}) + \left[\sum_{j=k+1}^{n} B_{n,j}(x) \right] (x_{n,k+1} - x_{n,k}) \right\}$$

$$= \frac{b-a}{n} ||f'||_\infty \left[\sum_{j=0}^{n} B_{n,j}(x) \right] = \frac{b-a}{n} ||f'||_\infty,$$

which is independent of k, and therefore implies the desired result (3.5.4).   ∎

3.6   Exercises

Exercise 3.1 For any sequence $\{ \triangle_n : n \in \mathbb{N} \}$, with

$$\triangle_n := \{ x_{n,0}, \ldots, x_{n,n} \} \subset [0, \pi], \quad n \in \mathbb{N},$$

let $\{ P_n^I : n \in \mathbb{N} \}$ denote the polynomial sequence in π_n such that, for each $n \in \mathbb{N}$, the polynomial P_n^I interpolates the function

$$f(x) = \sin x, \quad x \in [0, \pi],$$

at the points $\triangle_n$. Use the interpolation error estimate (2.1.14) in Theorem 2.1.4 to prove the uniform convergence result

$$\max_{0 \leqslant x \leqslant \pi} |\sin x - P_n^I(x)| \to 0, \quad n \to \infty.$$

Exercise 3.2 By arguing as in the derivation of (3.3.17), prove, for any $n \in \mathbb{N}$, the identity

$$\sum_{j=1}^{n} \frac{j^3}{n^3} \binom{n}{j} t^j (1-t)^{n-j} = t^3 + \frac{3n-2}{n^2} t(1-t) \left(t + \frac{1}{3n-2} \right), \quad t \in \mathbb{R},$$

and thereby extending the identities (3.3.18) - (3.3.20).

[*Hint:* Apply the identity $j^2 = (j-1)(j-2) + 3(j-1) + 1, \quad j \in \mathbb{N}.$]

Exercise 3.3 As a continuation of Exercise 3.2, prove that, analogously to (3.3.13) in Theorem 3.3.2(a), and for any bounded interval $[a,b] \subset \mathbb{R}$, the Bernstein polynomial approximation P_n^B in π_n of the function

$$f(x) = x^3, \quad x \in [a,b], \tag{$*$}$$

is given by

$$P_n^B(x) = x^3 + \frac{(x-a)(b-x)}{n} \left[\left(3 - \frac{2}{n} \right) x + \frac{a+b}{n} \right],$$

and then use this formula to verify that

$$P_n^B \in \begin{cases} \pi_n, & \text{if } n \in \{1,2\}; \\ \pi_3, & \text{if } n \geqslant 3. \end{cases}$$

Exercise 3.4 As a continuation of Exercise 3.3, prove that, analogously to (3.3.15) in Theorem 3.3.2(a),

$$\|f - P_n^B\|_\infty \leqslant \frac{(b-a)^2}{4n} K_n,$$

where

$$K_n := \max \left\{ \left| 3a + \frac{b-a}{n} \right|, \left| 3b - \frac{b-a}{n} \right| \right\}.$$

Exercise 3.5 For the function f given by $(*)$ in Exercise 3.3, and by using the results of Exercises 3.3 and 3.4, verify the results of (a) Theorem 3.3.1(a); (b) Theorem 3.3.1(b), with $a = 0$; (c) Theorem 3.3.3; (d) Theorem 3.4.3(a), with $a = 0$; (e) Theorem 3.5.1.

[*Hint:* In (e), consider separately the three cases $0 \leqslant a < b$; $a < 0 \leqslant b$; $a < b < 0$.]

Exercise 3.6 According to Theorem 3.2.1(d), there exists, for integers $k \geqslant 0$ and $n \geqslant k$, a (unique) coefficient sequence $\{\beta_{n,k,j} : j = 0, \ldots, n\}$ such that

$$t^k = \sum_{j=0}^{n} \beta_{n,k,j} \binom{n}{j} t^j (1-t)^{n-j}, \quad t \in \mathbb{R}.$$

By applying the identities (3.3.18) - (3.3.20), as well as the identity derived in Exercise 3.2, calculate the coefficient sequences $\{\beta_{n,k,j} : j = 0, \ldots, n\}$ for $k = 0, 1, 2$, and 3, and any $n \geqslant k$.

Exercise 3.7 Apply the results of Exercise 3.6 to obtain, in the form (3.2.12), the Bernstein representation in π_n with respect to the interval $[a,b]$ of the polynomial P if:

(a) $P(x) = 2x - 3$; $n = 5$; $[a,b] = [-1,2]$;   (b) $P(x) = x^2 + x + 1$; $n = 2$; $[a,b] = [0,3]$;

(c) $P(x) = x(x-1)(x-2)$; $n = 4$; $[a,b] = [0,2]$.

Exercise 3.8 Prove the Leibniz formula (3.4.5) by means of a proof by induction, together with the product rule for differentiation.

Exercise 3.9 For $n \in \mathbb{N}$, and any sequence $\{a_{j,k} : j,k = 0,\ldots,n\}$, verify the interchange of summation order result

$$\sum_{j=0}^{n}\sum_{k=0}^{j} a_{j,k} = \sum_{k=0}^{n}\sum_{j=k}^{n} a_{j,k},$$

as applied to establish (3.4.15) in the proof of Theorem 3.4.2.

Exercise 3.10 For the function

$$f(x) = \tan x, \quad x \in [0, \tfrac{\pi}{4}],$$

find the smallest value of the integer n for which, according to the convergence rate result (3.5.4) in Theorem 3.5.1, the Bernstein polynomial approximation P_n^B in π_n of f satisfies

$$\max_{0 \leqslant x \leqslant \frac{\pi}{4}} |\tan x - P_n^B(x)| < \frac{1}{10},$$

and explicitly write down the formula for the polynomial P_n^B for this specific value of n.

Chapter 4

Best Approximation

This chapter is concerned with the study of best polynomial approximation. For example, for any non-negative integer n and any function $f \in C[a,b]$, the problem to be considered is the existence of some polynomial $P^* \in \pi_n$, such that $||f - P^*||_\infty \leqslant ||f - P||_\infty$, for all polynomials $P \in \pi_n$. We shall study this problem in the more general setting of normed linear spaces.

4.1 Existence in normed linear spaces

For a linear (vector) space X, let $||\cdot|| : X \to \mathbb{R}$ denote a function such that the following conditions are satisfied:

(i)

$$||f|| \geqslant 0, \quad f \in X; \tag{4.1.1}$$

(ii)

$$||f|| = 0 \quad \text{if and only if} \quad f = 0; \tag{4.1.2}$$

(iii)

$$||\lambda f|| = |\lambda| \, ||f||, \quad \lambda \in \mathbb{R}, \quad f \in X; \tag{4.1.3}$$

(iv)

$$||f + g|| \leqslant ||f|| + ||g||, \quad f, g \in X \quad \text{(triangle inequality)}. \tag{4.1.4}$$

We then call $(X, ||\cdot||)$ a normed linear space, with corresponding norm $||\cdot||$.
For any $f, g \in X$, we see from (4.1.4) that

$$||f|| = ||(f - g) + g|| \leqslant ||f - g|| + ||g||,$$

71

and

$$||g|| = ||(g-f)+f|| \leqslant ||g-f|| + ||f|| = ||f-g|| + ||f||,$$

by using (4.1.3) with $\lambda = -1$, and thus

$$-||f-g|| \leqslant ||f|| - ||g|| \leqslant ||f-g||,$$

or equivalently,

$$|\ ||f|| - ||g||\ | \leqslant ||f-g||. \tag{4.1.5}$$

Example 4.1.1. For $n \in \mathbb{N}$, the Euclidean n-dimensional space $X = \mathbb{R}^n$, together with the associated Euclidean norm (or length)

$$||\mathbf{x}||_E := \sqrt{\sum_{j=1}^{n}(x_j)^2}, \quad \mathbf{x} = (x_1,\ldots,x_n) \in \mathbb{R}^n, \tag{4.1.6}$$

constitute the Euclidean normed linear space $(\mathbb{R}^n, ||\cdot||_E)$. ∎

Example 4.1.2. For a bounded interval $[a,b]$, the linear space $X = C[a,b]$, together with the maximum norm (or sup norm, or L^∞ norm, or Chebyshev norm)

$$||f||_\infty := \max_{a \leqslant x \leqslant b} |f(x)|, \quad f \in C[a,b], \tag{4.1.7}$$

as introduced in (2.1.13), constitute the normed linear space $(C[a,b], ||\cdot||_\infty)$ (see Exercise 4.1). ∎

For a given normed linear space $(X, ||\cdot||)$, suppose $f \in X$, and let $A \subset X$ denote an approximation set. If there exists an element $f^* \in A$ such that

$$||f-f^*|| \leqslant ||f-g||, \quad g \in A, \tag{4.1.8}$$

we say that f^* is a best approximation from A to f. Our following result establishes a sufficient condition on the approximation set A for the existence of f^*.

Theorem 4.1.1. *For a normed linear space $(X, ||\cdot||)$, let $f \in X$, and suppose $A \subset X$ is an approximation set such that A is a finite-dimensional subspace of X. Then there exists a best approximation f^* from A to f.*

Proof. Define $d := \dim(A)$, and let $\{f_1,\ldots,f_d\} \subset A$ denote a basis for A. Our first step is to prove that, for the function $w : \mathbb{R}^d \to \mathbb{R}$ defined by

$$w(\mathbf{x}) := \left|\left| \sum_{j=1}^{d} x_j f_j \right|\right|, \quad \mathbf{x} = (x_1,\ldots,x_d) \in \mathbb{R}^d, \tag{4.1.9}$$

there exists a positive constant m such that

$$w(\mathbf{x}) \geqslant m||\mathbf{x}||_E, \quad \mathbf{x} \in \mathbb{R}^d. \tag{4.1.10}$$

To this end, we first use consecutively (4.1.9), (4.1.5), (4.1.4), (4.1.3) and (4.1.6) to obtain, for any $\mathbf{x} = (x_1,\dots,x_d)$ and $\mathbf{y} = (y_1,\dots,y_d)$ in $\mathbb{R}^d$,

$$|w(\mathbf{x}) - w(\mathbf{y})| = \left| \left\| \sum_{j=1}^{d} x_j f_j \right\| - \left\| \sum_{j=1}^{d} y_j f_j \right\| \right| \tag{4.1.11}$$

$$\leqslant \left\| \sum_{j=1}^{d} (x_j - y_j) f_j \right\| \leqslant \sum_{j=1}^{d} |x_j - y_j| \, ||f_j||$$

$$= \sum_{j=1}^{d} \sqrt{(x_j - y_j)^2} \, ||f_j||$$

$$\leqslant \left[\sum_{j=1}^{d} ||f_j|| \right] ||\mathbf{x} - \mathbf{y}||_E,$$

and thus

$$\lim_{\mathbf{y} \to \mathbf{x}} w(\mathbf{y}) = w(\mathbf{x}), \quad \mathbf{x} \in \mathbb{R}^d, \tag{4.1.12}$$

that is, w is a continuous function on $\mathbb{R}^d$. Hence, since also

$$S := \{\mathbf{x} = (x_1,\dots,x_d) \in \mathbb{R}^d : ||\mathbf{x}||_E = 1\} \tag{4.1.13}$$

is a closed and bounded (or compact) subset of $\mathbb{R}^d$, a standard result from calculus guarantees that the function w attains its minimum value on S, that is, there exists a point $\mathbf{y}^* = (y_1^*,\dots,y_d^*) \in S$ such that

$$m := w(\mathbf{y}^*) \leqslant w(\mathbf{y}), \quad \mathbf{y} \in S. \tag{4.1.14}$$

Note from (4.1.14), (4.1.9) and (4.1.1) that $m \geqslant 0$. If $m = 0$, then (4.1.14), (4.1.9) and (4.1.2) imply

$$\sum_{j=1}^{d} y_j^* f_j = 0,$$

and thus, since $\{f_1,\dots,f_d\}$ is a basis for A, and therefore a linearly independent set, we must have $y_1^* = \cdots = y_d^* = 0$, which contradicts the fact that, since $\mathbf{y}^* = (y_1^*,\dots,y_d^*) \in S$, (4.1.13) and (4.1.6) give $\sum_{j=1}^{d} (y_j^*)^2 = 1$. Hence $m > 0$.

Let $\mathbf{x} \in \mathbb{R}^d \setminus \{\mathbf{0}\}$, and define

$$\mathbf{y} := \frac{\mathbf{x}}{||\mathbf{x}||_E},$$

according to which $||\mathbf{y}||_E = 1$, so that, from (4.1.13), $\mathbf{y} \in S$, and thus, by virtue of (4.1.14),

$$w\left(\frac{\mathbf{x}}{||\mathbf{x}||_E}\right) \geqslant m. \tag{4.1.15}$$

Now observe that (4.1.9) and (4.1.3) yield

$$w\left(\frac{\mathbf{x}}{||\mathbf{x}||_E}\right) = \frac{1}{||\mathbf{x}||_E} w(\mathbf{x}). \tag{4.1.16}$$

It follows from (4.1.15) and (4.1.16) that the inequality in (4.1.10) is satisfied for $\mathbf{x} \in \mathbb{R}^d \setminus \{\mathbf{0}\}$. Since (4.1.9) gives $w(\mathbf{0}) = 0$, we see that (4.1.10) holds with both sides equal to zero if $\mathbf{x} = \mathbf{0}$, and thereby completing our proof of (4.1.10).

Next, we define the function $v : \mathbb{R}^d \to \mathbb{R}$ by

$$v(\mathbf{x}) := \left\|f - \sum_{j=1}^{d} x_j f_j\right\|, \quad \mathbf{x} = (x_1, \ldots, x_d) \in \mathbb{R}^d, \tag{4.1.17}$$

for which, by applying the inequality (4.1.5), as well as (4.1.3) with $\lambda = -1$, we deduce that, for any $\mathbf{x} = (x_1, \ldots, x_d) \in \mathbb{R}^d$ and $\mathbf{y} = (y_1, \ldots, y_d) \in \mathbb{R}^d$,

$$|v(\mathbf{x}) - v(\mathbf{y})| = \left| \left\|f - \sum_{j=1}^{d} x_j f_j\right\| - \left\|f - \sum_{j=1}^{d} y_j f_j\right\| \right| \leqslant \left\|\sum_{j=1}^{d} (x_j - y_j) f_j\right\|. \tag{4.1.18}$$

It then follows as in the steps leading from (4.1.11) to (4.1.12) that v is a continuous function on $\mathbb{R}^d$. A standard result from calculus then guarantees that the function v attains its minimum value on the closed and bounded (or compact) subset

$$T := \left\{\mathbf{x} = (x_1, \ldots, x_d) \in \mathbb{R}^d : ||\mathbf{x}||_E \leqslant \frac{2||f||}{m}\right\} \tag{4.1.19}$$

of $\mathbb{R}^d$, that is, there exists a point $\mathbf{x}^* = (x_1^*, \ldots, x_d^*) \in T$ such that

$$v(\mathbf{x}^*) \leqslant v(\mathbf{x}), \quad \mathbf{x} \in T. \tag{4.1.20}$$

Let $A_0 \subset A$ be defined by

$$A_0 := \left\{g \in A : g = \sum_{j=1}^{d} x_j f_j; \, \mathbf{x} = (x_1, \ldots, x_d) \in T\right\}, \tag{4.1.21}$$

with the subset T of $\mathbb{R}^d$ given by (4.1.19), and define

$$f^* := \sum_{j=1}^{d} x_j^* f_j, \tag{4.1.22}$$

for which, since $\mathbf{x}^* = (x_1^*, \ldots, x_d^*) \in T$, it follows from (4.1.21) that $f^* \in A_0$.

Now let $g \in A_0$, according to which, from (4.1.21),

$$g = \sum_{j=1}^{d} x_j f_j, \tag{4.1.23}$$

for some $\mathbf{x} = (x_1, \ldots, x_d) \in T$. By using (4.1.23), (4.1.17), (4.1.20) and (4.1.22), we obtain

$$||f - g|| = v(\mathbf{x}) \geqslant v(\mathbf{x}^*) = ||f - f^*||,$$

according to which we have now shown that

$$||f - f^*|| \leqslant ||f - g||, \quad g \in A_0. \tag{4.1.24}$$

We proceed to prove that also

$$||f - f^*|| < ||f - g||, \quad g \in A \setminus A_0, \tag{4.1.25}$$

which, together with (4.1.24), then shows that (4.1.8) is satisfied by f^*, and would therefore complete our proof.

To prove (4.1.25), we let $g \in A \setminus A_0$, so that, from (4.1.21) and (4.1.19), g is given by (4.1.23) for some $\mathbf{x} = (x_1, \ldots, x_d) \in \mathbb{R}^d$ satisfying

$$||\mathbf{x}||_E > \frac{2||f||}{m}. \tag{4.1.26}$$

It follows from (4.1.23), (4.1.9), (4.1.10) and (4.1.26) that

$$||g|| = w(\mathbf{x}) \geqslant m||\mathbf{x}||_E > 2||f||,$$

and thus, by using also (4.1.3) and (4.1.5),

$$||f - g|| = ||g - f|| \geqslant |\,||g|| - |f|\,| \geqslant ||g|| - ||f|| > ||f||. \tag{4.1.27}$$

Now observe from (4.1.21) and (4.1.19) that the zero element 0 of the subspace $A \subset X$ satisfies $0 \in A_0$, and it follows that we may choose $g = 0$ in (4.1.24) to deduce that

$$||f|| = ||f - 0|| \geqslant ||f - f^*||. \tag{4.1.28}$$

By combining (4.1.27) and (4.1.28), we obtain the desired result (4.1.25). ∎

Since the polynomial space π_n, with the polynomial domains restricted to $[a,b]$, is a finite-dimensional subspace of $C[a,b]$, the following existence result is an immediate consequence of Theorem 4.1.1.

Theorem 4.1.2. *Let $f \in C[a,b]$. Then, for each non-negative integer n, there exists a best approximation P^* from π_n to f with respect to the maximum norm on $[a,b]$, that is,*

$$||f - P^*||_\infty \leqslant ||f - P||_\infty, \quad P \in \pi_n. \tag{4.1.29}$$

Before proceeding to investigate the issue of uniqueness in best approximation, as will be done in Section 4.2, we first prove, in the setting of Theorem 4.1.1, a property of the set

$$A_f^* := \{f^* \in A : \ f^* \text{ is a best approximation from } A \text{ to } f\}. \tag{4.1.30}$$

A non-empty subset Y of a linear space X is called a convex set if the condition

$$\{\lambda f + (1-\lambda)g : \ \lambda \in [0,1]\} \subset Y, \quad f, g \in Y, \tag{4.1.31}$$

is satisfied. Observe that a convex subset Y of X has either precisely one element, or infinitely many elements.

Theorem 4.1.3. *For a normed linear space* $(X, \|\cdot\|)$, *let* $f \in X$, *and suppose* $A \subset X$ *is an approximation set such that* A *is a subspace of* X, *and such that the set* A_f^*, *as defined by* (4.1.30), *is non-empty. Then* A_f^* *is a convex set.*

Proof. Suppose $f^*, g^* \in A_f^*$, define

$$d^* := \min\{\|f - g\| : \ g \in A\} = \|f - f^*\| = \|f - g^*\|, \tag{4.1.32}$$

and let $\lambda \in [0,1]$. It follows from (4.1.4), (4.1.3) and (4.1.32) that

$$\|f - [\lambda f^* + (1-\lambda)g^*]\| = \|\lambda(f - f^*) + (1-\lambda)(f - g^*)\|$$

$$\leqslant \|\lambda(f - f^*)\| + \|(1-\lambda)(f - g^*)\|$$

$$= \lambda\|f - f^*\| + (1-\lambda)\|f - g^*\| = \lambda d^* + (1-\lambda)d^* = d^*,$$

that is,

$$\|f - [\lambda f^* + (1-\lambda)g^*]\| \leqslant d^*. \tag{4.1.33}$$

Since A is a subspace of X, we have $\lambda f^* + (1-\lambda)g^* \in A$, and thus, from the definition in (4.1.32) of d^*,

$$\|f - [\lambda f^* + (1-\lambda)g^*]\| \geqslant d^*. \tag{4.1.34}$$

It follows from (4.1.33), (4.1.34) and (4.1.32) that

$$\|f - [\lambda f^* + (1-\lambda)g^*]\| = d^* = \min\{\|f - g\| : \ g \in A\},$$

and thus, from (4.1.30), $\lambda f^* + (1-\lambda)g^* \in A_f^*$, which proves that A_f^* is a convex set. ∎

We deduce that, in the setting of Theorem 4.1.3, there either exists a unique best approximation f^* from A to f, or there exist infinitely many best approximations f^* from A to f.

In the next section, we identify a class of normed linear spaces $(X, \|\cdot\|)$ for which f^* is indeed unique.

4.2   Uniqueness in inner product spaces

For a linear (vector) space X, let $\langle \cdot, \cdot \rangle : X \times X \to \mathbb{R}$ denote a function such that the following conditions are satisfied:

(i)

$$\langle f, f \rangle \geqslant 0, \quad f \in X; \tag{4.2.1}$$

(ii)

$$\langle f, f \rangle = 0 \quad \text{if and only if} \quad f = 0; \tag{4.2.2}$$

(iii)

$$\langle f + g, h \rangle = \langle f, h \rangle + \langle g, h \rangle, \quad f, g, h \in X; \tag{4.2.3}$$

(iv)

$$\langle \lambda f, g \rangle = \lambda \langle f, g \rangle, \quad \lambda \in \mathbb{R}, \quad f, g \in X; \tag{4.2.4}$$

(v)

$$\langle f, g \rangle = \langle g, f \rangle, \quad f, g \in X. \tag{4.2.5}$$

We then call $(X, \langle \cdot, \cdot \rangle)$ an inner product space, with corresponding inner product $\langle \cdot, \cdot \rangle$. Observe that the choice $\lambda = 0$ in (4.2.4), together with (4.2.5), yields

$$\langle f, 0 \rangle = \langle 0, g \rangle = 0, \quad f, g \in X. \tag{4.2.6}$$

Example 4.2.1. For $n \in \mathbb{N}$, the Euclidean n-dimensional space $X = \mathbb{R}^n$, together with the associated Euclidean inner product (or dot product, or scalar product)

$$\langle \mathbf{x}, \mathbf{y} \rangle_E := \mathbf{x} \cdot \mathbf{y} = \sum_{j=1}^{n} x_j y_j, \quad \mathbf{x} = (x_1, \ldots, x_n) \in \mathbb{R}^n, \quad \mathbf{y} = (y_1, \ldots, y_n) \in \mathbb{R}^n, \tag{4.2.7}$$

constitute the Euclidean inner product space $(\mathbb{R}^n, \langle \cdot, \cdot \rangle_E)$. ∎

Example 4.2.2. Let w denote a real-valued function that is integrable on a bounded interval $[a, b]$, and such that the conditions

(a)

$$\int_a^b w(x)dx > 0; \tag{4.2.8}$$

(b)

$$w(x) \geqslant 0, \quad x \in (a, b); \tag{4.2.9}$$

are satisfied,

in which case w is called a weight function on $[a,b]$. For any such weight function w on $[a,b]$, the linear space $X = C[a,b]$, together with the weighted inner product

$$\langle f,g \rangle_{2,w} := \int_a^b w(x)f(x)g(x)dx, \quad f,g \in C[a,b], \tag{4.2.10}$$

constitute the inner product space $(C[a,b], \langle \cdot, \cdot \rangle_{2,w})$ (see Exercise 4.8). In the special case where the weight function w is given by

$$w(x) = 1, \quad x \in [a,b], \tag{4.2.11}$$

we write

$$\langle f,g \rangle_2 := \int_a^b f(x)g(x)dx \tag{4.2.12}$$

for the corresponding inner product. ∎

The following fundamental inequality is satisfied in inner product spaces.

Theorem 4.2.1 (Cauchy-Schwarz inequality). *Let* $(X, \langle \cdot, \cdot \rangle)$ *be any inner product space. Then*

$$|\langle f,g \rangle| \leqslant \sqrt{\langle f,f \rangle}\sqrt{\langle g,g \rangle}, \quad f,g \in X. \tag{4.2.13}$$

Proof. If either $f = 0$ or $g = 0$, it follows from (4.2.6) that (4.2.13) is satisfied as an equality with both sides equal to zero.

Suppose next $f,g \in X$, with $f \neq 0$ and $g \neq 0$, and let $\lambda \in \mathbb{R}$. Then (4.2.1), (4.2.3), (4.2.4) and (4.2.5) yield

$$0 \leqslant \langle f + \lambda g, f + \lambda g \rangle = \langle f,f \rangle + 2\lambda \langle f,g \rangle + \lambda^2 \langle g,g \rangle,$$

that is,

$$\langle g,g \rangle \lambda^2 + 2\langle f,g \rangle \lambda + \langle f,f \rangle \geqslant 0, \quad \lambda \in \mathbb{R}, \tag{4.2.14}$$

where, since $g \neq 0$, (4.2.2) and (4.2.1) imply $\langle g,g \rangle > 0$. It follows from (4.2.14) that the corresponding discriminant is non-positive, that is,

$$[2\langle f,g \rangle]^2 - 4\langle g,g \rangle \langle f,f \rangle \leqslant 0,$$

and thus

$$[\langle f,g \rangle]^2 \leqslant \langle f,f \rangle \langle g,g \rangle,$$

which, together with (4.2.1), yields the desired inequality (4.2.13). ∎

The Cauchy-Schwarz inequality (4.2.13) is instrumental in proving the following result, according to which every inner product space generates a normed linear space.

Theorem 4.2.2. *Suppose* $(X, \langle \cdot, \cdot \rangle)$ *is an inner product space. Then* $(X, \|\cdot\|)$, *where*

$$\|f\| := \sqrt{\langle f,f \rangle}, \quad f \in X, \tag{4.2.15}$$

is a normed linear space.

Proof. First, note from (4.2.15) that (4.2.1) and (4.2.2) imply, respectively, (4.1.1) and (4.1.2), whereas (4.2.4) and (4.2.5) yield

$$||\lambda f|| = \sqrt{\langle \lambda f, \lambda f \rangle} = \sqrt{\lambda^2 \langle f, f \rangle} = |\lambda|\, ||f||,$$

for any $\lambda \in \mathbb{R}$ and $f \in X$, and thereby proving (4.1.3). It therefore remains to verify that the triangle inequality (4.1.4) is satisfied by the definition (4.2.15). To this end, we let $f, g \in X$, and use (4.2.15), (4.2.3) and (4.2.5), as well as the Cauchy-Schwarz inequality (4.2.13), to obtain

$$
\begin{aligned}
||f + g||^2 &= \langle f + g, f + g \rangle \\
&= \langle f, f \rangle + 2\langle f, g \rangle + \langle g, g \rangle \\
&= ||f||^2 + 2\langle f, g \rangle + ||g||^2 \\
&\leqslant ||f||^2 + 2|\langle f, g \rangle| + ||g||^2 \\
&\leqslant ||f||^2 + 2\sqrt{\langle f, f \rangle}\sqrt{\langle g, g \rangle} + ||g||^2 \\
&= ||f||^2 + 2||f||\, ||g|| + ||g||^2 = (||f|| + ||g||)^2,
\end{aligned}
$$

from which (4.1.4) then immediately follows.   ∎

Observe that if, in Theorem 4.2.2, we choose $(X, \langle \cdot, \cdot \rangle) = (\mathbb{R}^n, \langle \cdot, \cdot \rangle_E)$, then the definition (4.2.15), together with (4.1.6) and (4.2.7), shows that the normed linear space thus generated is the Euclidean n-dimensional space $(\mathbb{R}^n, ||\cdot||_E)$ of Example 4.1.1.

Next, following Example 4.2.2, we choose $(X, \langle \cdot, \cdot \rangle) = (C[a,b], \langle \cdot, \cdot \rangle_{2,w})$ in Theorem 4.2.2, to obtain the normed linear space $(C[a,b], ||\cdot||_{2,w})$, where, from (4.2.15) and (4.2.10), the weighted L^2 norm is given by

$$||f||_{2,w} := \sqrt{\int_a^b w(x)[f(x)]^2 dx}, \quad f \in C[a,b], \tag{4.2.16}$$

for some weight function w on $[a,b]$ satisfying the conditions (4.2.8) and (4.2.9). For the special case where the weight function w is given by (4.2.11), we obtain the normed linear space $(C[a,b], ||\cdot||_2)$, where, from (4.2.15) and (4.2.12), the L^2 norm is given by

$$||f||_2 := \sqrt{\int_a^b [f(x)]^2 dx}, \quad f \in C[a,b]. \tag{4.2.17}$$

Analogously to Theorem 4.1.2, the existence result of Theorem 4.1.1 now immediately implies the following.

Theorem 4.2.3. *Let $f \in C[a,b]$. Then, for each non-negative integer n, there exists a polynomial $\widetilde{P}^*$ such that*

$$||f - \widetilde{P}^*||_{2,w} \leqslant ||f - P||_{2,w}, \quad P \in \pi_n. \tag{4.2.18}$$

We proceed to prove a uniqueness result for best approximation in inner product spaces. We shall rely on the fact that, in the setting of Theorem 4.2.2, it follows from (4.2.15), (4.2.3), (4.2.4) and (4.2.5), that, for $f, g \in X$,

$$\|f+g\|^2 + \|f-g\|^2 = \langle f+g, f+g \rangle + \langle f-g, f-g \rangle$$

$$= [\langle f,f \rangle + 2\langle f,g \rangle + \langle g,g \rangle] + [\langle f,f \rangle - 2\langle f,g \rangle + \langle g,g \rangle]$$

$$= 2\|f\|^2 + 2\|g\|^2. \tag{4.2.19}$$

Our uniqueness result is then as follows.

Theorem 4.2.4. *For a normed linear space* $(X, \|\cdot\|)$ *as in Theorem* 4.2.2, *let* $f \in X$, *and suppose* $A \subset X$ *is an approximation set such that* A *is a subspace of* X, *and such that there exists a best approximation* f^* *from* A *to* f. *Then* f^* *is the only best approximation from* A *to* f.

Proof. Suppose $g^* \in A$ is such that

$$\|f - g^*\| = \|f - f^*\| = \min\{\|f - g\| : g \in A\} =: d^*. \tag{4.2.20}$$

Since A is a subspace of X, we know that $\frac{1}{2}f^* + \frac{1}{2}g^* \in A$, so that we may apply Theorem 4.1.3 to deduce that the convex combination $\frac{1}{2}f^* + \frac{1}{2}g^*$ is a best approximation from A to f, that is, from (4.2.20),

$$\|f - (\tfrac{1}{2}f^* + \tfrac{1}{2}g^*)\| = d^*. \tag{4.2.21}$$

By using (4.1.3), (4.2.19), (4.2.20) and (4.2.21), we obtain

$$\|f^* - g^*\|^2 = \|g^* - f^*\|^2$$

$$= \|(f - f^*) - (f - g^*)\|^2$$

$$= 2\|f - f^*\|^2 + 2\|f - g^*\|^2 - \|(f - f^*) + (f - g^*)\|^2$$

$$= 2(d^*)^2 + 2(d^*)^2 - 4\|f - (\tfrac{1}{2}f^* + \tfrac{1}{2}g^*)\|^2$$

$$= 2(d^*)^2 + 2(d^*)^2 - 4(d^*)^2 = 0,$$

which, together with (4.1.2), yields $g^* = f^*$, and thereby completing our proof. ∎

By combining Theorems 4.1.1 and 4.2.4, we immediately deduce the following result.

Theorem 4.2.5. *In Theorem* 4.1.1, *suppose* $(X, \|\cdot\|)$ *is a normed linear space of the type described in Theorem* 4.2.2. *Then* f^* *is the only best approximation from* A *to* f.

An application of Theorem 4.2.5 now immediately yields the following improved formulation of Theorem 4.2.3.

Theorem 4.2.6. *Let $f \in C[a,b]$. Then, for any non-negative integer n, there exists precisely one polynomial $\widetilde{P}_n^* \in \pi_n$ such that*

$$||f - \widetilde{P}_n^*||_{2,w} \leqslant ||f - P||_{2,w}, \quad P \in \pi_n. \tag{4.2.22}$$

The polynomial $\widetilde{P}_n^*$ of Theorem 4.2.6 is called the best weighted L^2 (or weighted least-squares) approximation on $[a,b]$ from π_n to f. For the special case where the weight function w is given by (4.2.11), we have

$$||f - \widetilde{P}_n^*||_2 \leqslant ||f - P||_2, \quad P \in \pi_n, \tag{4.2.23}$$

in which case we call $\widetilde{P}_n^*$ the best L^2 (or least-squares) approximation on $[a,b]$ from π_n to f.

Observe from Theorem 4.2.6 that the best weighted L^2 approximation $\widetilde{P}_n^*$ on $[a,b]$ from π_n to f satisfies the condition

$$||f - \widetilde{P}_n^*||_{2,w} < ||f - P||_{2,w}, \quad \text{for} \quad P \in \pi_n, \quad \text{with} \quad P \neq \widetilde{P}_n^*. \tag{4.2.24}$$

4.3 Exercises

Exercise 4.1 Verify, as stated in Example 4.1.2, that $(C[a,b], ||\cdot||_\infty)$, with $||\cdot||_\infty$ defined by (4.1.7), is a normed linear space, by showing that $||\cdot|| = ||\cdot||_\infty$ satisfies the properties (4.1.1) - (4.1.4).

Exercise 4.2 For any normed linear space $(X, ||\cdot||)$, let $f \in X$ be fixed, and define, for $r > 0$, the sets

$$B(f,r) := \{g \in X : ||f - g|| < r\};$$
$$\overline{B}(f,r) := \{g \in X : ||f - g|| \leqslant r\};$$
$$\partial B(f,r) := \{g \in X : ||f - g|| = r\}.$$

(a) Show, as used in the proof of Theorem 4.1.1, with $(X, ||\cdot||) = (\mathbb{R}^d, ||\cdot||_E)$ as in Example 4.1.1, that $\overline{B}(f,r)$ and $\partial B(f,r)$ are closed and bounded (or compact) subsets of X. [Recall that A is a bounded subset of X if

$$||g|| \leqslant M, \quad g \in A,$$

for some constant M, whereas A is a closed subset of X if the limit (in X) of any convergent sequence $\{g_n : n = 0, 1, \ldots\} \subset A$ belongs to A, that is,

$$||g - g_n|| \to 0, \quad n \to \infty \Rightarrow g \in A.]$$

(b) Show that $B(f,r)$ and $\overline{B}(f,r)$ are both convex sets, whereas $\partial B(f,r)$ is not a convex set.

Exercise 4.3 Prove that $(\mathbb{R}^2, ||\cdot||_E)$, $(\mathbb{R}^2, ||\cdot||_1)$, and $(\mathbb{R}^2, ||\cdot||_\infty)$, where

$$\left.\begin{array}{l} ||\mathbf{x}||_E := \sqrt{x^2 + y^2}, \\[2mm] ||\mathbf{x}||_1 := |x| + |y|, \\[2mm] ||\mathbf{x}||_\infty := \max\{|x|, |y|\}, \end{array}\right\} \quad \mathbf{x} = (x,y) \in \mathbb{R}^2,$$

are normed linear spaces, by showing that each of $||\cdot|| = ||\cdot||_E$, $||\cdot|| = ||\cdot||_1$, and $||\cdot|| = ||\cdot||_\infty$, satisfies the properties (4.1.1) - (4.1.4).

Exercise 4.4 As a continuation of Exercise 4.3, let $\mathbf{y} \in \mathbb{R}^2$ be given by $\mathbf{y} = (1, \sqrt{3})$. For any $r > 0$, make a sketch of the set $B(\mathbf{y}, r)$, as defined in the general setting of Exercise 4.2, for each of the three normed linear spaces $(\mathbb{R}^2, ||\cdot||_E)$, $(\mathbb{R}^2, ||\cdot||_1)$, and $(\mathbb{R}^2, ||\cdot||_\infty)$.

Exercise 4.5 As a continuation of Exercise 4.4, find, for each of the three normed linear spaces $(\mathbb{R}^2, ||\cdot||_E)$, $(\mathbb{R}^2, ||\cdot||_1)$, and $(\mathbb{R}^2, ||\cdot||_\infty)$, the set $A_\mathbf{y}^*$ of best approximations from A to $\mathbf{y}$ if

(a) $A := \{(x,0) : x \in \mathbb{R}\};$ (b) $A := \{\lambda(1,1) : \lambda \in \mathbb{R}\},$

as well as the corresponding minimum values. Explain the consistency of the results thus obtained with Theorems 4.1.1, 4.1.3 and 4.2.4.

Exercise 4.6 For $n = 0$ and $n = 1$, and by arguing in a heuristic manner, find polynomials $\{P_{n,k}^* : k = 0,1,2\} \subset \pi_n$ satisfying the best approximation condition

$$\max_{0 \leqslant x \leqslant 1} |x^k - P_{n,k}^*(x)| \leqslant \max_{0 \leqslant x \leqslant 1} |x^k - P(x)|, \quad P \in \pi_n, \quad k = 0,1,2,$$

and the existence of which is guaranteed by Theorem 4.1.2.

Exercise 4.7 Suppose $(X, ||\cdot||)$ is a normed linear space, and let $f \in X$. For any two subsets A_0 and A_1 of X satisfying $A_0 \subset A_1 \subset X$, suppose $f^* \in A_1$ and $f^{**} \in A_0$ are such that f^* is a best approximation from A_1 to f, whereas f^{**} is a best approximation from A_0 to f^*. Investigate whether it is true or false that f^{**} is then necessarily a best approximation from A_0 to f.

[*Hint:* Consider first the special case provided by Exercise 4.6.]

Exercise 4.8 Verify, as stated in Example 4.2.2, that $(C[a,b], \langle\cdot,\cdot\rangle_{2,w})$, with $\langle\cdot,\cdot\rangle_{2,w}$ defined as in (4.2.10) in terms of a weight function w satisfying (4.2.8), (4.2.9), is an inner product space, by showing that $\langle\cdot,\cdot\rangle = \langle\cdot,\cdot\rangle_{2,w}$ satisfies the properties (4.2.1) - (4.2.5).

Exercise 4.9 For any inner product space $(X, \langle\cdot,\cdot\rangle)$, and $f, g \in X$, with $f \neq 0$ and $g \neq 0$, prove that the Cauchy-Schwarz inequality (4.2.13) in Theorem 4.2.1 holds with equality, that is,

$$|\langle f,g \rangle| = \sqrt{\langle f,f \rangle}\,\sqrt{\langle g,g \rangle},$$

if and only if

$$f \in \{\lambda g : \lambda \in \mathbb{R} \setminus \{0\}\}.$$

[*Hint:* For the proof in the "only if" direction, let $h \in X$ be defined by

$$h := \begin{cases} \sqrt{\langle g,g \rangle}\,f - \sqrt{\langle f,f \rangle}\,g, \text{ if } \langle f,g \rangle \geqslant 0; \\[2ex] \sqrt{\langle g,g \rangle}\,f + \sqrt{\langle f,f \rangle}\,g, \text{ if } \langle f,g \rangle < 0, \end{cases}$$

and use the fact that, according to (4.2.2), $\langle h,h \rangle = 0$ implies $h = 0$.]

Exercise 4.10 By applying a minimization method based on differentiation, find, for $k \in \mathbb{N}$, the polynomial $P^*_{w,k} \in \pi_0$ satisfying the best approximation condition

$$\sqrt{\int_0^1 w(x)[x^k - P^*_{w,k}(x)]^2 dx} \leqslant \sqrt{\int_0^1 w(x)[x^k - P(x)]^2 dx}, \quad P \in \pi_0,$$

and the existence and uniqueness of which are guaranteed by Theorems 4.2.3 and 4.2.6, for each of the following weight functions:

(a) $w(x) = 1$, $x \in [0,1]$;   (b) $w(x) = x$, $x \in [0,1]$;   (c) $w(x) = \dfrac{1}{\sqrt{x}}$, $x \in (0,1]$.

Chapter 5

Approximation Operators

In the previous chapters, we have studied the existence and formulations of certain polynomials $P \in \pi_n$ for the approximation of a given function $f \in C[a,b]$. In other words, we may formulate such results in terms of some projection $\mathscr{P}$ from $C[a,b]$ to $\pi_n \subset C[a,b]$, in that for each $f \in C[a,b]$, $\mathscr{P}f = P \in \pi_n$. In this chapter, we proceed to consider a more general point of view by introducing the concept of approximation operators $\mathscr{P}$ defined on a normed linear space and study various properties of $\mathscr{P}$ and the norm of the error function $f - \mathscr{P}f$.

5.1 Linearity and exactness

For a given linear space X, any approximation procedure which assigns to each $f \in X$ a unique approximation g_f belonging to some fixed approximation set $A \subset X$, can be associated with the corresponding approximation operator $\mathscr{A} : X \to A$ defined by

$$\mathscr{A}f := g_f, \quad f \in X. \tag{5.1.1}$$

We have the following examples from previous chapters.

Example 5.1.1. For any non-negative integer n, and a sequence $\{x_0, \ldots, x_n\}$ of $n+1$ distinct points in a given bounded interval $[a,b]$, the Lagrange polynomial interpolation operator $\mathscr{P}_n^I : C[a,b] \to \pi_n$ is defined by

$$\mathscr{P}_n^I f := P_n^I, \quad f \in C[a,b], \tag{5.1.2}$$

with P_n^I denoting the interpolation polynomial of Theorem 1.1.2, and where, from Theorem 1.2.2, we have the explicit formulation

$$(\mathscr{P}_n^I f)(x) = \sum_{j=0}^{n} f(x_j) L_{n,j}(x) \tag{5.1.3}$$

in terms of the Lagrange fundamental polynomials $\{L_{n,j} : j = 0,\ldots,n\}$, as defined in (1.2.1).

∎

Example 5.1.2. For any non-negative integer n and bounded interval $[a,b]$, the Bernstein polynomial interpolation operator $\mathscr{P}_n^B : C[a,b] \to \pi_n$ is defined by

$$\mathscr{P}_n^B f := P_n^B, \quad f \in C[a,b], \tag{5.1.4}$$

with P_n^B denoting the Bernstein polynomial approximation in π_n on $[a,b]$ of f, as given in (3.3.1), that is,

$$\left.\begin{array}{c} (\mathscr{P}_n^B f)(x) = \displaystyle\sum_{j=0}^{n} f(x_{n,j})B_{n,j}(x), \\[2mm] \text{where} \\[2mm] x_{n,j} := a + j\left(\dfrac{b-a}{n}\right), \quad j = 0,\ldots,n, \end{array}\right\} \tag{5.1.5}$$

and where the Bernstein polynomial sequence $\{B_{n,j} : j = 0,\ldots,n\}$ is defined by (3.2.2).

∎

Example 5.1.3. For any non-negative integer n, bounded interval $[a,b]$, and weight function w on $[a,b]$ satisfying the conditions (4.2.8) and (4.2.9), the best weighted L^2 approximation operator $\widetilde{\mathscr{P}}_n^* : C[a,b] \to \pi_n$ is defined by

$$\widetilde{\mathscr{P}}_n^* f := \widetilde{P}_n^*, \quad f \in C[a,b], \tag{5.1.6}$$

with the polynomial $\widetilde{P}_n^*$ as in Theorem 4.2.6, so that, from (4.2.24),

$$\|f - \widetilde{\mathscr{P}}_n^* f\|_{2,w} < \|f - P\|_{2,w}, \quad \text{for} \quad P \in \pi_n, \quad \text{with} \quad P \neq \widetilde{\mathscr{P}}_n^* f. \tag{5.1.7}$$

∎

For a normed linear space $(X, \|\cdot\|)$, let $A \subset X$ denote a fixed approximation set. If an approximation operator $\mathscr{A} : X \to A$ satisfies the condition

$$\mathscr{A}(\lambda f + \mu g) = \lambda(\mathscr{A}f) + \mu(\mathscr{A}g), \quad \lambda, \mu \in \mathbb{R}, \quad f, g \in X, \tag{5.1.8}$$

we say that $\mathscr{A}$ is linear. By choosing $\lambda = \mu = 0$ in (5.1.8), we deduce that

$$\mathscr{A}0 = 0, \quad \text{if } \mathscr{A} \text{ is linear.} \tag{5.1.9}$$

It is immediately evident from (5.1.3) and (5.1.5) that the approximation operators $\mathscr{P}_n^I$ and $\mathscr{P}_n^B$ are both linear. We shall in fact prove in Chapter 7 that the best approximation operator $\widetilde{\mathscr{P}}_n^*$ of Example 5.1.3 is also linear.

If an approximation operator $\mathscr{A} : X \to A$ satisfies the condition

$$\mathscr{A}f = f, \quad f \in M, \tag{5.1.10}$$

for some subset $M \subset A$, we say that $\mathscr{A}$ is exact on M. Note that optimal exactness is achieved in the case $M = A$.

Observe from (5.1.3), together with the identity (1.2.7) in Theorem 1.2.3, that the approximation operator $\mathscr{P}_n^I$ is exact on π_n, whereas, according to (5.1.5) and Theorem 3.3.1(c) and Theorem 3.3.2(a), the approximation operator $\mathscr{P}_n^B$ is exact on π_1, but not exact on π_n for $n \geqslant 2$.

To investigate the property of exactness with respect to the best approximation operator $\widetilde{\mathscr{P}}_n^*$ of Example 5.1.3, let $f \in \pi_n$, so that

$$||f - f||_{2,w} = 0 \leqslant ||f - P||_{2,w}, \quad P \in \pi_n,$$

and thus

$$\widetilde{\mathscr{P}}_n^* f = f, \quad f \in \pi_n. \tag{5.1.11}$$

In summary, we have therefore now proved the following result.

Theorem 5.1.1. *For any non-negative integer n and bounded interval $[a,b]$, the approximation operators $\mathscr{P}_n^I$, $\mathscr{P}_n^B$ and $\widetilde{\mathscr{P}}_n^*$, as defined by, respectively, (5.1.2), (5.1.4) and (5.1.6), satisfy the following properties:*

(a) *$\mathscr{P}_n^I$ and $\mathscr{P}_n^B$ are linear;*

(b) *$\mathscr{P}_n^I$ and $\widetilde{\mathscr{P}}_n^*$ are exact on π_n;*

(c) *$\mathscr{P}_n^B$ is exact on π_n if and only if $n = 1$.*

5.2 Boundedness and Lebesgue constants

For a normed linear space $(X, || \cdot ||)$, if a non-empty subset $Y \subset X$ satisfies the condition

$$||g|| \leqslant K, \quad g \in Y,$$

for some constant K, we say that Y is a bounded set. We proceed to introduce the notion of boundedness for approximation operators on X. For an approximation set $A \subset X$, let $\mathscr{A} : X \to A$ be an approximation operator. If

$$\left\{ \frac{||\mathscr{A}f||}{||f||} : f \in X; \quad f \neq 0 \right\} \tag{5.2.1}$$

is a bounded set, we say that $\mathscr{A}$ is bounded with respect to the norm $||\cdot||$, with corresponding operator norm

$$||\mathscr{A}|| := \sup\left\{\frac{||\mathscr{A}f||}{||f||} : f \in X; \quad f \neq 0\right\}. \tag{5.2.2}$$

For any bounded operator $\mathscr{A}$, the operator norm $||\mathscr{A}||$ in (5.2.2) will be referred to as the Lebesgue constant of $\mathscr{A}$ with respect to the norm $||\cdot||$. If the set (5.2.1) is unbounded, we say that $\mathscr{A}$ is unbounded with respect to the norm $||\cdot||$.

As an immediate consequence of the definition (5.2.2), we observe that, for any bounded approximation operator $\mathscr{A} : X \to A$, we have

$$||\mathscr{A}f|| \leqslant ||\mathscr{A}||\,||f||, \quad f \in X. \tag{5.2.3}$$

If, moreover, $\mathscr{A}$ is linear, it follows from (5.1.8) and (5.2.3) that, for any $f, \tilde{f} \in X$,

$$||\mathscr{A}f - \mathscr{A}\tilde{f}|| = ||\mathscr{A}(f - \tilde{f})|| \leqslant ||\mathscr{A}||\,||f - \tilde{f}||. \tag{5.2.4}$$

Suppose we wish to approximate a given element $f \in X$ by $\mathscr{A}f \in A$, and suppose that, perhaps due to measuring errors, or computer rounding errors, we are instead actually computing $\mathscr{A}\tilde{f}$, where $\tilde{f} \in X$ and $||f - \tilde{f}||$ is "small" in some sense. If $\mathscr{A}$ is linear and bounded, it follows from (5.2.4) that, for $\tilde{f} \neq f$, the quotient $||\mathscr{A}f - \mathscr{A}\tilde{f}||/||f - \tilde{f}||$ is bounded above by the Lebesgue constant $||\mathscr{A}||$. Since we would ideally like $||\mathscr{A}f - \mathscr{A}\tilde{f}||$ to have at most the same order of "smallness" as $||f - \tilde{f}||$, it follows that a relatively small value of $||\mathscr{A}||$ reflects favourably on an approximation operator $\mathscr{A}$.

The following result on the size of approximation operator norms should however be kept in mind.

Theorem 5.2.1. *For a normed linear space* $(X, ||\cdot||)$ *and an approximation set* $A \subset X$, *suppose the approximation operator* $\mathscr{A} : X \to A$ *is bounded, and exact on* $M \subset A$, *in the sense of* (5.1.10), *with* $M \neq \{0\}$. *Then the corresponding Lebesgue constant* $||\mathscr{A}||$ *satisfies the inequality*

$$||\mathscr{A}|| \geqslant 1. \tag{5.2.5}$$

Proof. Since (5.1.10) gives

$$\frac{||\mathscr{A}f||}{||f||} = 1, \quad \text{for} \quad f \in M, \quad \text{with} \quad f \neq 0,$$

it follows from the definition (5.2.2) that the inequality (5.2.5) is satisfied. ∎

The Lebesgue constant with respect to the maximum norm on $[a,b]$ of the Bernstein approximation operator $\mathscr{P}_n^B$ can now be computed as follows.

Theorem 5.2.2. *For any non-negative integer n and bounded interval $[a,b]$, the Bernstein approximation operator $\mathscr{P}_n^B : C[a,b] \to \pi_n$, as given by (5.1.5), is bounded with respect to the maximum norm on $[a,b]$, and has corresponding Lebesgue constant*

$$||\mathscr{P}_n^B||_\infty = 1. \tag{5.2.6}$$

Proof. Let $f \in C[a,b]$, with $f \neq 0$, and choose $x \in [a,b]$. It follows from (5.1.5), together with Theorem 3.2.1(b) and (c), that

$$|(\mathscr{P}_n^B f)(x)| \leqslant \sum_{j=0}^n |f(x_{n,j})| B_{n,j}(x) \leqslant ||f||_\infty \sum_{j=0}^n B_{n,j}(x) = ||f||_\infty,$$

and thus

$$||\mathscr{P}_n^B f||_\infty \leqslant ||f||_\infty,$$

so that

$$\frac{||\mathscr{P}_n^B f||_\infty}{||f||_\infty} \leqslant 1. \tag{5.2.7}$$

Hence, by using (5.2.7) and the definition (5.2.2), we conclude that the approximation operator $\mathscr{P}_n^B$ is bounded with respect to the maximum norm on $[a,b]$, with corresponding Lebesgue constant satisfying

$$||\mathscr{P}_n^B||_\infty \leqslant 1. \tag{5.2.8}$$

Next, we apply Theorem 5.1.1(c) and Theorem 5.2.1 to deduce that $||\mathscr{P}_n^B||_\infty \geqslant 1$, which, together with (5.2.8), implies the desired result (5.2.6). ■

Note from (5.2.6) that, subject to the constraint implied by exactness on π_1, the Lebesgue constant $||\mathscr{P}_n^B||_\infty$ is optimally small.

In order to compute the Lebesgue constant of the Lagrange polynomial interpolation operator $\mathscr{P}_n^I$ with respect to the maximum norm on $[a,b]$, we first establish the following simplified operator norm formulations for linear approximation operators.

Theorem 5.2.3. *For a normed linear space $(X, ||\cdot||)$ and an approximation set $A \subset X$, suppose $\mathscr{A} : X \to A$ is a linear approximation operator. Then the following three statements are equivalent:*

(i) *$\mathscr{A}$ is bounded;*

(ii) *the set $\{||\mathscr{A}f|| : f \in X; ||f|| \leqslant 1\}$ is bounded;*

(iii) *the set $\{||\mathscr{A}f|| : f \in X; ||f|| = 1\}$ is bounded.*

Moreover, if any one of the statements (i), (ii) *or* (iii) *holds, then the Lebesgue constant* $||\mathscr{A}||$ *satisfies the formulations*

$$||\mathscr{A}|| = \sup\{||\mathscr{A}f|| : f \in X; ||f|| \leqslant 1\}; \qquad (5.2.9)$$

$$||\mathscr{A}|| = \sup\{||\mathscr{A}f|| : f \in X; ||f|| = 1\}. \qquad (5.2.10)$$

Proof. First, observe that

$$\left. \begin{array}{c} f \in X, \quad ||f|| \leqslant 1, \quad f \neq 0 \quad \text{imply} \quad ||\mathscr{A}f|| \leqslant \dfrac{||\mathscr{A}f||}{||f||}, \\[2mm] \text{and} \quad \mathscr{A}0 = 0, \end{array} \right\} \qquad (5.2.11)$$

from (5.1.9), since $\mathscr{A}$ is linear, whereas the linearity of $\mathscr{A}$ also gives

$$\left. \begin{array}{c} f \in X, \quad f \neq 0 \quad \text{imply} \quad \dfrac{||\mathscr{A}f||}{||f||} = ||\mathscr{A}g||, \\[2mm] \text{where} \quad g := \dfrac{f}{||f||}, \quad \text{and thus} \quad ||g|| = 1. \end{array} \right\} \qquad (5.2.12)$$

We shall show that (i) $\Rightarrow$ (ii) $\Rightarrow$ (iii) $\Rightarrow$ (i), which will then imply the equivalence of the three statements (i), (ii) and (iii). Suppose therefore that (i) holds, that is, the set (5.2.1) is bounded. But then (5.2.11) shows that (ii) holds, and thus (i) $\Rightarrow$ (ii). Next, since

$$\{||\mathscr{A}f|| : f \in X; ||f|| = 1\} \subset \{||\mathscr{A}f|| : f \in X; ||f|| \leqslant 1\}, \qquad (5.2.13)$$

we deduce that (ii) $\Rightarrow$ (iii). If (iii) holds, we deduce from (5.2.12) that the set (5.2.1) is bounded, that is, (i) holds, so that (iii) $\Rightarrow$ (i), and thereby completing our proof of the equivalence of (i), (ii) and (iii).

Suppose that any one of the statements (i),(ii) or (iii) holds. It follows from the equivalence of (i), (ii) and (iii) that the definitions

$$k := \sup\left\{ \dfrac{||\mathscr{A}f||}{||f||} : f \in X, f \neq 0 \right\}; \quad \ell := \sup\{||\mathscr{A}f|| : f \in X, ||f|| \leqslant 1\};$$

$$m := \sup\{||\mathscr{A}f|| : f \in X; ||f|| = 1\}$$

yield, $k, \ell, m \in \mathbb{R}$.

We shall show that $k = \ell = m$, which, together with (5.2.2), will then imply the formulas (5.2.9) and (5.2.10).

To this end, we first note from (5.2.11) that $\ell \leqslant k$, whereas (5.2.12) shows that $k \leqslant m$. Since (5.2.13) implies $m \leqslant \ell$, it follows that $\ell \leqslant k \leqslant m \leqslant \ell$, and thus $k = \ell = m$, as required. ∎

The formula (5.2.10) in Theorem 5.2.3 enables us to obtain the following boundedness result, and explicit formulation of the Lebesgue constant, with respect to the maximum norm on $[a, b]$ of the Lagrange polynomial interpolation operator $\mathscr{P}_n^I$.

Theorem 5.2.4. *For any non-negative integer* n, *and a sequence* $\{x_0, \ldots, x_n\}$ *of* $n+1$ *distinct points in a given bounded interval* $[a,b]$, *the Lagrange interpolation operator* $\mathscr{P}_n^I : C[a,b] \to \pi_n$, *as defined by (5.1.2), is bounded with respect to the maximum norm on* $[a,b]$, *and has corresponding Lebesgue constant*

$$||\mathscr{P}_n^I||_\infty = \max_{a \leqslant x \leqslant b} \sum_{j=0}^n |L_{n,j}(x)|, \tag{5.2.14}$$

with the Lagrange fundamental polynomials $\{L_{n,j} : j = 0, \ldots, n\}$ *defined as in (1.2.1).*

Proof. Let $f \in C[a,b]$, with $f \neq 0$, and choose $x \in [a,b]$. It follows from (5.1.3) that

$$|(\mathscr{P}_n^I f)(x)| \leqslant \sum_{j=0}^n |f(x_j)||L_{n,j}(x)| \leqslant ||f||_\infty \sum_{j=0}^n |L_{n,j}(x)| \leqslant ||f||_\infty \max_{a \leqslant x \leqslant b} \sum_{j=0}^n |L_{n,j}(x)|,$$

and thus

$$||\mathscr{P}_n^I f||_\infty \leqslant ||f||_\infty \max_{a \leqslant x \leqslant b} \sum_{j=0}^n |L_{n,j}(x)|,$$

so that

$$\frac{||\mathscr{P}_n^I f||_\infty}{||f||_\infty} \leqslant \max_{a \leqslant x \leqslant b} \sum_{j=0}^n |L_{n,j}(x)|. \tag{5.2.15}$$

According to (5.2.15), the set $\{||\mathscr{P}_n^I f||_\infty / ||f||_\infty : f \in C[a,b]; f \neq 0\}$ is bounded, that is, the approximation operator $\mathscr{P}_n^I$ is bounded with respect to the maximum norm on $[a,b]$, with, from the definition (5.2.2), corresponding Lebesgue constant satisfying

$$||\mathscr{P}_n^I||_\infty \leqslant \max_{a \leqslant x \leqslant b} \sum_{j=0}^n |L_{n,j}(x)|. \tag{5.2.16}$$

We shall show that also

$$||\mathscr{P}_n^I||_\infty \geqslant \max_{a \leqslant x \leqslant b} \sum_{j=0}^n |L_{n,j}(x)|, \tag{5.2.17}$$

which, together with (5.2.16), will then complete our proof of the formula (5.2.14).

To prove the inequality (5.2.17), we choose a fixed $x \in [a,b]$, and let f denote any (for example piecewise linear) function in $C[a,b]$ satisfying $||f||_\infty = 1$, and

$$f(x_j) = \begin{cases} 1, & \text{if } L_{n,j}(x) \geqslant 0; \\ -1, & \text{if } L_{n,j}(x) < 0, \end{cases} \tag{5.2.18}$$

according to which the choice of f depends on the chosen value of x.

By applying (5.2.18) and (5.1.3), and using the facts that $||f||_\infty = 1$ and $\mathscr{P}_n^I$ is bounded, we obtain

$$\sum_{j=0}^n |L_{n,j}(x)| = \left| \sum_{j=0}^n f(x_j) L_{n,j}(x) \right|$$

$$\leqslant \max_{a \leqslant t \leqslant b} \left| \sum_{j=0}^{n} f(x_j) L_{n,j}(t) \right|$$

$$= \max_{a \leqslant t \leqslant b} \left| \left(\mathscr{P}_n^I f \right)(t) \right|$$

$$= \|\mathscr{P}_n^I f\|_\infty$$

$$\leqslant \sup\{\|\mathscr{P}_n^I f\|_\infty : f \in C[a,b]; \|f\|_\infty = 1\} = \|\mathscr{P}_n^I\|_\infty, \tag{5.2.19}$$

from the formula (5.2.10) in Theorem 5.2.3, after having recalled also from Theorem 5.1.1(a) that the approximation operator $\mathscr{P}_n^I$ is linear. Since the right hand side of (5.2.19) is independent of x, it follows that the required inequality (5.2.17) is indeed satisfied. ∎

Example 5.2.1. In Example 5.1.1, let $n = 1, [a,b] = [0,1]$, and $\{x_0, x_1\} = \{0,1\}$, so that, for any $f \in C[0,1]$, the graph of the interpolation polynomial $\mathscr{P}_1^I f$ is the straight line joining the points $(0, f(0))$ and $(1, f(1))$. Moreover, the definition (1.2.1) yields the Lagrange fundamental polynomials

$$L_{1,0}(x) = 1 - x; \quad L_{1,1}(x) = x. \tag{5.2.20}$$

The formula (5.2.14), together with (5.2.20), gives the Lebesgue constant

$$\|\mathscr{P}_1^I\|_\infty = \max_{0 \leqslant x \leqslant 1} [(1-x) + x] = 1, \tag{5.2.21}$$

which, in view of Theorem 5.1.1(b) and Theorem 5.2.1, and subject to the constraint implied by exactness on π_1, is an optimally small Lebesgue constant. ∎

Let us also consider the question of whether the approximation operator $\mathscr{P}_1^I$ of Example 5.2.1 is also bounded with respect to the L^2 norm on $[0,1]$, as defined in (4.2.17). To this end, we define the polynomial sequence $\{e_j : j = 0, 1, \ldots\}$ by

$$e_j(x) := x^j, \quad j = 0, 1, \ldots. \tag{5.2.22}$$

Then, since for each $j = 0, 1, \ldots$, it holds that $\mathscr{P}_1^I e_j \in \pi_1$, with $(\mathscr{P}_1^I e_j)(0) = e_j(0) = 0$, and $(\mathscr{P}_1^I e_j)(1) = e_j(1) = 1$, we deduce that

$$\mathscr{P}_1^I e_j = e_1, \quad j = 0, 1, \ldots, \tag{5.2.23}$$

and thus, from (4.2.17) and (5.2.22),

$$\|\mathscr{P}_1^I e_j\|_2 = \|e_1\|_2 = \sqrt{\int_0^1 x^2 dx} = \frac{1}{\sqrt{3}}, \quad j = 0, 1, \ldots. \tag{5.2.24}$$

Since, moreover, (4.2.17) and (5.2.22) yield

$$\|e_j\|_2 = \sqrt{\int_0^1 x^{2j} dx} = \frac{1}{\sqrt{2j+1}}, \quad j = 0, 1, \ldots, \tag{5.2.25}$$

we may deduce from (5.2.24) and (5.2.25) that

$$\frac{\|\mathscr{P}_1^I e_j\|_2}{\|e_j\|_2} = \sqrt{\frac{2j+1}{3}} \to \infty, \quad j \to \infty. \tag{5.2.26}$$

We see from (5.2.26) that the set

$$\left\{ \frac{\|\mathscr{P}_1^I f\|_2}{\|f\|_2} : f \in C[0,1]; \ f \neq 0 \right\}$$

is unbounded, and it follows that the approximation operator $\mathscr{P}_1^I$ is unbounded with respect to the L^2 norm on $[0,1]$.

5.3   The approximation error

For a normed linear space $(X, \|\cdot\|)$ and an approximation set $A \subset X$, let $\mathscr{A} : X \to A$ denote an approximation operator. For any $f \in X$, the quantity

$$\|f - \mathscr{A}f\| \tag{5.3.1}$$

is then called the corresponding approximation error with respect to the norm $\|\cdot\|$. It is evident that $\mathscr{A}f$ will be considered to be a "good approximation" from A to f with respect to the norm $\|\cdot\|$ if the corresponding approximation error (5.3.1) is appropriately "small". The approximation error (5.3.1) evidently depends on the choice of the norm $\|\cdot\|$. Our following result establishes a fundamental inequality between the two norms we have thus far established for the linear space $C[a, b]$.

Theorem 5.3.1. *For a bounded interval $[a,b]$ and a weight function w on $[a,b]$ satisfying the conditons (4.2.8) and (4.2.9), let the norms $\|\cdot\|_\infty$ and $\|\cdot\|_{2,w}$ be as defined in, respectively, (4.1.7) and (4.2.16). Then*

$$\|f\|_{2,w} \leqslant \sqrt{\int_a^b w(x)dx}\, \|f\|_\infty, \quad f \in C[a,b]. \tag{5.3.2}$$

Proof. Let $f \in C[a,b]$. By using (4.2.16), (4.2.8) and (4.2.9), we obtain

$$(\|f\|_{2,w})^2 = \int_a^b w(x)[f(x)]^2 dx \leqslant (\|f\|_\infty)^2 \int_a^b w(x)dx,$$

which then implies (5.3.2). ∎

As an immediate consequence of (5.3.2) in Theorem 5.3.1, we observe that, for any approximation operator $\mathscr{A} : C[a,b] \to A$, with $A \subset C[a,b]$ denoting an arbitrary approximation set, the corresponding approximation error satisfies

$$||f - \mathscr{A}f||_{2,w} \leqslant \sqrt{\int_a^b w(x)dx}\, ||f - \mathscr{A}f||_\infty, \quad f \in C[a,b], \tag{5.3.3}$$

and thus, for the special case where the weight function w is given by (4.2.11),

$$||f - \mathscr{A}f||_2 \leqslant \sqrt{b-a}\, ||f - \mathscr{A}f||_\infty. \tag{5.3.4}$$

The inequality (5.3.4) may, for example, be applied to deduce from Theorem 3.3.3, together with (5.1.5), that, for any $f \in C[a,b]$,

$$||f - \mathscr{P}_n^B f||_2 \leqslant \sqrt{b-a}\, ||f - \mathscr{P}_n^B f||_\infty \to 0, \quad n \to \infty, \tag{5.3.5}$$

that is, by using also (4.2.17),

$$\sqrt{\int_a^b [f(x) - (\mathscr{P}_n^B f)(x)]^2 dx} \to 0, \quad n \to \infty. \tag{5.3.6}$$

In general, we deduce from (5.3.3) that, provided the value of $\sqrt{\int_a^b w(x)dx}$ is not "large" in some sense, then the approximation error $||f - \mathscr{A}f||_{2,w}$ has at least the same order of "smallness" as the approximation error $||f - \mathscr{A}f||_\infty$.

The converse does not hold, in the sense that there does not exist a positive constant K such that

$$||f||_\infty \leqslant K||f||_{2,w}, \quad f \in C[a,b],$$

as is evident by considering the polynomial sequence $\{e_j : j = 0,1,\ldots\}$ defined by (5.2.22), for which we calculate as in (5.2.25) that, for $[a,b] = [0,1]$,

$$\frac{||e_j||_\infty}{||e_j||_2} = \frac{1}{(2j+1)^{-\frac{1}{2}}} = \sqrt{2j+1} \to \infty, \quad j \to \infty.$$

For a certain class of approximation operators, the size of the corresponding approximation error can be bounded in terms of the best approximation error, as follows.

Theorem 5.3.2 (Lebesgue inequality). *For a normed linear space* $(X, ||\cdot||)$, *let* $A \subset X$ *denote an approximation set such that* A *is a subspace of* X, *and such that there exists a best approximation from* A *to each* $f \in X$. *Furthermore, suppose that* $\mathscr{A} : X \to A$ *is an approximation operator which is linear and bounded, and such that* $\mathscr{A}$ *is exact on* A. *Then the corresponding approximation error satisfies*

$$||f - \mathscr{A}f|| \leqslant (1 + ||\mathscr{A}||) \min_{g \in A} ||f - g||, \quad f \in X, \tag{5.3.7}$$

with $||\mathscr{A}||$ *denoting the corresponding Lebesgue constant, as defined by (5.2.2).*

Proof. Let $f \in X$, and suppose f^* is a best approximation from A to f, that is

$$||f - f^*|| = \min_{g \in A}||f - g||. \tag{5.3.8}$$

Observe that, since $f^* \in A$, and since $\mathscr{A}$ is exact on A, we have

$$\mathscr{A} f^* = f^*. \tag{5.3.9}$$

By using the triangle inequality (4.1.4), as well as (5.3.9) and the linearity of $\mathscr{A}$, and finally the inequality (5.2.3), we deduce that

$$||f - \mathscr{A}f|| \leqslant ||f - \mathscr{A}f^*|| + ||Af^* - \mathscr{A}f|| = ||f - f^*|| + ||\mathscr{A}(f^* - f)||$$

$$\leqslant ||f - f^*|| + ||\mathscr{A}|| \, ||f^* - f||$$

$$= (1 + ||\mathscr{A}||)||f - f^*||,$$

which, together with (5.3.8), then yields (5.3.7). ∎

Since the approximation operator $\mathscr{A}$ in Theorem 5.3.2 is exact on a subspace A of X, with $A \neq \{0\}$, we deduce from (5.2.5) in Theorem 5.2.1 that the constant $(1 + ||\mathscr{A}||)$ appearing in the Lebesgue inequality (5.3.7) satisfies

$$1 + ||\mathscr{A}|| \geqslant 2. \tag{5.3.10}$$

By combining Theorem 5.3.2, Theorem 4.1.2, Theorem 5.1.1(a) and (b), and Theorem 5.2.4, we deduce the following upper bound for the approximation error in polynomial interpolation.

Theorem 5.3.3. *For any non-negative integer n, and a sequence $\{x_0, \ldots, x_n\}$ of $n+1$ distinct points in a bounded interval $[a,b]$, the approximation error corresponding to the Lagrange polynomial interpolation operator $\mathscr{P}_n^I : C[a,b] \to \pi_n$, as defined by (5.1.2), satisfies the Lebesgue inequality*

$$||f - \mathscr{P}_n^I f||_\infty \leqslant \left(1 + \max_{a \leqslant x \leqslant b} \sum_{j=0}^{n} |L_{n,j}(x)|\right) \min_{P \in \pi_n} ||f - P||_\infty, \ f \in C[a,b], \tag{5.3.11}$$

where the Lagrange polynomials $\{L_{n,j} : j = 0, \ldots, n\}$ are given by (1.2.1).

Example 5.3.1. The approximation error corresponding to the Lagrange polynomial interpolation operator $\mathscr{P}_1^I : C[0,1] \to \pi_1$ of Example 5.2.1 satisfies, from (5.2.21) and (5.3.11), the Lebesgue inequality

$$||f - \mathscr{P}_1^I f||_\infty \leqslant 2 \min_{P \in \pi_1} ||f - P||_\infty, \ f \in C[0,1]. \tag{5.3.12}$$

It follows from (5.3.12) that, in the $||\cdot||_\infty$ norm, and for any $f \in C[0,1]$, the approximation error $||f - \mathscr{P}_1^I f||_\infty$ is at most twice the (minimum) approximation error corresponding to a best approximation in the $||\cdot||_\infty$ norm from π_1 to f.   ∎

Example 5.3.2. In Theorem 5.3.3, let $n = 2$, $\{x_0, x_1, x_2\} = \{-1, 1, 2\}$, and $[a,b] = [-1,2]$. Then (1.2.1) gives

$$L_{2,0}(x) = \frac{(x-1)(x-2)}{(-1-1)(-1-2)} = \frac{1}{6}(x^2 - 3x + 2) = \frac{1}{6}x^2 - \frac{1}{2}x + \frac{1}{3};$$

$$L_{2,1}(x) = \frac{(x-(-1))(x-2)}{(1-(-1))(1-2)} = -\frac{1}{2}(x^2 - x - 2) = -\frac{1}{2}x^2 + \frac{1}{2}x + 1;$$

$$L_{2,2}(x) = \frac{(x-(-1))(x-1)}{(2-(-1))(2-1)} = \frac{1}{3}(x^2 - 1) = \frac{1}{3}x^2 - \frac{1}{3},$$

and thus

$$\sum_{j=0}^{2} |L_{2,j}(x)| = \begin{cases} L_{2,0}(x) + L_{2,1}(x) - L_{2,2}(x) = u(x), \ x \in [-1,1]; \\ -L_{2,0}(x) + L_{2,1}(x) + L_{2,2}(x) = v(x), \ x \in [1,2], \end{cases} \qquad (5.3.13)$$

where

$$u(x) := -\frac{2}{3}x^2 + \frac{5}{3}; \quad v(x) := -\frac{1}{3}x^2 + x + \frac{1}{3}. \qquad (5.3.14)$$

But (5.3.14) implies

$$\max_{-1 \leqslant x \leqslant 1} u(x) = \max\{u(-1), u(0), u(1)\} = \max\{1, \tfrac{5}{3}, 1\} = \tfrac{5}{3};$$

$$\max_{1 \leqslant x \leqslant 2} v(x) = \max\{v(1), v(\tfrac{3}{2}), v(2)\} = \max\{1, \tfrac{13}{12}, 1\} = \tfrac{13}{12},$$

and thus, from (5.2.14) and (5.3.13), the corresponding Lagrange interpolation operator $\mathscr{P}_2^I : C[-1,2] \to \pi_2$ has Lebesgue constant given by

$$||\mathscr{P}_2^I||_\infty = \max_{-1 \leqslant x \leqslant 2} \sum_{j=0}^{2} |L_{2,j}(x)| = \max\left\{\frac{5}{3}, \frac{13}{12}\right\} = \frac{5}{3}. \qquad (5.3.15)$$

By using (5.3.15), it follows from (5.3.11) in Theorem 5.3.3 that the corresponding approximation error satisfies the Lebesgue inequality

$$||f - \mathscr{P}_2^I f||_\infty \leqslant \frac{8}{3} \min_{P \in \pi_2} ||f - P||_\infty, \ f \in C[-1,2]. \qquad (5.3.16)$$

∎

5.4   Exercises

Exercise 5.1 For a bounded interval $[a,b] \subset \mathbb{R}$, and any non-negative integer n, let $\{x_0,\ldots,x_n\}$ be a sequence of $n+1$ distinct points in $[a,b]$, and denote by $\mathbf{r} = \{r_0,\ldots,r_n\}$ a sequence of non-negative integers. Also, let the integers ν and k be defined by, respectively, (1.4.3) and

$$k := \max\{r_0,\ldots,r_n\}.$$

Prove that the Hermite interpolation operator $\mathscr{P}_{n,\mathbf{r}}^I : C^k[a,b] \to \pi_\nu$, as defined by

$$\mathscr{P}_{n,\mathbf{r}}^I f := P_{n,\mathbf{r}}^I, \quad f \in C^k[a,b],$$

with $P_{n,\mathbf{r}}^I$ denoting the Hermite interpolation polynomial of Theorem 1.4.2, is linear.

[*Hint:* Use Theorem 1.4.1 to derive a Hermite interpolation formula analogous to (1.1.16) in Theorem 1.1.2.]

Exercise 5.2 As a continuation of Exercise 5.1, prove that the Hermite interpolation operator $\mathscr{P}_{n,\mathbf{r}}^I$ is exact on π_ν.

Exercise 5.3 Calculate, for $n = 1$ and $n = 2$, the Lebesgue constant $||\mathscr{P}_n^I||_\infty$ of the Lagrange interpolation operator $\mathscr{P}_n^I$ defined by (5.1.2), with $[a,b] = [0,2]$, and where P_n^I is the interpolation polynomial of Exercise 2.2.

Exercise 5.4 As a continuation of Exercise 5.3, for $n = 1$ and $n = 2$, write down the corresponding Lebesgue inequality, and then use this inequality, together with the exact value of $||E_n^I||_\infty$, as obtained in Exercise 2.2, to obtain a lower bound on the minimum value

$$\min_{P \in \pi_n} \max_{0 \leqslant x \leqslant 2} |\ln(x+2) - P(x)|.$$

Exercise 5.5 Repeat Exercises 5.3 and 5.4, with the interpolation sequences $\triangle_n$ of Exercise 2.2 replaced, as in Exercise 2.6, by the Chebyshev interpolation points $\triangle_n^C$, as calculated in Exercise 2.5.

Exercise 5.6 Calculate the Lebesgue constant $||\mathscr{P}_2^I||_\infty$ with respect to the interval $[0,4]$ and the interpolation points $\{1,2,4\}$.

Exercise 5.7 Prove the existence of an interpolation operator

$$\mathscr{L} : C[0,2] \to A := \operatorname{span}\{1,x,x^3\}$$

with the defining property

$$(\mathscr{L}f)(j) = f(j), \quad j = 0,1,2, \quad f \in C[0,2].$$

[*Hint:* Argue analogously to the reasoning which led from (1.1.3) to (1.1.8), and show that the resulting determinant is non-zero, after which a definition analogous to (5.1.2) may be used.]

Exercise 5.8 As a continuation of Exercise 5.7, obtain a sequence $\{P_0, P_1, P_2\} \subset A$ satisfying, analogously to (1.2.3) in Theorem 1.2.1(a), the condition

$$P_j(k) = \delta_{j-k}, \quad j, k = 0, 1, 2,$$

and express $\mathscr{L}f$, for any $f \in C[0,2]$, and analogously to the Lagrange polynomial interpolation formula (1.2.5) in Theorem 1.2.2, in terms of the sequence $\{P_0, P_1, P_2\}$. Then use this formula to show that the interpolation operator $\mathscr{L}$ is linear, and exact on A.

Exercise 5.9 As a continuation of Exercise 5.8, argue as in the proof of Theorem 5.2.4 to prove that, with $||\cdot||_\infty$ denoting the maximum norm on $[0,2]$, the Lebesgue constant $||\mathscr{L}||_\infty$ of the interpolation operator $\mathscr{L}$ is given explicitly by

$$||\mathscr{L}||_\infty = \max_{0 \leqslant x \leqslant 2} \sum_{j=0}^{2} |P_j(x)|,$$

and then use this formula to calculate the value of $||\mathscr{L}||_\infty$.

Exercise 5.10 As a continuation of Exercise 5.9, write down the Lebesgue inequality for the interpolation operator $\mathscr{L}$, after first having explained why, for any $f \in C[0,2]$, there exists a best approximation with respect to the maximum norm on $[0,2]$ from A to f. Now use the interpolation formula derived in Exercise 5.8 to calculate, for the function

$$f(x) = x^2, \quad x \in [0,2],$$

its corresponding interpolant $\mathscr{L}f \in A$, and then calculate the exact value of the corresponding error $||f - \mathscr{L}f||_\infty$. By combining this result with the Lebesgue inequality for $\mathscr{L}$, derive a lower bound on the minimum value

$$\min_{\alpha,\beta,\gamma \in \mathbb{R}} \max_{0 \leqslant x \leqslant 2} |\alpha x^3 + x^2 + \beta x + \gamma|,$$

and give the corresponding optimal values of α, β and γ.

[*Hint:* Observe that

$$\min_{\alpha,\beta,\gamma \in \mathbb{R}} \max_{0 \leqslant x \leqslant 2} |\alpha x^3 + x^2 + \beta x + \gamma| = \min_{\alpha,\beta,\gamma \in \mathbb{R}} \max_{0 \leqslant x \leqslant 2} |x^2 - (\alpha x^3 + \beta x + \gamma)|.]$$

Chapter 6

Best Uniform Polynomial Approximation

This chapter is a continuation of Chapter 4, in that the best uniform polynomial approximation $P^* \in \pi_n$ of $f \in C[a,b]$, with existence of P^* guaranteed by Theorem 4.1.2, will be characterized in terms of the alternation properties of the error function $f - P^*$. As an application, the uniqueness of $P^* \in \pi_n$ as the only best uniform polynomial approximant of $f \in C[a,b]$ is assured.

6.1   A necessary condition

According to Theorem 4.1.2, there exists, for $f \in C[a,b]$ and any non-negative integer n, a polynomial $P^* \in \pi_n$ such that

$$||f - P^*||_\infty \leqslant ||f - P||_\infty, \quad P \in \pi_n. \tag{6.1.1}$$

The following necessary condition is satisfied by the error function $f - P^*$.

Theorem 6.1.1. *For $f \in C[a,b]$ and any non-negative integer n, let P^* denote a polynomial such that the best approximation property (6.1.1) holds. Then there exist points $\xi, \widetilde{\xi} \in [a,b]$, with $\xi \neq \widetilde{\xi}$, such that*

$$|f(\xi) - P^*(\xi)| = |f(\widetilde{\xi}) - P^*(\widetilde{\xi})| = ||f - P^*||_\infty, \tag{6.1.2}$$

and

$$f(\xi) - P^*(\xi) = -[f(\widetilde{\xi}) - P^*(\widetilde{\xi})]. \tag{6.1.3}$$

Proof. If $f \in \pi_n$, then $P^* = f$, and the theorem holds for arbitrary points $\xi, \widetilde{\xi} \in [a,b]$, with $\xi \neq \widetilde{\xi}$, and where both left hand and right hand sides of (6.1.2) and (6.1.3) are equal to zero.

Suppose next $f \notin \pi_n$, so that $||f - P^*||_\infty > 0$, and suppose that there do not exist points $\xi, \tilde{\xi} \in [a,b]$, with $\xi \neq \tilde{\xi}$, such that (6.1.2) and (6.1.3) are satisfied. We shall prove the existence of a polynomial $Q \in \pi_n$ such that

$$0 < ||f - Q||_\infty < ||f - P^*||_\infty, \tag{6.1.4}$$

which contradicts (6.1.1), and would therefore complete our proof.

Introducing the notation

$$m := \min_{a \leqslant x \leqslant b} [f(x) - P^*(x)]; \quad M := \max_{a \leqslant x \leqslant b} [f(x) - P^*(x)]; \quad d^* := ||f - P^*||_\infty, \tag{6.1.5}$$

we see that, since $f \neq P^*$, we have either $-m < M$ or $-m > M$, since $-m = M$ implies the existence of points $\xi, \tilde{\xi} \in [a,b]$, with $\xi \neq \tilde{\xi}$, such that (6.1.2) and (6.1.3) are satisfied.

(a) Suppose first $-m < M$. Then $M > 0$, for if not, that is, $M \leqslant 0$, then $-m < M \leqslant 0$, and thus $m > 0 \geqslant M$, so that $m > M$, which contradicts the definitions of m and M in (6.1.5). It then follows from the definitions in (6.1.5) that $d^* = M$, and thus

$$c := \frac{m + d^*}{2} = \frac{m + M}{2} > 0. \tag{6.1.6}$$

With the definition

$$Q(x) := P^*(x) + c, \tag{6.1.7}$$

according to which $Q \in \pi_n$, we have

$$f(x) - Q(x) = [f(x) - P^*(x)] - c,$$

and it follows from (6.1.5) and (6.1.6) that, for any $x \in [a,b]$,

$$m - c \leqslant f(x) - Q(x) \leqslant M - c = d^* - c. \tag{6.1.8}$$

Since also (6.1.6) gives $2c = m + d^*$, and thus $m - c = -(d^* - c)$, it follows from (6.1.8) that

$$-(d^* - c) \leqslant f(x) - Q(x) \leqslant d^* - c,$$

that is,

$$|f(x) - Q(x)| \leqslant d^* - c,$$

from which, together with (6.1.6) and (6.1.5), we deduce that

$$0 < ||f - Q||_\infty = d^* - c < d^* = ||f - P^*||_\infty,$$

and thereby yielding the desired result (6.1.4).

(b) Suppose next $-m > M$. But then, since (6.1.1), together with the fact that

$$\{-P : P \in \pi_n\} = \pi_n,$$

yields

$$||(-f) - (-P^*)||_\infty = ||f - P^*||_\infty = \min_{P \in \pi_n} ||f - P||_\infty$$

$$= \min_{P \in \pi_n} || - f + P||_\infty = \min_{P \in \pi_n} ||(-f) - P||_\infty,$$

and since

$$\min_{a \leqslant x \leqslant b} [(-f(x)) - (-P^*(x))] = \max_{a \leqslant x \leqslant b} [f(x) - P^*(x)];$$

$$\max_{a \leqslant x \leqslant b} [(-f(x)) - (-P^*(x))] = \min_{a \leqslant x \leqslant b} [f(x) - P^*(x)],$$

we may appeal to the proof in (a) to deduce the existence of a polynomial $\widetilde{Q} \in \pi_n$ such that

$$0 < ||(-f) - \widetilde{Q}||_\infty < ||(-f) - (-P^*)||_\infty,$$

according to which $Q := -\widetilde{Q}$ is a polynomial in π_n satisfying (6.1.4), as required. ∎

The necessary condition in Theorem 6.1.1 enables us to obtain the constant polynomial P^* satisfying (6.1.1) for $n = 0$, as follows.

Theorem 6.1.2. *Let $f \in C[a,b]$. Then the constant polynomial*

$$P^*(x) = P_0^*(x) := \frac{1}{2} \left[\max_{a \leqslant x \leqslant b} f(x) + \min_{a \leqslant x \leqslant b} f(x) \right], \qquad (6.1.9)$$

is the only polynomial in π_0 such that the best approximation condition (6.1.1) is satisfied for $n = 0$, with, moreover,

$$||f - P_0^*||_\infty = \frac{1}{2} \left[\max_{a \leqslant x \leqslant b} f(x) - \min_{a \leqslant x \leqslant b} f(x) \right]. \qquad (6.1.10)$$

Proof. According to Theorem 4.1.2, there exists a constant polynomial P^* that satisfies the condition (6.1.1) for $n = 0$. Suppose P^* is not given by (6.1.9). Then there do not exist points $\xi, \widetilde{\xi} \in [a,b]$, with $\xi \neq \widetilde{\xi}$, such that (6.1.2) and (6.1.3) hold. It follows from Theorem 6.1.1 that P^* does not satisfy the condition (6.1.1) for $n = 0$, which is a contradiction. Hence $P^* = P_0^*$, as given by (6.1.9), is the only constant polynomial satisfying (6.1.1) for $n = 0$. The maximum error value (6.1.10) is then an immediate consequence of (6.1.9). ∎

Example 6.1.1. In Theorem 6.1.2, if we choose $[a,b] = [0,1]$, and, for any $k \in \mathbb{N}$,

$$f(x) = x^k, \quad x \in [0,1], \qquad (6.1.11)$$

then the constant polynomial P_0^* is given by

$$P_0^*(x) = \frac{1}{2}. \qquad (6.1.12)$$

∎

6.2　The equi-oscillation property

Observe from Theorem 6.1.1 that the error function $f - P^*$ in polynomial best approxi-
mation attains each of the two extreme values $||f - P^*||_\infty$ and $-||f - P^*||_\infty$ at least once
on $[a,b]$. We proceed to show that, if $P^* \in \pi_n$ is such that $f - P^*$ equi-oscillates at least
$n+2$ times between $||f - P^*||_\infty$ and $-||f - P^*||_\infty$, then P^* satisfies the best approximation
condition (6.1.1).

Theorem 6.2.1. *For $f \in C[a,b]$ and any non-negative integer n, suppose the polynomial
$P^* \in \pi_n$ is such that there exists a sequence $\{\xi_0,\ldots,\xi_{n+1}\}$ of $n+2$ distinct points in $[a,b]$
satisfying the conditions*

$$a \leqslant \xi_0 < \xi_1 < \cdots < \xi_{n+1} \leqslant b; \tag{6.2.1}$$

$$|f(\xi_j) - P^*(\xi_j)| = ||f - P^*||_\infty, \quad j = 0,\ldots,n+1; \tag{6.2.2}$$

$$f(\xi_j) - P^*(\xi_j) = -[f(\xi_{j+1}) - P^*(\xi_{j+1})], \quad j = 0,\ldots,n. \tag{6.2.3}$$

Then P^* satisfies the best approximation condition (6.1.1).

Proof. Suppose that $f \in \pi_n$, and thus $f - P^* \in \pi_n$. If $f - P^*$ is not the zero polynomial,
then $||f - P^*||_\infty > 0$, and it follows from (6.2.1), (6.2.2) and (6.2.3), together with the
intermediate value theorem, that $f - P^*$ has at least one zero in each of the $n+1$ distinct
intervals $(\xi_0,\xi_1),\ldots,(\xi_n,\xi_{n+1})$. Hence $f - P^*$ is a polynomial in π_n with at least $n+1$
distinct real zeros, and thus $f - P^*$ is the zero polynomial, which is a contradiction. Hence
$P^* = f$, and it follows that P^* satisfies the condition (6.1.1).
Suppose next $f \notin \pi_n$, according to which $P^* \neq f$, and thus $||f - P^*||_\infty > 0$. Using a proof
by contradiction, we next suppose that P^* does not satisfy the condition (6.1.1), that is,
there exists a polynomial $Q \in \pi_n$ such that

$$0 < ||f - Q||_\infty < ||f - P^*||_\infty, \tag{6.2.4}$$

according to which it then also holds that $P^* \neq Q$. With the definition

$$R := P^* - Q, \tag{6.2.5}$$

it follows that $R \in \pi_n$, and R is not the zero polynomial.
Now observe from (6.2.2) that either $f(\xi_0) - P^*(\xi_0) = ||f - P^*||_\infty$, or $f(\xi_0) - P^*(\xi_0) =
-||f - P^*||_\infty$. If the first alternative holds, it follows from the assumption (6.2.4) that

$$f(\xi_0) - Q(\xi_0) < ||f - P^*||_\infty = f(\xi_0) - P^*(\xi_0),$$

and thus

$$R(\xi_0) = P^*(\xi_0) - Q(\xi_0) < 0. \tag{6.2.6}$$

Next, we use (6.2.3) and (6.2.4) to deduce that

$$f(\xi_1) - Q(\xi_1) > -||f - P^*||_\infty = f(\xi_1) - P^*(\xi_1),$$

so that

$$R(\xi_1) = P^*(\xi_1) - Q(\xi_1) > 0. \tag{6.2.7}$$

It follows from (6.2.6), (6.2.7), together with the intermediate value theorem, that the polynomial R has a zero $\eta_0 \in (\xi_0, \xi_1)$.

Repeated further applications of the same procedure yield the existence of zeros $\eta_j \in (\xi_j, \xi_{j+1})$, $j = 0, \ldots, n$, of the polynomial R. Hence R is a polynomial in π_n with at least $n+1$ real zeros, and thus R is the zero polynomial, which is a contradiction. The case $f(\xi_0) - P^*(\xi_0) = -||f - P^*||_\infty$ similarly yields the same contradiction (see Exercise 6.2). Hence there does not exist a polynomial $Q \in \pi_n$ such that (6.2.4) holds, from which we then deduce that P^* does indeed satisfy the best approximation condition (6.1.1). ∎

Observe that the result (2.2.23) of Theorem 2.2.2 can immediately be deduced from Theorem 6.2.1, as follows. For any $j \in \mathbb{N}$, we have

$$\min_{P \in \pi_j} \max_{-1 \leqslant x \leqslant 1} |P(x)| = \min_{c_0, \ldots, c_{j-1} \in \mathbb{R}} \max_{-1 \leqslant x \leqslant 1} \left| x^j + \sum_{k=0}^{j-1} c_k x^k \right|$$

$$= \min_{c_0, \ldots, c_{j-1} \in \mathbb{R}} \max_{-1 \leqslant x \leqslant 1} \left| x^j - \sum_{k=0}^{j-1} c_k x^k \right|$$

$$= \min_{P \in \pi_{j-1}} \max_{-1 \leqslant x \leqslant 1} \left| x^j - P(x) \right| = \max_{-1 \leqslant x \leqslant 1} \left| \widetilde{T}_j(x) \right| = 2^{1-j},$$

as deduced from (2.2.21), (2.2.22), together with Theorem 2.2.1(a), (c), (d), and an application of Theorem 6.2.1 with $[a,b] = [-1, 1]$, $f(x) = x^j$ and $n = j - 1$.

According to Theorem 6.2.1, the equi-oscillation property (6.2.1), (6.2.2), (6.2.3) is a sufficient condition on a polynomial $P^* \in \pi_n$ to satisfy the best approximation condition (6.1.1). We proceed to prove that the same equi-oscillation property is also a necessary condition, as follows.

Theorem 6.2.2. *For $f \in C[a,b]$ and any non-negative integer n, let P^* be a best approximation from π_n to f with respect to the maximum norm on $[a,b]$ as in Theorem 4.1.2. Then there exists a sequence $\{\xi_0, \ldots, \xi_{n+1}\}$ of $n+2$ distinct points in $[a,b]$ such that the conditions (6.2.1), (6.2.2), (6.2.3) are satisfied.*

Proof. If $f \in \pi_n$, then $P^* = f$, and it follows that (6.2.2) and (6.2.3) are satisfied, with both left hand and right hand sides equal to zero, for any choice of a sequence $\{\xi_0, \ldots, \xi_{n+1}\}$ in $[a,b]$ satisfying (6.2.1).

Next, for $f \notin \pi_n$, we denote by k the largest positive integer for which there exists a point sequence $\{\xi_0, \ldots, \xi_k\}$ in $[a,b]$ such that the conditions

$$a \leqslant \xi_0 < \xi_1 < \cdots < \xi_k \leqslant b; \tag{6.2.8}$$

$$|f(\xi_j) - P^*(\xi_j)| = \|f - P^*\|_\infty, \quad j = 0, \ldots, k; \tag{6.2.9}$$

$$f(\xi_j) - P^*(\xi_j) = -[f(\xi_{j+1}) - P^*(\xi_{j+1})], \quad j = 0, \ldots, k-1, \tag{6.2.10}$$

hold. Observe in particular from Theorem 6.1.1 that $k \geqslant 1$. Note also that if such a largest integer k does not exist, our result immediately follows. We shall prove that the assumption

$$1 \leqslant k \leqslant n \tag{6.2.11}$$

yields the existence of a polynomial $Q \in \pi_n$ such that

$$0 < \|f - Q\|_\infty < \|f - P^*\|_\infty, \tag{6.2.12}$$

which contradicts the best approximation property (6.1.1) of P^*, and from which it will then follow that $k \geqslant n+1$, and thereby completing our proof.

Suppose therefore that the integer k satisfies the inequalities (6.2.11). With the notation

$$E^* := f - P^*; \quad d^* := \|E^*\|_\infty, \tag{6.2.13}$$

it follows from $f \notin \pi_n$ that $d^* > 0$. Since $E^* \in C[a,b]$, we know that E^* is uniformly continuous on $[a,b]$, and thus there exists a positive constant δ such that

$$x,y \in [a,b], \text{ with } |x-y| < \delta \Rightarrow |E^*(x) - E^*(y)| < \frac{d^*}{2}. \tag{6.2.14}$$

Let μ be any positive integer such that

$$\mu > \frac{b-a}{\delta}, \tag{6.2.15}$$

and define

$$t_j := a + j\left(\frac{b-a}{\mu}\right), \quad j = 0, \ldots, \mu, \tag{6.2.16}$$

according to which

$$t_{j+1} - t_j = \frac{b-a}{\mu} < \delta, \quad j = 0, \ldots, \mu - 1. \tag{6.2.17}$$

Observe from (6.2.14) and (6.2.17) that

$$x,y \in [t_j, t_{j+1}] \Rightarrow |E^*(x) - E^*(y)| < \frac{d^*}{2}, \quad j = 0, \ldots, \mu - 1. \tag{6.2.18}$$

Next, we define the interval sets

$$M := \{[t_j, t_{j+1}] : j = 0, \ldots, \mu - 1\}; \tag{6.2.19}$$

$$M^+ := \{[t_j, t_{j+1}] \in M : \text{ there exists a point } t \in [t_j, t_{j+1}] \text{ such that } E^*(t) = d^*\}; \tag{6.2.20}$$

$$M^- := \{[t_j, t_{j+1}] \in M : \text{ there exists a point } \tau \in [t_j, t_{j+1}] \text{ such that } E^*(\tau) = -d^*\}. \tag{6.2.21}$$

It then follows from (6.2.18), (6.2.20) and (6.2.21) that

$$\left. \begin{array}{l} I \in M^+ \Rightarrow E^*(x) > \dfrac{d^*}{2} > 0, \qquad x \in I; \\[2mm] I \in M^- \Rightarrow E^*(x) < -\dfrac{d^*}{2} < 0, \qquad x \in I, \end{array} \right\} \tag{6.2.22}$$

and thus

$$M^+ \cap M^- = \emptyset. \tag{6.2.23}$$

Also, we deduce from (6.2.22), together with $E^* \in C[a,b]$, that

$$I \in M^+, \quad \tilde{I} \in M^- \Rightarrow I \cap \tilde{I} = \emptyset. \tag{6.2.24}$$

Consider now the interval sequence

$$\{I_1, \ldots, I_N\} := M^+ \cup M^-, \tag{6.2.25}$$

where, for each $j \in \{1, \ldots, N-1\}$, the interval I_{j+1} is situated, along $[a,b]$, to the right of the interval I_j. Suppose $I_1 \in M^+$. It then follows from the definition of the integer k, along with the definitions (6.2.19), (6.2.20) and (6.2.21), that the interval sequence $\{I_1, \ldots, I_N\}$ may be partitioned into the $k+1$ subsequences

$$\left. \begin{array}{ll} \{I_1, \ldots, I_{j_0}\} & \subset M^+; \\[1mm] \{I_{j_0+1}, \ldots, I_{j_1}\} & \subset M^-; \\ \quad \vdots & \quad \vdots \\ \{I_{j_{k-1}+1}, \ldots, I_{j_k}\} & \subset M^{(-)^k}, \end{array} \right\} \tag{6.2.26}$$

with $j_k = N$, and where

$$(-)^k := \begin{cases} +, & \text{if } k \text{ is even;} \\ -, & \text{if } k \text{ is odd.} \end{cases}$$

By using also (6.2.24), we deduce that

$$I_{j_\ell} \cap I_{j_\ell+1} = \emptyset, \quad \ell = 0, \ldots, k-1,$$

and hence there exists a point sequence $\{x_\ell : \ell = 0,\ldots,k-1\}$ such that, for each $\ell \in \{0,\ldots,k-1\}$,

$$x_\ell > x,\ x \in I_{j_\ell} \quad ; \quad x_\ell < x,\ x \in I_{j_\ell+1}. \tag{6.2.27}$$

For the polynomial

$$P(x) := \prod_{\ell=0}^{k-1}(x_\ell - x), \tag{6.2.28}$$

it then follows from the assumption (6.2.11) that $P \in \pi_k \subset \pi_n$. Moreover, we deduce from (6.2.28), (6.2.27) and (6.2.26), together with the assumption $I_1 \in M^+$, that

$$\left.\begin{array}{l} I \in M^+ \Rightarrow P(x) > 0, \quad x \in I; \\[2mm] I \in M^- \Rightarrow P(x) < 0, \quad x \in I, \end{array}\right\} \tag{6.2.29}$$

and thus also

$$P(x) \neq 0, \quad x \in I_j, \quad j = 1,\ldots,N. \tag{6.2.30}$$

Next, with the interval set M given as in (6.2.19), we define the interval set

$$S := M \setminus \bigcup_{j=1}^{N} I_j. \tag{6.2.31}$$

Then S is the union of a finite number of (closed) intervals, so that we may define

$$\widetilde{d} := \max_{x \in S} |E^*(x)|. \tag{6.2.32}$$

It then follows from (6.2.32), (6.2.31), (6.2.25), (6.2.19), (6.2.20), (6.2.21) and (6.2.13) that $\widetilde{d} < d^*$, according to which we may define the constant λ to be any real number satisfying

$$0 < \lambda < \frac{\min\{d^* - \widetilde{d}, \frac{d^*}{2}\}}{||P||_\infty}. \tag{6.2.33}$$

Let

$$Q := P^* + \lambda P, \tag{6.2.34}$$

so that $P^*, P \in \pi_n$ implies $Q \in \pi_n$. We proceed to show that Q satisfies the condition (6.2.12), which will then complete our proof for the case $I_1 \in M^+$.

To this end, we first observe from (6.2.34), (6.2.13), (6.2.32) and (6.2.33) that, for any $x \in S$,

$$|f(x) - Q(x)| = |E^*(x) - \lambda P(x)| \leqslant |E^*(x)| + \lambda |P(x)| < \widetilde{d} + \frac{d^* - \widetilde{d}}{||P||_\infty}||P||_\infty = d^*,$$

and thus

$$|f(x) - Q(x)| < d^*, \quad x \in S. \tag{6.2.35}$$

Next, since (6.2.33) yields

$$|\lambda P(x)| < \frac{d^*/2}{||P||_\infty} ||P||_\infty = \frac{d^*}{2}, \quad x \in [a,b],$$

it follows from (6.2.25), (6.2.20), (6.2.21) and (6.2.22) that, with the definition

$$J := \bigcup_{j=1}^{N} I_j, \tag{6.2.36}$$

we have

$$|E^*(x)| > \frac{d^*}{2} > |\lambda P(x)|, \quad x \in J. \tag{6.2.37}$$

Next, we note from (6.2.22), (6.2.29), (6.2.36) and (6.2.25), together with $\lambda > 0$, that $E^*(x)$ and $\lambda P(x)$ have the same sign for $x \in J$, according to which, by using also (6.2.33), (6.2.13) and (6.2.37), we obtain, for any $x \in J$,

$$|f(x) - Q(x)| = |E^*(x) - \lambda P(x)| = |E^*(x)| - |\lambda P(x)| \leqslant d^* - \lambda|P(x)|. \tag{6.2.38}$$

Finally, since the definition (6.2.36) implies that J is a closed subset of $[a,b]$, we use (6.2.30) and (6.2.36) to obtain

$$\min_{x \in J} |P(x)| > 0. \tag{6.2.39}$$

Since also $\lambda > 0$, we deduce from (6.2.38) and (6.2.39) that

$$|f(x) - Q(x)| < d^*, \quad x \in J. \tag{6.2.40}$$

Since (6.2.31), (6.2.19), (6.2.16) and (6.2.36) imply that $S \cup J = [a,b]$, it follows from (6.2.35) and (6.2.40) that

$$|f(x) - Q(x)| < d^*, \quad x \in [a,b], \tag{6.2.41}$$

which, together with (6.2.13), yields the desired result (6.2.12).

The proof for the case $I_1 \in M^-$ is similar (see Exercise 6.3). ∎

Together, Theorems 6.2.1 and 6.2.2 imply the following full characterisation of polynomial best approximation with respect to the maximum norm on $[a,b]$.

Theorem 6.2.3. *For $f \in C[a,b]$ and any non-negative integer n, a polynomial $P^* \in \pi_n$ satisfies the best approximation condition (6.1.1) if and only if there exists a sequence $\{\xi_0, \ldots, \xi_{n+1}\}$ of $n+2$ distinct points in $[a,b]$ such that the equi-oscillation property (6.2.1), (6.2.2), (6.2.3) is satisfied.*

6.3   Uniqueness

The necessary condition in Theorem 6.2.2 enables us to establish, analogously to Theorem 4.2.6 for the case of best weighted L^2 polynomial approximation, the following improved formulation, which includes uniqueness, of Theorem 4.1.2.

Theorem 6.3.1. *Let $f \in C[a,b]$. Then, for each non-negative integer n, there exists precisely one polynomial $P_n^* \in \pi_n$ such that*

$$||f - P_n^*||_\infty \leqslant ||f - P||_\infty, \quad P \in \pi_n. \tag{6.3.1}$$

Proof. According to Theorem 4.1.2, there exists a polynomial $P^* \in \pi_n$ such that the best approximation property (6.1.1) is satisfied.

To prove the uniqueness in π_n of P^*, suppose $Q^* \in \pi_n$ is such that

$$||f - Q^*||_\infty = ||f - P^*||_\infty = \min_{P \in \pi_n} ||f - P||_\infty =: d^*. \tag{6.3.2}$$

By applying Theorem 4.1.3, we deduce from (6.3.2) that

$$||f - (\tfrac{1}{2}P^* + \tfrac{1}{2}Q^*)||_\infty = d^*, \tag{6.3.3}$$

and thus, from Theorem 6.2.2, there exists a sequence $\{\xi_0, \ldots, \xi_{n+1}\}$ of $n+2$ distinct points in $[a,b]$ such that

$$a \leqslant \xi_0 < \xi_1 < \cdots < \xi_{n+1} \leqslant b; \tag{6.3.4}$$

$$|f(\xi_j) - [\tfrac{1}{2}P^*(\xi_j) + \tfrac{1}{2}Q^*(\xi_j)]| = d^*, \quad j = 0, \ldots, n+1; \tag{6.3.5}$$

$$f(\xi_j) - [\tfrac{1}{2}P^*(\xi_j) + \tfrac{1}{2}Q^*(\xi_j)] = -\{f(\xi_{j+1}) - [\tfrac{1}{2}P^*(\xi_{j+1}) + \tfrac{1}{2}Q^*(\xi_{j+1})]\}, \quad j = 0, \ldots, n. \tag{6.3.6}$$

Consider first the case where (6.3.5) gives

$$f(\xi_0) - [\tfrac{1}{2}P^*(\xi_0) + \tfrac{1}{2}Q^*(\xi_0)] = d^*, \tag{6.3.7}$$

or equivalently,

$$\tfrac{1}{2}[f(\xi_0) - P^*(\xi_0)] + \tfrac{1}{2}[f(\xi_0) - Q^*(\xi_0)] = d^*. \tag{6.3.8}$$

It follows from (6.3.2) and (6.3.8) that

$$\frac{d^*}{2} \geqslant \frac{1}{2}[f(\xi_0) - P^*(\xi_0)] = d^* - \frac{1}{2}[f(\xi_0) - Q^*(\xi_0)] \geqslant d^* - \frac{d^*}{2} = \frac{d^*}{2},$$

and thus

$$\frac{d^*}{2} = \frac{1}{2}[f(\xi_0) - P^*(\xi_0)] = d^* - \frac{1}{2}[f(\xi_0) - Q^*(\xi_0)],$$

from which it then follows that

$$d^* = f(\xi_0) - P^*(\xi_0) = f(\xi_0) - Q^*(\xi_0),$$

and thereby yielding

$$P^*(\xi_0) = Q^*(\xi_0). \tag{6.3.9}$$

Next, we use (6.3.7) and (6.3.6) to obtain

$$f(\xi_1) - [\tfrac{1}{2}P^*(\xi_1) + \tfrac{1}{2}Q^*(\xi_1)] = -d^*,$$

or equivalently,

$$\tfrac{1}{2}[f(\xi_1) - P^*(\xi_1)] + \tfrac{1}{2}[f(\xi_1) - Q^*(\xi_1)] = -d^*. \tag{6.3.10}$$

It follows from (6.3.2) and (6.3.10) that

$$-\frac{d^*}{2} \leqslant \frac{1}{2}[f(\xi_1) - P^*(\xi_1)] = -d^* - \frac{1}{2}[f(\xi_1) - Q^*(\xi_1)] \leqslant -d^* + \frac{d^*}{2} = -\frac{d^*}{2},$$

and thus

$$-\frac{d^*}{2} = \frac{1}{2}[f(\xi_1) - P^*(\xi_1)] = -d^* - \frac{1}{2}[f(\xi_1) - Q^*(\xi_1)],$$

from which it follows that

$$-d^* = f(\xi_1) - P^*(\xi_1) = f(\xi_1) - Q^*(\xi_1),$$

which gives

$$P^*(\xi_1) = Q^*(\xi_1). \tag{6.3.11}$$

Repeated applications of the procedure which led to (6.3.9) and (6.3.11) eventually yield

$$P^*(\xi_j) = Q^*(\xi_j), \quad j = 0, \dots, n+1. \tag{6.3.12}$$

For the case where (6.3.5) gives

$$f(\xi_0) - [\tfrac{1}{2}P^*(\xi_0) + \tfrac{1}{2}Q^*(\xi_0)] = -d^*, \tag{6.3.13}$$

it is similarly shown that (6.3.12) is also satisfied (see Exercise 6.4).

Hence, if we define the polynomial $R^* := P^* - Q^*$, so that $R^* \in \pi_n$, it follows from (6.3.12) and (6.3.4) that R^* has at least $n+2$ distinct real zeros, and thus R^* is the zero polynomial, that is, $Q^* = P^*$, and thereby completing our proof. ∎

The polynomial P_n^* of Theorem 6.3.1 is called the best uniform (or Chebyshev, or minimax) approximation on $[a, b]$ from π_n to f.

6.4   The approximation operator

Based on Theorem 6.3.1, we may now define, for any bounded interval $[a,b]$ and non-negative integer n, the best uniform polynomial approximation operator $\mathscr{P}_n^* : C[a,b] \to \pi_n$ by

$$\mathscr{P}_n^* f = P_n^*, \quad f \in C[a,b]. \tag{6.4.1}$$

The following example shows that $\mathscr{P}_n^*$ is not linear.

Example 6.4.1. Let $[a,b] = [-1,1]$, and

$$f(x) = x^2 \quad ; \quad g(x) = x,$$

so that, from (6.4.1), together with the formula (6.1.9) in Theorem 6.1.2, the approximation operator $\mathscr{P}_0^* : C[-1,1] \to \pi_0$ satisfies

$$(\mathscr{P}_0^* f)(x) = \frac{1}{2} \quad ; \quad (\mathscr{P}_0^* g)(x) = 0,$$

whereas, since also

$$\max_{-1\leqslant x\leqslant 1} (x^2+x) = 2 \; ; \quad \min_{-1\leqslant x\leqslant 1} (x^2+x) = (-\tfrac{1}{2})^2 + (-\tfrac{1}{2}) = -\tfrac{1}{4},$$

we have

$$\mathscr{P}_0^*(f+g) = \frac{1}{2}\left(2 - \frac{1}{4}\right) = \frac{7}{8}.$$

Hence

$$\mathscr{P}_0^* f + \mathscr{P}_0^* g = \frac{1}{2} + 0 = \frac{1}{2} \neq \frac{7}{8} = \mathscr{P}_0^*(f+g),$$

according to which the best approximation operator $\mathscr{P}_0^*$ is not linear. ∎

The approximation operator $\mathscr{P}_n^*$ possesses the following properties.

Theorem 6.4.1. *For any bounded interval $[a,b]$ and non-negative integer n, the best uniform polynomial approximation $\mathscr{P}_n^* : C[a,b] \to \pi_n$, as defined by (6.4.1), satisfies:*

(a) *$\mathscr{P}_n^*$ is exact on π_n.*

(b) *The uniform convergence property*

$$\|f - \mathscr{P}_n^* f\|_\infty \to 0, \quad n \to \infty, \quad f \in C[a,b], \tag{6.4.2}$$

holds.

(c) *If $f \in C^k[a,b]$ for an integer $k \in \mathbb{N}$, the convergence rate result*

$$||f - \mathscr{P}_n^* f||_\infty \leqslant (b-a)^k \frac{(n-k)!}{n!} ||f^{(k)}||_\infty, \quad n = k, k+1, \ldots, \tag{6.4.3}$$

holds, with, in particular,

$$||f - \mathscr{P}_n^* f||_\infty \leqslant \frac{b-a}{n} ||f'||_\infty, \quad n = 1, 2, \ldots, \; f \in C^1[a,b]. \tag{6.4.4}$$

Proof. (a) If $f \in \pi_n$, then, since

$$||f - f||_\infty = 0 \leqslant ||f - P||_\infty, \quad P \in \pi_n,$$

it follows from (6.4.1) and (6.1.1) that $\mathscr{P}_n^* f = f$.

(b) From (6.4.1) and (6.1.1), we deduce that

$$||f - \mathscr{P}_n^* f||_\infty \leqslant ||f - \mathscr{P}_n^B f||_\infty, \quad n = 1, 2, \ldots, \quad f \in C[a,b], \tag{6.4.5}$$

with $\mathscr{P}_n^B : C[a,b] \to \pi_n$ denoting the Bernstein polynomial approximation operator, as defined by (5.1.5). The uniform convergence result (6.4.2) is then an immediate consequence of (6.4.5) and Theorem 3.3.3.

(c) If $f \in C^1[a,b]$, it follows from (6.1.1), (6.4.1), (6.4.5) and (5.1.4), together with the inequality (3.5.4) in Theorem 3.5.1, that

$$\min_{P \in \pi_n} ||f - P||_\infty = ||f - \mathscr{P}_n^* f||_\infty \leqslant \frac{b-a}{n} ||f'||_\infty, \quad n = 1, 2, \ldots, \tag{6.4.6}$$

which shows that (6.4.3) holds for $k = 1$, and is indeed precisely (6.4.4).

Proceeding inductively, suppose next that (6.4.3) is satisfied for a fixed integer $k \in \mathbb{N}$, and let $f \in C^{k+1}[a,b]$. Then $f' \in C^k[a,b]$, and it follows from the inductive hypothesis (6.4.3), with n replaced by $n-1$, that

$$||f' - \mathscr{P}_{n-1}^*(f')||_\infty \leqslant (b-a)^k \frac{(n-1-k)!}{(n-1)!} ||f^{(k+1)}||_\infty, \quad n = k+1, k+2, \ldots. \tag{6.4.7}$$

Now define

$$Q(x) := \int_a^x (\mathscr{P}_{n-1}^*(f'))(t)dt, \tag{6.4.8}$$

according to which, since $\mathscr{P}_{n-1}^*(f') \in \pi_{n-1}$, we have $Q \in \pi_n$, and thus, by using also (6.1.1) and (6.4.1),

$$||f - \mathscr{P}_n^* f||_\infty = \min_{P \in \pi_n} ||f - P||_\infty = \min_{P \in \pi_n} ||f - (P+Q)||_\infty = \min_{P \in \pi_n} ||(f-Q) - P||_\infty. \tag{6.4.9}$$

By applying (6.4.9), (6.4.6), (6.4.8) and (6.4.7), we deduce that, for $n = k+1, k+2, \ldots$,

$$||f - \mathscr{P}_n^* f||_\infty \leqslant \frac{b-a}{n} ||f' - Q'||_\infty = \frac{b-a}{n} ||f' - \mathscr{P}_{n-1}^*(f')||_\infty$$

$$\leqslant \frac{b-a}{n}\left[(b-a)^k\frac{(n-1-k)!}{(n-1)!}||f^{(k+1)}||_\infty\right]$$

$$= (b-a)^{k+1}\frac{(n-(k+1))!}{n!}||f^{(k+1)}||_\infty,$$

that is, (6.4.3) is satisfied with k replaced by $k+1$, and thereby completing our inductive proof of (6.4.3). ∎

As an application of the uniform convergence result (6.4.2) in Theorem 6.4.1(b), and by using also Theorem 5.3.1, we prove next the following weighted L^2 convergence property of the best approximation operator $\widetilde{\mathscr{P}}_n^*$.

Theorem 6.4.2. *For any bounded interval $[a,b]$ and weight function w on $[a,b]$ such that (4.2.8) and (4.2.9) hold, let $\{\widetilde{\mathscr{P}}_n^* : n=0,1,\ldots\}$ denote the corresponding sequence of best weighted L^2 approximation operators, as defined by (5.1.6). Then*

$$||f - \widetilde{\mathscr{P}}_n^* f||_{2,w} \to 0, \quad n \to \infty, \quad f \in C[a,b], \tag{6.4.10}$$

with the norm $||\cdot||_{2,w}$ defined by (4.2.16).

Proof. Let $f \in C[a,b]$. Then, from the defining property (4.2.23) of the polynomial $\widetilde{\mathscr{P}}_n^* f := \widetilde{P}_n^*$, we have, for any non-negative integer n,

$$||f - \widetilde{\mathscr{P}}_n^* f||_{2,w} \leqslant ||f - \mathscr{P}_n^* f||_{2,w}, \tag{6.4.11}$$

with $\mathscr{P}_n^* : C[a,b] \to \pi_n$ denoting the best uniform approximation operator defined by (6.4.1), whereas also, from the inequality (5.3.2) in Theorem 5.3.1,

$$||f - \mathscr{P}_n^* f||_{2,w} \leqslant \sqrt{\int_a^b w(x)dx}\,||f - \mathscr{P}_n^* f||_\infty. \tag{6.4.12}$$

It follows from (6.4.11) and (6.4.12), together with the uniform convergence result (6.4.2) in Theorem 6.4.1(b), that

$$||f - \widetilde{\mathscr{P}}_n^* f||_{2,w} \leqslant \sqrt{\int_a^b w(x)dx}\,||f - \mathscr{P}_n^* f||_\infty \to 0,\ n \to \infty,$$

from which (6.4.10) then immediately follows. ∎

Let $f \in C[a,b]$, and observe from (6.4.1) and Theorem 6.2.3 that, according to the equi-oscillation condition (6.2.1), (6.2.2), (6.2.3), there exists, for any non-negative integer n, a sequence of $n+1$ distinct points $\{x_{n,0},\ldots,x_{n,n}\}$ in $[a,b]$, with $x_{n,j} \in (\xi_{n,j},\xi_{n,j+1})$, $j = 0,\ldots,n+1$, where we have replaced ξ_j by $\xi_{n,j}$ in (6.2.1), (6.2.2) and (6.2.3), such that

$$(\mathscr{P}_n^* f)(x_{n,j}) = f(x_{n,j}), \quad j = 0,\ldots,n. \tag{6.4.13}$$

Hence, if we denote by $\mathscr{P}_n^I : C[a,b] \to \pi_n$ the Lagrange polynomial interpolation operator with respect to the interpolation points $\{x_{n,0}, \ldots, x_{n,n}\}$, it follows from (5.1.2) and the uniqueness statement in Theorem 1.1.2 that

$$\mathscr{P}_n^I f = \mathscr{P}_n^* f, \quad n = 0, 1, \ldots. \tag{6.4.14}$$

Hence we may apply the uniform convergence result (6.4.2) in Theorem 6.4.1(b), together with (6.4.14), to deduce that

$$||f - \mathscr{P}_n^I f||_\infty \to 0, \quad n \to \infty. \tag{6.4.15}$$

We may therefore conclude that, for any $f \in C[a,b]$, there exists a sequence

$$\triangle_n := \{x_{n,0}, \ldots, x_{n,n}\} \subset [a,b], \quad n = 0, 1, \ldots, \tag{6.4.16}$$

of (distinct) interpolation point sequences such that the uniform convergence condition (6.4.15) is satisfied. On the other hand, as mentioned before in Section 3.1 of Chapter 3, for any given sequence $\{\triangle_n : n = 0, 1, \ldots\}$ of (distinct) interpolation point sequences as in (6.4.16), there exists a function $f \in C[a,b]$ such that the divergence result (3.1.71) holds.

6.5   An application in polynomial interpolation

In this section, we show how the convergence rate result (6.4.3) of Theorem 6.4.1(c), together with the Lebesgue inequality (5.3.11) in Theorem 5.3.3, may be applied to prove uniform convergence for polynomial interpolation with respect to the Chebyshev interpolation points (2.2.31), for a class of functions f which includes the Runge example (3.1.5) in Example 3.1.3. We shall rely on the following properties of the first derivative of a Chebyshev polynomial.

Theorem 6.5.1. *For an integer $j \in \mathbb{N}$, let the Chebyshev polynomial T_j be defined as in (2.2.1), and denote by $\{t_{j,k} : k = 0, \ldots, j-1\}$ the zeros, as described in (2.2.15), (2.2.16) and (2.2.17), of T_j. Then:*

(a)

$$T_j'(t_{j,k}) = (-1)^{j-k+1} \frac{j}{\sin\left(\frac{2j-1-2k}{2j}\pi\right)}, \quad k = 0, \ldots, j-1; \tag{6.5.1}$$

(b)

$$\left.\begin{array}{l} T_j'(-1) = (-1)^{j+1} j^2; \\ T_j'(1) = j^2; \end{array}\right\} \tag{6.5.2}$$

(c)

$$\max_{-1\leqslant x\leqslant 1}|T_j'(x)| = |T_j'(-1)| = |T_j'(1)| = j^2. \tag{6.5.3}$$

Proof. (a) By differentiating the formula (2.2.5), we obtain

$$T_j'(x) = \frac{j\sin(j\arccos x)}{\sqrt{1-x^2}}, \quad x\in(-1,1). \tag{6.5.4}$$

Now observe that (2.2.16) yields

$$\sin(j\arccos t_{j,k}) = \sin\left((2j-1-2k)\frac{\pi}{2}\right) = (-1)^{j-k+1}, \quad k=0,\ldots,j-1, \tag{6.5.5}$$

as well as

$$\sqrt{1-(t_{j,k})^2} = \sin\left(\frac{2j-1-2k}{2j}\pi\right), \quad k=0,\ldots,j-1. \tag{6.5.6}$$

The formula (6.5.1) is a consequence of (6.5.4), (6.5.5) and (6.5.6).

(b) In terms of the one-to-one mapping (2.2.11), (2.2.12) between the intervals $[0,\pi]$ and $[-1,1]$, the equation (6.5.4) has the equivalent formulation

$$T_j'(x) = \frac{j\sin(j\theta)}{\sin\theta}, \quad \theta\in(0,\pi). \tag{6.5.7}$$

Now observe that

$$\lim_{\theta\to 0^+}\left[\frac{\sin(j\theta)}{\sin\theta}\right] = j\lim_{\theta\to 0^+}\left[\frac{\sin(j\theta)}{j\theta}\frac{1}{\frac{\sin\theta}{\theta}}\right] = j, \tag{6.5.8}$$

and similarly, since

$$\sin(j(\pi-\theta)) = \sin(j\pi)\cos(j\theta) - \cos(j\pi)\sin(j\theta)$$
$$= (0)\cos(j\theta) - (-1)^j\sin(j\theta) = (-1)^{j+1}\sin(j\theta),$$

we have

$$\lim_{\theta\to\pi^-}\left[\frac{\sin(j\theta)}{\sin\theta}\right] = (-1)^{j+1}j\lim_{\theta\to\pi^-}\left[\frac{\sin(j(\pi-\theta))}{j(\pi-\theta)}\frac{1}{\frac{\sin(\pi-\theta)}{\pi-\theta}}\right] = (-1)^{j+1}j. \tag{6.5.9}$$

It follows from (6.5.7), (6.5.8) and (6.5.9), together with (2.2.11), (2.2.12), that (6.5.2) is satisfied.

(c) Since, for any integer $m\in\mathbb{N}$, and $\theta\in(0,\pi)$, we have

$$\left|\frac{\sin((m+1)\theta)}{\sin\theta}\right| = \left|\frac{\sin(m\theta)}{\sin\theta}\cos\theta + \cos(m\theta)\right| \leqslant \left|\frac{\sin(m\theta)}{\sin\theta}\right| + 1,$$

it follows inductively (see Exercise 6.13) that

$$\left|\frac{\sin(m\theta)}{\sin\theta}\right| \leqslant m, \quad \theta\in(0,\pi), \quad m\in\mathbb{N}. \tag{6.5.10}$$

We deduce from (6.5.7) and (6.5.10), together with (2.2.11), (2.2.12), that

$$|T_j'(x)| \leqslant j^2, \quad x \in (-1,1). \tag{6.5.11}$$

The result (6.5.3) is an immediate consequence of (6.5.11) and (6.5.2). ∎

By using Theorem 6.5.1, we proceed to establish the following bound on the Lebesgue constant of the Lagrange polynomial interpolation operator for the case when the Chebyshev interpolation points are used.

Theorem 6.5.2. *For a bounded interval $[a,b]$ and any integer $n \in \mathbb{N}$, let $\mathscr{P}_n^I : C[a,b] \to \pi_n$ denote the Lagrange polynomial interpolation operator defined by (5.1.2), and where the interpolation points are chosen as the Chebyshev interpolation points, that is,*

$$x_j = x_{n,j}^C, \quad j = 0,\ldots,n, \tag{6.5.12}$$

as defined by (2.2.31). Then the corresponding Lebesgue constant satisfies the bound

$$\|\mathscr{P}_n^I\|_\infty \leqslant 2(n+1)^2. \tag{6.5.13}$$

Proof. By using the formula (5.2.14) of Theorem 5.2.4, as well as (1.2.1), (6.5.12), and the one-to-one mapping (2.2.29), (2.2.30) between the intervals $[-1,1]$ and $[a,b]$, together with the definitions (2.2.31) and (2.2.16), we deduce that

$$\|\mathscr{P}_n^I\|_\infty = \max_{a \leqslant x \leqslant b} \sum_{j=0}^n \left| \prod_{j \neq k=0}^n \frac{x - x_{n,k}^C}{x_{n,j}^C - x_{n,k}^C} \right|$$

$$= \max_{-1 \leqslant t \leqslant 1} \left[\sum_{j=0}^n F_{n,j}(t) \right] \leqslant \sum_{j=0}^n \left[\max_{-1 \leqslant t \leqslant 1} F_{n,j}(t) \right], \tag{6.5.14}$$

where

$$F_{n,j}(t) := \left| \prod_{j \neq k=0}^n \frac{t - t_{n+1,k}}{t_{n+1,j} - t_{n+1,k}} \right|, \tag{6.5.15}$$

and with the sequence $\{t_{n+1,k} : k = 0,\ldots,n\}$ defined as in (2.2.16).

Let $j \in \{0,\ldots,n\}$ be fixed. It follows from (6.5.15) that

$$F_{n,j}(t_{n+1,k}) = \delta_{j-k}, \quad k = 0,\ldots,n. \tag{6.5.16}$$

Next, by noting from (2.2.9) in Theorem 2.2.1(f), together with (2.2.16), that

$$T_{n+1}(t) = 2^n \prod_{k=0}^n (t - t_{n+1,k}), \tag{6.5.17}$$

and thus

$$T_{n+1}'(t_{n+1,j}) = 2^n \prod_{j \neq k=0}^n (t_{n+1,j} - t_{n+1,k}), \tag{6.5.18}$$

we deduce from (6.5.15), together with the formula (6.5.1) in Theorem 6.5.1(a), that, for $t \in [-1,1] \setminus \{t_{n+1,j}\}$,

$$F_{n,j}(t) = \left| \frac{1}{T'_{n+1}(t_{n+1,j})} \frac{T_{n+1}(t)}{t - t_{n+1,j}} \right| = \frac{\sin\left(\frac{2n+1-2j}{2n+2} \pi \right)}{n+1} \left| \frac{T_{n+1}(t)}{t - t_{n+1,j}} \right|$$

$$\leqslant \frac{1}{n+1} \left| \frac{T_{n+1}(t)}{t - t_{n+1,j}} \right|. \tag{6.5.19}$$

Now observe that, for $t \in [-1,1] \setminus \{t_{n+1,j}\}$,

$$\frac{d}{dt} \left[\frac{T_{n+1}(t)}{t - t_{n+1,j}} \right] = \frac{(t - t_{n+1,j})T'_{n+1}(t) - T_{n+1}(t)}{(t - t_{n+1,j})^2},$$

and thus

$$\frac{d}{dt} \left[\frac{T_{n+1}(t)}{t - t_{n+1,j}} \right] = 0 \text{ if and only if } \frac{T_{n+1}(t)}{t - t_{n+1,j}} = T'_{n+1}(t). \tag{6.5.20}$$

Next, we use (2.2.7), (2.2.16) and (2.2.17) to obtain

$$\left| \frac{T_{n+1}(-1)}{-1 - t_{n+1,j}} \right| = \frac{1}{|-1 - t_{n+1,j}|}$$

$$\leqslant \frac{1}{|-1 - t_{n+1,0}|} = \frac{1}{1 + t_{n+1,0}}$$

$$= \frac{1}{1 + \cos\left(\frac{2n+1}{2n+2} \pi \right)}$$

$$= \frac{1 - \cos\left(\frac{2n+1}{2n+2} \pi \right)}{\sin^2\left(\frac{2n+1}{2n+2} \pi \right)}$$

$$\leqslant \frac{2}{\sin^2\left(\frac{2n+1}{2n+2} \pi \right)} = \frac{2}{\sin^2\left(\frac{\pi}{2(n+1)} \right)}, \tag{6.5.21}$$

and similarly,

$$\left| \frac{T_{n+1}(1)}{1 - t_{n+1,j}} \right| \leqslant \frac{1}{1 - t_{n+1,n}} = \frac{1}{1 + t_{n+1,0}} \leqslant \frac{2}{\sin^2\left(\frac{\pi}{2(n+1)} \right)}. \tag{6.5.22}$$

By applying (6.5.16), (6.5.19), (6.5.20), (6.5.21) and (6.5.22), we deduce that

$$\max_{-1 \leqslant t \leqslant 1} F_{n,j}(t) \leqslant \max \left\{ 1, \frac{1}{n+1} \max_{-1 \leqslant t \leqslant 1} |T'_{n+1}(t)|, \frac{2}{(n+1)\sin^2\left(\frac{\pi}{2(n+1)} \right)} \right\}. \tag{6.5.23}$$

Now observe from (6.5.3) in Theorem 6.5.1(c) that

$$\frac{1}{n+1} \max_{-1 \leqslant t \leqslant 1} |T'_{n+1}(t)| = n+1. \tag{6.5.24}$$

Next, we note that the definition

$$G(x) := \sin x - \frac{2x}{\pi}$$

yields

$$G'(x) = \cos x - \frac{2}{\pi} \quad ; \quad G''(x) = -\sin x,$$

and thus

$$\left. \begin{array}{c} G(0) = G(\tfrac{\pi}{2}) = 0; \quad G'(0) = 1 - \tfrac{2}{\pi} > 0; \\ G''(x) < 0, \quad x \in (0, \tfrac{\pi}{2}], \end{array} \right\}$$

according to which we may deduce that

$$G(x) \geqslant 0, \quad x \in [0, \tfrac{\pi}{2}],$$

that is,

$$\sin x \geqslant \frac{2x}{\pi}, \quad x \in [0, \tfrac{\pi}{2}]. \tag{6.5.25}$$

Hence we may apply (6.5.25) to obtain

$$\sin^2\left(\frac{\pi}{2(n+1)}\right) \geqslant \left[\frac{2}{\pi}\frac{\pi}{2(n+1)}\right]^2 = \frac{1}{(n+1)^2},$$

and thus

$$\frac{2}{(n+1)\sin^2\left(\frac{\pi}{2(n+1)}\right)} \leqslant 2(n+1). \tag{6.5.26}$$

It follows from (6.5.23), (6.5.24) and (6.5.26) that

$$\max_{-1\leqslant t\leqslant 1} F_{n,j}(t) \leqslant 2(n+1),$$

which, together with (6.5.14), yields the desired upper bound in (6.5.13). ∎

Finally, we combine the results of Theorem 6.5.2 and Theorem 6.4.1(c) to deduce the following result by means of the Lebesgue inequality (5.3.11) in Theorem 5.3.3.

Theorem 6.5.3. *Suppose* $f \in C^k[a,b]$ *for an integer* $k \in \mathbb{N}$, *and denote by* $\mathscr{P}_n^I$ *the Lagrange polynomial interpolation operator of Theorem 6.5.2. Then:*

(a)

$$\|f - \mathscr{P}_n^I f\|_\infty \leqslant (b-a)^k \frac{[1 + 2(n+1)^2](n-k)!}{n!} \|f^{(k)}\|_\infty, \quad n = k, k+1, \ldots; \tag{6.5.27}$$

(b) *if* $k \geqslant 3$, *then*

$$\|f - \mathscr{P}_n^I f\|_\infty \to 0, \quad n \to \infty, \tag{6.5.28}$$

with, in particular,

$$\|f - \mathscr{P}_n^I f\|_\infty \leqslant (b-a)^3 \frac{11/2}{n-2} \|f'''\|_\infty, \quad n = 3, 4, \ldots. \tag{6.5.29}$$

Proof. (a) According to Theorem 5.3.3, the Lebesgue inequality (5.3.7) in Theorem 5.3.2 may be applied to the Lagrange polynomial interpolation operator $\mathscr{P}_n^I$, as defined by (5.1.2), and thus, by using also (6.5.13) in Theorem 6.5.2, as well as (6.4.3) in Theorem 6.4.1(c), we obtain (6.5.27).

(b) If $f \in C^k[a,b]$ for an integer $k \geqslant 3$, then $f \in C^3[a,b]$, and it follows from (6.5.27) that

$$\|f - \mathscr{P}_n^I f\|_\infty \leqslant (b-a)^3 \frac{1+2(n+1)^2}{n(n-1)} \frac{1}{n-2} \|f'''\|_\infty, \quad n = 3,4,\ldots. \qquad (6.5.30)$$

The upper bound in (6.5.29) then follows by applying the inequality

$$1 + 2(n+1)^2 \leqslant \tfrac{11}{2} n(n-1), \quad n = 3,4,\ldots, \qquad (6.5.31)$$

(see Exercise 6.14) in (6.5.30). Finally observe that (6.5.29) implies the uniform convergence result (6.5.28).

∎

Observe that the Runge example (3.1.5) in Example 3.1.3 satisfies $f \in C^k[-5,5]$ for each integer $k \geqslant 3$, so that (6.5.28) in Theorem 6.5.3(b) verifies our statement in Example 3.1.3 that

$$\max_{-5 \leqslant x \leqslant 5} \left| \frac{1}{1+x^2} - P_n^I(x) \right| \to 0, \quad n \to \infty,$$

if the Chebyshev interpolation points (3.1.69) are chosen.

6.6 Explicit computation for specific cases

Finally in this chapter, we provide examples of functions $f \in C[a,b]$ for which the corresponding best uniform approximation $\mathscr{P}_n^* f = P_n^*$ can be computed explicitly.

First, for any non-negative integer n, let $f \in \pi_{n+1}$, that is,

$$f(x) = \sum_{j=0}^{n+1} c_j x^j \qquad (6.6.1)$$

for some coefficient sequence $\{c_0,\ldots,c_{n+1}\}$. It follows from (6.6.1) that the error function

$$E^*(x) := f(x) - P_n^*(x), \qquad (6.6.2)$$

with P_n^* denoting the best uniform approximation on $[a,b]$ from π_n to f, satisfies

$$E^*(x) = c_{n+1} x^{n+1} + Q^*(x), \qquad (6.6.3)$$

for some polynomial $Q^* \in \pi_n$.

Now observe from Theorem 2.2.1(a), (c) and (d), together with the one-to-one mapping (2.2.29), (2.2.30) between the intervals $[-1,1]$ and $[a,b]$, that the polynomial

$$P(x) := c_{n+1} \left(\frac{b-a}{2} \right)^{n+1} \frac{1}{2^n} T_{n+1} \left(\frac{2}{b-a}x - \frac{a+b}{b-a} \right) \qquad (6.6.4)$$

satisfies

$$P(x) = c_{n+1}x^{n+1} + \widetilde{Q}^*(x), \qquad (6.6.5)$$

for some polynomial $\widetilde{Q}^* \in \pi_n$, and

$$\|P\|_\infty = |c_{n+1}| \left(\frac{b-a}{2} \right)^{n+1} \frac{1}{2^n}, \qquad (6.6.6)$$

as well as, for the points

$$\xi_j := \left(\frac{b-a}{2} \right)^{n+1} \cos \left(\frac{n+1-j}{n+1}\pi \right) - \frac{a+b}{2}, \quad j=0,\ldots,n+1, \qquad (6.6.7)$$

$$a = \xi_0 < \xi_1 < \cdots < \xi_{n+1} = b, \qquad (6.6.8)$$

the conditions

$$|P(\xi_j)| = \|P\|_\infty, \quad j=0,\ldots,n+1, \qquad (6.6.9)$$

and

$$P(\xi_j) = -P(\xi_{j+1}), \quad j=0,\ldots,n. \qquad (6.6.10)$$

By applying the characterisation result of Theorem 6.2.3, as well as the uniqueness result of Theorem 6.3.1, we now deduce from (6.6.2) - (6.6.10) that we must have

$$E^*(x) = f(x) - P_n^*(x) = P(x), \qquad (6.6.11)$$

with the polynomial P defined by (6.6.4), from which we can solve for $P_n^*(x)$ to obtain the following results from (6.6.4) and (6.6.6).

Theorem 6.6.1. *For a non-negative integer n, let $f \in \pi_{n+1}$, with leading coefficient c_{n+1} in the formulation (6.6.1). Then, for any bounded interval $[a,b]$, the best uniform approximation on $[a,b]$ from π_n to f is given by the formula*

$$(\mathscr{P}_n^* f)(x) = P_n^*(x) = f(x) - c_{n+1} \left(\frac{b-a}{2} \right)^{n+1} \frac{1}{2^n} T_{n+1} \left(\frac{2}{b-a}x - \frac{a+b}{b-a} \right), \qquad (6.6.12)$$

with T_{n+1} denoting the Chebyshev polynomial of degree $n+1$, as defined in (2.2.1). Moreover,

$$\|f - \mathscr{P}_n^* f\|_\infty = |c_{n+1}| \left(\frac{b-a}{2} \right)^{n+1} \frac{1}{2^n}. \qquad (6.6.13)$$

Observe from (6.6.12) and (6.6.13) that the case $c_{n+1} = 0$ of Theorem 6.6.1, in which case $f \in \pi_n$, is consistent with Theorem 6.4.1(a).

Example 6.6.1. For the case $[a,b] = [0,1]$, $f(x) = x^2$ and $n = 1$, it follows from (6.6.12) and (2.2.2) that

$$(\mathscr{P}_1^* f)(x) = x^2 - 1 \left(\tfrac{1}{2}\right)^2 \tfrac{1}{2}[2(2x-1)^2 - 1] = x - \tfrac{1}{8}, \qquad (6.6.14)$$

with, moreover, from (6.6.13),

$$\|f - \mathscr{P}_1^* f\|_\infty = 1 \left(\tfrac{1}{2}\right)^2 \tfrac{1}{2} = \tfrac{1}{8}. \qquad (6.6.15)$$

Recall now, from (5.3.12), in Example 5.3.1, the Lebesgue inequality

$$\|f - \mathscr{P}_1^I f\|_\infty \leqslant 2\|f - \mathscr{P}_1^* f\|_\infty, \quad f \in C[0,1], \qquad (6.6.16)$$

with $\mathscr{P}_1^I : C[0,1] \to \pi_1$ denoting the (linear) Lagrange polynomial interpolation operator with respect to the interpolation points $\{x_0, x_1\} = \{0,1\}$.

For $f(x) = x^2$, it follows that $(\mathscr{P}_1^I f)(x) = x$, and thus

$$\|f - \mathscr{P}_1^I f\|_\infty = \max_{0 \leqslant x \leqslant 1} |x^2 - x| = \left|\left(\tfrac{1}{2}\right)^2 - \tfrac{1}{2}\right| = \tfrac{1}{4}. \qquad (6.6.17)$$

It follows from (6.6.15) and (6.6.17) that the Lebesgue inequality (6.6.16) is satisfied as an equation for the case $f(x) = x^2$. Hence the inequality (6.6.16) is sharp, in the sense that the constant 2 in the right hand side of (6.6.16) can not be replaced by a smaller constant. ∎

Example 6.6.2. Let $[a,b] = [0,2]$,

$$f(x) = \frac{3}{x+1}, \qquad (6.6.18)$$

and $n = 1$, and set

$$(\mathscr{P}_1^* f)(x) = P_1^*(x) = -\alpha x + \beta \qquad (6.6.19)$$

for some positive coefficients α and β. Based on the characterisation result of Theorem 6.2.3, and since (6.6.18) and (6.6.19) give

$$\frac{d}{dx}(f(x) - P_1^*(x)) = -\frac{3}{(x+1)^2} + \alpha,$$

according to which

$$\frac{d}{dx}(f(x) - P_1^*(x)) = 0 \text{ for } x > 0 \text{ if and only if } x = \sqrt{\frac{3}{\alpha}} - 1,$$

we choose

$$\xi_0 = 0; \quad \xi_1 = \sqrt{\frac{3}{\alpha}} - 1; \quad \xi_2 = 2, \qquad (6.6.20)$$

and, with the notation

$$h := ||f - \mathscr{P}_1^* f||_\infty, \tag{6.6.21}$$

proceed to solve for α, β and h from the system

$$\left.\begin{array}{l} f(\xi_0) - P_1^*(\xi_0) = h \\ f(\xi_1) - P_1^*(\xi_1) = -h \\ f(\xi_2) - P_1^*(\xi_2) = h \end{array}\right\}, \tag{6.6.22}$$

which, from (6.6.18) and (6.6.19), is equivalent to the non-linear system

$$\left.\begin{array}{l} 3 - \beta \qquad\qquad = h \\ 2\sqrt{3}\alpha - \alpha - \beta = -h \\ 1 + 2\alpha - \beta \qquad = h \end{array}\right\}. \tag{6.6.23}$$

It follows from the first and third equations in (6.6.23) that $\alpha = 1$, which, when substituted into the second equation of (6.6.23), yields

$$2\sqrt{3} - 1 - \beta = -h. \tag{6.6.24}$$

It then follows from the first equation of (6.6.23), together with (6.6.24), that $\beta = 1 + \sqrt{3}$ and $h = 2 - \sqrt{3}$. Hence

$$(\mathscr{P}_1^* f)(x) = P_1^*(x) = -x + (1 + \sqrt{3}),$$

with, moreover, from (6.6.21),

$$||f - \mathscr{P}_1^* f||_\infty = 2 - \sqrt{3},$$

since we can now verify that the sequence $\{\xi_0, \xi_1, \xi_2\} = \{0, \sqrt{3} - 1, 2\}$ and the polynomial $\mathscr{P}_1^* f$ then does indeed satisfy the required equi-oscillation property (6.2.1), (6.2.2), (6.2.3).

∎

Next, we consider a case where f is a continuous piecewise linear function, as follows:

Example 6.6.3. Let $[a, b] = [-1, 2]$,

$$f(x) = |x|, \tag{6.6.25}$$

and $n = 2$, and set

$$(\mathscr{P}_2^* f)(x) = P_2^*(x) = \alpha x^2 + \beta x + \gamma, \tag{6.6.26}$$

for some coefficients α, β and γ such that $\alpha > 0, \beta < 1$ and $\gamma > 0$. Based on the characterisation result of Theorem 6.2.3, and since (6.6.25) and (6.6.26) give, for $x > 0$,

$$\frac{d}{dx}[f(x) - P_2^*(x)] = 0 \text{ if and only if } x = \frac{1-\beta}{2\alpha},$$

we choose

$$\xi_0 = -1; \quad \xi_1 = 0; \quad \xi_2 = \frac{1-\beta}{2\alpha}; \quad \xi_3 = 2, \tag{6.6.27}$$

and, with the notation

$$h := \|f - \mathscr{P}_2^* f\|_\infty, \tag{6.6.28}$$

proceed to solve for α, β, γ and h from the system

$$\left. \begin{array}{l} f(\xi_0) - P_2^*(\xi_0) = h \\ f(\xi_1) - P_2^*(\xi_1) = -h \\ f(\xi_2) - P_2^*(\xi_2) = h \\ f(\xi_3) - P_2^*(\xi_3) = -h \end{array} \right\}, \tag{6.6.29}$$

which, from (6.6.25) and (6.6.26), is equivalent to the non-linear system

$$\left. \begin{array}{rl} 1 - \alpha + \beta - \gamma & = h \\ -\gamma & = -h \\ \frac{1-\beta}{2\alpha} - \alpha\left(\frac{1-\beta}{2\alpha}\right)^2 - \frac{\beta(1-\beta)}{2\alpha} - \gamma = h \\ 2 - 4\alpha - 2\beta - \gamma & = -h \end{array} \right\}. \tag{6.6.30}$$

It follows from the second and fourth equations of (6.6.30) that $\beta = 1 - 2\alpha$, and thus also $(1-\beta)/(2\alpha) = 1$, which, together with the second and third equations of (6.6.30), gives $\alpha = 2h$, and thus $\beta = 1 - 4h$. By substituting these values of α and β, together with $\gamma = h$, into the first equation of (6.6.30), we obtain $h = \frac{1}{4}$, from which it then follows that $\alpha = \frac{1}{2}, \beta = 0$ and $\gamma = \frac{1}{4}$. Hence, from (6.6.26),

$$(\mathscr{P}_2^* f)(x) = \tfrac{1}{2}x^2 + \tfrac{1}{4},$$

with, moreover, from (6.6.28),

$$\|f - \mathscr{P}_2^* f\|_\infty = \tfrac{1}{4},$$

since we can now verify that the sequence $\{\xi_0, \xi_1, \xi_2, \xi_3\} = \{-1, 0, 1, 2\}$ and the polynomial $\mathscr{P}_2^* f$ then does indeed satisfy the required equi-oscillation property (6.2.1), (6.2.2), (6.2.3).

■

Example 6.6.4. Let $[a,b] = [-\pi, \pi]$ and $f(x) = \sin(4x)$. Since then $||f||_\infty = 1$, with $f(x)$ attaining the values 1 and -1 with alternating sign at precisely 8 successive, and strictly increasing, points in $(-\pi, \pi)$, we deduce from Theorem 6.2.3 that $\mathscr{P}_n^* f$ is the zero polynomial if and only if $n \in \{0, \ldots, 6\}$. ∎

Example 6.6.5. Let $[a,b] = [-1,2]$, $f(x) = 2x^3 - x$ and $n = 2$. An application of the formula (6.6.12) in Theorem 6.6.1, together with (2.2.2), then yields

$$(\mathscr{P}_2^* f)(x) = P_2^*(x) = 2x^3 - x - 2(\tfrac{3}{2})^3 \tfrac{1}{4} \left[4 \left(\frac{2x-1}{3} \right)^3 - 3 \left(\frac{2x-1}{3} \right) \right]$$

$$= \frac{1}{16}(48x^2 + 14x - 23),$$

with, moreover, from (6.6.13),

$$||f - \mathscr{P}_2^* f||_\infty = 2 \left(\tfrac{3}{2} \right)^3 \tfrac{1}{4} = \tfrac{27}{16}.$$ ∎

In general, it is not possible, for an arbitrary choice of $f \in C[a,b]$, to explicitly compute the best uniform polynomial approximation $\mathscr{P}_n^* f$ on $[a,b]$ from π_n to f, as in the examples provided in this section. In such cases iterative methods like the exchange algorithm, or Remez algorithm, the presentation of which is outside the scope of this book, can be applied to approximate $\mathscr{P}_n^* f$ arbitrarily well within a finite number of iterations.

6.7 Exercises

Exercise 6.1 Investigate whether it is true or false that the polynomial

$$P^*(x) := (e-1)x + \frac{9}{10}$$

satisfies the best approximation condition

$$\max_{0 \leqslant x \leqslant 1} |e^x - P^*(x)| \leqslant \max_{0 \leqslant x \leqslant 1} |e^x - P(x)|, \quad P \in \pi_1.$$

[*Hint:* Consider first the applicability of Theorem 6.1.1.]

Exercise 6.2 In the proof of Theorem 6.2.1, show that the case

$$f(\xi_0) - P^*(\xi_0) = -||f - P^*||_\infty$$

also yields the contradiction that R is the zero polynomial.

Exercise 6.3 In the proof of Theorem 6.2.2, show that the case $I_1 \in M^-$ also yields the inequality (6.2.41).

Exercise 6.4 In the proof of Theorem 6.3.1, show that the case (6.3.13) also yields the result (6.3.12).

Exercise 6.5 Suppose $f \in C[-a,a]$, where $a > 0$, and, for any non-negative integer n, denote by P_n^* the best uniform polynomial approximation on $[-a,a]$ from π_n to f. Prove that:

(a) If f is an even function, then P_n^* is an even polynomial;

(b) If f is an odd function, then P_n^* is an odd polynomial.

[*Hint:* Apply the uniqueness result of Theorem 6.3.1.]

Exercise 6.6 For the function

$$f(x) = \ln(x+2), \quad x \in [0,2],$$

calculate the linear polynomial $\mathscr{P}_1^* f$, with $\mathscr{P}_1^* : C[0,2] \to \pi_1$ denoting the best uniform linear polynomial approximation operator on the interval $[0,2]$. Compute also the corresponding minimum value $||f - \mathscr{P}_1^* f||_\infty$, and verify that

$$||f - \mathscr{P}_1^* f||_\infty < ||E_1^l||_\infty,$$

with $||E_1^l||_\infty$ evaluated as in both Exercise 2.2 and Exercise 2.6.

Exercise 6.7 Calculate the following minimum values, as well as the corresponding optimal linear polynomials $P_1^* \in \pi_1$:

$$\text{(a)} \min_{P \in \pi_1} \max_{0 \leqslant x \leqslant 1} |e^x - P(x)|; \qquad \text{(b)} \min_{P \in \pi_1} \max_{0 \leqslant x \leqslant 4} |\sqrt{x} - P(x)|.$$

Also, verify that the result for (a) is consistent with the conclusion in Exercise 6.1.

Exercise 6.8 Find the minimum value

$$\min_{\alpha, \beta \in \mathbb{R}} \max_{0 \leqslant x \leqslant 1} |x^3 + \alpha x + \beta|,$$

as well as the corresponding optimal values of α and β.

Exercise 6.9 Find the best uniform polynomial approximation P_4^* on $[-1,2]$ from π_4 to the function

$$f(x) = x^2(x+1)(2x-1)(x-2), \quad x \in [-1,2],$$

as well as the corresponding minimum value.

[*Hint:* Consider first the applicability of Theorem 6.6.1.]

Exercise 6.10 Apply Example 6.6.1 to verify the polynomial $P_{1,2}^*$ in Exercise 4.6.

Exercise 6.11 Find the best uniform polynomial approximation P_2^* on $[-2,2]$ from π_2 to the function

$$f(x) = |x+1|, \quad x \in [-2,2],$$

as well as the corresponding minimum value $||f - P_2^*||_\infty$, with $||\cdot||_\infty$ denoting the maximum norm on $[-2,2]$.

Exercise 6.12 Let T_j denote the Chebyshev polynomial of degree $j \in \mathbb{N}$. Prove that the best uniform polynomial approximation $P_{n,j}^*$ from π_n to T_j with respect to the interval $[-1,1]$ satisfies

$$P_{n,j}^* \text{ is the zero polynomial } \Leftrightarrow n \in \{0, \dots, j-1\}.$$

Exercise 6.13 Provide the details of the inductive proof of the inequality (6.5.10).

Exercise 6.14 Prove the inequality (6.5.31).

Exercise 6.15 For $n \in \mathbb{N}$, denote by $\mathscr{P}_n^I$ the Lagrange polynomial interpolation operator with respect to the Chebyshev interpolation points $\{x_{n,0}^C, \dots, x_{n,n}^C\}$ for the interval $[0,10]$, as obtained from (2.2.31), and let

$$f(x) = \ln(x+2), \quad x \in [0,10].$$

(a) Apply Theorem 6.5.2 to deduce the uniform convergence result
$$||f - \mathscr{P}_n^I f||_\infty \to 0, \quad n \to \infty, \qquad (*)$$
with $|| \cdot ||_\infty$ denoting the maximum norm on $[0,10]$.

(b) Verify that the interpolation error estimate (2.2.36) in Theorem 2.2.4 does not yield the uniform convergence result $(*)$ in (a).

Chapter 7

Orthogonality

For $f \in C[a,b]$ and any non-negative integer n, the best weighted L^2 polynomial approximation $\widetilde{\mathscr{P}}_n^* f = \widetilde{P}_n^*$, as considered in Theorem 4.2.6 and (5.1.6), satisfies the condition

$$||f - \widetilde{\mathscr{P}}_n^* f||_{2,w} \leqslant ||f - P||_{2,w}, \quad P \in \pi_n,$$

or equivalently,

$$\sqrt{\int_a^b w(x)[f(x) - (\widetilde{\mathscr{P}}_n^* f)(x)]^2 dx} \leqslant \sqrt{\int_a^b w(x)[f(x) - P(x)]^2 dx}, \quad P \in \pi_n,$$

where w denotes any weight function on $[a,b]$ that satisfies the conditions (4.2.8), (4.2.9). In this chapter, we derive a full characterisation of $\widetilde{P}_n^*$ in terms of the orthogonality property in the general inner product setting, and study the special case of best weighted L^2 polynomial approximation $\widetilde{\mathscr{P}}_n^* f = \widetilde{P}_n^*$.

7.1 The fundamental characterising condition

Let $(X, \langle \cdot, \cdot \rangle)$ denote any inner product space. If $f, g \in X$ satisfy the condition $\langle f, g \rangle = 0$, we say that f and g are orthogonal with respect to the inner product $\langle \cdot, \cdot \rangle$, and we write $f \perp g$. Observe from (4.2.6) that the zero element 0 of X is orthogonal to each $f \in X$. For $f \in X$ and a subset Y of X, if it holds that

$$\langle f, g \rangle = 0, \quad g \in Y,$$

we say that f is orthogonal to Y, and write $f \perp Y$.

Consider the Euclidean inner product space $(\mathbb{R}^2, \langle \cdot, \cdot \rangle_E)$ and its corresponding normed linear space $(\mathbb{R}^n, ||\cdot||_E)$, as defined in, respectively, Examples 4.1.1 and 4.2.1. Let A denote the one-dimensional subspace of $\mathbb{R}^2$ consisting of all the points on a straight line through the origin $\mathbf{0} = (0,0)$. Then, for any $\mathbf{x} \in \mathbb{R}^2 \setminus A$, we may apply Theorem 4.1.1 and Theorem 4.2.5 to deduce the existence of a unique point $\mathbf{x}^* \in A$ such that

$$||\mathbf{x} - \mathbf{x}^*||_E \leqslant ||\mathbf{x} - \mathbf{y}||_E, \quad \mathbf{y} \in A. \tag{7.1.1}$$

It is geometrically evident that $\mathbf{x}^*$ is uniquely characterised by the orthogonality condition $\mathbf{x} - \mathbf{x}^* \perp A$, that is, from (4.2.7),

$$(\mathbf{x} - \mathbf{x}^*) \cdot \mathbf{y} = 0, \quad \mathbf{y} \in A. \tag{7.1.2}$$

Our following result shows that the characterisation property (7.1.2) of best approximation in $(\mathbb{R}^2, || \cdot ||_E)$ generalizes to all inner product spaces.

Theorem 7.1.1. *For an inner product space* $(X, \langle \cdot, \cdot \rangle)$, *let* $f \in X$, *and suppose* $A \subset X$ *is an approximation set such that* A *is a subspace of* X. *Then* $f^* \in A$ *is a best approximation from* A *to* f *with respect to the norm* $|| \cdot ||$ *defined by* (4.2.15), *if and only if* $f - f^* \perp A$, *that is,*

$$\langle f - f^*, g \rangle = 0, \quad g \in A. \tag{7.1.3}$$

Proof. Suppose $f^* \in A$ satisfies the orthogonality condition (7.1.3), and let $g \in A$. By using (4.2.15), (4.2.3) and (4.2.5), we obtain

$$||f - g||^2 = ||(f - f^*) + (f^* - g)||^2$$

$$= ||f - f^*||^2 + 2\langle f - f^*, f^* - g \rangle + ||f^* - g||^2. \tag{7.1.4}$$

Since A is a subspace of X, it follows that $f^* - g \in A$, according to which (7.1.3) yields

$$\langle f - f^*, f^* - g \rangle = 0. \tag{7.1.5}$$

By substituting (7.1.5) into (7.1.4), we obtain

$$||f - g||^2 = ||f - f^*||^2 + ||f^* - g||^2. \tag{7.1.6}$$

We deduce from (7.1.6) and (4.1.2) that f^* satisfies the condition

$$||f - f^*|| < ||f - g||, \quad \text{for} \quad g \in A, \quad \text{with} \quad g \neq f^*, \tag{7.1.7}$$

and thus f^* is the (unique, in accordance also with Theorem 4.2.4) best approximation from A to f with respect to the norm $|| \cdot ||$ defined by (4.2.15).

Conversely, suppose $f^* \in A$ satisfies the best approximation condition

$$||f - f^*|| \leqslant ||f - g||, \quad g \in A. \tag{7.1.8}$$

If $f \in A$, then $f^* = f$, and thus, from (4.2.6),

$$\langle f - f^*, g \rangle = \langle 0, g \rangle = 0, \quad g \in A,$$

which shows that (7.1.3) holds for this case.

Suppose next $f \notin A$, so that $f^* \neq f$, and thus, from (4.1.1) and (4.1.2), $\|f - f^*\| > 0$. Using a proof by contradiction, we proceed to prove that if the condition (7.1.3) is not satisfied then there exists an element $g^* \in A$ such that

$$0 < \|f - g^*\| < \|f - f^*\|, \tag{7.1.9}$$

which contradicts (7.1.8), and would thereby conclude our proof.

Suppose therefore that (7.1.3) does not hold, that is, there exists an element $h \in A$ such that

$$\delta := \langle f - f^*, h \rangle \neq 0, \tag{7.1.10}$$

and let the function $u : \mathbb{R} \to \mathbb{R}$ be defined by

$$u(\lambda) := \|f - (f^* + \lambda h)\|^2, \quad \lambda \in \mathbb{R}. \tag{7.1.11}$$

By using (7.1.11), (4.2.15), (4.2.3), (4.2.4) and (4.2.5), we deduce that

$$u(\lambda) = \langle (f - f^*) - \lambda h, (f - f^*) - \lambda h \rangle$$
$$= \|f - f^*\|^2 - 2\lambda \langle f - f^*, h \rangle + \lambda^2 \|h\|^2 = \|h\|^2 \lambda^2 - 2\delta\lambda + \|f - f^*\|^2,$$
$$\tag{7.1.12}$$

from (7.1.10). Since also (7.1.10) and (4.2.6) imply $h \neq 0$, so that (4.1.2) yields $\|h\|^2 > 0$, we deduce from (7.1.12) that u is a quadratic polynomial in λ, with minimum value

$$u\left(\frac{\delta}{\|h\|^2}\right) = \|f - f^*\|^2 - \frac{\delta^2}{\|h\|^2}. \tag{7.1.13}$$

Now define

$$g^* := f^* + \frac{\delta}{\|h\|^2} h, \tag{7.1.14}$$

according to which, since A is a subspace of X, we have $g^* \in A$. Moreover, it follows from (7.1.14), (7.1.11) and (7.1.13) that

$$0 < \|f - g^*\| = \sqrt{\|f - f^*\|^2 - \frac{\delta^2}{\|h\|^2}} < \|f - f^*\|,$$

which yields (7.1.9), as required.   ∎

We next deduce the following Pythagoras rule for inner product spaces from Theorem 7.1.1 and its proof.

Theorem 7.1.2. *For a normed linear space* $(X, \|\cdot\|)$ *as in Theorem 4.2.2, let* $f \in X$, *and suppose* $A \subset X$ *is an approximation set such that* A *is a subspace of* X, *and such that there exists a best approximation* f^* *from* A *to* f *with respect to the norm* $\|\cdot\|$ *defined by* (4.2.15). *Then*

$$\|f - g\|^2 = \|f - f^*\|^2 + \|f^* - g\|^2, \quad g \in A, \tag{7.1.15}$$

with, in particular,

$$\|f\|^2 = \|f - f^*\|^2 + \|f^*\|^2. \tag{7.1.16}$$

Proof. Let $f^* \in A$ be such that the best approximation condition (7.1.8) is satisfied. It follows from Theorem 7.1.1 that f^* satisfies the orthogonality condition (7.1.3). Hence we may follow the steps in the first part of the proof of Theorem 7.1.1 to deduce (7.1.6), which is precisely (7.1.15).

Since A is a subspace of X, we know that the zero element 0 belongs to A, so that we may set $g = 0$ in (7.1.15) to obtain (7.1.16). ∎

We proceed to show how the results of Theorem 7.1.1 and Theorem 7.1.2 may be applied to prove the following properties of the best approximation operator with respect to the norm generated by an inner product space as in Theorem 4.2.2.

Theorem 7.1.3. *For a normed linear space* $(X, \|\cdot\|)$ *as in Theorem 4.2.2, let* $A \subset X$ *be an approximation set such that* A *is a subspace of* X, *and such that, for each* $f \in X$, *there exists a (unique, according to Theorem 4.2.4) best approximation* f^* *from* A *to* f *with respect to the norm* $\|\cdot\|$ *defined by* (4.2.15). *Then the corresponding best approximation operator* $\mathscr{A}^* : X \to A$ *defined by*

$$\mathscr{A}^* f := f^*, \quad f \in X, \tag{7.1.17}$$

satisfies the following:

(a) $\mathscr{A}^*$ *is linear;*

(b) $\mathscr{A}^*$ *is exact on* A;

(c) $\mathscr{A}^*$ *is bounded, with corresponding operator norm*

$$\|\mathscr{A}^*\| = 1. \tag{7.1.18}$$

Proof. (a) For any $f, \widetilde{f} \in X$ and $\lambda, \mu \in \mathbb{R}$, define

$$h := \lambda f + \mu \widetilde{f} \quad ; \quad h^* := \lambda(\mathscr{A}^* f) + \mu(\mathscr{A}^* \widetilde{f}), \tag{7.1.19}$$

so that, since A is a subspace of X, and $\mathscr{A}^* : X \to A$, we have $h^* \in A$. It then follows from (7.1.19), (4.2.3) and (4.2.4), together with Theorem 7.1.1, that, for any $g \in A$,

$$\langle h - h^*, g \rangle = \langle \lambda (f - \mathscr{A}^* f) + \mu (\widetilde{f} - \mathscr{A}^* \widetilde{f}), g \rangle$$

$$= \lambda \langle f - \mathscr{A}^* f, g \rangle + \mu \langle \widetilde{f} - \mathscr{A}^* \widetilde{f}, g \rangle = 0 + 0 = 0,$$

and thus, again from Theorem 7.1.1,

$$h^* = \mathscr{A}^* h,$$

that is, from (7.1.19),

$$\lambda (\mathscr{A}^* f) + \mu (\mathscr{A}^* \widetilde{f}) = \mathscr{A}^* (\lambda f + \mu \widetilde{f}),$$

according to which $\mathscr{A}^*$ satisfies the characterising property (5.1.8) of linearity for approximation operators.

(b) If $f \in A$, then (4.2.1) and (4.2.2) yield

$$\|f - f\| = 0 \leqslant \|f - g\|, \quad g \in A,$$

and it follows that

$$\mathscr{A}^* f = f, \quad f \in A, \tag{7.1.20}$$

that is, $\mathscr{A}^*$ is exact on A.

(c) Let $f \in X$, with $f \neq 0$. By applying (7.1.17) and (7.1.16), we obtain

$$\|\mathscr{A}^* f\| = \sqrt{\|f\|^2 - \|f - \mathscr{A}^* f\|^2} \leqslant \|f\|,$$

and thus

$$\frac{\|\mathscr{A}^* f\|}{\|f\|} \leqslant 1, \tag{7.1.21}$$

according to which $\{\|\mathscr{A}^* f\|/\|f\| : f \in X, f \neq 0\}$ is a bounded set, that is, $\mathscr{A}^*$ is bounded. Moreover, the operator norm definition (5.2.2), together with (7.1.21), yields

$$\|\mathscr{A}^*\| \leqslant 1. \tag{7.1.22}$$

By applying also (7.1.20) and Theorem 5.2.1, we deduce that

$$\|\mathscr{A}^*\| \geqslant 1,$$

which, together with (7.1.22), then gives the desired result (7.1.18). ∎

Note from (7.1.18) and Theorem 5.2.1 that, subject to the constraint of exactness on a nontrivial subset of X, the best approximation operator $\mathscr{A}^*$ of Theorem 7.1.3 possesses the favourable property of having the optimally small operator norm value of one.

The choices $(X, ||\cdot||) = (C[a,b], ||\cdot||_{2,w})$ and $A = \pi_n$ in Theorem 7.1.3 yield $\mathscr{A}^* = \widetilde{\mathscr{P}}_n^*$, as defined by (5.1.6), from which we immediately deduce the following result, the first part of which extends Theorem 5.1.1(a).

Theorem 7.1.4. *For any bounded interval* $[a,b]$ *and non-negative integer* n, *the best weighted* L^2 *approximation operator* $\widetilde{\mathscr{P}}_n^* : C[a,b] \to \pi_n$, *as defined by (5.1.6), satisfies:*

(a) $\widetilde{\mathscr{P}}_n^*$ *is linear;*

(b) $\widetilde{\mathscr{P}}_n^*$ *is bounded, with corresponding Lebesgue constant*

$$||\widetilde{\mathscr{P}}_n^*||_{2,w} = 1. \tag{7.1.23}$$

The following formulations of the approximation error in the context of Theorem 7.1.2 are useful in applications.

Theorem 7.1.5. *In Theorem* 7.1.2, *the corresponding approximaton error satisfies*

$$||f - f^*|| = \sqrt{\langle f, f - f^* \rangle}, \tag{7.1.24}$$

and

$$||f - f^*|| = \sqrt{||f||^2 - ||f^*||^2}. \tag{7.1.25}$$

Proof. Let $f \in X$. It follows from (4.2.15), (4.2.3), (4.2.4) and (4.2.5) that

$$||f - f^*||^2 = \langle f - f^*, f - f^* \rangle = \langle f, f - f^* \rangle - \langle f^*, f - f^* \rangle$$

$$= \langle f, f - f^* \rangle - \langle f - f^*, f^* \rangle. \tag{7.1.26}$$

Since $f^* \in A$, it follows from Theorem 7.1.1 that

$$\langle f - f^*, f^* \rangle = 0,$$

which, together with (7.1.26), then implies (7.1.24). Finally, we note that (7.1.25) is an immediate consequence of (7.1.16) in Theorem 7.1.2. ∎

7.2   The Gram matrix

Let $(X, ||\cdot||)$ denote a normed linear space as in Theorem 4.2.2, that is, with norm $||\cdot||$ generated by an inner product as in (4.2.15), and suppose $A \subset X$ is an approximation set such that A is a finite-dimensional subspace of X, with $\dim(A) = d \in \mathbb{N}$. As was done in Theorem 4.2.5, it follows from Theorem 4.1.1 and Theorem 4.2.4 that we may define, as in Theorem 7.1.3, the corresponding best approximation operator $\mathscr{A}^* : X \to A$ by means of (7.1.17). We proceed to demonstrate how an application of the characterisation result of Theorem 7.1.1 yields, for any $f \in X$, a matrix-inversion method for the construction of $\mathscr{A}^* f$.

We shall rely on the following result.

Theorem 7.2.1. *For an inner product space* $(X, \langle \cdot, \cdot \rangle)$, *suppose A is a finite-dimensional subspace of X, with* $\dim(A) = d \in \mathbb{N}$. *Let* $f \in X$, *and suppose* $\{f_1, \ldots, f_d\}$ *is a basis for A. Then* $f \perp A$, *that is,*

$$\langle f, g \rangle = 0, \quad g \in A, \tag{7.2.1}$$

if and only if

$$\langle f, f_j \rangle = 0, \quad j = 1, \ldots, d. \tag{7.2.2}$$

Proof. Suppose (7.2.1) holds. Since $f_j \in A, j = 1, \ldots, d$, it then follows that (7.2.2) is satisfied.

Conversely, suppose (7.2.2) holds, and let $g \in A$. Since $\{f_1, \ldots, f_d\}$ is a basis for A, it follows that

$$g = \sum_{j=1}^{d} c_j f_j \tag{7.2.3}$$

for some coefficient sequence $\{c_1, \ldots, c_d\} \subset \mathbb{R}$. By applying (7.2.3), (4.2.3), (4.2.4) and (7.2.2), we obtain

$$\langle f, g \rangle = \sum_{j=1}^{d} c_j \langle f, f_j \rangle = 0,$$

that is, (7.2.1) is satisfied. ∎

For any basis $\{f_1, \ldots, f_d\}$ of A, as in Theorem 7.2.1, there exists a unique coefficient sequence $\{c_1^*, \ldots, c_d^*\} \subset \mathbb{R}$ such that

$$\mathscr{A}^* f = \sum_{k=1}^{d} c_k^* f_k. \tag{7.2.4}$$

Moreover, Theorem 7.1.1 shows that $f - \mathscr{A}^* f \perp A$, which, according to (7.2.4), together with Theorem 7.2.1, is equivalent to

$$\langle f - \sum_{k=1}^{d} c_k^* f_k, f_j \rangle = 0, \quad j = 1, \ldots, d,$$

or equivalently, from (4.2.3), (4.2.4) and (4.2.5),

$$\sum_{k=1}^{d} \langle f_j, f_k \rangle c_k^* = \langle f, f_j \rangle, \quad j = 1, \ldots, d. \tag{7.2.5}$$

Hence $\mathbf{c}^* := (c_1^*, \ldots, c_d^*)$ is the unique solution in $\mathbb{R}^d$ of the linear system

$$G\mathbf{c}^* = \mathbf{f}, \tag{7.2.6}$$

where G is the $d \times d$ matrix

$$G := \begin{bmatrix} \langle f_1, f_1 \rangle & \langle f_1, f_2 \rangle & \cdots & \langle f_1, f_d \rangle \\ \langle f_2, f_1 \rangle & \langle f_2, f_2 \rangle & \cdots & \langle f_2, f_d \rangle \\ \vdots & \vdots & & \vdots \\ \langle f_d, f_1 \rangle & \langle f_d, f_2 \rangle & \cdots & \langle f_d, f_d \rangle \end{bmatrix}, \tag{7.2.7}$$

and with $\mathbf{f} \in \mathbb{R}^d$ denoting the (column) vector

$$\mathbf{f} := [\langle f, f_1 \rangle, \ldots, \langle f, f_d \rangle]^T. \tag{7.2.8}$$

The matrix G in (7.2.7) is called the Gram matrix corresponding to the sequence $\{f_1, \ldots, f_d\}$. Note from (7.2.7) and (4.2.5) that G is a symmetrix matrix.

Since the $d \times d$ linear system (7.2.6) is uniquely solved by $\mathbf{c}^* = (c_1^*, \ldots, c_d^*)$, a standard result from linear algebra implies that the Gram matrix G in (7.2.7) is invertible, which, together with (7.2.4), then yields the following formulation of the best approximation $\mathscr{A}^* f$.

Theorem 7.2.2. *For a normed linear space $(X, ||\cdot||)$ as in Theorem 4.2.2, let $A \subset X$ be an approximation set such that A is a finite-dimensional subspace of X, with $\dim(A) = d \in \mathbb{N}$. Also, denote by $\{f_1, \ldots, f_d\}$ any basis for A, and let the best approximation operator $\mathscr{A}^*$: $X \to A$, with respect to the norm defined by (4.2.15), be given by (7.1.17) in Theorem 7.1.3. Then the Gram matrix G, as given by (7.2.7), is invertible, and*

$$\mathscr{A}^* f = \sum_{j=1}^{d} (G^{-1}\mathbf{f})_j f_j, \quad f \in X, \tag{7.2.9}$$

where the (column) vector $\mathbf{f} \in \mathbb{R}^d$ is defined by (7.2.8).

An advantage of best approximation from a finite-dimensional subspace A of X with respect to a norm generated by an inner product as in (4.2.15), is that the system (7.2.6) to be solved is a linear one, whereas the use of specifically the maximum norm was shown in Examples 6.6.2 and 6.6.3 to depend on solving non-linear systems.

Example 7.2.1. Consider the case where, in Theorem 7.2.2, we choose $X = C[0,1]$, $||\cdot|| = ||\cdot||_2$ as in (4.2.17), $A = \pi_1$, $\{f_1, f_2\} = \{1, x\}$ and $f(x) = x^2$. Observe that here $\mathscr{A}^* =$

$\widetilde{\mathscr{P}}_1^* : C[a,b] \to \pi_1$, as defined in (5.1.6). By using (4.2.12) and (4.2.5), we obtain the inner products

$$\langle f_1, f_1 \rangle_2 = \int_0^1 (1)(1)dx = \int_0^1 dx = 1;$$

$$\langle f_1, f_2 \rangle_2 = \int_0^1 (1)(x)dx = \int_0^1 xdx = \tfrac{1}{2};$$

$$\langle f_2, f_1 \rangle_2 = \langle f_1, f_2 \rangle_2 = \tfrac{1}{2};$$

$$\langle f_2, f_2 \rangle_2 = \int_0^1 (x)(x)dx = \int_0^1 x^2 dx = \tfrac{1}{3},$$

which can be substituted into (7.2.7) to give the Gram matrix

$$G = \begin{bmatrix} 1 & \tfrac{1}{2} \\ \tfrac{1}{2} & \tfrac{1}{3} \end{bmatrix}. \tag{7.2.10}$$

Also, from (4.2.12),

$$\langle f, f_1 \rangle_2 = \int_0^1 (x^2)(1)dx = \int_0^1 x^2 dx = \tfrac{1}{3};$$

$$\langle f, f_2 \rangle_2 = \int_0^1 (x^2)(x)dx = \int_0^1 x^3 dx = \tfrac{1}{4},$$

which, together with (7.2.8), yield

$$\mathbf{f} = \begin{bmatrix} \tfrac{1}{3} \\ \tfrac{1}{4} \end{bmatrix}. \tag{7.2.11}$$

From (7.2.10) we now calculate the inverse matrix

$$G^{-1} = \begin{bmatrix} 4 & -6 \\ -6 & 12 \end{bmatrix}. \tag{7.2.12}$$

It follows from (7.2.12) and (7.2.11) that

$$G^{-1}\mathbf{f} = \begin{bmatrix} 4 & -6 \\ -6 & 12 \end{bmatrix} \begin{bmatrix} \tfrac{1}{3} \\ \tfrac{1}{4} \end{bmatrix} = \begin{bmatrix} -\tfrac{1}{6} \\ 1 \end{bmatrix}. \tag{7.2.13}$$

Finally, we use (7.2.13) in the formula (7.2.9) to obtain

$$(\widetilde{\mathscr{P}}_1^* f)(x) = -\tfrac{1}{6} + x. \tag{7.2.14}$$

To compute the corresponding best approximation error, it is convenient to use the formula (7.1.24), together with (4.2.12), according to which

$$\|f - \widetilde{\mathscr{P}}_1^* f\|_2 = \sqrt{\langle f, f - \widetilde{\mathscr{P}}_1^* f \rangle_2} = \sqrt{\int_0^1 x^2 \left[x^2 - (x - \tfrac{1}{6}) \right] dx}$$

$$= \sqrt{\int_0^1 \left(x^4 - x^3 + \tfrac{1}{6}x^2\right) dx} = \sqrt{\tfrac{1}{5} - \tfrac{1}{4} + \tfrac{1}{18}} = \tfrac{1}{6\sqrt{5}}.$$

$$(7.2.15)$$

∎

It is interesting to compare the two best approximation polynomials $\mathscr{P}_1^* f$ and $\widetilde{\mathscr{P}}_1^* f$ from Examples 6.6.1 and 7.2.1, as given, according to (6.6.14) and (7.2.14), by

$$(\mathscr{P}_1^* f)(x) = x - \tfrac{1}{8} \quad ; \quad (\widetilde{\mathscr{P}}_1^* f)(x) = x - \tfrac{1}{6}. \tag{7.2.16}$$

According to the formula (7.2.9) in Theorem 7.2.2, the calculation of the best approximation $\mathscr{A}^* f$ involves the computation of the inverse G^{-1} of the Gram matrix G, as given in
(7.2.7). We proceed in the next section to show that it is always possible to choose a basis
$\{f_1, \ldots, f_d\}$ for A such that G is a diagonal matrix, in which case the calculation of the
inverse matrix G^{-1} is trivial.

7.3 Orthogonal bases

For an inner product space $(X, \langle \cdot, \cdot \rangle)$, let $A \subset X$ denote a finite-dimensional subspace of X,
with $\dim(A) = d \in \mathbb{N}$. If, for $d \geqslant 2$, a basis $\{f_1, \ldots, f_d\}$ of A satisfies the condition

that is,
$$\left. \begin{array}{l} f_j \perp f_k, \quad j, k = 1, \ldots, d, \, j \neq k, \\[2mm] \langle f_j, f_k \rangle = 0, \, j, k = 1, \ldots, d, \, j \neq k, \end{array} \right\} \tag{7.3.1}$$

we say that $\{f_1, \ldots, f_d\}$ is an orthogonal basis for A, in which case it follows that the
corresponding Gram matrix G in (7.2.7) is a diagonal matrix, with inverse

$$G^{-1} = \begin{bmatrix} \dfrac{1}{||f_1||^2} & 0 & \cdots & 0 \\[3mm] 0 & \dfrac{1}{||f_2||^2} & \cdots & 0 \\[3mm] \vdots & \vdots & & \vdots \\[3mm] 0 & 0 & & \dfrac{1}{||f_d||^2} \end{bmatrix}, \tag{7.3.2}$$

in terms of the norm $|| \cdot ||$ defined by (4.2.15). If $d = 1$, any non-zero element $f_1 \in A$ yields
the orthogonal basis $\{f_1\}$ of the one-dimensional subspace A of X, and G^{-1} is the 1×1

matrix given by $[||f_1||^{-2}]$. Hence, for the case where, in Theorem 7.2.2, $\{f_1,\ldots,f_d\}$ is an orthogonal basis for A, the following explicit formulation for $\mathscr{A}^*f$ is a direct consequence of (7.2.9), (7.3.2) and (7.2.8).

Theorem 7.3.1. *In Theorem 7.2.2, suppose that $\{f_1,\ldots,f_d\}$ is an orthogonal basis for A. Then*

$$\mathscr{A}^*f = \sum_{j=1}^{d} \frac{\langle f,f_j\rangle}{||f_j||^2}\, f_j, \quad f \in X. \tag{7.3.3}$$

We proceed to show that, for any finite-dimensional subspace A of an inner product space $(X,\langle\cdot,\cdot\rangle)$, there exists an orthogonal basis for A. We shall rely on the following preliminary result.

Theorem 7.3.2. *For an inner product space $(X,\langle\cdot,\cdot\rangle)$, suppose $\{f_1,\ldots,f_d\} \subset X$, with $d \geqslant 2$, is a sequence such that the orthogonality condition (7.3.1) is satisfied, and $f_j \neq 0, j = 1,\ldots,d$. Then $\{f_1,\ldots,f_d\}$ is a linearly independent set.*

Proof. Let the coefficient sequence $\{c_1,\ldots,c_d\}$ be such that

$$\sum_{j=1}^{d} c_j f_j = 0. \tag{7.3.4}$$

For any integer $k \in \{1,\ldots,d\}$, it follows from (7.3.4), (4.2.6), (4.2.3), (4.2.4) and (7.3.1) that

$$0 = \left\langle \sum_{j=1}^{d} c_j f_j, f_k \right\rangle = \sum_{j=1}^{d} c_j \langle f_j, f_k\rangle = c_k \langle f_k, f_k\rangle,$$

and thus, since also $f_k \neq 0$, so that (4.2.2) implies $\langle f_k, f_k\rangle \neq 0$, we have $c_k = 0$. Hence $c_j = 0, j = 1,\ldots,d$, which completes our proof. ∎

For any inner product space $(X,\langle\cdot,\cdot\rangle)$, let $A \subset X$ denote a finite-dimensional subspace of X, with $\dim(A) = d \geqslant 2$, and suppose $\{g_1,\ldots,g_d\}$ is any basis for A. We proceed to show how the results of Theorems 7.3.1 and 7.3.2 can be applied to explicitly construct from $\{g_1,\ldots,g_d\}$, by means of a recursive method called the Gram-Schmidt procedure, an orthogonal basis $\{f_1,\ldots,f_d\}$ for A.

Since $\{g_1,\ldots,g_d\}$ is a basis for A, we know that $\{g_1,\ldots,g_d\}$ is a linearly independent set, and thus $\{g_1,\ldots,g_j\}$ is a linearly independent set for any $j \in \{1,\ldots,d\}$. Hence, if we define

$$A_j := \text{span}\{g_1,\ldots,g_j\}, \quad j = 1,\ldots,d, \tag{7.3.5}$$

it follows that

$$\dim(A_j) = j, \quad j = 1, \ldots, d, \tag{7.3.6}$$

and, for each $j = 1, \ldots, d$, $\{g_1, \ldots, g_j\}$ is a basis for A_j. Moreover, the definition (7.3.5) implies the nesting property

$$A_j \subset A_{j+1}, \quad j = 1, \ldots, d-1, \tag{7.3.7}$$

as well as

$$A_d = \mathrm{span}\{g_1, \ldots, g_d\} = A. \tag{7.3.8}$$

Moreover,

$$g_j \notin A_{j-1}, \quad j = 2, \ldots, d, \tag{7.3.9}$$

since if not, that is $g_j \in A_{j-1}$ for some integer $j \in \{2, \ldots, d\}$, then, according to (7.3.5), g_j is a linear combination of the sequence $\{g_1, \ldots, g_{j-1}\}$, which contradicts the linear independence of the sequence $\{g_1, \ldots, g_j\}$.

Now define

$$f_1 := g_1, \tag{7.3.10}$$

which, together with (7.3.5) and (7.3.6), implies that $\{f_1\}$ is an orthogonal basis for the one-dimensional space A_1. By applying the formula (7.3.3) in Theorem 7.3.1, we deduce that the (unique) best approximation f_1^* from A_1 to g_2, with respect to the norm $\|\cdot\|$ in (4.2.15), is given by

$$f_1^* = \frac{\langle g_2, f_1 \rangle}{\|f_1\|^2} f_1, \tag{7.3.11}$$

with, moreover, from Theorem 7.1.1,

$$\langle g_2 - f_1^*, g \rangle = 0, \quad g \in A_1,$$

and thus

$$\langle g_2 - f_1^*, f_1 \rangle = 0. \tag{7.3.12}$$

Let

$$f_2 := g_2 - f_1^*, \tag{7.3.13}$$

according to which (7.3.12) gives

$$\langle f_2, f_1 \rangle = 0. \tag{7.3.14}$$

Observe that $f_2 \neq 0$, since if not, then (7.3.13) yields $g_2 = f_1^* \in A_1$, which contradicts
(7.3.9). Also, $f_1 \neq 0$, as follows from (7.3.10), together with the fact that $g_1 \neq 0$, since
$\{g_1, \ldots, g_d\}$ is a basis for A. Hence, since the orthogonality condition (7.3.14) is satisfied,
with, moreover, $f_1 \neq 0$ and $f_2 \neq 0$, we may apply Theorem 7.3.2 to deduce that $\{f_1, f_2\}$ is a
linearly independent set. Since, moreover, $\{f_1, f_2\} \subset A_2$, as follows from (7.3.10), (7.3.13),
(7.3.11) and (7.3.5), and since also $\dim(A_2) = 2$, from (7.3.6), a standard result from linear
algebra implies that $\{f_1, f_2\}$ is a basis for A_2. Hence the definitions (7.3.10) and (7.3.13)
have yielded an orthogonal basis $\{f_1, f_2\}$ for A_2, and thus

$$A_2 = \text{span}\{f_1, f_2\}. \tag{7.3.15}$$

Also,

$$f_2 \in A_2 \setminus A_1, \tag{7.3.16}$$

since the assumption $f_2 \in A_1 = \text{span}\{f_1\}$ contradicts the linear independence of the set
$\{f_1, f_2\}$. Moreover, the orthogonality condition

$$f_2 \perp A_1, \tag{7.3.17}$$

is satisfied, by virtue of (7.3.14), together with the fact that $A_1 = \text{span}\{f_1\}$.

If $d \geqslant 3$, we continue similarly, by first deducing from the fact that $\{f_1, f_2\}$ is an orthogonal
basis for A_2, together with (7.3.3) in Theorem 7.3.1, that the best approximation f_2^* from
A_2 to g_3, with respect to the norm $\| \cdot \|$ defined by (4.2.15), is given by

$$f_2^* = \frac{\langle g_3, f_1 \rangle}{\|f_1\|^2} f_1 + \frac{\langle g_3, f_2 \rangle}{\|f_2\|^2} f_2, \tag{7.3.18}$$

with, moreover, from Theorem 7.1.1,

$$\langle g_3 - f_2^*, g \rangle = 0, \quad g \in A_2,$$

according to which the definition

$$f_3 := g_3 - f_2^*, \tag{7.3.19}$$

together with (7.3.15), yields

$$\langle f_3, f_1 \rangle = 0 \quad ; \quad \langle f_3, f_2 \rangle = 0. \tag{7.3.20}$$

Also, $f_3 \neq 0$, for if not, then (7.3.19) yields $g_3 - f_2^* \in A_2$, from (7.3.18) and (7.3.15), which
contradicts (7.3.9). Hence, from (7.3.20) and (7.3.14), we have

$$\langle f_j, f_k \rangle = 0, \quad j,k = 1,2,3; \quad j \neq k, \tag{7.3.21}$$

with also $f_j \neq 0, j = 1,2,3$, so that we may appeal to Theorem 7.3.2 to deduce that
$\{f_1, f_2, f_3\}$ is a linearly independent set. Moreover, since $f_3 \in A_3$, as follows from

(7.3.19), (7.3.5), (7.3.18) and (7.3.15), together with $A_2 \subset A_3$, as given in (7.3.7), with also $\{f_1, f_2\} \subset A_2 \subset A_3$, from (7.3.15), and $\dim(A_3) = 3$, from (7.3.6), it follows that $\{f_1, f_2, f_3\}$ is a basis for A_3. Hence the definitions (7.3.19), (7.3.13) and (7.3.10) have yielded an orthogonal basis $\{f_1, f_2, f_3\}$ for A_3, and thus

$$A_3 = \text{span}\{f_1, f_2, f_3\}. \tag{7.3.22}$$

Also,

$$f_3 \in A_3 \setminus A_2, \tag{7.3.23}$$

since the assumption $f_3 \in A_2 = \text{span}\{f_1, f_2\}$, as in (7.3.15), contradicts the linear independence of the set $\{f_1, f_2, f_3\}$. Also (7.3.20), together with (7.3.14) and (7.3.15), implies the orthogonality condition

$$f_3 \perp A_2. \tag{7.3.24}$$

Repeated applications for $d \geqslant 4$, $d \geqslant 5, \ldots$, of the above procedure establish the following result, as based on (7.3.5)–(7.3.24).

Theorem 7.3.3. (Gram-Schmidt procedure) *For any inner product space* $(X, \langle \cdot, \cdot \rangle)$*, let A denote a finite-dimensional subspace of X, with* $\dim(A) = d \geqslant 2$*. Suppose, moreover, that* $\{g_1, \ldots, g_d\}$ *is a basis for A, and let*

$$A_j := \text{span}\{g_1, \ldots, g_j\}, \quad j = 1, \ldots, d. \tag{7.3.25}$$

Then the sequence $\{f_1, \ldots, f_d\}$ *defined recursively by*

$$\left.\begin{array}{l} f_1 := g_1; \\[2mm] f_j := g_j - \displaystyle\sum_{k=1}^{j-1} \frac{\langle g_j, f_k \rangle}{\|f_k\|^2} f_k, \quad j = 2, \ldots, d, \end{array}\right\} \tag{7.3.26}$$

with $\| \cdot \|$ *denoting the norm defined by (4.2.15), satisfies the following:*

(a)

$$f_j \in \begin{cases} A_1, & \text{if } j = 1; \\[2mm] A_j \setminus A_{j-1}, & \text{if } j = 2, \ldots, d. \end{cases} \tag{7.3.27}$$

(b)

$$f_j \perp A_{j-1}, \quad j = 2, \ldots, d. \tag{7.3.28}$$

(c) *For* $j = 1, \ldots, d$*, the sequence* $\{f_1, \ldots, f_j\}$ *is an orthogonal basis for* A_j*.*

(d) *The sequence* $\{f_1, \ldots, f_d\}$ *is an orthogonal basis for A.*

Hence the Gram-Schmidt procedure (7.3.26) uses any basis $\{g_1,\ldots,g_d\}$ of a finite-dimensional subspace A of an inner product space $(X,\langle\cdot,\cdot\rangle)$ to recursively construct an orthogonal basis $\{f_1,\ldots,f_d\}$ of A, and thereby proving the existence of an orthogonal basis for any such finite-dimensional subspace A of X.

Example 7.3.1. In Theorem 7.3.3, choose $X = C[0,1], ||\cdot|| = ||\cdot||_2$, as given in (4.2.17), $A = \pi_2$, so that $d = 3$, and $\{g_1,g_2,g_3\} = \{1,x,x^2\}$. By applying the first line in the Gram-Schmidt procedure (7.3.26), we obtain

$$f_1(x) = 1.$$

Next, we apply (4.2.12) and (4.2.17) to calculate

$$\langle g_2,f_1\rangle_2 = \int_0^1 (x)(1)dx = \int_0^1 xdx = \tfrac{1}{2},$$

and

$$(||f_1||_2)^2 = \int_0^1 (1)^2 dx = \int_0^1 dx = 1,$$

and thus, from the second line of (7.3.26),

$$f_2(x) = x - \tfrac{1/2}{1}(1) = x - \tfrac{1}{2}.$$

By using again (4.2.12) and (4.2.17), we obtain

$$\langle g_3,f_1\rangle_2 = \int_0^1 (x^2)(1)dx = \int_0^1 x^2 dx = \tfrac{1}{3};$$

$$\langle g_3,f_2\rangle_2 = \int_0^1 (x^2)\left(x-\tfrac{1}{2}\right)dx = \int_0^1 \left(x^3 - \tfrac{1}{2}x^2\right)dx = \tfrac{1}{4} - \tfrac{1}{6} = \tfrac{1}{12};$$

$$(||f_2||_2)^2 = \int_0^1 \left(x-\tfrac{1}{2}\right)^2 dx = \int_0^1 \left(x^2 - x + \tfrac{1}{4}\right)dx = \tfrac{1}{3} - \tfrac{1}{2} + \tfrac{1}{4} = \tfrac{1}{12},$$

and thus, from (7.3.26),

$$f_2(x) = x^2 - \tfrac{1/3}{1}(1) - \tfrac{1/12}{1/12}\left(x-\tfrac{1}{2}\right) = x^2 - \tfrac{1}{3} - \left(x-\tfrac{1}{2}\right) = x^2 - x + \tfrac{1}{6}.$$

It follows from Theorem 7.3.3 that

$$\{f_1,f_2\} = \{1, x-\tfrac{1}{2}\} \tag{7.3.29}$$

is an orthogonal basis for π_1, whereas

$$\{f_1,f_2,f_3\} = \{1, x-\tfrac{1}{2}, x^2 - x + \tfrac{1}{6}\} \tag{7.3.30}$$

is an orthogonal basis for π_2, with respect to the L^2 norm $||\cdot||_2$ on the interval $[0,1]$.

Hence we may use the orthogonal basis (7.3.29) to calculate, for $f(x) = x^2$, the best approximation $\widetilde{\mathscr{P}}_1^* f$ of Example 7.2.1 by means of the formula (7.3.3) of Theorem 7.3.1 . Since (4.2.12) and (7.3.29) yield the inner products

$$\langle f, f_1 \rangle_2 = \int_0^1 (x^2)(1)dx = \int_0^1 x^2 dx = \tfrac{1}{3};$$

$$\langle f, f_2 \rangle_2 = \int_0^1 (x^2)\left(x - \tfrac{1}{2}\right)dx = \int_0^1 \left(x^3 - \tfrac{1}{2}x^2\right)dx = \tfrac{1}{4} - \tfrac{1}{6} = \tfrac{1}{12},$$

we may apply (7.3.3) and (7.3.29) to obtain

$$(\widetilde{\mathscr{P}}_1^* f)(x) = \tfrac{1/3}{1}(1) + \tfrac{1/12}{1/12}\left(x - \tfrac{1}{2}\right) = \tfrac{1}{3} + \left(x - \tfrac{1}{2}\right) = x - \tfrac{1}{6},$$

which agrees with (7.2.14) in Example 7.2.1. ∎

Next we prove, analogously to Theorem 7.1.5, a formulation of the approximation error $\|f - \mathscr{A}^* f\|$ in terms of an orthogonal basis.

Theorem 7.3.4. *In Theorem* 7.3.1, *the corresponding approximation error satisfies*

$$\|f - \mathscr{A}^* f\| = \sqrt{\|f\|^2 - \sum_{j=1}^{d}\left[\frac{\langle f, f_j \rangle}{\|f_j\|}\right]^2}, \quad f \in X. \tag{7.3.31}$$

Proof. Let $f \in X$. By applying (4.2.15), the formula (7.3.3) in Theorem 7.3.1, as well as (4.2.3) and (4.2.4), and finally (7.3.1), we obtain

$$\|\mathscr{A}^* f\|^2 = \left\langle \sum_{j=1}^{d}\frac{\langle f, f_j \rangle}{\|f_j\|^2}f_j, \sum_{k=1}^{d}\frac{\langle f, f_k \rangle}{\|f_k\|^2}f_k \right\rangle = \sum_{j=1}^{d}\frac{\langle f, f_j \rangle}{\|f_j\|^2}\left[\sum_{k=1}^{d}\frac{\langle f, f_k \rangle}{\|f_k\|^2}\langle f_j, f_k \rangle\right]$$

$$= \sum_{j=1}^{d}\frac{\langle f, f_j \rangle}{\|f_j\|^2}\left[\frac{\langle f, f_j \rangle}{\|f_j\|^2}\langle f_j, f_j \rangle\right]$$

$$= \sum_{j=1}^{d}\left[\frac{\langle f, f_j \rangle}{\|f_j\|}\right]^2. \tag{7.3.32}$$

The formula (7.3.31) is then an immediate consequence of (7.1.25) in Theorem 7.1.5, as well as (7.1.17) in Theorem 7.1.3, together with (7.3.32). ∎

Example 7.3.2. As in Examples 7.2.1 and 7.3.1, let $f(x) = x^2$. Then, by using (4.2.17), we obtain

$$(\|f\|_2)^2 = \int_0^1 (x^2)^2 dx = \int_0^1 x^4 dx = \tfrac{1}{5},$$

whereas, from the calculations in Example 7.3.1,

$$\left[\frac{\langle f, f_1 \rangle_2}{||f_1||_2}\right]^2 = \left[\frac{1/3}{1}\right]^2 = \frac{1}{9};$$

$$\left[\frac{\langle f, f_2 \rangle_2}{||f_2||_2}\right]^2 = \left[\frac{1/12}{1/\sqrt{12}}\right]^2 = \frac{1}{12},$$

and it follows from (7.3.31) that

$$||f - \widetilde{\mathscr{P}_1^*} f||_2 = \sqrt{\frac{1}{5} - \frac{1}{9} - \frac{1}{12}} = \frac{1}{6\sqrt{5}},$$

which agrees with (7.2.15) in Example 7.2.1.   ∎

Finally in this section, we prove that the orthogonal basis $\{f_1, \ldots, f_d\}$ in Theorem 7.3.3, as constructed from the basis $\{g_1, \ldots, g_d\}$ by means of the Gram-Schmidt procedure (7.3.26), is unique up to a multiplicative constant, as follows.

Theorem 7.3.5. *In Theorem 7.3.3, suppose the sequence $\{h_1, \ldots, h_d\}$ is an orthogonal basis for A, and such that*

(a)

$$h_j \in \begin{cases} A_1, & \text{if } j = 1; \\ A_j \setminus A_{j-1}, & \text{if } j = 2, \ldots, d; \end{cases} \tag{7.3.33}$$

(b)

$$h_j \perp A_{j-1}, \quad j = 2, \ldots, d. \tag{7.3.34}$$

Then

$$h_j = c_j f_j, \quad j = 1, \ldots, d, \tag{7.3.35}$$

for some coefficients $\{c_1, \ldots, c_d\} \subset \mathbb{R}$, with $c_j \neq 0$, $j = 1, \ldots, d$.

Proof. For any fixed $j \in \{1, \ldots, d\}$, it follows from (7.3.33) and Theorem 7.3.3(c) that

$$h_j = \sum_{k=1}^{j} c_{j,k} f_k, \tag{7.3.36}$$

for some coefficient sequence $\{c_{j,k} : k = 1, \ldots, j\} \subset \mathbb{R}$. If $j = 1$, it follows from (7.3.36) that $h_1 = c_{1,1} f_1$. Since $\{h_1, \ldots, h_d\}$ is a basis for A, we have $h_1 \neq 0$, so that also $c_{1,1} \neq 0$, and thereby proving (7.3.35), for $j = 1$, with $c_1 := c_{1,1}$.

Suppose next $j \in \{2, \ldots, d\}$, and let $\ell \in \{1, \ldots, j-1\}$. Since $\{f_1, \ldots, f_d\}$ is an orthogonal basis for A, and therefore satisfies (7.3.1), we may now deduce from (7.3.34), Theorem 7.3.3(c), (7.3.36), (4.2.3), (4.2.4), and (4.2.15) that

$$0 = \langle h_j, f_\ell \rangle = \left\langle \sum_{k=1}^{j} c_{j,k} f_k, f_\ell \right\rangle = \sum_{k=1}^{j} c_{j,k} \langle f_k, f_\ell \rangle = c_{j,\ell} \langle f_\ell, f_\ell \rangle = c_{j,\ell} ||f_\ell||^2. \tag{7.3.37}$$

Since $\{f_1,\ldots,f_d\}$ is a basis for A, we know that $f_\ell \neq 0$, and thus, from (4.2.2), $\|f_\ell\|^2 \neq 0$, which, together with (7.3.37), yields $c_{j,\ell} = 0$. Hence we have shown that

$$c_{j,k} = 0, \quad k = 1,\ldots,j-1,$$

which we may now insert into (7.3.36) to obtain $h_j = c_{j,j}f_j$. Since $\{h_1,\ldots,h_d\}$ is a basis for A, we have $h_j \neq 0$, and thus also $c_{j,j} \neq 0$, so that the definition $c_j := c_{j,j}$ yields the desired result (7.3.35), and thereby completing our proof. ∎

We proceed in Section 7.4 to specialize the results of this section to the case $X = C[a,b]$, $\langle \cdot,\cdot \rangle = \langle \cdot,\cdot \rangle_{2,w}$ and $A = \pi_n$, for any positive integer n.

7.4 Orthogonal polynomials

By applying Theorem 7.3.3, Theorem 7.3.5 and Theorem 7.3.1 for the case $X = C[a,b]$, $\|\cdot\| = \|\cdot\|_{2,w}$ as in (4.2.16), $A = \pi_n$, with $n \in \{0,1,\ldots\}$, so that $d := \dim(A) = n+1$, and $\{g_1,\ldots,g_d\} = \{g_1,\ldots,g_{n+1}\} = \{1,x,\ldots,x^n\}$ if $n \geqslant 1$, we immediately deduce the following result.

Theorem 7.4.1. *For a given bounded interval $[a,b]$, and any weight function w on $[a,b]$ satisfying the conditions (4.2.8) and (4.2.9), let the polynomial sequence $\{P_j^\perp : j = 0,1,\ldots\}$ be defined recursively by means of the Gram-Schmidt procedure*

$$\left. \begin{array}{l} P_0^\perp(x) := 1; \\[2ex] P_j^\perp(x) := x^j - \displaystyle\sum_{k=0}^{j-1} \left[\frac{\displaystyle\int_a^b w(x)x^j P_k^\perp(x)dx}{\displaystyle\int_a^b w(x)[P_k^\perp(x)]^2 dx} \right] P_k^\perp(x), \quad j = 1,2,\ldots. \end{array} \right\} \quad (7.4.1)$$

Then:

(a) *For $j = 0,1,\ldots,P_j^\perp$ is a monic polynomial, with, more precisely,*

$$P_j^\perp \in \tilde{\pi}_j, \quad j = 0,1,\ldots. \quad (7.4.2)$$

(b) *$\{P_j^\perp : j = 0,1,\ldots\}$ is an orthogonal sequence, with respect to the inner product $\langle \cdot,\cdot \rangle_{2,w}$ as in (4.2.10), that is,*

$$\int_a^b w(x)P_j^\perp(x)P_k^\perp(x)dx = 0, \quad for \quad j,k \in \{0,1,\ldots\}, with \ j \neq k. \quad (7.4.3)$$

(c)

$$P_j^\perp \perp \pi_{j-1}, \quad j = 1,2,\ldots. \quad (7.4.4)$$

(d) *For any non-negative integer n, the sequence $\{P_j^\perp : j = 0,\ldots,n\}$ is an orthogonal basis with respect to the inner product $\langle \cdot,\cdot \rangle_{2,w}$, as given by (4.2.10), for the polynomial space π_n, and $\{P_j^\perp : j = 0,\ldots,n\}$ is, moreover, the only orthogonal basis for π_n with respect to the inner product $\langle \cdot,\cdot \rangle_{2,w}$ such that (7.4.2) and (7.4.4) hold.*

(e) *For any non-negative integer n, the best weighted L^2 polynomial approximation operator $\widetilde{\mathscr{P}}_n^* : C[a,b] \to \pi_n$, as defined by (5.1.6), satisfies the explicit formulation*

$$\widetilde{\mathscr{P}}_n^* f = \sum_{j=0}^{n} \left[\frac{\int_a^b w(x)f(x)P_j^\perp(x)dx}{\int_a^b w(x)[P_j^\perp(x)]^2 dx} \right] P_j^\perp. \tag{7.4.5}$$

The sequence $\{P_j^\perp : j = 0,1,\ldots\}$ of Theorem 7.4.1 will be called the orthogonal polynomials with respect to the weight function w on the interval $[a,b]$.

Although the Gram-Schmidt formula (7.4.1) provides an explicit construction method for the orthogonal polynomials $\{P_j^\perp : j = 0,1,\ldots\}$, it has the disadvantage of increasing in length with the degree j of $P_j^\perp$. We proceed to show that the sequence $\{P_j^\perp : j = 0,1,\ldots\}$ satisfies a three-term recursion formula, which, for $j \geqslant 3$, then yields a more efficient construction method than the Gram-Schmidt procedure (7.4.1).

To this end, we define the polynomials

$$Q_j(x) := xP_j^\perp(x), \quad j = 0,1,\ldots, \tag{7.4.6}$$

with $\{P_j^\perp : j = 0,1,\ldots\}$ denoting the orthogonal polynomials of Theorem 7.4.1, as obtained by means of the Gram-Schmidt procedure (7.4.1). It then follows from (7.4.6) and (7.4.2) that, for each $j \in \{0,1,\ldots\}, Q_j$ is a monic polynomial, with, more precisely,

$$Q_j \in \widetilde{\pi}_{j+1}, \quad j = 0,1,\ldots. \tag{7.4.7}$$

According to Theorem 7.4.1(d), the sequence $\{P_k^\perp : k = 0,\ldots,j\}$ is, for each non-negative integer j, a basis for π_j, so that we may deduce from (7.4.7) that

$$Q_j = \sum_{k=0}^{j+1} c_{j,k} P_k^\perp, \quad j = 0,1,\ldots, \tag{7.4.8}$$

for some (unique for each j) coefficient sequence $\{c_{j,k} : k = 0,1,\ldots,j+1\} \subset \mathbb{R}$. Moreover, it follows from (7.4.7) and (7.4.2) that, in (7.4.8), we have $c_{j,j+1} = 1$, and thus

$$Q_j = \sum_{k=0}^{j} c_{j,k} P_k^\perp + P_{j+1}^\perp, \quad j = 0,1,\ldots. \tag{7.4.9}$$

For any fixed $j \in \{0,1,\ldots\}$, let $\ell \in \{0,\ldots,j\}$. It follows from (7.4.9) and Theorem 7.4.1(b), together with (4.2.3), (4.2.4) and (4.2.15), that

$$\langle Q_j, P_\ell^\perp \rangle_{2,w} = \left\langle \sum_{k=0}^{j} c_{j,k} P_k^\perp + P_{j+1}^\perp, P_\ell^\perp \right\rangle_{2,w}$$

$$= \sum_{k=0}^{j} c_{j,k} \langle P_k^\perp, P_\ell^\perp \rangle_{2,w} + \langle P_{j+1}^\perp, P_\ell^\perp \rangle_{2,w}$$

$$= c_{j,\ell} \langle P_\ell^\perp, P_\ell^\perp \rangle_{2,w} + 0 = c_{j,\ell} \left(\|P_\ell^\perp\|_{2,w} \right)^2,$$

and thus

$$c_{j,\ell} = \frac{\langle Q_j, P_\ell^\perp \rangle_{2,w}}{(\|P_\ell^\perp\|_{2,w})^2}, \quad \ell = 0, \ldots, j. \tag{7.4.10}$$

We may now use (7.4.9), (7.4.10), (7.4.6), (4.2.10) and (4.2.16) to obtain, for $j = 0$,

$$P_1^\perp(x) = \left(x - \frac{\displaystyle\int_a^b w(x)x[P_0^\perp(x)]^2 dx}{\displaystyle\int_a^b w(x)[P_0^\perp(x)]^2 dx} \right) P_0^\perp(x), \tag{7.4.11}$$

whereas, for $j = 1$,

$$P_2^\perp(x) = \left(x - \frac{\displaystyle\int_a^b w(x)x[P_1^\perp(x)]^2 dx}{\displaystyle\int_a^b w(x)[P_1^\perp(x)]^2 dx} \right) P_1^\perp(x) - \left[\frac{\displaystyle\int_a^b w(x)xP_1^\perp(x)P_0^\perp(x)dx}{\displaystyle\int_a^b w(x)[P_0^\perp(x)]^2 dx} \right] P_0^\perp(x),$$
$$\tag{7.4.12}$$

and, for $j \geqslant 2$,

$$P_{j+1}^\perp(x) = \left(x - \frac{\displaystyle\int_a^b w(x)x[P_j^\perp(x)]^2 dx}{\displaystyle\int_a^b w(x)[P_j^\perp(x)]^2 dx} \right) P_j^\perp(x) - \left[\frac{\displaystyle\int_a^b w(x)xP_j^\perp(x)P_{j-1}^\perp(x)dx}{\displaystyle\int_a^b w(x)[P_{j-1}^\perp(x)]^2 dx} \right] P_{j-1}^\perp(x)$$

$$- \left[\sum_{k=0}^{j-2} \frac{\displaystyle\int_a^b w(x)xP_j^\perp(x)P_k^\perp(x)dx}{\displaystyle\int_a^b w(x)[P_k^\perp(x)]^2 dx} \right] P_k^\perp(x). \tag{7.4.13}$$

Now observe from (7.4.6), (7.4.7) and (4.2.10), together with (7.4.4) in Theorem 7.4.1, that, for $j \geqslant 2$ and $k \in \{0, 1, \ldots, j-2\}$,

$$\int_a^b w(x)xP_j^\perp(x)P_k^\perp(x)dx = \int_a^b w(x)P_j^\perp(x)Q_k(x)dx = \langle P_j^\perp, Q_k \rangle_{2,w} = 0, \tag{7.4.14}$$

since $Q_k \in \tilde{\pi}_{k+1} \subset \pi_{j-1}$. Hence we may substitute (7.4.14) into (7.4.13) to obtain, for $j \geqslant 2$,

$$P_{j+1}^\perp(x) = \left(x - \frac{\displaystyle\int_a^b w(x)x[P_j^\perp(x)]^2 dx}{\displaystyle\int_a^b w(x)[P_j^\perp(x)]^2 dx} \right) P_j^\perp(x) - \left[\frac{\displaystyle\int_a^b w(x)xP_j^\perp(x)P_{j-1}^\perp(x)dx}{\displaystyle\int_a^b w(x)[P_{j-1}^\perp(x)]^2 dx} \right] P_{j-1}^\perp(x).$$
$$\tag{7.4.15}$$

Next, with the definition

$$\tilde{Q}_j(x) := P_j^\perp(x) - xP_{j-1}^\perp(x), \quad j = 1, 2, \ldots, \tag{7.4.16}$$

we deduce from (7.4.2) in Theorem 7.4.1(a), that

$$\widetilde{Q}_j \in \pi_{j-1}, \quad j = 1, 2, \ldots. \tag{7.4.17}$$

It follows from (7.4.17), and (7.4.4) in Theorem 7.4.1(c), together with (4.2.10) and (7.4.16), that, for any $j \in \{1, 2, \ldots\}$,

$$0 = \langle P_j^\perp, \widetilde{Q}_j \rangle_{2,w} = \int_a^b w(x) P_j^\perp(x)[P_j^\perp(x) - x P_{j-1}^\perp(x)]dx$$

$$= \int_a^b w(x)[P_j^\perp(x)]^2 dx - \int_a^b w(x) x P_j^\perp(x) P_{j-1}^\perp(x)dx,$$

and thus

$$\int_a^b w(x) x P_j^\perp(x) P_{j-1}^\perp(x)dx = \int_a^b w(x)[P_j^\perp(x)]^2 dx. \tag{7.4.18}$$

By substituting (7.4.18) into (7.4.15), we obtain, for $j \geqslant 2$, the formula

$$P_{j+1}^\perp(x) = \left(x - \frac{\int_a^b w(x) x [P_j^\perp(x)]^2 dx}{\int_a^b w(x)[P_j^\perp(x)]^2 dx} \right) P_j^\perp(x) - \left[\frac{\int_a^b w(x)[P_j^\perp(x)]^2 dx}{\int_a^b w(x)[P_{j-1}^\perp(x)]^2 dx} \right] P_{j-1}^\perp(x).$$

$$\tag{7.4.19}$$

According to (7.4.11), (7.4.12) and (7.4.19), together with the first line of (7.4.1), we have therefore established the following three-term recursion formulation.

Theorem 7.4.2. *The orthogonal polynomials* $\{P_j^\perp : j = 0, 1, \ldots\}$ *of Theorem 7.4.1 satisfy the recursion formulation*

$$\left. \begin{array}{l} P_0^\perp(x) = 1; \\[2mm] P_1^\perp(x) = (x - \alpha_0)P_0^\perp(x); \\[2mm] P_{j+1}^\perp(x) = (x - \alpha_j)P_j^\perp(x) - \beta_j P_{j-1}^\perp(x), \quad j = 1, 2, \ldots, \end{array} \right\} \tag{7.4.20}$$

where

$$\alpha_j := \frac{\int_a^b w(x) x [P_j^\perp(x)]^2 dx}{\int_a^b w(x)[P_j^\perp(x)]^2 dx}, \quad j = 0, 1, \ldots; \tag{7.4.21}$$

$$\beta_j := \frac{\int_a^b w(x)[P_j^\perp(x)]^2 dx}{\int_a^b w(x)[P_{j-1}^\perp(x)]^2 dx}, \quad j = 1, 2, \ldots. \tag{7.4.22}$$

In the next section, we analyze the special case of Theorem 7.4.2 where the weight function w on $[a,b]$ is given by (4.2.11).

Analogously to Theorem 6.6.1, we proceed to obtain an explicit formulation of the best weighted L^2 polynomial approximation on $[a,b]$ from π_n to any $f \in \pi_{n+1}$. To this end, for any non-negative integer n, we let

$$f(x) = \sum_{j=0}^{n+1} c_j x^j, \tag{7.4.23}$$

for some coefficient sequence $\{c_0, \ldots, c_{n+1}\}$. For any bounded interval $[a,b]$, let $\widetilde{\mathscr{P}}_n^* : C[a,b] \to \pi_n$ denote the corresponding best weighted L^2 polynomial approximation operator. It then follows from (7.4.23) that

$$f(x) - (\widetilde{\mathscr{P}}_n^* f)(x) = c_{n+1} x^{n+1} + \widetilde{Q}(x), \tag{7.4.24}$$

for some polynomial $\widetilde{Q} \in \pi_n$. But, from the characterisation result of Theorem 7.1.1, we must have

$$\langle f - \widetilde{\mathscr{P}}_n^* f, \, P \rangle_{2,w} = 0, \quad P \in \pi_n. \tag{7.4.25}$$

Hence, by recalling also (7.4.2) in Theorem 7.4.1(a), as well as (7.4.4) in Theorem 7.4.1(c), it follows from (7.4.24) that we must have

$$f - \widetilde{\mathscr{P}}_n^* f = c_{n+1} P_{n+1}^\perp, \tag{7.4.26}$$

with $P_{n+1}^\perp$ denoting the orthogonal polynomial of degree $n+1$ as in Theorem 7.4.1. We can therefore solve for $\widetilde{\mathscr{P}}_n^* f$ from the equation (7.4.26) to obtain the following result.

Theorem 7.4.3. *For any non-negative integer n, let $f \in \pi_{n+1}$, with leading coefficient c_{n+1} in the formulation (7.4.23). Then, for any bounded interval $[a,b]$ and weight function w on $[a,b]$ satisfying the conditions (4.2.8) and (4.2.9), the best weighted L^2 polynomial approximation on $[a,b]$ from π_n to f is given by the formula*

$$\widetilde{\mathscr{P}}_n^* f = f - c_{n+1} P_{n+1}^\perp, \tag{7.4.27}$$

with $P_{n+1}^\perp$ denoting the orthogonal polynomial of degree $n+1$ in Theorem 7.4.1, and with corresponding approximation error

$$\| f - \widetilde{\mathscr{P}}_n^* f \|_{2,w} = |c_{n+1}| \, \| P_{n+1}^\perp \|_{2,w}. \tag{7.4.28}$$

7.5 The Legendre polynomials

For the case where, in Theorem 7.4.1, we choose $[a,b] = [-1,1]$ and the weight function

$$w(x) = 1, \quad x \in [-1,1], \tag{7.5.1}$$

the resulting orthogonal polynomials are called the Legendre polynomials, and will be denoted by $\{L_j^\perp : j = 0,1,\ldots\}$, for which we therefore have, from (7.4.3) in Theorem 7.4.1(b),

$$\int_{-1}^{1} L_j^\perp(x) L_k^\perp(x) dx = 0, \quad \text{for} \quad j,k = 0,1,\ldots, \quad \text{with} \quad j \neq k. \tag{7.5.2}$$

Also, the three-term recursion formulation (7.4.20) of Theorem 7.4.2 yields

$$\left.\begin{aligned}
L_0^\perp(x) &= 1; \\
L_1^\perp(x) &= (x - \alpha_0) L_0^\perp(x); \\
L_{j+1}^\perp(x) &= (x - \alpha_j) L_j^\perp(x) - \beta_j L_{j-1}^\perp(x), \quad j = 1,2,\ldots,
\end{aligned}\right\} \tag{7.5.3}$$

where, from (7.4.21) and (7.4.22),

$$\alpha_j := \frac{\displaystyle\int_{-1}^{1} x[L_j^\perp(x)]^2 dx}{\displaystyle\int_{-1}^{1} [L_j^\perp(x)]^2 dx}, \quad j = 0,1,\ldots; \tag{7.5.4}$$

$$\beta_j := \frac{\displaystyle\int_{-1}^{1} [L_j^\perp(x)]^2 dx}{\displaystyle\int_{-1}^{1} [L_{j-1}^\perp(x)]^2 dx}, \quad j = 1,2,\ldots. \tag{7.5.5}$$

We proceed to explicitly compute the coefficient sequences $\{\alpha_j : j = 0,1,\ldots\}$ and $\{\beta_j : j = 1,2,\ldots\}$ in (7.5.4) and (7.5.5), and thereby yielding an explicit formulation for the corresponding recursion relations in (7.5.3).

Our first step is to prove inductively that

$$\left.\begin{aligned}
L_{2j}^\perp &\ \text{is an even polynomial,} \\
L_{2j+1}^\perp &\ \text{is an odd polynomial,}
\end{aligned}\right\} \quad j = 0,1,\ldots. \tag{7.5.6}$$

To this end, we first note from the first line of (7.5.3) that

$$\int_{-1}^{1} x[L_0^\perp(x)]^2 dx = \int_{-1}^{1} x dx = 0,$$

which, together with (7.5.4), implies $\alpha_0 = 0$, and thus, by using also the second line of (7.5.3), we have

$$L_1^\perp(x) = x, \tag{7.5.7}$$

according to which, by recalling also from the first line of (7.5.3) that $L_0^\perp(x) = 1$, we have now shown that (7.5.6) holds for $j = 0$. Suppose next that (7.5.6) is satisfied for a fixed non-negative integer j. Our inductive proof of (7.5.6) will be complete if we can show that then

$$\left.\begin{array}{l} L_{2j+2}^\perp \text{ is an even polynomial} \\ \\ L_{2j+3}^\perp \text{ is an odd polynomial} \end{array}\right\}. \tag{7.5.8}$$

To prove (7.5.8), we note that $2j+2 \geqslant 2$, so that we may apply the third line of (7.5.3) to obtain

$$L_{2j+2}^\perp(x) = (x - \alpha_{2j+1})L_{2j+1}^\perp(x) - \beta_{2j+1}L_{2j}^\perp(x). \tag{7.5.9}$$

From the inductive hypothesis that $L_{2j+1}^\perp$ is an odd polynomial, it follows that $x[L_{2j+1}^\perp(x)]^2$ is an odd polynomial, and thus

$$\int_{-1}^1 x[L_{2j+1}^\perp(x)]^2 dx = 0,$$

which, together with (7.5.4), gives $\alpha_{2j+1} = 0$, and it follows from (7.5.9) that

$$L_{2j+2}^\perp(x) = xL_{2j+1}^\perp(x) - \beta_{2j+1}L_{2j}^\perp(x). \tag{7.5.10}$$

Hence, from (7.5.10) and the inductive assumption (7.5.6), we deduce that the first line of (7.5.8) is satisfied.

Next, we apply the third line of (7.5.3) to obtain

$$L_{2j+3}^\perp(x) = (x - \alpha_{2j+2})L_{2j+2}^\perp(x) - \beta_{2j+2}L_{2j+1}^\perp(x). \tag{7.5.11}$$

Since, as already shown, $L_{2j+2}^\perp$ is an even polynomial, we see that $x[L_{2j+2}^\perp(x)]^2$ is an odd polynomial, and thus

$$\int_{-1}^1 x\left[L_{2j+2}^\perp(x)\right]^2 dx = 0,$$

which, together with (7.5.4), yields $\alpha_{2j+2} = 0$, and it follows from (7.5.11) that

$$L_{2j+3}^\perp(x) = xL_{2j+2}^\perp(x) - \beta_{2j+2}L_{2j+1}^\perp(x). \tag{7.5.12}$$

Since, as already shown, $L_{2j+2}^\perp$ is an even polynomial, whereas $L_{2j+1}^\perp$ is an odd polynomial from the inductive assumption (7.5.6), it follows from (7.5.12) that $L_{2j+3}^\perp$ is an odd polynomial. Hence the second line of (7.5.8) also holds, so that we have now completed our inductive proof of the statement (7.5.6).

It follows from (7.5.6) that $x[L_j^\perp(x)]^2$ is an odd polynomial for each $j = 0, 1, \ldots$, and thus

$$\int_{-1}^1 x[L_j^\perp(x)]^2 dx = 0, \quad j = 0, 1, \ldots,$$

which we can now use in (7.5.4) to obtain

$$\alpha_j = 0, \quad j = 0, 1, \dots. \tag{7.5.13}$$

By inserting (7.5.13) into the third line of (7.5.3), we obtain

$$L_{j+1}^{\perp}(x) = xL_j^{\perp}(x) - \beta_j L_{j-1}^{\perp}(x), \quad j = 1, 2, \dots. \tag{7.5.14}$$

In order to find an explicit formulation for the sequence $\{\beta_j : j = 1, 2, \dots\}$ in (7.5.14), we first use integration by parts to obtain, for any $j \in \{0, 1, \dots\}$,

$$\int_{-1}^{1} \left[L_j^{\perp}(x)\right]^2 dx = \left[\left\{L_j^{\perp}(x)\right\}^2 x\right]_{-1}^{1} - \int_{-1}^{1} 2L_j^{\perp}(x)(L_j^{\perp})'(x)x\,dx$$

$$= 2\left[L_j^{\perp}(1)\right]^2 - 2\int_{-1}^{1} xL_j^{\perp}(x)(L_j^{\perp})'(x)dx, \tag{7.5.15}$$

after having used also the fact that, according to (7.5.6), $[L_j^{\perp}(x)]^2$ is an even polynomial, and thus $[L_j^{\perp}(-1)]^2 = [L_j^{\perp}(1)]^2$. Now observe that, for any $j \in \{0, 1, \dots\}$, the polynomial

$$\widetilde{P}_j(x) := x(L_j^{\perp})'(x) - jL_j^{\perp}(x) \tag{7.5.16}$$

satisfies, from (7.4.2) in Theorem 7.4.1(a), $\widetilde{P}_j \in \pi_{j-1}$, and thus, from (7.4.4) in Theorem 7.4.1(c), together with (4.2.12) and (7.5.16),

$$0 = \langle L_j^{\perp}, \widetilde{P}_j \rangle_2 = \int_{-1}^{1} L_j^{\perp}(x)\left[x(L_j^{\perp})'(x) - jL_j^{\perp}(x)\right]dx$$

$$= \int_{-1}^{1} xL_j^{\perp}(x)(L_j^{\perp})'(x)dx - j\int_{-1}^{1} [L_j^{\perp}(x)]^2 dx,$$

and hence

$$\int_{-1}^{1} xL_j^{\perp}(x)(L_j^{\perp})'(x)dx = j\int_{-1}^{1} [L_j^{\perp}(x)]^2 dx,$$

which can now be substituted into (7.5.15) to obtain

$$\int_{-1}^{1} [L_j^{\perp}(x)]^2 dx = 2[L_j^{\perp}(1)]^2 - 2j\int_{-1}^{1} [L_j^{\perp}(x)]^2 dx,$$

from which it then follows that

$$\int_{-1}^{1} [L_j^{\perp}(x)]^2 dx = \frac{2}{2j+1}[L_j^{\perp}(1)]^2, \quad j = 0, 1, \dots. \tag{7.5.17}$$

Observe from (7.5.5) and (7.5.17) that

$$\beta_j = \frac{2j-1}{2j+1}(\gamma_j)^2, \quad j = 1, 2, \dots, \tag{7.5.18}$$

where

$$\gamma_j := \frac{L_j^{\perp}(1)}{L_{j-1}^{\perp}(1)}, \quad j = 1, 2, \dots, \tag{7.5.19}$$

after having also noted from (7.5.17) that

$$L_j^\perp(1) \neq 0, \quad j = 0,1,\ldots. \tag{7.5.20}$$

In order to find an explicit formula for γ_j in (7.5.19), we first set $x = 1$ in (7.5.14), and use (7.5.18) to obtain, for $j = 1,2,\ldots,$

$$L_{j+1}^\perp(1) = L_j^\perp(1) - \frac{2j-1}{2j+1}(\gamma_j)^2\, L_{j-1}^\perp(1),$$

which we now divide by $L_j^\perp(1)$ to deduce, by using also the definition (7.5.19), the formula

$$\gamma_{j+1} = 1 - \frac{2j-1}{2j+1}\,\gamma_j, \quad j = 1,2,\ldots, \tag{7.5.21}$$

with also

$$\gamma_1 = 1, \tag{7.5.22}$$

from (7.5.19), together with the first line of (7.5.3), and (7.5.7).

We proceed to prove inductively from (7.5.21) and (7.5.22) that

$$\gamma_j = \frac{j}{2j-1}, \quad j = 1,2,\ldots. \tag{7.5.23}$$

After first noting from (7.5.22) that (7.5.23) holds for $j = 1$, we suppose next that (7.5.23) is true for a fixed integer $j \in \mathbb{N}$. But then (7.5.21) yields

$$\gamma_{j+1} = 1 - \frac{2j-1}{2j+1}\left(\frac{j}{2j-1}\right) = 1 - \frac{j}{2j+1} = \frac{j+1}{2j+1},$$

which shows that (7.5.23) also holds with j replaced by $j+1$, and thereby completing our inductive proof of (7.5.23).

By substituting (7.5.23) into (7.5.18), we obtain

$$\beta_j = \frac{j^2}{(2j-1)(2j+1)}, \quad j = 1,2,\ldots. \tag{7.5.24}$$

According to the first line of (7.5.3), together with (7.5.7), (7.5.14) and (7.5.24), we have now established the following explicit three-term recursion formulation for Legendre polynomials.

Theorem 7.5.1. *The Legendre polynomials* $\{L_j^\perp : j = 0,1,\ldots\}$, *which are precisely the orthogonal polynomials* $\{P_j^\perp : j = 0,1,\ldots\}$ *of Theorem 7.4.1 for the case* $[a,b] = [-1,1]$ *and weight function w on* $[-1,1]$ *given by* (7.5.1), *satisfy the recursion formulation*

$$\left.\begin{array}{l} L_0^\perp(x) = 1 \quad ; \quad L_1^\perp(x) = x; \\[2mm] L_{j+1}^\perp(x) = xL_j^\perp(x) - \dfrac{j^2}{(2j-1)(2j+1)}\, L_{j-1}^\perp(x), \quad j = 1,2,\ldots \end{array}\right\} \tag{7.5.25}$$

By using (7.5.25), we calculate the Legendre polynomials

$$L_2^\perp(x) = x^2 - \tfrac{1}{3}; \quad L_3^\perp(x) = x^3 - \tfrac{3}{5}x; \quad L_4^\perp(x) = x^4 - \tfrac{6}{7}x^2 + \tfrac{3}{35};$$
$$L_5^\perp(x) = x^5 - \tfrac{10}{9}x^3 + \tfrac{5}{21}x; \quad L_6^\perp(x) = x^6 - \tfrac{15}{11}x^4 + \tfrac{5}{11}x^2 - \tfrac{5}{231}. \quad \left.\right\} \tag{7.5.26}$$

By considering the one-to-one mapping (2.2.29), (2.2.30) between the intervals $[-1,1]$ and $[a,b]$, we now define the polynomials

$$P_j(x) := \left(\frac{b-a}{2}\right)^j L_j^\perp \left(\frac{2}{b-a}x - \frac{a+b}{b-a}\right), \quad j = 0,1,\ldots, \tag{7.5.27}$$

for which it then follows, since $L_j^\perp \in \tilde{\pi}_j$, $j = 0,1,\ldots$, that

$$P_j \in \tilde{\pi}_j, \quad j = 0,1,\ldots. \tag{7.5.28}$$

Moreover, for $j,k \in \{0,1,\ldots\}$, with $j \neq k$, (4.2.12), (2.2.29), (2.2.30) and (7.5.27) yield

$$\langle P_j P_k \rangle_2 = \int_a^b P_j(x)P_k(x)dx = \left(\frac{b-a}{2}\right)^{j+k+1} \int_{-1}^1 L_j^\perp(t)L_k^\perp(t)dt = 0, \tag{7.5.29}$$

from (7.5.2). It follows from (7.5.29) and Theorem 7.3.2 that, for each $j = 0,1,\ldots, \{P_0,\ldots,P_j\}$ is a linearly independent set, and therefore an orthogonal basis for π_j. Hence, if for any $j \in \mathbb{N}$ we let $P \in \pi_{j-1}$, there exists a (unique) coefficient sequence $\{c_0,\ldots,c_{j-1}\} \subset \mathbb{R}$ such that $P = \sum_{k=1}^{j-1} c_k P_k$, and thus, from (4.2.3), (4.2.4), (4.2.5) and (7.5.29),

$$\langle P_j, P \rangle_2 = \sum_{k=1}^{j-1} c_k \langle P_j, P_k \rangle_2 = 0,$$

according to which

$$P_j \perp \pi_{j-1}, \quad j = 1,2,\ldots. \tag{7.5.30}$$

It then follows from (7.5.28), (7.5.29) and (7.5.30) that we may apply the uniqueness statement in Theorem 7.4.1, together with the definition (7.5.27), to deduce the following result.

Theorem 7.5.2. *In Theorem 7.4.1, let the weight function w on $[a,b]$ be given by (4.2.11), that is,*

$$w(x) = 1, \quad x \in [a,b]. \tag{7.5.31}$$

Then the corresponding orthogonal polynomials $\{P_j^\perp : j = 0,1,\ldots\}$ are given by

$$P_j^\perp(x) = \left(\frac{b-a}{2}\right)^j L_j^\perp \left(\frac{2}{b-a}x - \frac{a+b}{b-a}\right), \quad j = 0,1,\ldots, \tag{7.5.32}$$

with $\{L_j^\perp : j = 0,1,\ldots\}$ denoting the Legendre polynomials as in Theorem 7.5.1.

According to the formula (7.4.5) in Theorem 7.4.1(e), for the case when the weight function w on $[a,b]$ is given by (7.5.31), the corresponding best L^2 polynomial approximation operator $\widetilde{\mathscr{P}}_n^* : C[a,b] \to \pi_n$ is given by

$$\widetilde{\mathscr{P}}_n^* f = \sum_{j=0}^{n} \left[\frac{\int_a^b f(x)P_j^\perp(x)dx}{\int_a^b [P_j^\perp(x)]^2 dx} \right] P_j^\perp, \quad f \in C[a,b], \tag{7.5.33}$$

with $\{P_j^\perp : j = 0,1,\dots\}$ denoting the orthogonal polynomials in Theorem 7.5.2. Now observe from (2.2.29), (2.2.30), (7.5.32) and (7.5.17) that, for $j = 0,1,\dots$,

$$\int_a^b [P_j^\perp(x)]^2 dx = \left(\frac{b-a}{2}\right)^{2j+1} \int_{-1}^1 [L_j^\perp(t)dt]^2 dt = \left(\frac{b-a}{2}\right)^{2j+1} \frac{2}{2j+1} \left[L_j^\perp(1)\right]^2. \tag{7.5.34}$$

By using (7.5.19) and (7.5.23), as well as $L_0^\perp(x) = 1$, we obtain, for any $j \in \mathbb{N}$,

$$L_j^\perp(1) = \frac{j}{2j-1} L_{j-1}^\perp(1) = \frac{j}{2j-1} \frac{j-1}{2j-3} L_{j-2}^\perp(1) \quad (\text{if } j \geq 2)$$

$$= \cdots$$

$$= \frac{j}{2j-1} \frac{j-1}{2j-3} \cdots \frac{1}{1} L_0^\perp(1)$$

$$= \frac{j!(2j)(2j-2)\dots(2)}{(2j)!} = \frac{2^j(j!)^2}{(2j)!} = \frac{2^j}{\binom{2j}{j}}. \tag{7.5.35}$$

By inserting (7.5.35) into (7.5.34), we obtain the explicit formula

$$\int_a^b [P_j^\perp(x)]^2 dx = \frac{(b-a)^{2j+1}}{2j+1} \frac{1}{\binom{2j}{j}^2}, \quad j = 0,1,\dots. \tag{7.5.36}$$

It follows from (7.5.33), (7.5.36), and (4.2.17), that we have now established the following result.

Theorem 7.5.3. *For a bounded interval $[a,b]$ and any non-negative integer n, the best L^2 polynomial approximation operator $\widetilde{\mathscr{P}}_n^* : C[a,b] \to \pi_n$, as defined uniquely by the condition*

$$\sqrt{\int_a^b [f(x) - (\widetilde{\mathscr{P}}_n^* f)(x)]^2 dx} \leq \sqrt{\int_a^b [f(x) - P(x)]^2 dx}, \quad P \in \pi_n, \tag{7.5.37}$$

satisfies the explicit formulation

$$\widetilde{\mathscr{P}}_n^* f = \sum_{j=0}^{n} \left[\frac{2j+1}{(b-a)^{2j+1}} \binom{2j}{j}^2 \int_a^b f(x)P_j^\perp(x)dx \right] P_j^\perp, \quad f \in C[a,b], \tag{7.5.38}$$

with $\{P_j^\perp : j = 0,1,\dots\}$ denoting the orthogonal polynomials of Theorem 7.5.2. Moreover,

$$\|P_j^\perp\|_2 := \sqrt{\int_a^b [P_j^\perp(x)]^2 dx} = \sqrt{\frac{(b-a)^{2j+1}}{2j+1} \frac{1}{\binom{2j}{j}}}, \quad j = 0,1,\dots. \tag{7.5.39}$$

The formula (7.5.38) holds for arbitrary $f \in C[a,b]$. Recall from Theorem 7.4.3 the simple formulation of $\widetilde{\mathscr{P}}_n^* f$ for the case when $f \in \pi_{n+1}$. In the following example, we apply the explicit formulation (7.5.32) of Theorem 7.5.2 to solve the following best approximation problem in the settting of Theorem 7.4.3.

Example 7.5.1. As in Example 6.6.5, let $[a,b] = [-1,2]$, $f(x) = 2x^3 - x$ and $n = 2$. Then, for the weight function

$$w(x) = 1, \quad x \in [-1,2],$$

it follows from (7.4.27) in Theorem 7.4.3, together with (7.5.32) and (7.5.26), that

$$(\widetilde{\mathscr{P}}_2^* f)(x) = 2x^3 - x - 2\left(\frac{3}{2}\right)^2\left[\left(\frac{2}{3}x - \frac{1}{3}\right)^3 - \frac{3}{5}\left(\frac{2}{3}x - \frac{1}{3}\right)\right]$$

$$= 3x^2 + \frac{1}{5}x - \frac{11}{10},$$

with, from (7.4.28) and (7.5.39),

$$\|f - \widetilde{\mathscr{P}}_2^* f\|_2 = 2\sqrt{\frac{37}{7}}\frac{1}{20} = \frac{27\sqrt{3}}{10\sqrt{7}}. \qquad \blacksquare$$

By combining Theorem 5.1.1(b), according to which the best approximation operator $\widetilde{\mathscr{P}}_n^*$ in Theorem 7.5.3 is exact on π_n, with the formula (7.5.38) in Theorem 7.5.3, we immediately deduce the following polynomial identity.

Theorem 7.5.4. *For a bounded interval $[a,b]$ and any non-negative integer n,*

$$P = \sum_{j=0}^{n}\left[\frac{2j+1}{(b-a)^{2j+1}}\binom{2j}{j}^2\int_a^b P(x)P_j^\perp(x)dx\right]P_j^\perp, \quad P \in \pi_n, \qquad (7.5.40)$$

with $\{P_j^\perp : j = 0, 1, \ldots\}$ denoting the orthogonal polynomials of Theorem 7.5.2.

For any polynomial $P \in \pi_n$, the finite sum in the right hand side of (7.5.40) is called the Legendre expansion with respect to the interval $[a,b]$ in π_n of P.

7.6 Chebyshev polynomials and orthogonality

In this section, we let $[a,b] = [-1,1]$, and consider the weight function w on $[-1,1]$ as given by

$$w(x) = \frac{1}{\sqrt{1-x^2}}, \quad x \in (-1,1), \qquad (7.6.1)$$

for which, since

$$\int_{-1+\varepsilon}^{1-\delta} \frac{1}{\sqrt{1-x^2}} dx = \arcsin(1-\delta) - \arcsin(-1+\varepsilon) \to \frac{\pi}{2} - \left(-\frac{\pi}{2}\right) = \pi,$$

$$\text{for } \varepsilon \to 0^+, \quad \delta \to 0^+,$$

and thus

$$\int_{-1}^{1} \frac{1}{\sqrt{1-x^2}} dx = \pi,$$

we see that w is integrable on $[-1,1]$, and with w also satisfying the required conditions (4.2.8) and (4.2.9) to qualify as a weight function on $[-1,1]$.

Let $\{T_j : j = 0,1,\ldots\}$ denote the Chebyshev polynomials, as defined by (2.2.1). By using the one-to-one mapping (2.2.11), (2.2.12) between the intervals $[0,\pi]$ and $[-1,1]$, it follows from (4.2.10) and (7.6.1), together with the formula (2.2.5) in Theorem 2.2.1(b), that, for $j,k \in \{0,1,\ldots\}$, with $j \neq k$,

$$\langle T_j, T_k \rangle_{2,w} = \int_{-1}^{1} \frac{1}{\sqrt{1-x^2}} T_j(x) T_k(x) dx$$

$$= \int_{0}^{\pi} T_j(\cos\theta) T_k(\cos\theta) d\theta$$

$$= \int_{0}^{\pi} \cos(j\theta)\cos(k\theta) d\theta = \frac{1}{2} \int_{0}^{\pi} [\cos((j+k)\theta) + \cos((j-k)\theta))] d\theta$$

$$= \frac{1}{2} \left[\frac{\sin((j+k)\theta)}{j+k} + \frac{\sin((j-k)\theta)}{j-k} \right]_{0}^{\pi}$$

$$= \frac{1}{2}[0-0] = 0. \tag{7.6.2}$$

Hence, if we define

$$\widetilde{P}_0 := T_0 \quad ; \quad \widetilde{P}_j = 2^{1-j} T_j, \quad j = 1,2,\ldots, \tag{7.6.3}$$

we may deduce from (7.6.2) that

$$\langle \widetilde{P}_j, \widetilde{P}_k \rangle_{2,w} = 0, \quad \text{for} \quad j,k = 0,1,\ldots, \quad \text{with} \quad j \neq k. \tag{7.6.4}$$

It follows from (7.6.4) and Theorem 7.3.2 that, for each $j = 0,1,\ldots$, the polynomial sequence $\{\widetilde{P}_0,\ldots,\widetilde{P}_j\}$ is a linearly independent set, and therefore an orthogonal basis for π_j. Hence, if for any $j \in \mathbb{N}$ we let $P \in \pi_{j-1}$, then there exists a (unique) coefficient sequence $\{c_0,\ldots,c_{j-1}\} \subset \mathbb{R}$ such that $P = \sum_{k=0}^{j-1} c_k \widetilde{P}_k$, and thus, from (4.2.3), (4.2.4), (4.2.5) and (7.6.4),

$$\langle \widetilde{P}_j, P \rangle_{2,w} = \sum_{k=0}^{j-1} c_k \langle \widetilde{P}_j, \widetilde{P}_k \rangle_{2,w} = 0,$$

from which we deduce that

$$\widetilde{P}_j \perp \pi_{j-1}, \quad j = 1, 2, \ldots . \tag{7.6.5}$$

Also, observe from (7.6.3), together with Theorem 2.2.1(a), that

$$\widetilde{P}_j \in \widetilde{\pi}_j, \quad j = 0, 1, \ldots . \tag{7.6.6}$$

It follows from (7.6.6), (7.6.4) and (7.6.5), together with (7.6.3), that we may apply the uniqueness statement in Theorem 7.4.1 to deduce the following analogue of Theorem 7.5.2.

Theorem 7.6.1. *In Theorem 7.4.1, let $[a,b] = [-1,1]$, and let the weight function w on $[-1,1]$ be defined by (7.6.1). Then the corresponding orthogonal polynomials $\{P_j^\perp : j = 0, 1, \ldots\}$ are given by*

$$P_0^\perp = T_0; \quad P_j^\perp = 2^{1-j} T_j, \quad j = 1, 2, \ldots, \tag{7.6.7}$$

with $\{T_j : j = 0, 1, \ldots\}$ denoting the Chebyshev polynomials as defined in (2.2.1).

According to the formula (7.4.5) in Theorem 7.4.1(e), for the case $[a,b] = [-1,1]$, and with the weight function w on $[-1,1]$ defined by (7.6.1), we may apply (7.6.7) in Theorem 7.6.1 to deduce that the corresponding best weighted L^2 polynomial approximation operator is given by

$$\widetilde{\mathscr{P}}_n^* f = \sum_{j=0}^{n} \left[\frac{\displaystyle\int_{-1}^{1} \frac{1}{\sqrt{1-x^2}} f(x) T_j(x) dx}{\displaystyle\int_{-1}^{1} \frac{1}{\sqrt{1-x^2}} [T_j(x)]^2 dx} \right] T_j, \quad f \in C[-1,1], \tag{7.6.8}$$

with $\{T_j : j = 0, 1, \ldots\}$ denoting the Chebyshev polynomials as defined in (2.2.1).
By using (2.2.11), (2.2.12), as well as the formula (2.2.5) in Theorem 2.2.1(b), we obtain, for any $f \in C[-1,1]$,

$$\int_{-1}^{1} \frac{1}{\sqrt{1-x^2}} f(x) T_j(x) dx = \int_{0}^{\pi} f(\cos\theta) \cos(j\theta) d\theta, \quad j = 0, 1, \ldots, \tag{7.6.9}$$

and

$$\int_{-1}^{1} \frac{1}{\sqrt{1-x^2}} [T_j(x)]^2 dx = \int_{0}^{\pi} \cos^2(j\theta) d\theta. \tag{7.6.10}$$

But

$$\int_{0}^{\pi} \cos^2(j\theta) d\theta = \begin{cases} \pi, & \text{if } j = 0; \\ \displaystyle\int_{0}^{\pi} \frac{1 - \sin(2j\theta)}{2} d\theta = \frac{\pi}{2}, & \text{if } j = 1, 2, \ldots . \end{cases} \tag{7.6.11}$$

According to (7.6.7)–(7.6.10), together with (2.2.1), we have now established the following analogue of Theorem 7.5.3.

Theorem 7.6.2. *For any non-negative integer n, the best weighted L^2 polynomial approximation operator $\widetilde{\mathscr{P}}_n^* : C[-1,1] \to \pi_n$, as defined uniquely by the condition*

$$\sqrt{\int_{-1}^{1} \frac{1}{\sqrt{1-x^2}} [f(x) - (\widetilde{\mathscr{P}}_n^* f)(x)]^2 dx} \leqslant \sqrt{\int_{-1}^{1} \frac{1}{\sqrt{1-x^2}} [f(x) - P(x)]^2 dx}, \quad P \in \pi_n,$$

(7.6.12)

satisfies the explicit formulation

$$\widetilde{\mathscr{P}}_n^* f = \frac{1}{\pi} \int_0^{\pi} f(\cos \theta) d\theta + \frac{2}{\pi} \sum_{j=1}^{n} \left[\int_0^{\pi} f(\cos \theta) \cos(j\theta) d\theta \right] T_j, \quad f \in C[-1,1], \quad (7.6.13)$$

with $\{T_j : j = 1, 2, \ldots\}$ denoting the Chebyshev polynomials defined in (2.2.1).

By combining Theorem 5.1.1(b), according to which the best approximation operator $\widetilde{\mathscr{P}}_n^*$ in Theorem 7.6.2 is exact on π_n, with the formula (7.6.13) in Theorem 7.6.2, we immediately deduce the following analogue of Theorem 7.5.4.

Theorem 7.6.3. *For any non-negative integer n,*

$$P = \frac{1}{\pi} \int_0^{\pi} P(\cos \theta) d\theta + \frac{2}{\pi} \sum_{j=1}^{n} \left[\int_0^{\pi} P(\cos \theta) \cos(j\theta) d\theta \right] T_j, \quad P \in \pi_n, \quad (7.6.14)$$

with $\{T_j : j = 0, 1, \ldots\}$ denoting the Chebyshev polynomials defined in (2.2.1).

For any polynomial $P \in \pi_n$, the right hand side of (7.6.13) is called the Chebyshev expansion in π_n of P.

We proceed to show, as will also be relied on in Section 8.3, that the integrals in the right hand side of the Chebyshev expansion (7.6.14) in Theorem 7.6.3 can be expressed as finite sums involving Chebyshev polynomial evaluations at the points where the Chebyshev polynomial T_n attains its extreme values.

We shall rely on the following result, in which we adopt, for $n \in \mathbb{N}$, the notation

$$\sideset{}{'}\sum_{j=0}^{n} a_j := \tfrac{1}{2} a_0 + \sum_{j=1}^{n-1} a_j + \tfrac{1}{2} a_n. \quad (7.6.15)$$

Theorem 7.6.4. *For any positive integer n, let the point sequence $\{\xi_{n,j} : j = 0, \ldots, n\}$ be as in (2.2.19), (2.2.20), that is,*

$$\xi_{n,\ell} := \cos\left(\frac{n-\ell}{n}\pi\right), \quad \ell = 0, \ldots, n, \quad (7.6.16)$$

according to which

$$-1 = \xi_{n,0} < \xi_{n,1} < \cdots < \xi_{n,n} = 1. \quad (7.6.17)$$

Then

$$\sideset{}{'}\sum_{\ell=0}^{n} T_j(\xi_{n,\ell})T_k(\xi_{n,\ell}) = 0, \ for \ j,k=0,\ldots,n, \ with \ j \neq k; \qquad (7.6.18)$$

$$\sideset{}{'}\sum_{\ell=0}^{n} \left[T_j(\xi_{n,\ell})\right]^2 = \begin{cases} n, & for \ j=0, \ or \ j=n, \\ \frac{1}{2}n, & for \ j=1,\ldots,n-1 \quad (if \ n \geqslant 2), \end{cases} \qquad (7.6.19)$$

where $\{T_j : j=0,1,\ldots\}$ are the Chebyshev polynomials as in (2.2.1), and with the symbol Σ' defined by (7.6.15).

Proof. Let $j,k \in \{0,\ldots,n\}$, with $j \geqslant k$. It follows from (7.6.16), (7.6.17), the formula (2.2.5) in Theorem 2.2.1(b), as well as the definition (7.6.15), that

$$\sideset{}{'}\sum_{\ell=0}^{n} T_j(\xi_{n,\ell})T_k(\xi_{n,\ell}) = \sideset{}{'}\sum_{\ell=0}^{n} \cos\left(\frac{j(n-\ell)}{n}\pi\right)\cos\left(\frac{k(n-\ell)}{n}\pi\right)$$

$$= \sideset{}{'}\sum_{\ell=0}^{n} \cos\left(\frac{j\ell}{n}\pi\right)\cos\left(\frac{k\ell}{n}\pi\right)$$

$$= \frac{1}{2}\left[\sideset{}{'}\sum_{\ell=0}^{n} \left\{\cos\left(\frac{(j+k)\ell}{n}\pi\right)+\cos\left(\frac{(j-k)\ell}{n}\pi\right)\right\}\right]$$

$$= \frac{1}{2}\left[\sideset{}{'}\sum_{\ell=0}^{n} \cos\left(\frac{(j+k)\ell}{n}\pi\right)+\sideset{}{'}\sum_{\ell=0}^{n} \cos\left(\frac{(j-k)\ell}{n}\pi\right)\right]. \qquad (7.6.20)$$

Note that (7.6.20) also holds for the case $j < k$.

Now observe that, if $m = 2\mu n$ for some $\mu \in \mathbb{Z}$, then

$$\sideset{}{'}\sum_{\ell=0}^{n} \cos\left(\frac{m\ell}{n}\pi\right) = \begin{cases} \frac{1}{2}+\frac{1}{2}=1, & if \ n=1; \\ \frac{1}{2}+(n-1)+\frac{1}{2}=n, & if \ n \geqslant 2, \end{cases}$$

and thus

$$\sideset{}{'}\sum_{\ell=0}^{n} \cos\left(\frac{m\ell}{n}\pi\right) = n, \quad if \quad m=2\mu n, \ for \ some \ \mu \in \mathbb{Z}. \qquad (7.6.21)$$

Next, if $m \neq 2\mu n$ for $\mu \in \mathbb{Z}$, with the standard notation $\text{Re}(\alpha+i\beta)=\alpha$, where $i=\sqrt{-1}$, for the real part of a complex number $\alpha+i\beta$, we apply De Moivre's theorem, together with the geometric series summation formula

$$\sum_{\ell=0}^{n-1}(\alpha+i\beta)^\ell = \frac{1-(\alpha+i\beta)^n}{1-(\alpha+i\beta)}, \quad if \quad \alpha+i\beta \neq 1,$$

to deduce that

$$\sum_{\ell=0}^{n-1}\cos\left(\frac{m\ell}{n}\pi\right) = \sum_{\ell=0}^{n-1}\text{Re}\left(\cos\left(\frac{m\ell}{n}\pi\right)+i\sin\left(\frac{m\ell}{n}\right)\right)$$

$$= \mathrm{Re} \sum_{\ell=0}^{n-1} \left[\cos\left(\ell\left(\frac{m\pi}{n}\right)\right) + i\sin\left(\ell\left(\frac{m\pi}{n}\right)\right) \right]$$

$$= \mathrm{Re} \sum_{\ell=0}^{n-1} \left[\cos\left(\frac{m\pi}{n}\right) + i\sin\left(\frac{m\pi}{n}\right) \right]^{\ell}$$

$$= \mathrm{Re} \left[\frac{1 - \left(\cos\left(\frac{m\pi}{n}\right) + i\sin\left(\frac{m\pi}{n}\right) \right)^{n}}{1 - \left(\cos\left(\frac{m\pi}{n}\right) + i\sin\left(\frac{m\pi}{n}\right) \right)} \right]$$

$$= \mathrm{Re} \left[\frac{1 - (\cos(m\pi) + i\sin(m\pi))}{\left(1 - \cos\left(\frac{m\pi}{n}\right)\right) - i\sin\left(\frac{m\pi}{n}\right)} \right]$$

$$= \mathrm{Re} \left[\frac{(1 - (-1)^{m}) \left(\left(1 - \cos\left(\frac{m\pi}{n}\right)\right) + i\sin\left(\frac{m\pi}{n}\right) \right)}{\left(1 - \cos\left(\frac{m\pi}{n}\right)\right)^{2} + \sin^{2}\left(\frac{m\pi}{n}\right)} \right]$$

$$= \frac{(1 - (-1)^{m}) \left(1 - \cos\left(\frac{m\pi}{n}\right)\right)}{2 \left(1 - \cos\left(\frac{m\pi}{n}\right)\right)} = \frac{1}{2}(1 - (-1)^{m}),$$

that is,

$$\sum_{\ell=0}^{n-1} \cos\left(\frac{m\ell}{n}\pi\right) = \frac{1}{2}(1 - (-1)^{m}), \quad \text{if} \quad m \neq 2\mu n \quad \text{for} \quad \mu \in \mathbb{Z}. \tag{7.6.22}$$

By using (7.6.22), together with the definition (7.6.15) of the symbol Σ', we obtain

$$\sum_{\ell=0}^{n}{}' \cos\left(\frac{m\ell}{n}\pi\right) = \frac{1}{2} + \left[\sum_{\ell=0}^{n-1} \cos\left(\frac{m\ell}{n}\pi\right) - 1 \right] + \frac{1}{2}(-1)^{m}$$

$$= \frac{1}{2} + \left[\frac{1}{2}(1 - (-1)^{m}) - 1 \right] + \frac{1}{2}(-1)^{m} = 0,$$

and thus

$$\sum_{\ell=0}^{n}{}' \cos\left(\frac{m\ell}{n}\pi\right) = 0, \text{ if } m \neq 2\mu n \text{ for } \mu \in \mathbb{Z}. \tag{7.6.23}$$

If $j \neq k$, then $1 \leqslant j+k \leqslant 2n-1$ and $1 \leqslant j-k \leqslant n$, so that we may use (7.6.20) and (7.6.23) to deduce that

$$\sum_{\ell=0}^{n}{}' T_j(\xi_{n,\ell}) T_k(\xi_{n,\ell}) = \tfrac{1}{2}[0+0] = 0,$$

which proves (7.6.18).

Next, if $j = k$, with either $j = 0$ or $j = n$, then (7.6.21) and (7.6.20) yield

$$\sum_{\ell=0}^{n}{}' [T_j(\xi_{n,\ell})]^{2} = \tfrac{1}{2}[n+n] = n,$$

which proves the first line of (7.6.19).

Finally, if $j = k$, with $1 \leqslant j \leqslant n - 1$ (for $n \geqslant 2$), it follows from (7.6.23) that

$$\sideset{}{'}\sum_{\ell=0}^{n} \cos\left(\frac{(j+k)\ell}{n}\pi\right) = \sideset{}{'}\sum_{\ell=0}^{n} \cos\left(\frac{2j\ell}{n}\pi\right) = 0. \tag{7.6.24}$$

Hence, by using (7.6.24) and (7.6.21) in (7.6.20), we obtain

$$\sideset{}{'}\sum_{\ell=0}^{n} [T_j(\xi_{n,\ell})]^2 = \tfrac{1}{2}[0+n] = \frac{n}{2},$$

which proves the second line of (7.6.19). ∎

Suppose now $P \in \pi_n$, for a non-negative integer n. Since, according to Theorem 7.6.1 and Theorem 7.4.1(d), the Chebyshev polynomial sequence $\{T_j : j = 0, \ldots, n\}$ is a basis for π_n we know that there exists a unique coefficient sequence $\{c_0, \ldots, c_n\} \subset \mathbb{R}$ such that

$$P = \sum_{j=0}^{n} c_j T_j. \tag{7.6.25}$$

Let $k \in \{0, \ldots, n\}$. It then follows from (7.6.25) that, with $\{\xi_{n,\ell} : \ell = 0, \ldots, n\}$ denoting the point sequence (7.6.16) in Theorem 7.6.4, we have

$$\sideset{}{'}\sum_{\ell=0}^{n} P(\xi_{n,\ell}) T_k(\xi_{n,\ell}) = \sideset{}{'}\sum_{\ell=0}^{n} \left[\sum_{j=0}^{n} c_j T_j(\xi_{n,\ell}) \right] T_k(\xi_{n,\ell})$$

$$= \sum_{j=0}^{n} c_j \sideset{}{'}\sum_{\ell=0}^{n} T_j(\xi_{n,\ell}) T_k(\xi_{n,\ell})$$

$$= c_k \sideset{}{'}\sum_{\ell=0}^{n} [T_k(\xi_{n,\ell})]^2,$$

by virtue of (7.6.18) in Theorem 7.6.4, and thus

$$c_k = \frac{\displaystyle\sideset{}{'}\sum_{\ell=0}^{n} P(\xi_{n,\ell}) T_k(\xi_{n,\ell})}{\displaystyle\sideset{}{'}\sum_{\ell=0}^{n} [T_k(\xi_{n,\ell})]^2}, \quad k = 1, \ldots, n,$$

since (7.6.19) assures that the denominator is non-zero, and which can now be substituted into (7.6.25) to yield the formula

$$P = \sum_{j=0}^{n} \left[\frac{\displaystyle\sideset{}{'}\sum_{\ell=0}^{n} P(\xi_{n,\ell}) T_j(\xi_{n,\ell})}{\displaystyle\sideset{}{'}\sum_{\ell=0}^{n} [T_j(\xi_{n,\ell})]^2} \right] T_j. \tag{7.6.26}$$

It follows from (7.6.26), (7.6.19), (7.6.16), the formula (2.2.5) in Theorem 2.2.1(b), together with the definition (7.6.15) of the symbol Σ', that we have now established the following alternative formulation of the Chebyshev expansion (7.6.14) in Theorem 7.6.3.

Theorem 7.6.5. *For any non-negative integer* n,

$$P = \frac{2}{n} \sum_{j=0}^{n} {}' \left[\sum_{\ell=0}^{n} {}' P\left(\cos\left(\frac{\ell}{n}\pi \right) \right) \cos\left(\frac{j\ell}{n}\pi \right) \right] T_j, \quad P \in \pi_n, \tag{7.6.27}$$

with $\{T_j : j = 0,1,\dots\}$ *denoting the Chebyshev polynomials in* (2.2.1), *and with the symbol* Σ' *defined by* (7.6.15).

7.7 Exercises

Exercise 7.1 By applying Theorem 7.2.2, find the polynomial

$$f^* \in A := \text{span}\{1, x^2\}$$

for which it holds that

$$d^* := \sqrt{\int_1^2 [\sqrt{x} - f^*(x)]^2 dx} < \sqrt{\int_1^2 [\sqrt{x} - g(x)]^2 dx}, \quad g \in A, \quad g \neq f^*,$$

and then use the error expression (7.1.24) in Theorem 7.1.5 to evaluate d^*.

Exercise 7.2 For the constant polynomials $\{P^*_{w,k} : k \in \mathbb{N}\}$ of Exercise 4.10, apply Theorem 7.2.2 to prove the explicit formulation

$$\mathscr{P}^*_{w,k}(x) = c_{w,k} := \int_0^1 t^k w(t) dt, \quad x \in \mathbb{R}, \quad k \in \mathbb{N},$$

and then verify that this formula yields, for each $k \in \mathbb{N}$, the same constant polynomials as those obtained in (a), (b) and (c) of Exercise 4.10.

Exercise 7.3 As a continuation of Exercise 7.2, apply the error expression (7.1.25) in Theorem 7.1.5 to obtain, for any $k \in \mathbb{N}$, an explicit formulation of the error

$$\sqrt{\int_0^1 w(x)[x^k - P^*_{w,k}(x)]^2 dx},$$

and then use this formula to evaluate the corresponding approximation error for each of the cases (a), (b) and (c) of Exercise 4.10.

Exercise 7.4 For any inner product space $(X, \langle \cdot, \cdot \rangle)$, prove that a set $\{f_1, \dots, f_d\} \subset X$ is linearly independent if and only if the corresponding Gram matrix G, as given by (7.2.7), is invertible.

[*Hint:* Observe that the proof in the "only if" direction may be obtained directly from Theorem 7.2.2.]

Exercise 7.5 Prove that $\{1, e^x, e^{-x}\}$ is a linearly independent set, to deduce that the linear space

$$A := \text{span}\{1, e^x, e^{-x}\}$$

satisfies $\dim(A) = 3$, and $\{1, e^x, e^{-x}\}$ is a basis of A.

[*Hint:* As in a standard result from linear algebra based on the Wronskian determinant for the set $\{1, e^x, e^{-x}\}$, differentiate the identity

$$\alpha + \beta e^x + \gamma e^{-x} = 0, \quad x \in \mathbb{R},$$

with $\{\alpha, \beta, \gamma\} \subset \mathbb{R}$ denoting a coefficient sequence, twice, and show that the determinant of the 3×3 coefficient matrix obtained by setting $x = 0$ in the resulting system of three identities, is not equal to zero.]

Exercise 7.6 As a continuation of Exercise 7.5, use the Gram-Schmidt procedure of Theorem 7.3.3, together with the basis $\{1, e^x, e^{-x}\}$ of A, to obtain an orthogonal basis for A with respect to the inner product space $(C[-1, 1], \langle \cdot, \cdot \rangle)$, where

$$\langle f, g \rangle := \int_{-1}^{1} (1 - x^2) f(x) g(x) dx, \quad f, g \in C[-1, 1].$$

Exercise 7.7 As a continuation of Exercise 7.6, apply Theorem 7.3.1 to find the function $f^* \in A$ satisfying the best approximation condition

$$d^* := \sqrt{\int_{-1}^{1} (1 - x^2)[x - f^*(x)]^2 dx} < \sqrt{\int_{-1}^{1} (1 - x^2)[x - g(x)]^2 dx}, \quad g \in A, \quad g \neq f^*,$$

and then use Theorem 7.3.4 to evaluate the minimum value d^*.

Exercise 7.8 For an inner product space $(X, \langle \cdot, \cdot \rangle)$, suppose A is a finite-dimensional subspace of X, with $\dim(A) = d \in \mathbb{N}$, and let $\{f_1, \ldots, f_d\}$ denote an orthogonal basis for A. By applying Theorem 7.3.4, show that, for any $f \in X$, the infinite series

$$\sum_{j=1}^{\infty} \left[\frac{\langle f, f_j \rangle}{\|f_j\|} \right]^2,$$

with the norm $\| \cdot \|$ given by (4.2.15), is convergent, and, moreover, prove the Bessel inequality

$$\sum_{j=1}^{\infty} \left[\frac{\langle f, f_j \rangle}{\|f_j\|} \right]^2 \leq \|f\|^2, \quad f \in X.$$

Exercise 7.9 Apply the three-term recursion formulation of Theorem 7.4.2 to obtain the sequence $\{P_j^{\perp} : j = 0, 1, 2, 3\}$ of (monic) orthogonal polynomials, with $P_j^{\perp} \in \tilde{\pi}_j, j = 0, 1, 2, 3$, such that

$$\int_{-1}^{1} x^2 P_j^{\perp}(x) P_k^{\perp}(x) dx = 0, \quad j, k = 0, \ldots, 3; \quad j \neq k.$$

Exercise 7.10 As a continuation of Exercise 7.9, apply Theorems 7.4.1 and 7.3.4 to calculate the minimum value

$$\min_{P \in \pi_3} \sqrt{\int_{-1}^{1} x^2 [|x| - P(x)]^2 dx},$$

as well as the corresponding optimal polynomial $P = \widetilde{P}_3^* \in \pi_3$.

Exercise 7.11 By using the recursion formulation in Theorem 7.5.1, extend the formulas in (7.5.26) by computing the Legendre polynomials $L_7^{\perp}$ and $L_8^{\perp}$.

Exercise 7.12 For the function

$$f(x) = \ln(x+2), \quad x \in [0,2],$$

by applying (7.5.38) in Theorem 7.5.3, calculate the linear polynomial $\widetilde{\mathscr{P}}_1^* f$, with $\widetilde{\mathscr{P}}_1^*$: $C[0,2] \to \pi_1$ denoting the best L^2 polynomial approximation operator on the interval $[0,2]$. Compute also the corresponding minimum value

$$\tilde{d}^* := \sqrt{\int_0^2 [f(x) - (\widetilde{\mathscr{P}}_1^* f)(x)]^2 dx},$$

and verify that, with $\mathscr{P}_1^* f$ denoting the linear polynomial of Exercise 6.6, the inequalities

$$\tilde{d}^* < \sqrt{\int_0^2 [f(x) - (\mathscr{P}_1^* f)(x)]^2 dx}; \quad ||f - \mathscr{P}_1^* f||_\infty < ||f - \widetilde{\mathscr{P}}_1^* f||_\infty,$$

are satisfied, where $||\cdot||_\infty$ denotes the maximum norm on $[0,2]$.

Exercise 7.13 Find the minimum value

$$\min_{P \in \pi_1} \sqrt{\int_0^1 [e^x - P(x)]^2 dx},$$

as well as the corresponding optimal linear polynomial $\widetilde{P}_1^*$.

Exercise 7.14 By applying Theorem 7.4.3, find the best L^2 polynomial approximation $\widetilde{P}_6^*$ on $[0,2]$ from π_6 to the polynomial function

$$f(x) = x^3(1 - 2x^4), \quad x \in [0,2],$$

as well as the corresponding minimum value

$$\sqrt{\int_0^2 [f(x) - \widetilde{P}_6^*(x)]^2 dx}.$$

Exercise 7.15 Prove the following analogue of Theorem 2.2.2 : For $n \in \mathbb{N}$, and any weight function w on $[a,b]$, it holds that

$$\min_{P \in \pi_n} \sqrt{\int_a^b w(x)[P(x)]^2 dx} = \sqrt{\int_a^b w(x)[P_n^\perp(x)]^2 dx},$$

with $P_n^\perp$ denoting the orthogonal polynomial of Theorem 7.4.1.

[*Hint:* Apply Theorem 7.4.3.]

Exercise 7.16 By using Exercise 7.15, find the minimum values

(a) $\min_{P \in \tilde{\pi}_3} \sqrt{\int_{-1}^{2} [P(x)]^2 dx}$;   (b) $\min_{P \in \tilde{\pi}_3} \sqrt{\int_{-1}^{1} \frac{1}{\sqrt{1-x^2}} [P(x)]^2 dx}$,

as well as the corresponding optimal polynomials.

[*Hint:* Use also Theorems 7.5.1, 7.5.2 and 7.6.1, as well as (7.6.10), (7.6.11).]

Exercise 7.17 Prove that the Legendre polynomials $\{L_j^{\perp} : j = 0, 1, \ldots\}$ satisfy the Rodrigues formula

$$L_j^{\perp}(x) = \frac{j!}{(2j)!} \left(\frac{d}{dx}\right)^j (x^2 - 1)^j, \quad j = 0, 1, \ldots.$$

[*Hint:* Show that the polynomial sequence

$$P_j(x) := \frac{j!}{(2j)!} \left(\frac{d}{dx}\right)^j (x^2 - 1)^j, \quad j = 0, 1, \ldots,$$

satisfies $P_j \in \tilde{\pi}_j, j = 0, 1, \ldots$, with also, by using integration by parts, $P_j \perp \pi_{j-1}, j = 1, 2, \ldots$, before applying the uniqueness result implied by Theorem 7.4.1(d).]

Exercise 7.18 Verify that the Legendre polynomial formulation in Exercise 7.17 yields, for $j = 0, \ldots, 4$, the formulas in the first lines of (7.5.25) and (7.5.26).

Exercise 7.19 Find the minimum value

$$\min_{a,b,c \in \mathbb{R}} \sqrt{\int_{-1}^{1} \frac{1}{\sqrt{1-x^2}} \left[|x| + ax^2 + bx + c\right]^2 dx},$$

as well as the corresponding optimal values of a, b and c.

[*Hint:* Apply Theorem 7.6.2.]

Exercise 7.20 Find (a) the Legendre expansion with respect to the interval $[-1, 1]$ in π_6; (b) the Chebyshev expansion in π_6, of the polynomial

$$P(x) = x^3,$$

by applying, respectively, Theorems 7.5.4 and 7.6.3. Verify the result of (b) by an application of Theorem 7.6.5.

Chapter 8

Interpolatory Quadrature

It is well-known that if the anti-derivative F of a given function $f \in C[a,b]$ is available, then the definite integral of f on $[a,b]$ can be evaluated by applying the fundamental theorem of calculus, namely:

$$\int_a^b f(x)dx = F(b) - F(a).$$

Unfortunately, with the exception of only a handful of functions f, it is not feasible to find their anti-derivatives F. Examples of functions that do not have simple expressions of anti-derivatives include

$$f(x) = e^{x^2}; \quad f(x) = \frac{\sin x}{x}; \quad f(x) = \frac{1}{\ln x}; \quad f(x) = \sqrt{1+x^3}.$$

For this reason, only numerical methods can be used to approximate $\int_a^b f(x)dx$ within certain desirable error tolerance. In this chapter, we investigate the method of interpolatory quadrature, by considering definite integrals of the polynomial interpolant of $f \in C[a,b]$, as numerical approximation of the definite integral of f.

8.1 General formulation, exactness and convergence

For a bounded interval $[a,b]$ and a weight function w on $[a,b]$ satisfying the conditions (4.2.8) and (4.2.9), we define the functional $\mathscr{I} : C[a,b] \to \mathbb{R}$ as the weighted integral

$$\mathscr{I}[f] := \int_a^b w(x)f(x)dx, \quad f \in C[a,b] \tag{8.1.1}$$

Next, for any non-negative integer n, we denote by

$$\triangle_n := \{x_{n,0}, \dots, x_{n,n}\} \tag{8.1.2}$$

a sequence of $n+1$ distinct points in $[a,b]$ such that

$$a \leqslant x_{n,0} < x_{n,1} < \cdots < x_{n,n} \leqslant b, \tag{8.1.3}$$

in terms of which we introduce the numerical approximation method

$$\mathscr{I}[f] \approx \mathscr{I}[\mathscr{P}_n^I f], \quad f \in C[a,b], \tag{8.1.4}$$

where $\mathscr{P}_n^I : C[a,b] \to \pi_n$ is the Lagrange polynomial interpolation operator with respect to the interpolation points $\triangle_n$, as defined in (5.1.2). Hence, with the functional $\mathscr{Q}_n : C[a,b] \to \mathbb{R}$ defined by

$$\mathscr{Q}_n[f] := \mathscr{I}[\mathscr{P}_n^I f], \quad f \in C[a,b], \tag{8.1.5}$$

the numerical approximation method (8.1.4) has the equivalent formulation

$$\mathscr{I}[f] \approx \mathscr{Q}_n[f], \quad f \in C[a,b]. \tag{8.1.6}$$

The functional $\mathscr{Q}_n$ defined by (8.1.5) is called an interpolatory quadrature rule (or formula). Observe from (8.1.5) and (8.1.1), together with (5.1.2) and the Lagrange interpolation formula (1.2.5) in Theorem 1.2.2, that, for any $f \in C[a,b]$,

$$\mathscr{Q}_n[f] = \int_a^b w(x) \left[\sum_{j=0}^n f(x_{n,j}) L_{n,j}(x) \right] dx = \sum_{j=0}^n f(x_{n,j}) \left[\int_a^b w(x) L_{n,j}(x) dx \right],$$

and thus

$$\mathscr{Q}_n[f] = \sum_{j=0}^n w_{n,j} f(x_{n,j}), \tag{8.1.7}$$

where

$$w_{n,j} := \int_a^b w(x) L_{n,j}(x) dx, \quad j = 0, \ldots, n, \tag{8.1.8}$$

with $\{L_{n,j} : j = 0, \ldots, n\}$ denoting, as given in (1.2.1), the Lagrange fundamental polynomials with respect to the sequence $\triangle_n$. The real numbers $\{w_{n,0}, \ldots, w_{n,n}\}$, as defined by (8.1.8), are called the weights of the interpolatory quadrature rule $\mathscr{Q}_n$.

Observe from (8.1.1) and (8.1.7) that the functionals $\mathscr{I}$ and $\mathscr{Q}_n$ are both linear, that is,

$$\mathscr{I}[\lambda f + \mu g] = \lambda \mathscr{I}[f] + \mu \mathscr{I}[g], \quad \lambda, \mu \in \mathbb{R}, \quad f, g \in C[a,b]; \tag{8.1.9}$$

$$\mathscr{Q}_n[\lambda f + \mu g] = \lambda \mathscr{Q}_n[f] + \mu \mathscr{Q}_n[g], \quad \lambda, \mu \in \mathbb{R}, \quad f, g \in C[a,b]. \tag{8.1.10}$$

By using the definition (8.1.8), together with the identity (1.2.8) in Theorem 1.2.3, we deduce that

$$\sum_{j=0}^n w_{n,j} = \sum_{j=0}^n \left[\int_a^b w(x) L_{n,j}(x) \right] dx = \int_a^b w(x) \left[\sum_{j=0}^n L_{n,j}(x) \right] dx = \int_a^b w(x) dx. \tag{8.1.11}$$

Suppose, in an application of the quadrature rule $\mathscr{Q}_n$ as in (8.1.5), as a result of measuring errors, or computer rounding errors, we are instead actually computing $\mathscr{Q}_n[\tilde{f}]$, where

$$|f(x_{n,j}) - \tilde{f}(x_{n,j})| \leq \varepsilon, \quad j = 0, \ldots, n, \tag{8.1.12}$$

for a given tolerance $\varepsilon > 0$, and thus, from (8.1.10) and (8.1.7),

$$|\mathcal{Q}_n[f] - \mathcal{Q}_n[\tilde{f}]| = |\mathcal{Q}_n[f - \tilde{f}]| = \left| \sum_{j=0}^{n} w_{n,j}[f(x_{n,j}) - \tilde{f}(x_{n,j})] \right|$$

$$\leqslant \sum_{j=0}^{n} |w_{n,j}| \, |f(x_{n,j}) - \tilde{f}(x_{n,j})| \leqslant \varepsilon \sum_{j=0}^{n} |w_{n,j}|. \qquad (8.1.13)$$

Since, ideally, we would like the absolute difference $|\mathcal{Q}_n[f] - \mathcal{Q}_n[\tilde{f}]|$ to be "small", we see from (8.1.13) that it reflects favourably on an interpolatory quadrature rule $\mathcal{Q}_n$ if the points $\triangle_n$ are chosen in such a way that the quantity $\sum_{j=0}^{n} |w_{n,j}|$ is minimized. Now observe from (8.1.11) and (4.2.8) that

$$\sum_{j=0}^{n} |w_{n,j}| \geqslant \sum_{j=0}^{n} w_{n,j} = \int_{a}^{b} w(x)dx > 0,$$

that is,

$$\sum_{j=0}^{n} |w_{n,j}| \geqslant \int_{a}^{b} w(x)dx, \qquad (8.1.14)$$

for any choice of $\triangle_n$ as in (8.1.2) and (8.1.3). Moreover,

$$w_{n,j} \geqslant 0, \; j = 0, \ldots, n \Rightarrow \sum_{j=0}^{n} |w_{n,j}| = \sum_{j=0}^{n} w_{n,j} = \int_{a}^{b} w(x)dx, \qquad (8.1.15)$$

which, together with (8.1.14), and the fact that $\int_{a}^{b} w(x)dx$ is independent of $\triangle_n$, implies that, if the points $\triangle_n$ are chosen in such a way that the weight sequence is non-negative, that is,

$$w_{n,j} \geqslant 0, \quad j = 0, \ldots, n, \qquad (8.1.16)$$

then the quantity $\sum_{j=0}^{n} |w_{n,j}|$ in the right hand side of the inequality (8.1.13) is indeed minimized.

Next, we note from (8.1.5), together with the exactness on π_n of the approximation operator $\mathcal{P}_n^I$, as given in Theorem 5.1.1(b), that

$$\mathcal{Q}_n[P] = \mathcal{I}[P], \quad P \in \pi_n. \qquad (8.1.17)$$

The degree of exactness of an interpolatory quadrature rule $\mathcal{Q}_n$ is now defined as the largest non-negative integer m for which it holds that

$$\mathcal{Q}_n[P] = \mathcal{I}[P], \quad P \in \pi_m. \qquad (8.1.18)$$

As an immediate consequence of the fact that (8.1.17) is satisfied, we then have the following minimum degree of exactness in interpolatory quadrature.

Theorem 8.1.1. *For any non-negative integer n, the degree of exactness m of an interpolatory quadrature rule $\mathcal{Q}_n$ as in (8.1.5) satisfies the inequality*

$$m \geqslant n. \tag{8.1.19}$$

We proceed to apply Theorem 6.4.1 and Theorem 8.1.1 in order to show that the non-negative weight condition (8.1.16) guarantees interpolatory quadrature rule convergence on $C[a,b]$, as follows.

Theorem 8.1.2. *Suppose the sequence $\{\triangle_n : n = 0, 1, \ldots\}$, as in (8.1.2), (8.1.3), is chosen in such a way that the corresponding sequence $\{\mathcal{Q}_n : n = 0, 1, \ldots\}$ of interpolatory quadrature rules, as defined by (8.1.5), and as given in terms of weights in (8.1.7), (8.1.8), satisfies the non-negative weight condition (8.1.16) for $n = 0, 1, \ldots$.. Then:*

(a) *Convergence on $C[a,b]$ holds, that is,*

$$|\mathscr{I}[f] - \mathcal{Q}_n[f]| \to 0, \quad n \to \infty, \quad f \in C[a,b], \tag{8.1.20}$$

with $\mathscr{I}[f]$ denoting the weighted integral in (8.1.1).

(b) *If $f \in C^k[a,b]$ for an integer $k \in \mathbb{N}$, the convergence rate result*

$$|\mathscr{I}[f] - \mathcal{Q}_n[f]| \leqslant 2 \left[\int_a^b w(x)dx \right] (b-a)^k \frac{(m_n-k)!}{(m_n)!} ||f^{(k)}||_\infty, \ n = k, k+1, \ldots, \tag{8.1.21}$$

holds, where, for $n = 0, 1, \ldots$, the non-negative integer m_n denotes the degree of exactness of $\mathcal{Q}_n$, so that, according to Theorem 8.1.1,

$$m_n \geqslant n, \quad n = 0, 1, \ldots, \tag{8.1.22}$$

and with, in particular,

$$|\mathscr{I}[f] - \mathcal{Q}_n[f]| \leqslant 2 \left[\int_a^b w(x)dx \right] \frac{b-a}{m_n} ||f'||_\infty, \quad n = 1, 2, \ldots, \ f \in C^1[a,b]. \tag{8.1.23}$$

Proof. (a) Let $f \in C[a,b]$, and, for $n = 0, 1, \ldots$, and with m_n denoting the degree of exactness of $\mathcal{Q}_n$, let $\mathscr{P}^*_{m_n} : C[a,b] \to \pi_{m_n}$ denote the sequence of best uniform polynomial approximation operators, as obtained by replacing n by m_n in the definition (6.4.1). Then

$$\mathscr{P}^*_{m_n} f \in \pi_{m_n}, \quad n = 0, 1, \ldots,$$

and thus, since $\mathcal{Q}_n$ has degree of exactness m_n,

$$\mathscr{I}[\mathscr{P}^*_{m_n} f] = \mathcal{Q}_n[\mathscr{P}^*_{m_n} f], \quad n = 0, 1, \ldots. \tag{8.1.24}$$

By using, consecutively, (8.1.24), (8.1.9), (8.1.10), (8.1.1), (8.1.7), the non-negativity (4.2.9) on (a,b) of the weight function w, (8.1.16) and (8.1.11), we deduce that, for $n = 0, 1, \ldots,$

$$|\mathscr{I}[f] - \mathscr{Q}_n[f]| = |(\mathscr{I}[f] - \mathscr{I}[\mathscr{P}^*_{m_n}f]) + (\mathscr{Q}_n[\mathscr{P}^*_{m_n}f] - \mathscr{Q}_n[f])|$$

$$\leqslant |\mathscr{I}[f] - \mathscr{I}[\mathscr{P}^*_{m_n}f]| + |\mathscr{Q}_n[\mathscr{P}^*_{m_n}f] - \mathscr{Q}_n[f]|$$

$$= |\mathscr{I}[f - \mathscr{P}^*_{m_n}f]| + |\mathscr{Q}_n[\mathscr{P}^*_{m_n}f - f]|$$

$$= \left| \int_a^b w(x)[f(x) - (\mathscr{P}^*_{m_n}f)(x)]dx \right| + \left| \sum_{j=0}^{n} w_{n,j}[(\mathscr{P}^*_{m_n}f)(x_{n,j}) - f(x_{n,j})] \right|$$

$$\leqslant \int_a^b w(x)|f(x) - (\mathscr{P}^*_{m_n}f)(x)|dx + \sum_{j=0}^{n} w_{n,j}|(\mathscr{P}^*_{m_n}f)(x_{n,j}) - f(x_{n,j})|$$

$$\leqslant \|f - \mathscr{P}^*_{m_n}f\|_\infty \left[\int_a^b w(x)dx + \sum_{j=0}^{n} w_{n,j} \right]$$

$$= 2 \left[\int_a^b w(x)dx \right] \|f - \mathscr{P}^*_{m_n}f\|_\infty. \tag{8.1.25}$$

Now observe from (8.1.22) that

$$\pi_n \subset \pi_{m_n}, \quad n = 0, 1, \ldots, \tag{8.1.26}$$

and thus

$$\|f - \mathscr{P}^*_{m_n}f\|_\infty = \min_{P \in \pi_{m_n}} \|f - P\|_\infty \leqslant \min_{P \in \pi_n} \|f - P\|_\infty = \|f - \mathscr{P}^*_n f\|_\infty, \tag{8.1.27}$$

for $n = 0, 1, \ldots$. It follows from (8.1.25) and (8.1.27) that

$$|\mathscr{I}[f] - \mathscr{Q}_n[f]| \leqslant 2 \left[\int_a^b w(x)dx \right] \|f - \mathscr{P}^*_n f\|_\infty, \, n = 0, 1, \ldots. \tag{8.1.28}$$

The convergence result (8.1.20) is now an immediate consequence of (8.1.28), together with the uniform convergence result (6.4.2) in Theorem 6.4.1(b).

(b) Suppose $f \in C^k[a,b]$ for an integer $k \in \mathbb{N}$, and let $n \in \{k, k+1, \ldots\}$. But then (8.1.22) implies $m_n \in \{k, k+1, \ldots\}$, so that we may apply the convergence rate result (6.4.3) in Theorem 6.4.1(c) to deduce that

$$\|f - \mathscr{P}^*_{m_n}f\|_\infty \leqslant (b-a)^k \frac{(m_n - k)!}{(m_n)!} \|f^{(k)}\|_\infty,$$

which, together with (8.1.25), then yields (8.1.21). Finally observe that (8.1.23) is obtained by setting $k = 1$ in (8.1.21). ∎

Observe from (8.1.22) that, in the right hand side of the convergence rate result (8.1.21) in Theorem 8.1.2(b), for any $k \in \mathbb{N}$, and $n = k, k+1, \ldots$,

$$\frac{(m_n - k)!}{(m_n)!} = \frac{1}{(m_n)(m_n - 1) \ldots (m_n - k + 1)}$$

$$\leqslant \frac{1}{n(n-1) \ldots (n-k+1)} = \frac{(n-k)!}{n!}, \qquad (8.1.29)$$

which, together with (8.1.21), yields the inequality

$$|\mathscr{I}[f] - \mathscr{Q}_n[f]| \leqslant 2 \left[\int_a^b w(x) dx \right] (b-a)^k \frac{(n-k)!}{n!} ||f^{(k)}||_\infty, \ n = k, k+1, \ldots. \qquad (8.1.30)$$

We proceed in the next section to construct an interpolatory quadrature rule sequence $\{\mathscr{Q}_n : n = 0, 1, \ldots\}$ satisfying the conditions of Theorem 8.1.2, and with, moreover, corresponding degrees of exactness $m_n, n = 0, 1, \ldots$, which significantly exceed the lower bound (8.1.22) in Theorem 8.1.2(b).

8.2   Gauss quadrature

According to (8.1.17), and as stated in Theorem 8.1.1, the degree of exactness m of an interpolatory quadrature formula $\mathscr{Q}_n$ satisfies the inequality $m \geqslant n$ for any choice of the sequence $\triangle_n$ in (8.1.2), (8.1.3). Noting from (8.1.2) that the sequence $\triangle_n$ has precisely $n+1$ (distinct) points, we proceed to show that there exists an optimal choice of $\triangle_n$ for which the corresponding interpolatory quadrature formula $\mathscr{Q}_n$ has degree of exactness

$$m = n + (n+1) = 2n + 1. \qquad (8.2.1)$$

We shall rely on the following result on the zeros of the orthogonal polynomials studied in Chapter 7.

Theorem 8.2.1. *For $j \in \mathbb{N}$, the orthogonal polynomial $P_j^\perp$ in Theorem 7.4.1 has precisely j distinct real zeros in (a, b), each of which corresponds to a sign change of $P_j^\perp$.*

Proof. Let the non-negative integer r denote the number of distinct real zeros in (a, b) of $P_j^\perp$, such that each such zero corresponds to a sign change of $P_j^\perp$. Our proof will be complete if we can show that

$$r = j. \qquad (8.2.2)$$

Using a proof by contradiction, suppose

$$r \leqslant j - 1, \qquad (8.2.3)$$

and, if $r \geqslant 1$, suppose that the sign changes in (a,b) of $P_j^{\perp}$ occur at the points $\{x_1, \ldots, x_r\}$, where

$$a < x_1 < \cdots < x_r < b. \tag{8.2.4}$$

For the polynomial

$$Q(x) := \begin{cases} 1, & \text{if } r = 0; \\ \displaystyle\prod_{k=1}^{r}(x_k - x), & \text{if } r \geqslant 1, \end{cases} \tag{8.2.5}$$

it follows from the assumption (8.2.3) that $Q \in \pi_r \subset \pi_{j-1}$, so that $Q \in \pi_{j-1}$, and thus, from (7.4.4) in Theorem 7.4.1(c),

$$\int_a^b w(x)P_j^{\perp}(x)Q(x)dx = 0. \tag{8.2.6}$$

Next, we observe from (8.2.5), together with the definitions of the non-negative integer r and the sequence $\{x_1, \ldots, x_r\}$ if $r \geqslant 1$, that the product $P_j^{\perp}(x)Q(x)$ is of one sign on the set

$$\left.\begin{array}{ll} (a,b), & \text{if } r = 0; \\ \\ (a,b) \setminus \{x_1, \ldots, x_r\}, & \text{if } r \geqslant 1, \end{array}\right\}$$

and thus, by recalling also the properties (4.2.8) and (4.2.9) of the weight function w, we deduce that

$$\int_a^b w(x)P_j^{\perp}(x)Q(x)dx \neq 0,$$

which contradicts (8.2.6). Hence (8.2.3) is not true, that is, $r \geqslant j$. But, according to (7.4.2) in Theorem 7.4.1(a), the polynomial $P_j^{\perp}$ can have at most j distinct zeros in (a,b), and thus $r \leqslant j$, which, together with $r \geqslant j$, then yields the desired result (8.2.2). ∎

Our next result now specifies the choice of $\triangle_n$ which yields the degree of exactness (8.2.1).

Theorem 8.2.2. *For any non-negative integer n, let the sequence $\triangle_n$ in (8.1.2) be chosen, as guaranteed by Theorem 8.2.1, as the $n+1$ distinct zeros in (a,b), ordered as in (8.1.3), of the orthogonal polynomial $P_{n+1}^{\perp}$ in Theorem 7.4.1, with weight function w on $[a,b]$ as in (8.1.1). Then the corresponding interpolatory quadrature rule $\mathcal{Q}_n =: \mathcal{Q}_n^G$, as given by (8.1.5), has degree of exactness*

$$m = 2n+1. \tag{8.2.7}$$

Proof. Our first step is to show that $m \geqslant 2n+1$, for which, according to Theorem 8.1.1, it will suffice to prove that, if $P \in \pi_{2n+1}$, with $\deg(P) \geqslant n+1$, then

$$\mathscr{I}[P] = \mathscr{Q}_n^G[P]. \tag{8.2.8}$$

Suppose therefore that P is a polynomial, with

$$n+1 \leqslant \deg(P) \leqslant 2n+1. \tag{8.2.9}$$

Since (7.4.2) in Theorem 7.4.1(a) gives

$$\deg\left(P_{n+1}^{\perp}\right) = n+1, \tag{8.2.10}$$

we may now deduce from the polynomial division theorem that there exist polynomials Q and R, with $\deg(R) < \deg(P_{n+1}^{\perp}) = n+1$, or R is the zero polynomial, that is,

$$R \in \pi_n, \tag{8.2.11}$$

such that

$$P = QP_{n+1}^{\perp} + R. \tag{8.2.12}$$

Observe from (8.2.12) that the assumption that Q is the zero polynomial yields $P = R$, which is not possible, by virtue of the first inequality in (8.2.9), together with (8.2.11). Hence Q is not the zero polynomial, according to which we may deduce from the second inequality in (8.2.9), together with (8.2.12), (8.2.10) and (8.2.11), that

$$2n+1 \geqslant \deg(P) = \deg(QP_{n+1}^{\perp} + R)$$

$$= \deg(QP_{n+1}^{\perp}) = \deg(Q) + \deg(P_{n+1}^{\perp}) = \deg(Q) + n+1,$$

and thus

$$\deg(Q) \leqslant n,$$

that is,

$$Q \in \pi_n. \tag{8.2.13}$$

Next, we apply (8.2.12), (8.1.9) and (8.1.1) to obtain

$$\mathscr{I}[P] = \mathscr{I}[QP_{n+1}^{\perp}] + \mathscr{I}[R] = \int_a^b w(x)P_{n+1}^{\perp}(x)Q(x)dx + \mathscr{I}[R]. \tag{8.2.14}$$

But, from (4.2.10) and (8.2.13), together with (7.4.4) in Theorem 7.4.1(c),

$$\int_a^b w(x)P_{n+1}^{\perp}(x)Q(x)dx = \langle P_{n+1}^{\perp}, Q \rangle_{2,w} = 0,$$

which, together with (8.2.14), yields

$$\mathscr{I}[P] = \mathscr{I}[R]. \tag{8.2.15}$$

By using (8.2.12), (8.1.10) and (8.1.7), we furthermore deduce that

$$\mathscr{Q}_n^G[P] = \mathscr{Q}_n^G[QP_{n+1}^\perp] + \mathscr{Q}_n^G[R] = \sum_{j=0}^{n} w_{n,j} Q(x_{n,j}) P_{n+1}^\perp(x_{n,j}) + \mathscr{Q}_n^G[R]. \tag{8.2.16}$$

Since $\{x_{n,0}, \ldots, x_{n,n}\}$ are the zeros of the polynomial $P_{n+1}^\perp$, it follows that

$$\sum_{j=0}^{n} w_{n,j} Q(x_{n,j}) P_{n+1}^\perp(x_{n,j}) = 0,$$

which, together with (8.2.16), gives

$$\mathscr{Q}_n^G[P] = \mathscr{Q}_n^G[R]. \tag{8.2.17}$$

Finally, observe from (8.2.11) and (8.1.17) that

$$\mathscr{Q}_n^G[R] = \mathscr{I}[R]. \tag{8.2.18}$$

The desired result (8.2.8) then follows from (8.2.15), (8.2.17) and (8.2.18).

We have therefore now shown that the exactness condition (8.1.18) holds for $m = 2n+1$, and thus $m \geqslant 2n+1$. Hence, to establish the fact that the degree of exactness result (8.2.7) is satisfied, it remains to find a polynomial $\widetilde{P}$ of degree $2n+2$ such that

$$\mathscr{Q}_n^G[\widetilde{P}] \neq \mathscr{I}[\widetilde{P}]. \tag{8.2.19}$$

To this end, we let

$$\widetilde{P} := (P_{n+1}^\perp)^2,$$

according to which (7.4.2) in Theorem 7.4.1(a) then implies that $\deg(\widetilde{P}) = 2n+2$. Also, from (8.1.7), and the fact that $\{x_{n,0}, \ldots, x_{n,n}\}$ are the zeros of $P_{n+1}^\perp$, we have

$$\mathscr{Q}_n^G[\widetilde{P}] = \sum_{j=0}^{n} w_{n,j} [P_{n+1}^\perp(x_{n,j})]^2 = 0. \tag{8.2.20}$$

Moreover, (8.1.1) and (4.2.9) yield

$$\mathscr{I}[\widetilde{P}] = \int_a^b w(x) [P_{n+1}^\perp(x)]^2 dx > 0,$$

which, together with (8.2.20), then implies the desired result (8.2.19). ∎

The interpolatory quadrature rule $\mathscr{Q}_n^G$ in Theorem 8.2.2 is known as the Gauss quadrature rule of degree n for the weight function w on the interval $[a,b]$.

We proceed to show that Gauss quadrature rules satisfy the condition (8.1.16) of non-negative weights, as required in Theorem 8.1.2.

Theorem 8.2.3. *For any non-negative integer n, let $\mathscr{Q}_n^G$ denote the Gauss quadrature rule in Theorem 8.2.2. Then the corresponding weights $\{w_{n,0}, \ldots, w_{n,n}\}$, as in (8.1.7), (8.1.8), satisfy the positivity condition*

$$w_{n,j} > 0, \quad j = 0, \ldots, n. \tag{8.2.21}$$

Proof. For any $k \in \{0, \ldots, n\}$, define the polynomial

$$P(x) := [L_{n,k}(x)]^2, \qquad (8.2.22)$$

where $L_{n,k}$ denotes the Lagrange fundamental polynomial with respect to the sequence $\{x_{n,0}, \ldots, x_{n,n}\}$ in Theorem 8.2.2, as defined, according to (1.2.1), by

$$L_{n,k}(x) := \prod_{\substack{k \neq \ell = 0}}^{n} \frac{x - x_{n,\ell}}{x_{n,k} - x_{n,\ell}}.$$

Since $\deg(L_{n,k}) = n$, it follows from (8.2.22) that $\deg(P) = 2n$, and thus

$$P \in \pi_{2n}. \qquad (8.2.23)$$

Also, from (1.2.3) in Theorem 1.2.1(a), together with (8.2.22), we obtain

$$P(x_{n,j}) = \delta_{k-j}, \quad j = 0, \ldots, n. \qquad (8.2.24)$$

From the degree of exactness $m = 2n + 1$ of the Gauss quadrature rule $\mathcal{Q}_n^G$, as given in (8.2.7) of Theorem 8.2.2, and since $\pi_{2n} \subset \pi_{2n+1}$, we deduce from (8.2.23) that

$$\mathcal{Q}_n^G[P] = \mathcal{I}[P]. \qquad (8.2.25)$$

It then follows from (4.2.9), (8.1.1), (8.2.22), (8.2.25), (8.1.7) and (8.2.24) that

$$0 < \int_a^b w(x)[L_{n,k}(x)]^2 dx = \sum_{j=0}^{n} w_{n,j} P(x_{n,j}) = \sum_{j=0}^{n} w_{n,j} \delta_{k-j} = w_{n,k},$$

and thereby completing our proof of (8.2.21). ∎

Observe from Theorems 8.2.2 and 8.2.3 that the Gauss quadrature formula $\mathcal{Q}_n^G$ satisfies the conditions of Theorem 8.1.2, with

$$m_n = 2n + 1, \quad n = 0, 1, \ldots, \qquad (8.2.26)$$

and thus we immediately have the following result as a special case of Theorem 8.1.2.

Theorem 8.2.4. *Let $\{\mathcal{Q}_n^G : n = 0, 1, \ldots\}$ denote the sequence of Gauss quadrature rules of Theorem 8.2.2. Then:*

(a) *Convergence on $C[a,b]$ holds, that is,*

$$|\mathcal{I}[f] - \mathcal{Q}_n^G[f]| \to 0, \quad n \to \infty, \quad f \in C[a,b], \qquad (8.2.27)$$

with $\mathcal{I}[f]$ denoting the weighted integral in (8.1.1).

(b) *If $f \in C^k[a,b]$ for an integer $k \in \mathbb{N}$, the convergence rate result*

$$|\mathcal{I}[f] - \mathcal{Q}_n^G[f]| \leqslant 2 \left[\int_a^b w(x) dx \right] (b-a)^k \frac{(2n+1-k)!}{(2n+1)!} ||f^{(k)}||_\infty, \ n = k, k+1, \ldots, \qquad (8.2.28)$$

holds, with, in particular,

$$|\mathcal{I}[f] - \mathcal{Q}_n^G[f]| \leqslant 2 \left[\int_a^b w(x) dx \right] \frac{b-a}{2n+1} ||f'||_\infty, \ n = 1, 2, \ldots, \ f \in C^1[a,b]. \quad (8.2.29)$$

A remarkable implication of the convergence result (8.2.27) in Theorem 8.2.4(a) is the following. Let the sequence $\{\triangle_n : n = 0, 1, \ldots\}$ be chosen as in Theorems 8.2.2 and 8.2.4. As mentioned immediately after Example 3.1.3, there exists a function $f \in C[a,b]$ such that

$$||f - \mathscr{P}_n^I f||_\infty \to \infty, \quad n \to \infty, \tag{8.2.30}$$

with $\{\mathscr{P}_n^I : n = 0, 1, \ldots\}$ denoting the sequence of Lagrange interpolation operators, as defined in (5.1.2), with respect to the interpolation point sequence $\{\triangle_n : n = 0, 1, \ldots\}$. Nevertheless, according to (8.2.27) in Theorem 8.2.4(a), the corresponding quadrature formula sequence converges for this function f, that is, by using also (8.1.5),

$$|\mathscr{I}[f] - \mathscr{I}[\mathscr{P}_n^I f]| = |\mathscr{I}[f] - \mathscr{Q}_n^G[f]| \to 0, \quad n \to \infty. \tag{8.2.31}$$

The specific Gauss quadrature formula $\mathscr{Q}_n^G$ obtained by choosing the weight function w on $[a,b]$ as given by (7.5.31), is known as the Gauss-Legendre quadrature rule $\mathscr{Q}_n^{GL}$, that is,

$$\mathscr{Q}_n^{GL} := \mathscr{Q}_n^G, \quad \text{if } w(x) = 1, \quad x \in [a,b]. \tag{8.2.32}$$

Example 8.2.1. Consider the problem of designing the Gauss-Legendre quadrature rule $\mathscr{Q}_1^{GL}$ for the integral

$$\mathscr{I}[f] = \int_0^2 f(x)dx, \tag{8.2.33}$$

where $f \in C[0,2]$. It follows from Theorem 8.2.2, together with (8.1.7), that

$$\mathscr{Q}_1^{GL}[f] = w_{1,0}f(x_{1,0}) + w_{1,1}f(x_{1,1}), \tag{8.2.34}$$

where $\{x_{1,0}, x_{1,1}\}$ are the zeros in $(0,2)$ of the orthogonal polynomial $P_2^\perp$ of Theorem 7.5.2, as guaranteed by Theorem 8.2.1, and where, from (8.1.8), (8.2.32) and (1.2.1), the weights $\{w_{1,0}, w_{1,1}\}$ are given by the formulas

$$w_{1,0} = \int_0^2 \frac{x - x_{1,1}}{x_{1,0} - x_{1,1}}dx; \tag{8.2.35}$$

$$w_{1,1} = \int_0^2 \frac{x - x_{1,0}}{x_{1,1} - x_{1,0}}dx. \tag{8.2.36}$$

By using the formula (7.5.32) in Theorem 7.5.2, together with (7.5.26), we obtain the orthogonal polynomial

$$P_2^\perp(x) = L_2^\perp(x-1) = (x-1)^2 - \tfrac{1}{3} = x^2 - 2x + \tfrac{2}{3},$$

the zeros of which are given by

$$x_{1,0} = 1 - \tfrac{1}{\sqrt{3}}; \quad x_{1,1} = 1 + \tfrac{1}{\sqrt{3}}, \tag{8.2.37}$$

both of which are indeed in the interval $(0,2)$, as guaranteed by Theorem 8.2.1.

Next, we substitute (8.2.37) into (8.2.35) and (8.2.36) to obtain

$$w_{1,0} = \int_0^2 \frac{x - (1 + \frac{1}{\sqrt{3}})}{(1 - \frac{1}{\sqrt{3}}) - (1 + \frac{1}{\sqrt{3}})} dx = -\frac{\sqrt{3}}{2} \left[\frac{(x - (1 + \frac{1}{\sqrt{3}}))^2}{2} \right]_0^2 = 1; \qquad (8.2.38)$$

$$w_{1,1} = \int_0^2 \frac{x - (1 - \frac{1}{\sqrt{3}})}{(1 + \frac{1}{\sqrt{3}}) - (1 - \frac{1}{\sqrt{3}})} dx = \frac{\sqrt{3}}{2} \left[\frac{(x - (1 - \frac{1}{\sqrt{3}}))^2}{2} \right]_0^2 = 1. \qquad (8.2.39)$$

By inserting (8.2.37), (8.2.38) and (8.2.39) into (8.2.34), we obtain the Gauss-Legendre quadrature rule

$$\mathcal{Q}_1^{GL}[f] = f\left(1 - \frac{1}{\sqrt{3}}\right) + f\left(1 + \frac{1}{\sqrt{3}}\right). \qquad (8.2.40)$$

According to (8.2.7) in Theorem 8.2.2, the quadrature rule $\mathcal{Q}_1^{GL}$ has degree of exactness $m = 3$, that is,

$$\mathcal{Q}_1^{GL}[P] = \int_0^2 P(x)dx, \quad P \in \pi_3, \qquad (8.2.41)$$

an illustration of which is provided by the choice $P(x) = x^3$, in which case

$$\int_0^2 P(x)dx = \int_0^2 x^3 dx = 4,$$

whereas (8.2.40) gives

$$\mathcal{Q}_1^{GL}[P] = \left(1 - \frac{1}{\sqrt{3}}\right)^3 + \left(1 + \frac{1}{\sqrt{3}}\right)^3$$

$$= \left(1 - \sqrt{3} + 1 - \frac{1}{3\sqrt{3}}\right) + \left(1 + \sqrt{3} + 1 + \frac{1}{3\sqrt{3}}\right) = 4,$$

which verifies (8.2.41) for this case.

Next, for the choice $f(x) = e^x$, we obtain

$$\int_0^2 f(x)dx = \int_0^2 e^x dx = e^2 - 1 \approx 6.389056, \qquad (8.2.42)$$

whereas (8.2.40) yields

$$\mathcal{Q}_1^{GL}[f] = e^{1 - \frac{1}{\sqrt{3}}} + e^{1 + \frac{1}{\sqrt{3}}} \approx 6.368108, \qquad (8.2.43)$$

and thus the corresponding quadrature error is given by

$$\left| \int_0^2 f(x)dx - \mathcal{Q}_1^{GL}[f] \right| \approx 2.095 \times 10^{-2}, \qquad (8.2.44)$$

which illustrates the accuracy of the quadrature rule $\mathcal{Q}_1^{GL}$ for this choice of f.   ∎

8.3   The Clenshaw-Curtis quadrature rule

The Gauss-Legendre quadrature rule (8.2.32) for the numerical approximation of the integral $\mathscr{I}[f]$ in (8.1.1), for the weight function w as given by (7.5.31), that is,

$$\mathscr{I}[f] := \int_a^b f(x)dx, \quad f \in C[a,b],$$

is an interpolatory quadrature rule such that the corresponding weights satisfy the non-negativity condition (8.1.16), and for which the convergence results of Theorem 8.1.2 therefore hold. As a further example of this kind, we present in this section, for the numerical approximation of the integral $\mathscr{I}[f]$, the Clenshaw-Curtis quadrature rule $\mathscr{Q}_n^{CC}$, as defined for any positive integer n by

$$\mathscr{Q}_n^{CC}[f] := \mathscr{I}[\mathscr{P}_n^I f] = \int_a^b (\mathscr{P}_n^I f)(x)dx, \quad f \in C[a,b], \tag{8.3.1}$$

where $\mathscr{P}_n^I : C[a,b] \to \pi_n$ is the Lagrange polynomial interpolation operator with respect to the interpolation points

$$x_j = x_{n,j}^{CC} := \tfrac{1}{2}(b-a)\xi_{n,j} + \tfrac{1}{2}(a+b), \quad j = 0,\ldots,n, \tag{8.3.2}$$

where, as in (7.6.16),

$$\xi_{n,j} := \cos\left(\frac{n-j}{n}\pi\right), \quad j = 0,\ldots,n, \tag{8.3.3}$$

and thus

$$a = x_{n,0}^{CC} < x_{n,1}^{CC} < \cdots < x_{n,n}^{CC} = b. \tag{8.3.4}$$

Recall from (2.2.18), (2.2.19), (2.2.20) that $\{\xi_{n,j} : j = 0,\ldots,n\}$ are the points in $[-1,1]$ where the Chebyshev polynomial T_n alternatively attains its extreme values 1 and -1. It follows from (8.3.1) and (8.3.2), together with (8.1.7) and (8.1.8), that

$$\mathscr{Q}_n^{CC}[f] = \sum_{j=0}^n w_{n,j} f(x_{n,j}^{CC}), \quad f \in C[a,b], \tag{8.3.5}$$

where

$$w_{n,j} := \int_a^b L_{n,j}(x)dx, \quad j = 0,\ldots,n, \tag{8.3.6}$$

with $\{L_{n,j} : j = 0,\ldots,n\}$ denoting the Lagrange fundamental polynomials in π_n, as obtained by setting $x_j = x_{n,j}^{CC}$, $j = 0,\ldots,n$, in (1.2.1), so that, according to (1.2.3) in Theorem 1.2.1,

$$L_{n,j}(x_{n,k}^{CC}) = \delta_{j-k}, \quad j,k = 0,\ldots,n. \tag{8.3.7}$$

We proceed to find explicit formulations of the Clenshaw-Curtis weights $\{w_{n,j} : j = 0,\ldots,n\}$ in (8.3.5), (8.3.6). To this end, we first observe from (8.3.6), together with the one-to-one mapping (2.2.29), (2.2.30) between the intervals $[-1,1]$ and $[a,b]$, that

$$w_{n,j} = \frac{1}{2}(b-a)\int_{-1}^{1} \tilde{L}_{n,j}(t)dt, \quad j = 0,\ldots,n, \tag{8.3.8}$$

where

$$\tilde{L}_{n,j}(t) := L_{n,j}\left(\tfrac{1}{2}(b-a)t + \tfrac{1}{2}(a+b)\right), \quad j = 0,\ldots,n, \tag{8.3.9}$$

and thus, by using also (8.3.7) and (8.3.2),

$$\tilde{L}_{n,j}(\xi_{n,k}) = \delta_{j-k}, \quad j,k = 0,\ldots,n. \tag{8.3.10}$$

Since $L_{n,j} \in \pi_n$, $j = 0,\ldots,n$, it follows from (8.3.9) that $\tilde{L}_{n,j} \in \pi_n$, $j = 0,\ldots,n$, so that we may apply the Chebyshev expansion (7.6.27) in Theorem 7.6.5, together with (8.3.3) and (8.3.10), to deduce that, for $j = 0,\ldots,n$,

$$\tilde{L}_{n,j} = \frac{2}{n}\sum_{k=0}^{n}{}' \left[\sum_{\ell=0}^{n}{}' \tilde{L}_{n,j}\left(\cos\left(\frac{\ell}{n}\pi\right)\right)\cos\left(\frac{k\ell}{n}\pi\right)\right]T_k$$

$$= \frac{2}{n}\sum_{k=0}^{n}{}' \left[\sum_{\ell=0}^{n}{}' \tilde{L}_{n,j}\left(\cos\left(\frac{n-\ell}{n}\pi\right)\right)\cos\left(\frac{k(n-\ell)}{n}\pi\right)\right]T_k$$

$$= \frac{2}{n}\sum_{k=0}^{n}{}' \left[\sum_{\ell=0}^{n}{}' \tilde{L}_{n,j}(\xi_{n,\ell})\cos\left(\frac{k(n-\ell)}{n}\pi\right)\right]T_k$$

$$= \frac{2}{n}\sum_{k=0}^{n}{}' \left[\sum_{\ell=0}^{n}{}' \delta_{j-\ell}\cos\left(\frac{k(n-\ell)}{n}\pi\right)\right]T_k. \tag{8.3.11}$$

By recalling the definition (7.6.15) of the symbol Σ', as well as the fact that the first line of (2.2.1) gives $T_0(x) := 1$, we deduce from (8.3.11) that

$$\tilde{L}_{n,0} = \frac{1}{n}\sum_{k=0}^{n}{}' (-1)^k T_k = \frac{1}{2n}\left[1 + 2\sum_{k=1}^{n-1}(-1)^k T_k + (-1)^n T_n\right]; \tag{8.3.12}$$

$$\tilde{L}_{n,j} = \frac{2}{n}\sum_{k=0}^{n}{}' \cos\left(\frac{k(n-j)}{n}\pi\right)T_k = \frac{1}{n}\left[1 + 2\sum_{k=1}^{n-1}\cos\left(\frac{k(n-j)}{n}\pi\right)T_k + (-1)^{n-j}T_n\right],$$

$$j = 1,\ldots,n-1 \quad (\text{if } n \geqslant 2); \tag{8.3.13}$$

$$\tilde{L}_{n,n} = \frac{1}{n}\sum_{k=0}^{n}{}' T_k = \frac{1}{2n}\left[1 + 2\sum_{k=1}^{n-1}T_k + T_n\right]. \tag{8.3.14}$$

Next, we use the formula (2.2.5) in Theorem 2.2.1(b), together with the one-to-one mapping (2.2.11), (2.2.12) between the intervals $[0, \pi]$ and $[-1, 1]$, to obtain by means of integration by parts, for any integer $k \geqslant 2$,

$$\int_{-1}^{1} T_k(t)dt = \int_{0}^{\pi} \cos(k\theta) \sin\theta \, d\theta$$

$$= \left[\cos(k\theta)(-\cos\theta)\right]_{0}^{\pi} - k\int_{0}^{\pi} \sin(k\theta)\cos\theta \, d\theta$$

$$= [(-1)^k + 1] - k\left\{\left[\sin(k\theta)\sin\theta\right]_{0}^{\pi} - k\int_{0}^{\pi} \cos(k\theta)\sin\theta \, d\theta\right\}$$

$$= 1 + (-1)^k + k^2 \int_{-1}^{1} T_k(t)dt,$$

and thus

$$\int_{-1}^{1} T_k(t)dt = \frac{1 + (-1)^k}{1 - k^2}, \quad k = 2, 3, \dots. \tag{8.3.15}$$

Also, observe from the first line of (2.2.1) that

$$\int_{-1}^{1} T_0(t)dt = \int_{-1}^{1} dt = 2; \quad \int_{-1}^{1} T_1(t)dt = \int_{-1}^{1} t \, dt = 0,$$

which, together with (8.3.15), yields

$$\int_{-1}^{1} T_k(t)dt = \begin{cases} \dfrac{2}{1 - k^2}, & \text{if } k \text{ is even,} \\ 0, & \text{if } k \text{ is odd,} \end{cases} \quad k = 0, 1, \dots. \tag{8.3.16}$$

It follows from (8.3.12), (8.3.13), (8.3.14) and (8.3.16) that

$$\int_{-1}^{1} \tilde{L}_{n,0}(t)dt = \frac{1}{n}\left[1 - 2\sum_{k=1}^{\lfloor \frac{1}{2}(n-1)\rfloor} \frac{1}{4k^2 - 1} + \frac{1}{2}(-1)^n \int_{-1}^{1} T_n(t)dt\right]; \tag{8.3.17}$$

$$\int_{-1}^{1} \tilde{L}_{n,j}(t)dt = \frac{2}{n}\left[1 - 2\sum_{k=1}^{\lfloor \frac{1}{2}(n-1)\rfloor} \frac{1}{4k^2 - 1}\cos\left(\frac{2jk}{n}\pi\right) + \frac{1}{2}(-1)^{n-j}\int_{-1}^{1} T_n(t)dt\right],$$

$$j = 1, \dots, n - 1 \ (\text{if } n \geqslant 2); \tag{8.3.18}$$

$$\int_{-1}^{1} \tilde{L}_{n,n}(t)dt = \frac{1}{n}\left[1 - 2\sum_{k=1}^{\lfloor \frac{1}{2}(n-1)\rfloor} \frac{1}{4k^2 - 1} + \frac{1}{2}\int_{-1}^{1} T_n(t)dt\right]. \tag{8.3.19}$$

Now observe that, for any positive integer μ,

$$\sum_{k=1}^{\mu} \frac{1}{4k^2 - 1} = \frac{1}{2}\sum_{k=1}^{\mu}\left[\frac{1}{2k - 1} - \frac{1}{2k + 1}\right]$$

$$= \frac{1}{2}\left[\left(1 - \frac{1}{3}\right) + \left(\frac{1}{3} - \frac{1}{5}\right) + \cdots + \left(\frac{1}{2\mu - 1} - \frac{1}{2\mu + 1}\right)\right]$$

$$= \frac{1}{2}\left[1 - \frac{1}{2\mu+1}\right] = \frac{\mu}{2\mu+1},$$

that is,

$$\sum_{k=1}^{\mu} \frac{1}{4k^2-1} = \frac{\mu}{2\mu+1}, \quad \mu \in \mathbb{N}, \tag{8.3.20}$$

according to which

$$\sum_{k=1}^{\lfloor \frac{1}{2}(n-1)\rfloor} \frac{1}{4k^2-1} = \begin{cases} \dfrac{\frac{1}{2}(n-2)}{(n-2)+1} = \dfrac{1}{2}\dfrac{n-2}{n-1}, & \text{if } n \text{ is even, } n \geqslant 4; \\[3mm] \dfrac{\frac{1}{2}(n-1)}{(n-1)+1} = \dfrac{1}{2}\dfrac{n-1}{n}, & \text{if } n \text{ is odd, } n \geqslant 3. \end{cases} \tag{8.3.21}$$

By combining (8.3.8) and (8.3.16)–(8.3.21), we obtain the following explicit formulation for the weights of the Clenshaw-Curtis quadrature rule, where we adopt once again the convention $\sum_{j=j_0}^{j_1} \alpha_j := 0$, if $j_1 < j_0$.

Theorem 8.3.1. *For any positive integer n, the Clenshaw-Curtis quadrature rule $\mathcal{Q}_n^{CC}$, as given by (8.3.5), (8.3.6), (8.3.2), (8.3.3), has weights $\{w_{n,j} : j = 0,\ldots,n\}$ given explicitly by*

(a) *if n is even,*

$$w_{n,0} = w_{n,n} = \frac{b-a}{2(n^2-1)}; \tag{8.3.22}$$

$$w_{n,j} = \frac{b-a}{n}\left[1 - 2\sum_{k=1}^{\frac{1}{2}(n-2)} \frac{1}{4k^2-1}\cos\left(\frac{2jk}{n}\pi\right) + \frac{(-1)^j}{1-n^2}\right],$$

$$j = 1,\ldots,n-1 \ (if \ n \geqslant 2); \tag{8.3.23}$$

(b) *if n is odd,*

$$w_{n,0} = w_{n,n} = \frac{b-a}{2n^2}; \tag{8.3.24}$$

$$w_{n,j} = \frac{b-a}{n}\left[1 - 2\sum_{k=1}^{\frac{1}{2}(n-1)} \frac{1}{4k^2-1}\cos\left(\frac{2jk}{n}\pi\right)\right],$$

$$j = 1,\ldots,n-1 \ (if \ n \geqslant 3). \tag{8.3.25}$$

We proceed to show that the weights $\{w_{n,j} : j = 0,\ldots,n\}$ in Theorem 8.3.1 are positive. First, observe from (8.3.22) and (8.3.24) that

$$w_{n,0} > 0, \quad w_{n,n} > 0, \quad \text{for } n = 1,2,\ldots. \tag{8.3.26}$$

Next, if n is even, with $n \geqslant 4$, we use (8.3.21) to deduce from (8.3.23) that, for $j = 1, \ldots, n-1$,

$$w_{n,j} \geqslant \frac{b-a}{n}\left[1 - 2\sum_{k=1}^{\frac{1}{2}(n-2)}\frac{1}{4k^2-1} - \frac{1}{n^2-1}\right]$$

$$= \frac{b-a}{n}\left[1 - \frac{n-2}{n-1} - \frac{1}{n^2-1}\right] = \frac{b-a}{n^2-1} > 0, \qquad (8.3.27)$$

whereas

$$w_{2,1} = \frac{b-a}{2}\left[1 + \frac{1}{3}\right] = \frac{2(b-a)}{3} > 0. \qquad (8.3.28)$$

Similarly, if n is odd, with $n \geqslant 3$, we use (8.3.21) to deduce from (8.3.25) that

$$w_{n,j} \geqslant \frac{b-a}{n}\left[1 - 2\sum_{k=1}^{\frac{1}{2}(n-1)}\frac{1}{4k^2-1}\right] = \frac{b-a}{n}\left[1 - \frac{n-1}{n}\right] = \frac{b-a}{n^2} > 0. \qquad (8.3.29)$$

By combining (8.3.26)–(8.3.29), it follows that we have proved the following result.

Theorem 8.3.2. *For any positive integer n, the weights $\{w_{n,j} : j = 0, \ldots, n\}$ of the Clenshaw-Curtis quadrature rule $\mathcal{Q}_n^{CC}$, as given in Theorem 8.3.1, are positive, that is,*

$$w_{n,j} > 0, \quad j = 0, \ldots, n. \qquad (8.3.30)$$

Our next step is to find, for any $n \in \mathbb{N}$, the degree of exactness m_n of the Clenshaw-Curtis quadrature rule. Since $\mathcal{Q}_n^{CC}$ is, according to (8.3.1), an interpolatory quadrature rule, we deduce from Theorem 8.1.1 that

$$m_n \geqslant n, \quad n = 1, 2, \ldots. \qquad (8.3.31)$$

Hence we proceed to investigate the validity of the statement

$$\int_a^b P(x)dx = \mathcal{Q}_n^{CC}[P], \quad P \in \pi_{n+1}. \qquad (8.3.32)$$

To this end, for any positive integer n, let $P \in \pi_{n+1}$, with $\deg(P) = n+1$. By using the one-to-one mapping (2.2.29), (2.2.30) between the intervals $[-1,1]$ and $[a,b]$, we deduce that

$$\int_a^b P(x)dx = \frac{b-a}{2}\int_1^1 \widetilde{P}(t)dt, \qquad (8.3.33)$$

where

$$\widetilde{P}(t) := P\left(\tfrac{1}{2}(b-a)t + \tfrac{1}{2}(a+b)\right), \qquad (8.3.34)$$

or equivalently,

$$P(x) = \widetilde{P}\left(\frac{2}{b-a}x - \frac{a+b}{b-a}\right). \qquad (8.3.35)$$

Since $P \in \pi_{n+1}$, it follows from (8.3.34) that $\widetilde{P} \in \pi_{n+1}$, so that we may deduce from Theorem 7.6.3 that

$$\widetilde{P} = \sum_{k=0}^{n+1} c_k T_k, \tag{8.3.36}$$

where the coefficient sequence $\{c_j : j = 0, \ldots, n+1\}$ can be obtained from the Chebyshev expansion (7.6.14). Note from (8.3.35) and (8.3.36), together with Theorem 2.2.1(a), that, since $\deg(P) = n+1$, we have $c_{n+1} \neq 0$. It follows from (8.3.33) and (8.3.36), together with (2.2.29), (2.2.30), that

$$\int_a^b P(x)dx = \int_a^b Q(x)dx + c_{n+1}\frac{b-a}{2}\int_{-1}^1 T_{n+1}(t)dt, \tag{8.3.37}$$

where

$$Q(x) := \sum_{j=0}^n c_j T_j\left(\frac{2}{b-a}x - \frac{a+b}{b-a}\right), \tag{8.3.38}$$

or equivalently,

$$\sum_{j=0}^n c_j T_j(t) = Q\left(\tfrac{1}{2}(b-a)t + \tfrac{1}{2}(a+b)\right). \tag{8.3.39}$$

Next, we use (8.3.5), (8.3.2), (8.3.34), (8.3.36) and (8.3.39) to obtain

$$\mathcal{Q}_n^{CC}[P] = \sum_{j=0}^n w_{n,j} P\left(\tfrac{1}{2}(b-a)\xi_{n,j} + \tfrac{1}{2}(a+b)\right)$$

$$= \sum_{j=0}^n w_{n,j}\widetilde{P}(\xi_{n,j})$$

$$= \sum_{j=0}^n w_{n,j} Q\left(\tfrac{1}{2}(b-a)\xi_{n,j} + \tfrac{1}{2}(a+b)\right) + c_{n+1}\sum_{j=0}^n w_{n,j} T_{n+1}(\xi_{n,j})$$

$$= \mathcal{Q}_n^{CC}[Q] + c_{n+1}\sum_{j=0}^n w_{n,j} T_{n+1}(\xi_{n,j}). \tag{8.3.40}$$

Since (8.3.38) and Theorem 2.2.1(a) imply $Q \in \pi_n$, we may now apply (8.1.17) to obtain

$$\int_a^b Q(x)dx = \mathcal{Q}_n^{CC}[Q]. \tag{8.3.41}$$

It follows from (8.3.37), (8.3.40) and (8.3.41), together with $c_{n+1} \neq 0$, that (8.3.32) holds if and only if

$$\frac{b-a}{2}\int_{-1}^1 T_{n+1}(t)dt = \sum_{j=0}^n w_{n,j} T_{n+1}(\xi_{n,j}),$$

or equivalently, from (8.3.3) and the formula (2.2.5) in Theorem 2.2.1(b),

$$\frac{b-a}{2}\int_{-1}^1 T_{n+1}(t)dt = \sum_{j=0}^n w_{n,j} \cos\left(\frac{(n+1)(n-j)}{n}\pi\right). \tag{8.3.42}$$

Note from the formulas (8.3.22)–(8.3.25) in Theorem 8.3.1 that the Clenshaw-Curtis weights $\{w_{n,j} : j = 0,\ldots,n\}$ satisfy the symmetry condition

$$w_{n,n-j} = w_{n,j}, \quad j = 0,\ldots,n. \tag{8.3.43}$$

It then follows from (8.3.43) that the condition (8.3.42) has the equivalent formulation

$$\frac{b-a}{2} \int_{-1}^{1} T_{n+1}(t)dt = \sum_{j=0}^{n} w_{n,j} \cos\left(\frac{(n+1)j}{n}\pi\right). \tag{8.3.44}$$

Suppose first n is even, that is, $n = 2v$ for a positive integer v. By using (8.3.43), we obtain

$$\sum_{j=0}^{2v} w_{2v,j} \cos\left(\frac{(2v+1)j}{2v}\pi\right)$$

$$= \sum_{j=0}^{v-1} w_{2v,j} \cos\left(\frac{(2v+1)j}{2v}\pi\right) + w_{2v,v} \cos\left(\left(v+\frac{1}{2}\right)\pi\right)$$

$$+ \sum_{j=v+1}^{2v} w_{2v,2v-j} \cos\left(\frac{(2v+1)j}{2v}\pi\right)$$

$$= \sum_{j=0}^{v-1} w_{2v,j} \left[\cos\left(\frac{(2v+1)j}{2v}\pi\right) + \cos\left(\frac{(2v+1)(2v-j)}{2v}\pi\right)\right]$$

$$= \sum_{j=0}^{v-1} w_{2v,j} \left[\cos\left(\frac{(2v+1)j}{2v}\pi\right) - \cos\left(\frac{(2v+1)j}{2v}\pi\right)\right] = 0. \tag{8.3.45}$$

Also, since $n+1$ is odd, it follows from (8.3.16) that

$$\int_{-1}^{1} T_{n+1}(t)dt = 0,$$

which, together with (8.3.45), shows that (8.3.44), and therefore also (8.3.32), do indeed hold. Hence the exactness condition (8.3.32) is satisfied if n is even, and thus

$$m_n \geqslant n+1, \quad \text{if } n \text{ is even.} \tag{8.3.46}$$

Suppose next n is odd, and thus $n+1$ is even, so that we may apply the formulas (8.3.24) and (8.3.25) in Theorem 8.3.1, as well as (7.6.22) and (8.3.21), to obtain, for $n \geqslant 3$,

$$\sum_{j=0}^{n} w_{n,j} \cos\left(\frac{(n+1)j}{n}\pi\right)$$

$$= \frac{b-a}{n^2}\left[1 + n\sum_{j=1}^{n-1}\left(1 - 2\sum_{k=1}^{\frac{1}{2}(n-1)}\frac{1}{4k^2-1}\cos\left(\frac{2jk}{n}\pi\right)\right)\cos\left(\frac{(n+1)j}{n}\pi\right)\right]$$

$$= \frac{b-a}{n^2}\left[1 + n\sum_{j=1}^{n-1}\cos\left(\frac{(n+1)j}{n}\pi\right)\right.$$

$$\left. -2n\sum_{k=1}^{\frac{1}{2}(n-1)}\frac{1}{4k^2-1}\sum_{j=1}^{n-1}\cos\left(\frac{2jk}{n}\pi\right)\cos\left(\frac{(n+1)j}{n}\pi\right)\right]$$

$$= \frac{b-a}{n^2}\left[1 + n\left\{\sum_{j=0}^{n-1}\cos\left(\frac{(n+1)j}{n}\pi\right) - 1\right\}\right.$$

$$-n\sum_{k=1}^{\frac{1}{2}(n-1)}\frac{1}{4k^2-1}\left\{\sum_{j=0}^{n-1}\cos\left(\frac{(2k+n+1)j}{n}\pi\right) - 1\right\}$$

$$\left. -n\sum_{k=1}^{\frac{1}{2}(n-1)}\frac{1}{4k^2-1}\left\{\sum_{j=0}^{n-1}\cos\left(\frac{(2k-n-1)j}{n}\pi\right) - 1\right\}\right]$$

$$= \frac{b-a}{n^2}\left[1 + n(0-1) - n\left\{\frac{1}{4(\frac{1}{2}(n-1))^2-1}(n) - \frac{n-1}{2n}\right\} - n\left\{0 - \frac{n-1}{2n}\right\}\right]$$

$$= -\frac{b-a}{n(n-2)}, \tag{8.3.47}$$

whereas, if $n = 1$, it follows from (8.3.24) that

$$\sum_{j=0}^{1}w_{1,j}\cos(2j\pi) = b - a,$$

from which we then deduce that the formula (8.3.47) is valid for any odd integer $n \geqslant 1$. Since $n+1$ is even, with $n+1 \geqslant 2$, it follows from (8.3.16) that

$$\frac{b-a}{2}\int_{-1}^{1}T_{n+1}(t)dt = \frac{b-a}{2}\left[\frac{2}{1-(n+1)^2}\right] = -\frac{b-a}{n(n+2)}. \tag{8.3.48}$$

It follows from (8.3.47) and (8.3.48) that

$$\frac{b-a}{2}\int_{-1}^{1}T_{n+1}(t)dt - \sum_{j=0}^{n}w_{n,j}\cos\left(\frac{(n+1)j}{n}\pi\right) = \frac{4(b-a)}{n(n^2-4)}, \tag{8.3.49}$$

according to which (8.3.44), and therefore also (8.3.32), are not satisfied if n is odd, which, together with (8.3.31), implies that

$$m_n = n, \quad \text{if} \quad n \quad \text{is odd.} \tag{8.3.50}$$

Hence it remains to establish the value of m_n if n is even. Suppose therefore that n is even, and let $P \in \pi_{n+2}$, with $\deg(P) = n+2$. Analogously to the argument which led from

(8.3.32) to (8.3.44), and by also using the fact that m_n satisfies the inequality (8.3.46), we deduce that the exactness condition

$$\int_a^b P(x)dx = \mathcal{Q}_n^{CC}[P] \tag{8.3.51}$$

holds if and only if

$$\frac{b-a}{2}\int_{-1}^1 T_{n+2}(t)dt = \sum_{j=0}^n w_{n,j}\cos\left(\frac{(n+2)j}{n}\pi\right). \tag{8.3.52}$$

First, since $n+2$ is even, we deduce from (8.3.16) that

$$\frac{b-a}{2}\int_{-1}^1 T_{n+2}(t)dt = \frac{b-a}{2}\left[\frac{2}{1-(n+2)^2}\right] = -\frac{b-a}{(n+1)(n+3)}. \tag{8.3.53}$$

Next, we apply the formulas (8.3.22) and (8.3.23) in Theorem 8.3.1, as well as (7.6.22) and (8.3.21), together with the fact that $n+2$ is even, whereas $n+1$ and $n+3$ are odd, to obtain, for $n \geqslant 4$,

$$\sum_{j=0}^n w_{n,j}\cos\left(\frac{(n+2)j}{n}\pi\right)$$

$$= (b-a)\left[\frac{1}{n^2-1} + \frac{1}{n}\sum_{j=1}^{n-1}\left\{1 - 2\sum_{k=1}^{\frac{1}{2}(n-2)}\frac{1}{4k^2-1}\cos\left(\frac{2jk}{n}\pi\right) + \frac{\cos(j\pi)}{1-n^2}\right\}\right.$$

$$\left. \times \cos\left(\frac{(n+2)j}{n}\pi\right)\right]$$

$$= (b-a)\left[\frac{1}{n^2-1} + \frac{1}{n}\left\{\sum_{j=0}^{n-1}\cos\left(\frac{(n+2)j}{n}\pi\right) - 1\right\}\right.$$

$$-\frac{1}{n}\sum_{k=1}^{\frac{1}{2}(n-2)}\frac{1}{4k^2-1}\left\{\sum_{j=0}^{n-1}\cos\left(\frac{(2k+n+2)j}{n}\pi\right) - 1\right\}$$

$$-\frac{1}{n}\sum_{k=1}^{\frac{1}{2}(n-2)}\frac{1}{4k^2-1}\left\{\sum_{j=0}^{n-1}\cos\left(\frac{(2k-n-2)j}{n}\pi\right) - 1\right\}$$

$$\left. -\frac{1}{2n(n^2-1)}\left\{\sum_{j=0}^{n-1}\cos\left(\frac{(n+3)j}{n}\pi\right) + \sum_{j=0}^{n-1}\cos\left(\frac{(n+1)j}{n}\pi\right) - 2\right\}\right]$$

$$= (b-a)\left[\frac{1}{n^2-1} + \frac{1}{n}\{0-1\} - \frac{1}{n}\left\{\frac{1}{4(\frac{1}{2}(n-2))^2-1}(n) - \frac{1}{2}\frac{n-2}{n-1}\right\}\right.$$

$$\left. -\frac{1}{n}\left\{0 - \frac{1}{2}\frac{n-2}{n-1}\right\} - \frac{1}{2n(n^2-1)}\{1+1-2\}\right]$$

$$= -(b-a)\frac{n+3}{n(n+1)(n-3)}. \tag{8.3.54}$$

It follows from (8.3.53) and (8.3.54) that, for $n \geqslant 4$,

$$\frac{b-a}{2}\int_{-1}^{1} T_{n+2}(t)dt - \sum_{j=0}^{n} w_{n,j}\cos\left(\frac{(n+2)j}{n}\pi\right) = \frac{9(b-a)}{n(n^2-9)}, \tag{8.3.55}$$

whereas, if $n=2$, it follows from (8.3.53), (8.3.22) and (8.3.23) that

$$\frac{b-a}{2}\int_{-1}^{1} T_4(t)dt - \sum_{j=0}^{2} w_{2,j}\cos(2j\pi) = -(b-a)\left[\frac{1}{15}+\left\{\frac{1}{3}+\frac{1}{2}\left(1+\frac{1}{3}\right)\right\}\right]$$

$$= -\frac{16(b-a)}{15}. \tag{8.3.56}$$

It follows from (8.3.55) and (8.3.56) that (8.3.52), and therefore also (8.3.51), are not satisfied if n is even, from which, together with (8.3.46), we deduce that

$$m_n = n+1, \quad \text{if } n \text{ is even.} \tag{8.3.57}$$

Hence, according to (8.3.50) and (8.3.57), together with (8.3.37), (8.3.40), (8.3.41) and (8.3.49) if n is odd, and, analogously, (8.3.55) and (8.3.56) if n is even, we have now established the following.

Theorem 8.3.3. *For any positive integer n, the degree of exactness m_n of the Clenshaw-Curtis quadrature rule $\mathscr{Q}_n^{CC}$ in Theorem 8.3.1 satisfies*

$$m_n = \begin{cases} n+1, & \text{if } n \text{ is even;} \\ n, & \text{if } n \text{ is odd.} \end{cases} \tag{8.3.58}$$

Also,

(a) *if n is odd, and $P \in \pi_{n+1}$, with*

$$P(x) = \sum_{j=0}^{n+1} c_j x^j,$$

then

$$\int_a^b P(x)dx - \mathscr{Q}_n^{CC}[P] = c_{n+1}(b-a)\frac{4}{n(n^2-4)}; \tag{8.3.59}$$

(b) *if n is even, and $P \in \pi_{n+2}$, with*

$$P(x) = \sum_{j=0}^{n+2} c_j x^j,$$

then

$$\int_a^b P(x)dx - \mathscr{Q}_n^{CC}[P] = c_{n+2}(b-a)\begin{cases} -\dfrac{16}{15}, & \text{if } n=2; \\ \dfrac{9}{n(n^2-9)}, & \text{if } n \geqslant 4. \end{cases} \tag{8.3.60}$$

Observe from Theorems 8.3.2 and 8.3.3 that the Clenshaw-Curtis quadrature rule $\mathscr{Q}_n^{CC}$ satisfies the conditions of Theorem 8.1.2, with degree of exactness m_n given by (8.3.58) in Theorem 8.3.3, according to which

$$m = 2\lfloor n/2 \rfloor + 1, \quad n = 1, 2, \ldots, \tag{8.3.61}$$

and thus we immediately have the following result as a special case of Theorem 8.1.2.

Theorem 8.3.4. *Let* $\{\mathscr{Q}_n^{CC} : n = 1, 2, \ldots\}$ *denote the sequence of Clenshaw-Curtis quadrature rules of Theorem 8.3.1. Then:*

(a) *Convergence on* $C[a,b]$ *holds, that is,*

$$\left| \int_a^b f(x)dx - \mathscr{Q}_n^{CC}[f] \right| \to 0, \quad n \to \infty, \quad f \in C[a,b]. \tag{8.3.62}$$

(b) *If* $f \in C^k[a,b]$ *for an integer* $k \in \mathbb{N}$, *the convergence rate result*

$$\left| \int_a^b f(x)dx - \mathscr{Q}_n^{CC}[f] \right| \leqslant 2(b-a)^{k+1} \frac{(2\lfloor n/2 \rfloor + 1 - k)!}{(2\lfloor n/2 \rfloor + 1)!} ||f^{(k)}||_\infty, \ n = k, k+1, \ldots, \tag{8.3.63}$$

holds, with, in particular,

$$\left| \int_a^b f(x)dx - \mathscr{Q}_n^{CC}[f] \right| \leqslant \frac{2(b-a)^2}{2\lfloor n/2 \rfloor + 1} ||f'||_\infty, \ n = 1, 2, \ldots, \ f \in C^1[a,b], \tag{8.3.64}$$

that is,

$$\left| \int_a^b f(x)dx - \mathscr{Q}_{2n}^{CC}[f] \right| \leqslant \frac{2(b-a)^2}{2n+1} ||f'||_\infty, \ n = 1, 2, \ldots, \ f \in C^1[a,b]; \tag{8.3.65}$$

$$\left| \int_a^b f(x)dx - \mathscr{Q}_{2n-1}^{CC}[f] \right| \leqslant \frac{2(b-a)^2}{2n-1} ||f'||_\infty, \ n = 1, 2, \ldots, \ f \in C^1[a,b]. \tag{8.3.66}$$

Example 8.3.1. To obtain the Clenshaw-Curtis quadrature rule $\mathscr{Q}_6^{CC}$ for the numerical approximation of the integral $\int_0^2 f(x)dx$, $f \in C[0,2]$, we first set $[a,b] = [0,2]$ and $n = 6$ in the formulas (8.3.22) and (8.3.23) of Theorem 8.3.1, to obtain

$$w_{6,0} = w_{6,6} = \frac{1}{35};$$

$$w_{6,1} = \frac{1}{3}\left[1 - 2\sum_{k=1}^{2} \frac{1}{4k^2 - 1}\cos\left(\frac{k\pi}{3}\right) + \frac{1}{35}\right] = \frac{16}{63};$$

$$w_{6,2} = \frac{1}{3}\left[1 - 2\sum_{k=1}^{2} \frac{1}{4k^2 - 1}\cos\left(\frac{2k\pi}{3}\right) - \frac{1}{35}\right] = \frac{16}{35};$$

$$w_{6,3} = \frac{1}{3}\left[1 - 2\sum_{k=1}^{2} \frac{1}{4k^2 - 1}\cos(k\pi) + \frac{1}{35}\right] = \frac{164}{315},$$

and thus, by using also the symmetry condition (8.3.43),

$$w_{6,4} = w_{6,2} = \frac{16}{35}; \qquad w_{6,5} = w_{6,1} = \frac{16}{63}.$$

Next, we use (8.3.2) and (8.3.3) to obtain

$$\{x_{6,0}^{CC}, x_{6,1}^{CC}, x_{6,2}^{CC}, x_{6,3}^{CC}, x_{6,4}^{CC}, x_{6,5}^{CC}, x_{6,6}^{CC}\} = \left\{0, 1 - \frac{\sqrt{3}}{2}, \frac{1}{2}, 1, \frac{3}{2}, 1 + \frac{\sqrt{3}}{2}, 2\right\}.$$

Hence, according to (8.3.5), we have the formulation

$$\mathscr{Q}_6^{CC}[f] = \frac{1}{35}f(0) + \frac{16}{63}f\left(1 - \frac{\sqrt{3}}{2}\right) + \frac{16}{35}f\left(\frac{1}{2}\right) + \frac{164}{315}f(1) + \frac{16}{35}f\left(\frac{3}{2}\right)$$

$$+ \frac{16}{63}f\left(1 + \frac{\sqrt{3}}{2}\right) + \frac{1}{35}f(2). \quad (8.3.67)$$

For the case $f(x) = e^x$, we calculate by means of (8.3.67), together with (8.2.42), that the corresponding quadrature error is given by

$$\left|\int_0^2 f(x)dx - \mathscr{Q}_6^{CC}[f]\right| \approx 5.597 \times 10^{-8}. \qquad \blacksquare$$

8.4 Newton-Cotes quadrature

In this section, we consider the numerical approximation of the integral

$$\mathscr{I}[f] := \int_a^b f(x)dx, \quad f \in C[a,b], \qquad (8.4.1)$$

by means of the Newton-Cotes quadrature rule $\mathscr{Q}_n^{NC}$, which, for any non-negative integer n, is defined by

$$\mathscr{Q}_n^{NC}[f] := \mathscr{I}[\mathscr{P}_n^I f] = \int_a^b (\mathscr{P}_n^I f)(x)dx, \quad f \in C[a,b], \qquad (8.4.2)$$

where $\mathscr{P}_n^I : C[a,b] \to \pi_n$ is the Lagrange interpolation operator, as defined by (5.1.2), with respect to the equispaced interpolation points

$$\left.\begin{array}{l} x_{0,0} := a; \\ \\ x_{n,j} := a + j\left(\dfrac{b-a}{n}\right), \quad j = 0,\ldots,n, \text{ if } n \geqslant 1. \end{array}\right\} \qquad (8.4.3)$$

Note that, if $n \geqslant 1$, the points $\{x_{n,j} : j = 0,\ldots,n\}$ in the second line of (8.4.3) are equispaced, with

$$x_{n,j+1} - x_{n,j} = \frac{b-a}{n}, \quad j = 0,\ldots,n-1, \qquad (8.4.4)$$

and such that

$$a = x_{n,0} < x_{n,1} < \cdots < x_{n,n} = b. \qquad (8.4.5)$$

According to (8.4.2), $\mathscr{Q}_n^{NC}$ is an interpolatory quadrature rule, as given by (8.1.7), that is,

$$\mathscr{Q}_n^{NC}[f] = \sum_{j=0}^{n} w_{n,j} f(x_{n,j}), \qquad (8.4.6)$$

where

$$w_{n,j} := \int_a^b L_{n,j}(x)dx, \quad j = 0,\ldots,n, \qquad (8.4.7)$$

with the Lagrange fundamental polynomials $\{L_{n,j} : j = 0,\ldots,n\}$ given as in (1.2.1).

We proceed to obtain explicit formulations for the Newton-Cotes weights $\{w_{n,j} : j = 0,\ldots,n\}$ in (8.4.6) and (8.4.7). To this end, we first observe from (8.4.7) and (1.2.1) that

$$w_{0,0} = \int_a^b dx = b - a, \qquad (8.4.8)$$

which, together with (8.4.6) and the first line of (8.4.3), yields the quadrature formula

$$\mathscr{Q}_0^{NC}[f] = (b-a)f(a). \qquad (8.4.9)$$

Suppose next that $n \in \mathbb{N}$, and let the integer $j \in \{0,\ldots,n\}$ be fixed. By using (8.4.7), (1.2.1), and the one-to-one mapping between the intervals $[0,n]$ and $[a,b]$ given by

$$t = n\frac{x-a}{b-a}, \quad a \leqslant x \leqslant b, \qquad (8.4.10)$$

or equivalently,

$$x = a + \left(\frac{b-a}{n}\right)t, \quad 0 \leqslant t \leqslant n, \qquad (8.4.11)$$

we deduce that

$$w_{n,j} = \frac{b-a}{n} \int_0^n \tilde{L}_{n,j}(t)dt, \qquad (8.4.12)$$

where

$$\tilde{L}_{n,j}(t) := L_{n,j}\left(a + \left(\frac{b-a}{n}\right)t\right). \qquad (8.4.13)$$

Observe from (8.4.13), together with (1.2.3) in Theorem 1.2.1, as well as (8.4.11), that

$$\tilde{L}_{n,j}(k) = \delta_{j-k}, \quad k = 0,\ldots,n. \qquad (8.4.14)$$

Now apply Theorem 1.1.2 to deduce that there exists a (unique) polynomial $g_j \in \pi_{n+1}$ such that the interpolation conditions

$$g_j(k) = \begin{cases} -1, & k = 0,\ldots,j; \\ 0, & k = j+1,\ldots,n+1, \end{cases} \qquad (8.4.15)$$

are satisfied, and in terms of which we define

$$h_j(t) := g_j(t+1) - g_j(t). \qquad (8.4.16)$$

Since $g_j \in \pi_{n+1}$, we deduce from (8.4.16) that $h_j \in \pi_n$.

Observe from (8.4.16) and (8.4.15) that

$$h_j(k) = \begin{cases} -1-(-1) = 0, & \text{if } k=0,\ldots,j-1 \quad (\text{if } j \geqslant 1); \\ 0-(-1) = 1, & \text{if } k=j; \\ 0-0 = 0, & \text{if } k=j+1,\ldots,n, \quad (\text{if } j \leqslant n-1), \end{cases}$$

that is,

$$h_j(k) = \delta_{j-k}, \quad k=0,\ldots,n. \tag{8.4.17}$$

Since $L_{n,j} \in \pi_n$, we see from (8.4.13) that $\widetilde{L}_{n,j} \in \pi_n$. But also $h_j \in \pi_n$, so that (8.4.14) and (8.4.17), together with the uniqueness statement in Theorem 1.1.2 with respect to the interpolation points $\{0,\ldots,n\}$, imply that

$$h_j = \widetilde{L}_{n,j}. \tag{8.4.18}$$

By combining (8.4.12), (8.4.18) and (8.4.16), we deduce that

$$w_{n,j} = \frac{b-a}{n} \int_0^n [g_j(t+1) - g_j(t)]dt. \tag{8.4.19}$$

Now observe that

$$\int_0^n [g_j(t+1) - g_j(t)]dt - \int_0^1 [g_j(n+1-t) - g_j(t)]dt$$

$$= -\left[\int_0^n g_j(t)dt - \int_0^1 g_j(t)dt \right] + \left[\int_0^n g_j(t+1)dt - \int_0^1 g_j(n+1-t)dt \right]$$

$$= -\int_1^n g_j(t)dt + \left[\int_1^{n+1} g_j(t)dt - \int_n^{n+1} g_j(t)dt \right]$$

$$= -\int_1^n g_j(t)dt + \int_1^n g_j(t)dt = 0,$$

and thus

$$\int_0^n [g_j(t+1) - g_j(t)]dt = \int_0^1 [g_j(n+1-t) - g_j(t)]dt,$$

which, together with (8.4.19), yields

$$w_{n,j} = \frac{b-a}{n} \int_0^1 [g_j(n+1-t) - g_j(t)]dt. \tag{8.4.20}$$

Since $g_j \in \pi_{n+1}$, we may apply the identity (1.3.25) in Theorem 1.3.5, together with the definition (1.3.11), and the first definition in (1.3.2), to deduce that

$$g_j(t) = g_j(0) + \sum_{k=1}^{n+1} g_j[0,\ldots,k] \prod_{\ell=0}^{k-1} (t-\ell) = -1 + \sum_{k=1}^{n+1} g_j[0,\ldots,k] \prod_{\ell=0}^{k-1}(t-\ell), \tag{8.4.21}$$

since the first line of (8.4.15) gives $g_j(0) = -1$, and similarly, with the definition

$$\widetilde{g}_j(t) := g_j(n+1-t), \tag{8.4.22}$$

according to which $g_j \in \pi_{n+1}$ implies $\widetilde{g}_j \in \pi_{n+1}$, we have

$$\widetilde{g}_j(t) = \widetilde{g}_j(0) + \sum_{k=1}^{n+1} \widetilde{g}_j[0,\ldots,k] \prod_{\ell=0}^{k-1}(t-\ell),$$

and thus, from (8.4.22),

$$g_j(n+1-t) = g_j(n+1) + \sum_{k=1}^{n+1} \widetilde{g}_j[0,\ldots,k] \prod_{\ell=0}^{k-1}(t-\ell)$$

$$= \sum_{k=1}^{n+1} \widetilde{g}_j[0,\ldots,k] \prod_{\ell=0}^{k-1}(t-\ell), \tag{8.4.23}$$

since the second line of (8.4.15) yields $g_j(n+1) = 0$.

Next, we prove that

$$g_j[m,\ldots,m+k] = \frac{1}{k}h_j[m,\ldots,m+k-1], \quad m \in \mathbb{Z}, \quad k = 1,2,\ldots, \tag{8.4.24}$$

where h_j is defined by (8.4.16), and similarly,

$$\widetilde{g}_j[m,\ldots,m+k] = \frac{1}{k}\widetilde{h}_j[m,\ldots,m+k-1], \quad m \in \mathbb{Z}, \quad k = 1,2,\ldots, \tag{8.4.25}$$

where

$$\widetilde{h}_j(t) := \widetilde{g}_j(t+1) - \widetilde{g}_j(t). \tag{8.4.26}$$

In order to prove (8.4.24) inductively, we first note from (1.3.4) and (8.4.16) that (8.4.24) is satisfied for $k = 1$. Moreover, if (8.4.24) holds for a fixed $k \in \mathbb{N}$, we may apply the recursion formula (1.3.21) in Theorem 1.3.4, together with the inductive hypothesis (8.4.24), to deduce that, for any $m \in \mathbb{Z}$,

$$h[m,\ldots,m+k] = \frac{h[m+1,\ldots,m+k] - h[m,\ldots,m+k-1]}{k}$$

$$= \frac{1}{k}\left\{kg_j[m+1,\ldots,m+k+1] - kg_j[m,\ldots,m+k]\right\}$$

$$= (k+1)\left\{\frac{g_j[m+1,\ldots,m+k+1] - g_j[m,\ldots,m+k]}{k+1}\right\}$$

$$= (k+1)g_j[m,\ldots,m+k+1],$$

according to which (8.4.24) also holds with k replaced by $k+1$, and thereby completing our inductive proof of (8.4.24). Similarly, it follows inductively from (8.4.26) that (8.4.25) holds (see Exercise 8.11).

Hence we may now use the case $m = 0$ of (8.4.24) and (8.4.25) in, respectively, (8.4.21) and (8.4.23), to obtain

$$g_j(t) = -1 + \sum_{k=1}^{n+1} \frac{h_j[0,\ldots,k-1]}{k} \prod_{\ell=0}^{k-1}(t-\ell); \tag{8.4.27}$$

$$g_j(n+1-t) = \sum_{k=1}^{n+1} \frac{\widetilde{h}_j[0,\ldots,k-1]}{k} \prod_{\ell=0}^{k-1}(t-\ell). \tag{8.4.28}$$

By noting from (8.4.22) and (8.4.15) that

$$\widetilde{g}_j(k) = \begin{cases} 0, & k=0,\ldots,n-j; \\ -1, & k=n-j+1,\ldots,n+1, \end{cases} \tag{8.4.29}$$

we may now deduce from (8.4.26) that

$$\widetilde{h}_j(k) = \begin{cases} 0-0 = 0, & \text{if } k=0,\ldots,n-j-1 \text{ (if } j \leqslant n-1); \\ -1-0 = -1, & \text{if } k=n-j; \\ -1-(-1) = 0, & \text{if } k=n-j+1,\ldots,n \text{ (if } j \geqslant 1), \end{cases}$$

that is,

$$\widetilde{h}_j(k) = -\delta_{n-j-k}, \quad k=0,\ldots,n. \tag{8.4.30}$$

Next, for $k = 1,\ldots,n$, we apply the formula (3.4.2) in Theorem 3.4.1, with, respectively, $f = h_j$ and $f = \widetilde{h}_j$, and with $\mu = 0$, $v = k-1$, $\{x_\mu,\ldots,x_v\} = \{0,\ldots,k-1\}$, and $h = 1$, to obtain

$$\left.\begin{aligned} h_j[0,\ldots,k-1] &= \frac{(-1)^{k-1}}{(k-1)!} \sum_{m=0}^{k-1}(-1)^m \binom{k-1}{m} h_j(m), \\ \widetilde{h}_j[0,\ldots,k-1] &= \frac{(-1)^{k-1}}{(k-1)!} \sum_{m=0}^{k-1}(-1)^m \binom{k-1}{m} \widetilde{h}_j(m), \end{aligned}\right\} \quad k=1,\ldots n. \tag{8.4.31}$$

It then follows from (8.4.27) and the first line of (8.4.31), together with (8.4.17), and keeping in mind also the second line of (3.2.1), that

$$g_j(t) = -1 + \sum_{k=1}^{n+1} \frac{(-1)^{k-1}}{k!} \sum_{m=0}^{k-1}(-1)^m \binom{k-1}{m} \delta_{j-m} \prod_{\ell=0}^{k-1}(t-\ell)$$

$$= -1 + (-1)^j \sum_{k=1}^{n+1} \frac{(-1)^{k-1}}{k!} \binom{k-1}{j} \prod_{\ell=0}^{k-1}(t-\ell), \tag{8.4.32}$$

and similarly, from (8.4.28), and the second line of (8.4.31), together with (8.4.30), we obtain

$$g_j(n+1-t) = \sum_{k=1}^{n+1} \frac{(-1)^{k-1}}{k!} \sum_{m=0}^{k-1}(-1)^m \binom{k-1}{m}(-\delta_{n-j-m}) \prod_{\ell=0}^{k-1}(t-\ell)$$

$$= (-1)^{n-j+1} \sum_{k=1}^{n+1} \frac{(-1)^{k-1}}{k!} \binom{k-1}{n-j} \prod_{\ell=0}^{k-1}(t-\ell). \tag{8.4.33}$$

Let the polynomial sequence $\left\{ \binom{t}{k} : k = 0, 1, \dots \right\}$ be defined by

$$\binom{t}{k} := \begin{cases} 1, & k = 0; \\ \frac{1}{k!} \prod_{\ell=0}^{k-1}(t-\ell), & k = 1, 2, \dots. \end{cases} \tag{8.4.34}$$

Observe that if we set $t = m \in \{0, 1, \dots\}$ in (8.4.34), then, for $k = 0, \dots, m$,

$$\binom{m}{k} = \frac{m(m-1)\dots(m-k+1)}{k!} = \frac{m!}{k!(m-k)!},$$

which is consistent with the binomial coefficient definition in (3.2.1). It follows from (8.4.32), (8.4.33) and (8.4.34) that

$$\left. \begin{aligned} g_j(t) &= -1 + (-1)^j \sum_{k=1}^{n+1}(-1)^{k-1}\binom{k-1}{j}\binom{t}{k}; \\ g_j(n+1-t) &= (-1)^{n-j+1} \sum_{k=1}^{n+1}(-1)^{k-1}\binom{k-1}{n-j}\binom{t}{k}. \end{aligned} \right\} \tag{8.4.35}$$

By substituting (8.4.35) into (8.4.20), we obtain

$$w_{n,j} = \frac{b-a}{n} \left[1 - (-1)^j \left\{ \sum_{k=1}^{n+1}(-1)^{k-1}\binom{k-1}{j} \int_0^1 \binom{t}{k} dt \right. \right.$$

$$\left. \left. + (-1)^n \sum_{k=1}^{n+1}(-1)^{k-1}\binom{k-1}{n-j} \int_0^1 \binom{t}{k} dt \right\} \right]. \tag{8.4.36}$$

The real numbers $\{\Lambda_k : 0, 1, \dots\}$ defined by

$$\Lambda_k := (-1)^{k-1} \int_0^1 \binom{t}{k} dt, \quad k = 0, 1, \dots, \tag{8.4.37}$$

with the polynomial sequence $\left\{ \binom{t}{k} : k = 0, 1, \dots \right\}$ given in (8.4.34), are known as the Laplace coefficients.

According to (8.4.8), (8.4.36) and (8.4.37), we have therefore now established the following formulation of the Newton-Cotes weights.

Theorem 8.4.1. *For any non-negative integer n, the Newton-Cotes quadrature rule $\mathcal{Q}_n^{NC}$, as given by (8.4.6), (8.4.7), (8.4.3), has weights $\{w_{n,j} : j = 0, \dots, n\}$ given by*

$$w_{0,0} = b - a;$$

$$\left. w_{n,j} = \frac{b-a}{n} \left[1 - (-1)^j \sum_{k=0}^{n} \Lambda_{k+1} \left\{ \binom{k}{j} + (-1)^n \binom{k}{n-j} \right\} \right], \, j = 0, \dots, n \; (if\, n \geqslant 1), \right\} \tag{8.4.38}$$

with the Laplace coefficients $\{\Lambda_1, \dots, \Lambda_{n+1}\}$ defined as in (8.4.37).

We proceed to derive a recursive formulation for the Laplace coefficients $\{\Lambda_k : k = 0, 1, \ldots\}$ in (8.4.37), for which we shall require the following polynomial identity.

Theorem 8.4.2. *For any non-negative integer k,*

$$\sum_{\ell=0}^{k} \binom{s}{\ell} \binom{t}{k-\ell} = \binom{s+t}{k}, \quad s, t \in \mathbb{R}, \tag{8.4.39}$$

with the polynomials $\left\{ \binom{t}{\ell} : \ell = 0, \ldots, k \right\}$ defined as in (8.4.34).

Proof. Our first step is to prove that (8.4.39) is satisfied for all non-negative integer values of s and t, that is,

$$\sum_{\ell=0}^{k} \binom{\mu}{\ell} \binom{\nu}{k-\ell} = \binom{\mu+\nu}{k}, \quad \mu, \nu = 0, 1, \ldots. \tag{8.4.40}$$

To this end, for any non-negative integers μ and ν, by keeping in mind also the second line of (3.2.1), we obtain, for any $x \in \mathbb{R}$,

$$\sum_{k=0}^{\mu+\nu} \left[\sum_{\ell=0}^{k} \binom{\mu}{\ell} \binom{\nu}{k-\ell} \right] x^k = \sum_{\ell=0}^{\mu+\nu} \binom{\mu}{\ell} \left[\sum_{k=\ell}^{\mu+\nu} \binom{\nu}{k-\ell} x^k \right]$$

$$= \sum_{\ell=0}^{\mu} \binom{\mu}{\ell} \left[\sum_{k=0}^{\mu+\nu-\ell} \binom{\nu}{k} x^{k+\ell} \right]$$

$$= \sum_{\ell=0}^{\mu} \binom{\mu}{\ell} \left[\sum_{k=0}^{\nu} \binom{\nu}{k} x^k 1^{\nu-k} \right] x^\ell$$

$$= \sum_{\ell=0}^{\mu} \binom{\mu}{\ell} (x+1)^\nu x^\ell$$

$$= (x+1)^\nu \sum_{\ell=0}^{\mu} \binom{\mu}{\ell} x^\ell 1^{\mu-\ell}$$

$$= (x+1)^\nu (x+1)^\mu = (x+1)^{\mu+\nu} = \sum_{k=0}^{\mu+\nu} \binom{\mu+\nu}{k} x^k,$$

and thus

$$\sum_{k=0}^{\mu+\nu} \left[\sum_{\ell=0}^{k} \binom{\mu}{\ell} \binom{\nu}{k-\ell} - \binom{\mu+\nu}{k} \right] x^k = 0, \quad x \in \mathbb{R},$$

from which (8.4.40) then immediately follows.

Hence, if we define the bivariate polynomial

$$F(s,t) := \sum_{\ell=0}^{k} \binom{s}{\ell} \binom{t}{k-\ell} - \binom{s+t}{k}, \quad s, t \in \mathbb{R}, \tag{8.4.41}$$

it follows from (8.4.40) that

$$F(\mu, \nu) = 0, \quad \mu, \nu = 0, 1, \ldots. \tag{8.4.42}$$

Now observe from (8.4.41) and (8.4.34) that

$$F(s,t) = \sum_{i=0}^{k} \sum_{j=0}^{k} a_{ij} s^i t^i, \tag{8.4.43}$$

for some coefficient sequence $\{a_{ij} : i, j = 0, \ldots, k\}$. It follows from (8.4.43) and (8.4.42) that

$$\sum_{i=0}^{k} m^i \sum_{j=0}^{k} a_{ij} n^j = 0, \quad m, n = 0, \ldots, k. \tag{8.4.44}$$

Let the $(k+1) \times (k+1)$ matrix V be defined by

$$V := \begin{bmatrix} 1 & x_0 & x_0^2 & \cdots & x_0^k \\ 1 & x_1 & x_1^2 & \cdots & x_1^k \\ \vdots & \vdots & \vdots & & \vdots \\ 1 & x_k & x_k^2 & \cdots & x_k^k \end{bmatrix}, \tag{8.4.45}$$

where

$$x_\mu := \mu, \quad \mu = 0, \ldots, k. \tag{8.4.46}$$

By comparing (8.4.45) and (1.1.7), we see that V is a Vandermonde matrix, and thus, since (8.4.46) shows that $\{x_\mu : \mu = 0, \ldots k\}$ is a sequence of $k+1$ distinct points in $\mathbb{R}$, we may apply Theorem 1.1.1 to deduce that V is an invertible matrix. Moreover, with the $(k+1) \times (k+1)$ matrix A defined by

$$A := [a_{ij}]_{1 \leqslant i,j \leqslant n}, \tag{8.4.47}$$

we deduce from (8.4.45), (8.4.46) and (8.4.47) that (8.4.44) is equivalent to the matrix equation

$$VAV^T = 0, \tag{8.4.48}$$

the zero matrix. Since V is invertible, we know from a standard result in linear algebra that V^T is invertible, so that we may deduce from (8.4.48) that

$$0 = V^{-1}(0)(V^T)^{-1} = V^{-1}(VAV^T)(V^T)^{-1} = (V^{-1}V)A(V^T(V^T)^{-1}) = IAI = A,$$

and thus $A = 0$, the zero matrix, and it follows from (8.4.47) that $a_{ij} = 0$, $i, j = 0, \ldots, n$, which, together with (8.4.43), implies that

$$F(s,t) = 0, \quad s, t \in \mathbb{R}. \tag{8.4.49}$$

The desired result (8.4.39) is then an immediate consequence of (8.4.41) and (8.4.49). ∎

The computation of the Laplace coefficients is facilitated by the following recursive formulation, the proof of which is based on the polynomial identity (8.4.39) in Theorem 8.4.2.

Theorem 8.4.3. *The Laplace coefficients* $\{\Lambda_k : k = 0, 1, \ldots\}$, *as defined by* (8.4.37), *satisfy the recursive formulation*

$$\left.\begin{aligned}
\Lambda_0 &= -1; \\
\Lambda_k &= \sum_{\ell=0}^{k-1} \frac{\Lambda_\ell}{\ell - k - 1}, \quad k = 1, 2, \ldots
\end{aligned}\right\} \tag{8.4.50}$$

Proof. First, observe from (8.4.37) and the first line of (8.4.34) that

$$\Lambda_0 = -\int_0^1 dt = -1,$$

which proves the first line of (8.4.50).

Next, by using (8.4.39) in Theorem 8.4.2, as well as (8.4.34), we obtain, for any $k \in \mathbb{N}$, $t \in \mathbb{R}$ and $h \in \mathbb{R} \setminus \{0\}$,

$$\frac{1}{h}\left[\binom{t+h}{k} - \binom{t}{k}\right] = \frac{1}{h}\left[\sum_{\ell=0}^{k}\binom{h}{\ell}\binom{t}{k-\ell} - \binom{t}{k}\right]$$

$$= \frac{1}{h}\sum_{\ell=1}^{k}\binom{h}{\ell}\binom{t}{k-\ell}$$

$$= \frac{1}{h}\sum_{\ell=1}^{k}\frac{h(h-1)\ldots(h-\ell+1)}{\ell!}\binom{t}{k-\ell}$$

$$= \sum_{\ell=1}^{k}\frac{(h-1)\ldots(h-(\ell-1))}{\ell!}\binom{t}{k-\ell},$$

and thus

$$\frac{d}{dt}\binom{t}{k} := \lim_{h\to 0}\frac{1}{h}\left[\binom{t+h}{k} - \binom{t}{k}\right] = \sum_{\ell=1}^{k}\frac{(-1)^{\ell-1}}{\ell}\binom{t}{k-\ell}. \tag{8.4.51}$$

It follows from (8.4.51) and (8.4.37) that, for $k \in \mathbb{N}$,

$$\int_0^1 \frac{d}{dt}\binom{t}{k}dt = \sum_{\ell=1}^{k}\frac{(-1)^{\ell-1}}{\ell}\int_0^1\binom{t}{k-\ell}dt = \sum_{\ell=0}^{k-1}\frac{(-1)^{k-1-\ell}}{k-\ell}\int_0^1\binom{t}{\ell}dt$$

$$= (-1)^k\sum_{\ell=0}^{k-1}\frac{\Lambda_\ell}{k-\ell}. \tag{8.4.52}$$

Moreover, the fundamental theorem of calculus gives

$$\int_0^1 \frac{d}{dt}\binom{t}{k}dt = \binom{1}{k} - \binom{0}{k} = \begin{cases} 1-0 = 1, & \text{if } k = 1; \\ 0-0 = 0, & \text{if } k = 2, 3, \ldots, \end{cases} \tag{8.4.53}$$

by virtue of (8.4.34).

By combining (8.4.52) and (8.4.53), we deduce that

$$(-1)^{k+1} \sum_{\ell=0}^{k} \frac{\Lambda_\ell}{k+1-\ell} = \begin{cases} 1, & \text{if } k=0; \\[2mm] 0, & \text{if } k=1,2,\ldots, \end{cases}$$

and thus

$$(-1)^{k+1} \left[\Lambda_k + \sum_{\ell=0}^{k-1} \frac{\Lambda_\ell}{k+1-\ell} \right] = 0, \quad k=1,2,\ldots,$$

which is equivalent to the second line of (8.4.50). ∎

By applying the recursive formulation (8.4.50), we obtain (see Exercise 8.12) the Laplace coefficients

$$\left. \begin{aligned} &\Lambda_1 = \tfrac{1}{2}; \quad \Lambda_2 = \tfrac{1}{12}; \quad \Lambda_3 = \tfrac{1}{24}; \quad \Lambda_4 = \tfrac{19}{720}; \quad \Lambda_5 = \tfrac{3}{160}; \\ &\Lambda_6 = \tfrac{863}{60480}; \; \Lambda_7 = \tfrac{275}{24192}; \; \Lambda_8 = \tfrac{33953}{3628800}; \; \Lambda_9 = \tfrac{8183}{1036800}; \; \Lambda_{10} = \tfrac{3250433}{479001600}. \end{aligned} \right\} \quad (8.4.54)$$

By using the formula (8.4.38), together with the Laplace coefficient values in (8.4.54), we obtain (see Exercise 8.12) the Newton-Cotes weights as given in Table 8.4.1.

Table 8.4.1 The Newton-Cotes weights $w_{n,j} = [(b-a)/n]\sigma_{n,j}$, $j=0,\ldots,n$, for $n=0,\ldots,8$.

n	$\{\sigma_{n,j}\}$
0	$\{1\}$
1	$\left\{ \dfrac{1}{2}, \dfrac{1}{2} \right\}$
2	$\left\{ \dfrac{1}{3}, \dfrac{4}{3}, \dfrac{1}{3} \right\}$
3	$\left\{ \dfrac{3}{8}, \dfrac{9}{8}, \dfrac{9}{8}, \dfrac{3}{8} \right\}$
4	$\left\{ \dfrac{14}{45}, \dfrac{64}{45}, \dfrac{8}{15}, \dfrac{64}{45}, \dfrac{14}{45} \right\}$
5	$\left\{ \dfrac{95}{288}, \dfrac{125}{96}, \dfrac{125}{144}, \dfrac{125}{144}, \dfrac{125}{96}, \dfrac{95}{288} \right\}$
6	$\left\{ \dfrac{41}{140}, \dfrac{54}{35}, \dfrac{27}{140}, \dfrac{68}{35}, \dfrac{27}{140}, \dfrac{54}{35}, \dfrac{41}{140} \right\}$
7	$\left\{ \dfrac{5257}{17280}, \dfrac{25039}{17280}, \dfrac{343}{640}, \dfrac{20923}{17280}, \dfrac{20923}{17280}, \dfrac{343}{640}, \dfrac{25039}{17280}, \dfrac{5257}{17280} \right\}$
8	$\left\{ \dfrac{3956}{14175}, \dfrac{23552}{14175}, -\dfrac{3712}{14175}, \dfrac{41984}{14175}, -\dfrac{3632}{2835}, \dfrac{41984}{14175}, -\dfrac{3712}{14175}, \dfrac{23552}{14175}, \dfrac{3956}{14175} \right\}$

Observe from the formula in the second line of (8.4.38) in Theorem 8.4.1 that the Newton-Cotes weights satisfy, for $j = 0,\ldots,n$,

$$w_{n,n-j} = \frac{b-a}{n}\left[1-(-1)^{n-j}\sum_{k=0}^{n}\Lambda_{k+1}\left\{\binom{k}{n-j}+(-1)^n\binom{k}{j}\right\}\right]$$

$$= \frac{b-a}{n}\left[1-(-1)^{j}\sum_{k=0}^{n}\Lambda_{k+1}\left\{(-1)^n\binom{k}{n-j}+\binom{k}{j}\right\}\right] = w_{n,j},$$

that is, the symmetry condition

$$w_{n,n-j} = w_{n,j}, \quad j = 0,\ldots,n, \tag{8.4.55}$$

is satisfied, as illustrated for $n = 0,\ldots,8$ in Table 8.4.1.

Example 8.4.1. To obtain the Newton-Cotes quadrature rules $\mathcal{Q}_1^{NC}$, $\mathcal{Q}_2^{NC}$, $\mathcal{Q}_3^{NC}$ and $\mathcal{Q}_4^{NC}$ for the numerical approximation of the integral (8.4.1), we apply (8.4.6) and (8.4.3), together with (8.4.38) in Theorem 8.4.1, to deduce from Table 8.4.1 that, for any $f \in C[a,b]$,

$$\mathcal{Q}_1^{NC}[f] = \frac{b-a}{2}[f(a)+f(b)]; \tag{8.4.56}$$

$$\mathcal{Q}_2^{NC}[f] = \frac{b-a}{6}\left[f(a)+4f\left(\frac{a+b}{2}\right)+f(b)\right]; \tag{8.4.57}$$

$$\mathcal{Q}_3^{NC}[f] = \frac{b-a}{8}\left[f(a)+3f\left(\frac{2a+b}{3}\right)+3f\left(\frac{a+2b}{3}\right)+f(b)\right]; \tag{8.4.58}$$

$$\mathcal{Q}_4^{NC}[f] = \frac{b-a}{90}\left[7f(a)+32f\left(\frac{3a+b}{4}\right)+12f\left(\frac{a+b}{2}\right)+32f\left(\frac{a+3b}{4}\right)+7f(b)\right]. \tag{8.4.59}$$

∎

8.5 Error analysis for Newton-Cotes quadrature

In the general setting of Section 8.1, for any non-negative integer n, let the functional $\mathscr{E}_n : C[a,b] \to \mathbb{R}$ be defined by

$$\mathscr{E}_n[f] := \mathscr{I}[f] - \mathcal{Q}_n[f], \quad f \in C[a,b], \tag{8.5.1}$$

with the functionals $\mathscr{I}$ and $\mathcal{Q}_n$ defined as in (8.1.1) and (8.1.5). For any $f \in C[a,b]$ and non-negative integer n, we call $\mathscr{E}_n[f]$ the corresponding quadrature error. Observe from (8.5.1), (8.1.9) and (8.1.10) that $\mathscr{E}_n$ is a linear functional, that is,

$$\mathscr{E}_n[\lambda f+\mu g] = \lambda\mathscr{E}_n[f]+\mu\mathscr{E}_n[g], \quad \lambda,\mu \in \mathbb{R}, \quad f,g \in C[a,b], \tag{8.5.2}$$

for $n = 0, 1, \ldots$.

The principal aim of this section is to investigate the Newton-Cotes quadrature error

$$\mathcal{E}_n^{NC}[f] := \int_a^b f(x)dx - \mathcal{Q}_n^{NC}[f], \quad f \in C[a,b], \tag{8.5.3}$$

with $\mathcal{Q}_n^{NC}$ as given in (8.4.6), (8.4.7), (8.4.3).

Our first step in this direction is to prove the following general result for linear functionals on $C[a,b]$.

Theorem 8.5.1. *Suppose $\mathcal{L} : C[a,b] \to \mathbb{R}$ is a linear functional such that, for some nonnegative integer k, it holds that either:*

(a)

$$f \in C^k[a,b], \quad \text{with} \quad f^{(k)}(x) > 0, x \in [a,b] \Rightarrow \mathcal{L}[f] > 0; \tag{8.5.4}$$

or

(b)

$$f \in C^k[a,b], \quad \text{with} \quad f^{(k)}(x) > 0, \ x \in [a,b] \Rightarrow \mathcal{L}[f] < 0. \tag{8.5.5}$$

Then, for $f \in C^k[a,b]$, and any monic polynomial $P \in \tilde{\pi}_k$, there exists a number $\xi \in [a,b]$ such that

$$\mathcal{L}[f] = \frac{f^{(k)}(\xi)}{k!} \mathcal{L}[P]. \tag{8.5.6}$$

Proof. Suppose first that $\mathcal{L}$ satisfies the condition (8.5.4). Let $f \in C^k[a,b]$, $P \in \tilde{\pi}_k$, and, for any $\varepsilon > 0$, with the definitions

$$m := \min_{a \leqslant x \leqslant b} f^{(k)}(x); \quad M := \max_{a \leqslant x \leqslant b} f^{(k)}(x), \tag{8.5.7}$$

define the polynomials

$$g := \frac{1}{k!}(m - \varepsilon)P; \quad h := \frac{1}{k!}(M + \varepsilon)P. \tag{8.5.8}$$

Since $P \in \tilde{\pi}_k$ implies

$$P^{(k)}(x) = k!, \quad x \in \mathbb{R}, \tag{8.5.9}$$

it follows from (8.5.8), (8.5.7), and $\varepsilon > 0$, that

$$\left. \begin{array}{l} g^{(k)}(x) = m - \varepsilon \leqslant f^{(k)}(x) - \varepsilon < f^{(k)}(x), \\ h^{(k)}(x) = M + \varepsilon \geqslant f^{(k)}(x) + \varepsilon > f^{(k)}(x), \end{array} \right\} \quad a \leqslant x \leqslant b,$$

and thus

$$\left.\begin{array}{l} (f-g)^{(k)}(x) > 0, \\[2mm] (h-f)^{(k)}(x) > 0, \end{array}\right\} \quad a \leqslant x \leqslant b. \tag{8.5.10}$$

Hence, from (8.5.10) and (8.5.4), together with the linearity of $\mathscr{L}$, we have

$$\left.\begin{array}{l} \mathscr{L}[f] - \mathscr{L}[g] = \mathscr{L}[f-g] > 0; \\[2mm] \mathscr{L}[h] - \mathscr{L}[f] = \mathscr{L}[h-f] > 0, \end{array}\right\}$$

and thus

$$\mathscr{L}[g] < \mathscr{L}[f] < \mathscr{L}[h]. \tag{8.5.11}$$

By using again the linearity of $\mathscr{L}$, we deduce from (8.5.8) and (8.5.11) that

$$\frac{m-\varepsilon}{k!}\mathscr{L}[P] < \mathscr{L}[f] < \frac{(M+\varepsilon)}{k!}\mathscr{L}[P]. \tag{8.5.12}$$

Now observe from (8.5.9) that

$$P^{(k)}(x) > 0, \quad x \in [a,b],$$

according to which (8.5.4) gives $\mathscr{L}[P] > 0$, so that we may divide the inequalities (8.5.12) by $\frac{1}{k!}\mathscr{L}[P]$ to deduce that

$$m - \varepsilon < \frac{k!\mathscr{L}[f]}{\mathscr{L}[P]} < M + \varepsilon. \tag{8.5.13}$$

Since ε is an arbitrary positive number, it follows from (8.5.13), together with (8.5.7), that

$$\min_{a \leqslant x \leqslant b} f^{(k)}(x) = m \leqslant \frac{k!\mathscr{L}[f]}{\mathscr{L}[P]} \leqslant M = \max_{a \leqslant x \leqslant b} f^{(k)}(x). \tag{8.5.14}$$

Hence, since also $f^{(k)} \in C[a,b]$, we may apply the intermediate value theorem, to deduce from (8.5.14) that there exists a number $\xi \in [a,b]$ such that

$$\frac{k!\mathscr{L}[f]}{\mathscr{L}[P]} = f^{(k)}(\xi), \tag{8.5.15}$$

which is equivalent to (8.5.6).

If $\mathscr{L}$ satisfies the condition (8.5.5), we may apply the argument which led from (8.5.7) to (8.5.14) with $\mathscr{L}$ replaced by $-\mathscr{L}$, to obtain, by using also the linearity of $\mathscr{L}$,

$$\frac{k!\mathscr{L}[f]}{\mathscr{L}[P]} = \frac{k!(-\mathscr{L}[f])}{-\mathscr{L}[P]} = \frac{k!(-\mathscr{L})[f]}{(-\mathscr{L})[P]} = f^{(k)}(\xi),$$

as in (8.5.15), and thereby completing our proof. ∎

Before proceeding to prove that the Newton-Cotes error functional $\mathscr{E}_n^{NC}$ in (8.5.3) belongs to the class of linear functionals $\mathscr{L}$ of Theorem 8.5.1, it is first necessary to establish the following property of divided differences. Here, and in the rest of the chapter, we shall rely on the mean value theorem for integrals which states that, if $f, g \in C[a,b]$, and $g(x)$ does not change sign on $[a,b]$, then there exists a point $\xi \in (a,b)$ such that

$$\int_a^b f(x)g(x)dx = f(\xi) \int_a^b g(x)dx, \qquad (8.5.16)$$

and similarly for iterated integrals.

Theorem 8.5.2. *Suppose $f \in C^{k+2}[a,b]$ for a non-negative integer k, and let $\{x_0, \ldots, x_k\}$ denote a sequence of $k+1$ distinct points in $[a,b]$. Then, for any $x \in [a,b]$, there is a point $\xi \in [a,b]$ such that*

$$\frac{d}{dx}(f[x, x_0, \ldots, x_k]) = \frac{f^{(k+2)}(\xi)}{(k+2)!}, \qquad (8.5.17)$$

with the divided difference $f[x, x_0, \ldots, x_k]$ defined as in (1.4.14).

Proof. Our first step is to prove that, for $g \in C^1[a,b]$, and any x, α and β such that

$$a < t(x - \alpha) + \beta < b, \quad t \in (0, \alpha),$$

we have

$$\frac{d}{dx}\left[\int_0^\alpha g(t(x-\alpha)+\beta)dt \right] = \int_0^\alpha t g'(t(x-\alpha)+\beta)dt, \quad x \neq \alpha. \qquad (8.5.18)$$

To prove (8.5.18), we suppose $x \neq \alpha$, and use the fundamental theorem of calculus to obtain

$$\frac{d}{dx}\left[\int_0^\alpha g(t(x-\alpha)+\beta)dt \right] = \frac{d}{dx}\left[\frac{\int_0^{\alpha(x-\alpha)} g(s+\beta)ds}{x-\alpha} \right]$$

$$= \frac{\alpha g(\alpha(x-\alpha)+\beta)}{x-\alpha} - \frac{\int_0^{\alpha(x-\alpha)} g(s+\beta)ds}{(x-\alpha)^2}$$

$$= \frac{\alpha g(\alpha(x-\alpha)+\beta)}{x-\alpha} - \frac{\int_0^\alpha g(t(x-\alpha)+\beta)dt}{x-\alpha}, \qquad (8.5.19)$$

whereas integration by parts yields

$$\int_0^\alpha t g'(t(x-\alpha)+\beta)dt = \left[\frac{t g(t(x-\alpha)+\beta)}{x-\alpha} \right]_{t=0}^{t=\alpha} - \int_0^\alpha \frac{g(t(x-\alpha)+\beta)}{x-\alpha}dt$$

$$= \frac{\alpha g(\alpha(x-\alpha)+\beta)}{x-\alpha} - \frac{\int_0^\alpha g(t(x-\alpha)+\beta)dt}{x-\alpha}, \qquad (8.5.20)$$

and it follows from (8.5.19) and (8.5.20) that (8.5.18) holds for $x \neq \alpha$.

By first noting that Theorem 1.3.6 is also valid for the more general definition (1.4.14) of divided differences, we now apply Theorem 1.3.6, the definition (1.4.14), (8.5.18), the remark following the statement of Theorem 1.4.3, as well as the mean value theorem for iterated integrals, as given similarly to (8.5.16), to deduce that, for $x \in [a,b] \setminus \{x_k\}$, there exists a point $\xi \in [a,b]$ such that, with $t_0 := 1$,

$$\frac{d}{dx}(f[x,x_0,\ldots,x_k]) = \frac{d}{dx}(f[x_0,\ldots,x_k,x])$$

$$= \frac{d}{dx}\left[\int_0^{t_0} \cdots \int_0^{t_{k-1}} \int_0^{t_k} f^{(k+1)}(t_{k+1}(x-x_k)+t_k(x_k-x_{k-1})+\ldots\right.$$

$$\left. +t_1(x_1-x_0)+x_0)dt_{k+1}dt_k\ldots dt_1\right]$$

$$= \int_0^{t_0} \cdots \int_0^{t_{k-1}} \frac{d}{dx}\left[\int_0^{t_k} f^{(k+1)}(t_{k+1}(x-x_k)+t_k(x_k-x_{k-1})+\cdots\right.$$

$$\left. +t_1(x_1-x_0)+x_0)dt_{k+1}\right]dt_k\ldots dt_1$$

$$= \int_0^{t_0} \cdots \int_0^{t_{k-1}} \int_0^{t_k} t_{k+1}f^{(k+2)}(t_{k+1}(x-x_k)+t_k(x_k-x_{k-1})+\cdots$$

$$+t_1(x_1-x_0)+x_0)dt_{k+1}dt_k\ldots dt_1$$

$$= f^{(k+2)}(\xi)\int_0^{t_0} \cdots \int_0^{t_{k-1}} \int_0^{t_k} t_{k+1}dt_{k+1}dt_k\ldots dt_1$$

$$= \frac{f^{(k+2)}(\xi)}{2}\int_0^{t_0} \cdots \int_0^{t_{k-1}} t_k^2 dt_k\ldots dt_1$$

$$= \cdots$$

$$= \frac{f^{(k+2)}(\xi)}{(k+1)!}\int_0^{t_0} t_1^{k+1}dt_1 = \frac{f^{(k+2)}(\xi)}{(k+1)!}\int_0^1 t_1^{k+1}dt_1 = \frac{f^{(k+2)}(\xi)}{(k+2)!},$$

which proves (8.5.17) for $x \in [a,b] \setminus \{x_k\}$.

If $x = x_k$, we use the fact that

$$f[x,x_0,\ldots,x_{k-1},x_k] = f[x_0,\ldots,x_k,x_{k-1},x],$$

and argue as above, with $x \in [a,b] \setminus \{x_{k-1}\}$, to deduce that (8.5.17) is also satisfied for $x = x_k$, and thereby completing our proof. ∎

By using Theorem 8.5.2, we can now prove that the Newton-Cotes error functional $\mathscr{E}_n^{NC}$ is a linear functional as in Theorem 8.5.1. Note that the case $n = 0$ is not included in

Theorem 8.5.3 below, but will be covered later in Theorem 8.6.4, in the context of the rectangle rule $\mathcal{Q}_n^{RE}$, with $n = 1$.

Theorem 8.5.3. *For any positive integer n, the Newton-Cotes quadrature error functional $\mathcal{E}_n^{NC}$, as defined by (8.5.3), satisfies the following:*

(a) *if n is even, then*

$$f \in C^{n+2}[a,b], \text{ with } f^{(n+2)}(x) > 0, \ x \in [a,b] \Rightarrow \mathcal{E}_n^{NC}[f] < 0; \tag{8.5.21}$$

(b) *if n is odd, then*

$$f \in C^{n+1}[a,b], \text{ with } f^{(n+1)}(x) > 0, \ x \in [a,b] \Rightarrow \mathcal{E}_n^{NC}[f] < 0. \tag{8.5.22}$$

Proof. First, we apply (8.5.3), (8.4.2) and (2.1.3), together with the formulation (2.1.5), (2.1.6) of Theorem 2.1.1, to deduce that, for $n = 0, 1, \ldots$,

$$\mathcal{E}_n^{NC}[f] = \int_a^b f[x, x_{n,0}, \ldots, x_{n,n}] \prod_{j=0}^n (x - x_{n,j}) dx, \qquad f \in C[a,b], \tag{8.5.23}$$

where the divided difference $f[x, x_{n,0}, \ldots, x_{n,n}]$ is defined by (1.4.14), and with the sequence $\{x_{n,j} : j = 0, \ldots, n\}$ given by (8.4.3).

(a) Suppose n is even, with $n \geqslant 2$, and let $f \in C^{n+2}[a,b]$, with

$$f^{(n+2)}(x) > 0, \quad x \in [a,b]. \tag{8.5.24}$$

By applying integration by parts, we deduce from (8.5.23) that

$$\mathcal{E}_n^{NC}[f] = \left[f[x, x_{n,0}, \ldots, x_{n,n}] G_n(x) \right]_{x=a}^{x=b} - \int_a^b \frac{d}{dx} (f[x, x_{n,0}, \ldots, x_{n,n}]) G_n(x) dx, \tag{8.5.25}$$

where we define

$$G_n(x) := \int_a^x \prod_{j=0}^n (\sigma - x_{n,j}) d\sigma, \quad n = 0, 1, \ldots. \tag{8.5.26}$$

Now use the one-to-one mapping (8.4.10), (8.4.11) between the intervals $[0, n]$ and $[a, b]$, together with (8.4.3), to deduce that, with the definition

$$\widetilde{G}_n(t) := \left(\frac{n}{b-a} \right)^{n+2} G_n \left(a + \left(\frac{b-a}{n} \right) t \right), \tag{8.5.27}$$

we have

$$\widetilde{G}_n(t) = \int_0^t \prod_{j=0}^n (\tau - j) d\tau. \tag{8.5.28}$$

It follows from (8.5.28) that

$$\widetilde{G}_n(n) = \int_0^{\frac{n}{2}} \prod_{j=0}^n (\tau - j) d\tau + \int_{\frac{n}{2}}^n \prod_{j=0}^n (\tau - j) d\tau. \tag{8.5.29}$$

But, by using the fact that $n+1$ is odd, we note that

$$\int_{\frac{n}{2}}^{n}\prod_{j=0}^{n}(\tau-j)d\tau = \int_{0}^{\frac{n}{2}}\prod_{j=0}^{n}[(n-s)-j]ds$$

$$= (-1)^{n+1}\int_{0}^{\frac{n}{2}}\prod_{j=0}^{n}[s-(n-j)]ds$$

$$= -\int_{0}^{\frac{n}{2}}\prod_{j=0}^{n}[s-(n-j)]ds = -\int_{0}^{\frac{n}{2}}\prod_{j=0}^{n}(s-j)ds,$$

which, together with (8.5.29), and (8.5.28), yields

$$\widetilde{G}_n(0) = 0; \quad \widetilde{G}_n(n) = 0, \tag{8.5.30}$$

so that, from (8.5.27),

$$G_n(a) = 0; \quad G_n(b) = 0. \tag{8.5.31}$$

Hence we may use (8.5.31) to deduce from (8.5.25) that

$$\mathscr{E}_n^{NC}[f] = -\int_a^b \frac{d}{dx}\left(f[x,x_{n,0},\ldots,x_{n,n}]\right)G_n(x)dx. \tag{8.5.32}$$

Our next step is to show that

$$G_n(x) \geqslant 0, \quad x \in [a,b], \tag{8.5.33}$$

or equivalently, from (8.5.27),

$$\widetilde{G}_n(t) \geqslant 0, \quad t \in [0,n].$$

Since n is even, we shall use an inductive proof to show that

$$\widetilde{G}_{2\nu}(t) \geqslant 0, \quad t \in [0,2\nu], \quad \nu = 1,2,\ldots. \tag{8.5.34}$$

First, note from (8.5.28) that

$$\widetilde{G}_2(t) = \int_0^t \tau(\tau-1)(\tau-2)d\tau = \tfrac{1}{4}t^2(t-2)^2, \tag{8.5.35}$$

(see Exercise 8.16), according to which (8.5.34) holds for $\nu = 1$. Suppose next that (8.5.34) is satisfied for a fixed positive integer ν. It follows from (8.5.28), together with integration by parts, as well as (8.5.30), that

$$\widetilde{G}_{2\nu+1}(t) = \int_0^t (\tau-2\nu-1)\prod_{j=0}^{2\nu}(\tau-j)d\tau$$

$$= \left[(\tau-2\nu-1)\widetilde{G}_{2\nu}(\tau)\right]_0^t - \int_0^t \widetilde{G}_{2\nu}(\tau)d\tau$$

$$= (t - 2v - 1)\widetilde{G}_{2v}(t) - \int_0^t \widetilde{G}_{2v}(\tau)d\tau, \tag{8.5.36}$$

and similarly,

$$\widetilde{G}_{2v+2}(t) = \left[(\tau - 2v - 2)\widetilde{G}_{2v+1}(\tau)\right]_0^t - \int_0^t \widetilde{G}_{2v+1}(\tau)d\tau$$

$$= (t - 2v - 2)\widetilde{G}_{2v+1}(t) - \int_0^t \widetilde{G}_{2v+1}(\tau)d\tau. \tag{8.5.37}$$

It follows from (8.5.36) and the inductive hypothesis (8.5.34) that

$$\widetilde{G}_{2v+1}(t) \leqslant 0, \quad t \in [0, 2v], \tag{8.5.38}$$

and hence, from (8.5.37) and (8.5.38),

$$\widetilde{G}_{2v+2}(t) \geqslant 0, \quad t \in [0, 2v]. \tag{8.5.39}$$

Next, we use (8.5.28) to obtain, for any $t \in [0, v+1]$,

$$\widetilde{G}_{2v+2}(2v + 2 - t) - \widetilde{G}_{2v+2}(t)$$

$$= \int_t^{2v+2-t} \prod_{j=0}^{2v+2} (\tau - j)d\tau$$

$$= \int_t^{v+1} \prod_{j=0}^{2v+2} (\tau - j)d\tau + \int_{v+1}^{2v+2-t} \prod_{j=0}^{2v+2} (\tau - j)d\tau$$

$$= \int_t^{v+1} \prod_{j=0}^{2v+2} (\tau - j)d\tau + \int_t^{v+1} \prod_{j=0}^{2v+2} [(2v + 2 - s) - j]ds$$

$$= \int_t^{v+1} \prod_{j=0}^{2v+2} (\tau - j)d\tau + (-1)^{2v+3} \int_t^{v+1} \prod_{j=0}^{2v+2} [s - (2v + 2 - j)]ds$$

$$= \int_t^{v+1} \prod_{j=0}^{2v+2} (\tau - j)d\tau - \int_t^{v+1} \prod_{j=0}^{2v+2} (s - j)ds = 0,$$

and thus, from (8.5.39) and $2v + 2 \geqslant 2$,

$$\widetilde{G}_{2v+2}(t) = \widetilde{G}_{2v+2}(2v + 2 - t) \geqslant 0, \quad t \in [2v, 2v + 2],$$

which, together with (8.5.39), shows that (8.5.34) holds with v replaced by $v + 1$, and thereby completing our inductive proof of (8.5.34).

Since (8.5.34), and therefore also (8.5.33) hold, we may apply the mean value theorem (8.5.16) for integrals to deduce from (8.5.32) that there exists a point $\tilde{x} \in [a, b]$ such that

$$\mathscr{E}_n^{NC}[f] = -\left[\frac{d}{dx}\left(f[x, x_{n,0}, \dots, x_{n,n}]\right)\right]_{x=\tilde{x}} \int_a^b G_n(x)dx. \tag{8.5.40}$$

But then, from $f \in C^{n+2}[a,b]$ and $\tilde{x} \in [a,b]$, we may apply Theorem 8.5.2 with $k = n$ to deduce that there exists a point $\xi \in [a,b]$ such that

$$\left[\frac{d}{dx}(f[x,x_{n,0},\ldots,x_{n,n}])\right]\bigg|_{x=\tilde{x}} = \frac{f^{(n+2)}(\xi)}{(n+2)!}, \tag{8.5.41}$$

which can now be substituted into (8.5.40) to yield

$$\mathscr{E}_n^{NC}[f] = -\frac{f^{(n+2)}(\xi)}{(n+2)!}\int_a^b G_n(x)dx. \tag{8.5.42}$$

By observing from (8.5.33) and (8.5.26) that

$$\int_a^b G_n(x)dx > 0, \tag{8.5.43}$$

we may now deduce from (8.5.42), (8.5.24) and (8.5.43) that $\mathscr{E}_n^{NC}[f] < 0$, which completes our proof of (8.5.21).

(b) Suppose n is odd, and $f \in C^{n+1}[a,b]$, with

$$f^{(n+1)}(x) > 0, \quad x \in [a,b].$$

It follows from (8.5.23), since also $n \geqslant 1$, that

$$\mathscr{E}_n^{NC}[f] = \int_a^b (x-x_{n,n})f[x,x_{n,0},\ldots,x_{n,n}]\prod_{j=0}^{n-1}(x-x_{n,j})dx. \tag{8.5.44}$$

Now observe from the recursion formula (1.3.21) in Theorem 1.3.4 that

$$(x-x_{n,n})f[x,x_{n,0},\ldots,x_{n,n}] = f[x,x_{n,0},\ldots,x_{n,n-1}] - f[x_{n,0},\ldots,x_{n,n}], \tag{8.5.45}$$

for $x \in [a,b] \setminus \{x_{n,0},\ldots,x_{n,n}\}$. But, since $f \in C^{n+1}[a,b]$, we deduce from Theorem 1.4.4 that both sides of (8.5.45) are continuous for $x \in [a,b]$, and it follows that (8.5.45) holds for each $x \in [a,b]$.

Hence we may substitute (8.5.45) into (8.5.44) to obtain

$$\mathscr{E}_n^{NC}[f] = \int_a^b (f[x,x_{n,0},\ldots,x_{n,n-1}] - f[x_{n,0},\ldots,x_{n,n}])\prod_{j=0}^{n-1}(x-x_{n,j})dx,$$

to which we may now apply integration by parts to obtain

$$\mathscr{E}_n^{NC}[f] = \left[(f[x,x_{n,0},\ldots,x_{n,n-1}] - f[x_{n,0},\ldots,x_{n,n}])G_{n-1}(x)\right]_{x=a}^{x=b}$$
$$- \int_a^b \frac{d}{dx}(f[x,x_{n,0},\ldots,x_{n,n-1}])G_{n-1}(x)dx, \tag{8.5.46}$$

where G_{n-1} is defined as in (8.5.26), with n replaced by $n-1$.

If $n = 1$, it follows from (8.4.3) that

$$|f[x,x_{n,0},\ldots,x_{n,n-1}] - f[x_{n,0},\ldots,x_{n,n}]|_{x=b} = f[b,a] - f[a,b] = 0,$$

from Theorem 1.3.6, whereas (8.5.26) gives $G_0(a) = 0$, and it follows from (8.5.46) that (8.5.32) holds with n replaced by $n-1$, if $n = 1$. Moreover, (8.5.26) and (8.4.3) yield

$$G_0(x) = \int_a^x (\sigma - a)d\sigma = \frac{(x-a)^2}{2} \geqslant 0, \quad x \in [a,b],$$

and thus also $\int_a^b G_0(x)dx > 0$.

Since $n-1$ is an even positive integer for $n \geqslant 3$, we may now argue as in (8.5.27)–(8.5.43) for $n \geqslant 3$, where Theorem 8.5.2 is applied with $k = n-1$, and, similarly, argue as in (8.5.40)–(8.5.43) for $n = 1$, to deduce that $\mathcal{E}_n^{NC}[f] < 0$, which then proves (8.5.22). ∎

It follows from Theorem 8.5.3 that the Newton-Cotes quadrature error functional $\mathcal{E}_n^{NC}$ belongs to the class of linear functionals $\mathcal{L}$ in Theorem 8.5.1, based on which we can now prove the following result.

Theorem 8.5.4. *For any positive integer n, the Newton-Cotes quadrature error functional $\mathcal{E}_n^{NC}$, as defined by (8.5.3), satisfies*

$$\mathcal{E}_n^{NC}[f] = \begin{cases} -\left(\dfrac{b-a}{n}\right)^{n+3}(2\Lambda_{n+3} - \Lambda_{n+2})f^{(n+2)}(\xi), \ f \in C^{n+2}[a,b], \ \text{if } n \text{ is even}; \\ -2\left(\dfrac{b-a}{n}\right)^{n+2}\Lambda_{n+2}f^{(n+1)}(\xi), \qquad f \in C^{n+1}[a,b], \ \text{if } n \text{ is odd}, \end{cases}$$

$$(8.5.47)$$

for some $\xi \in [a,b]$, and where the Laplace coefficients $\{\Lambda_k : k = 0,1,\ldots\}$ are defined by (8.4.37).

Proof. According to (8.5.21) and (8.5.22) in Theorem 8.5.3, the linear functional $\mathcal{L} = \mathcal{E}_n^{NC}$ satisfies the condition (8.5.5) of Theorem 8.5.1, with $k = n+2$ if n is even, and $k = n+1$ if n is odd. Hence, with the polynomial P_n defined for any non-negative integer n by

$$P_n(x) := \begin{cases} \displaystyle\prod_{j=0}^{n+1}(x - x_{n,j}), \ \text{if } n \text{ is even}; \\ \displaystyle\prod_{j=0}^{n}(x - x_{n,j}), \ \text{if } n \text{ is odd}, \end{cases}$$

$$(8.5.48)$$

with the points $\{x_{n,j} : j = 0,\ldots,n\}$ as in (8.4.3), and where

$$x_{n,n+1} := a + (n+1)\left(\frac{b-a}{n}\right),$$

$$(8.5.49)$$

according to which

$$P_n \in \begin{cases} \tilde{\pi}_{n+2}, \ \text{if } n \text{ is even}; \\ \tilde{\pi}_{n+1}, \ \text{if } n \text{ is odd}, \end{cases}$$

$$(8.5.50)$$

we may apply (8.5.6) in Theorem 8.5.1 to obtain, for $n = 0, 1, \ldots,$

$$\mathscr{E}_n^{NC}[f] = \begin{cases} \dfrac{f^{(n+2)}(\xi)}{(n+2)!}\mathscr{E}_n^{NC}[P_n], \; f \in C^{n+2}[a,b], \; \text{if } n \text{ is even}; \\ \dfrac{f^{(n+1)}(\xi)}{(n+1)!}\mathscr{E}_n^{NC}[P_n], \; f \in C^{n+1}[a,b], \; \text{if } n \text{ is odd}, \end{cases} \quad (8.5.51)$$

for some $\xi \in [a,b]$. By observing from (8.4.6) and (8.5.48) that, for $n = 0, 1, \ldots,$

$$\mathscr{Q}_n^{NC}[P_n] = \sum_{j=0}^{n} w_{n,j} P_n(x_{n,j}) = \sum_{j=0}^{n} w_{n,j}(0) = 0,$$

we deduce from (8.5.3) that

$$\mathscr{E}_n^{NC}[P_n] = \int_a^b P_n(x)dx, \quad n = 0, 1, \ldots. \quad (8.5.52)$$

Now use the one-to-one mapping (8.4.10), (8.4.11) between the intervals $[0,n]$ and $[a,b]$, together with (8.5.48), (8.4.3) and (8.5.49), to deduce that, for $n = 0, 1, \ldots,$

$$\int_a^b P_n(x)dx = \begin{cases} \left(\dfrac{b-a}{n}\right)^{n+3} \displaystyle\int_0^n \prod_{j=0}^{n+1}(t-j)dt, \; \text{if } n \text{ if even}; \\ \left(\dfrac{b-a}{n}\right)^{n+2} \displaystyle\int_0^n \prod_{j=0}^{n}(t-j)dt, \; \text{if } n \text{ is odd}, \end{cases}$$

and thus, by recalling the definition (8.4.34),

$$\int_a^b P_n(x)dx = \begin{cases} \left(\dfrac{b-a}{n}\right)^{n+3} (n+2)! \displaystyle\int_0^n \binom{t}{n+2}dt, \; \text{if } n \text{ is even}; \\ \left(\dfrac{b-a}{n}\right)^{n+2} (n+1)! \displaystyle\int_0^n \binom{t}{n+1}dt, \; \text{if } n \text{ is odd}. \end{cases} \quad (8.5.53)$$

It follows from the definition (8.4.34) that, for $t \in \mathbb{R}$ and $k \in \mathbb{N}$,

$$\binom{t}{k} + \binom{t}{k+1} = \frac{t(t-1)\ldots(t-k+1)}{k!} + \frac{t(t-1)\ldots(t-k)}{(k+1)!}$$

$$= \frac{t(t-1)\ldots(t-k+1)}{(k+1)!}[(k+1)+(t-k)]$$

$$= \frac{(t+1)t(t-1)\ldots((t+1)-(k+1)+1)}{(k+1)!}$$

$$= \binom{t+1}{k+1}, \quad (8.5.54)$$

which is consistent with the standard combinatorial identity obtained by setting $t = m \in \{0, 1, \ldots\}$ in (8.5.54). Hence we may use (8.5.54) in (8.5.53) to deduce that, for $n = 0, 1, \ldots,$

$$\int_a^b P_n(x)dx = \begin{cases} \left(\dfrac{b-a}{n}\right)^{n+3} (n+2)! \displaystyle\int_0^n \left[\binom{t+1}{n+3} - \binom{t}{n+3}\right]dt, \; \text{if } n \text{ is even}; \\ \left(\dfrac{b-a}{n}\right)^{n+2} (n+1)! \displaystyle\int_0^n \left[\binom{t+1}{n+2} - \binom{t}{n+2}\right]dt, \; \text{if } n \text{ is odd}, \end{cases}$$

from which, as in the argument which led from (8.4.19) to (8.4.20), we deduce that

$$\int_a^b P_n(x)dx = \begin{cases} \left(\dfrac{b-a}{n}\right)^{n+3}(n+2)!\displaystyle\int_0^1\left[\binom{n+1-t}{n+3}-\binom{t}{n+3}\right]dt, & \text{if } n \text{ is even;} \\ \left(\dfrac{b-a}{n}\right)^{n+2}(n+1)!\displaystyle\int_0^1\left[\binom{n+1-t}{n+2}-\binom{t}{n+2}\right]dt, & \text{if } n \text{ is odd.} \end{cases}$$
$$(8.5.55)$$

Suppose now that n is an even non-negative integer. It then follows from (8.4.34) and (8.5.54), and the fact that $n+3$ is odd, that, for $t \in \mathbb{R}$,

$$\binom{n+1-t}{n+3} = \frac{(n+1-t)(n-t)\ldots(-1-t)}{(n+3)!}$$

$$= (-1)^{n+3}\frac{(t+1)t\ldots((t+1)-(n+3)+1)}{(n+3)!}$$

$$= -\binom{t+1}{n+3} = -\binom{t}{n+3}-\binom{t}{n+2},$$

and thus

$$\int_0^1\left[\binom{n+1-t}{n+3}-\binom{t}{n+3}\right]dt = -\left[2\int_0^1\binom{t}{n+3}dt+\int_0^1\binom{t}{n+2}dt\right]. \quad (8.5.56)$$

Now observe from the definition (8.4.37) of the Laplace coefficients $\{\Lambda_k : k = 0,1,\ldots\}$ that

$$2\int_0^1\binom{t}{n+3}dt+\int_0^1\binom{t}{n+2}dt = 2(-1)^{n+2}\Lambda_{n+3}+(-1)^{n+1}\Lambda_{n+2}$$

$$= 2\Lambda_{n+3}-\Lambda_{n+2}, \quad (8.5.57)$$

since $n+2$ is even and $n+1$ is odd. The result in the first line of (8.5.47) is now obtained from the first line of (8.5.51), (8.5.52), and the first line of (8.5.55), together with (8.5.56) and (8.5.57).

Suppose next that n is an odd positive integer. It then follows from (8.4.34), and the fact that $n+2$ is odd, that, for $t \in \mathbb{R}$,

$$\binom{n+1-t}{n+2} = \frac{(n+1-t)(n-t)\ldots(-t)}{(n+2)!}$$

$$= (-1)^{n+2}\frac{t(t-1)\ldots(t-(n+2)+1)}{(n+2)!} = -\binom{t}{n+2},$$

and thus, by using also (8.4.37),

$$\int_0^1\left[\binom{n+1-t}{n+2}-\binom{t}{n+2}\right]dt = -2\int_0^1\binom{t}{n+2}dt = -2(-1)^{n+1}\Lambda_{n+2}$$

$$= -2\Lambda_{n+2}, \quad (8.5.58)$$

since $n+1$ is even. The result in the second line of (8.5.47) is now obtained by using the second line of (8.5.51), (8.5.52), and the second line of (8.5.55), together with (8.5.58), which then completes our proof.   ∎

We may now apply Theorems 8.5.3 and 8.5.4 to establish the following result.

Theorem 8.5.5. *For any non-negative integer n,*

(a) *the Laplace coefficients Λ_{n+2} and Λ_{n+3}, as appearing in (8.5.47) of Theorem 8.5.4, satisfy*

$$\left.\begin{array}{r} 2\Lambda_{n+3} - \Lambda_{n+2} > 0; \\[2mm] \Lambda_{n+2} > 0; \end{array}\right\} \tag{8.5.59}$$

(b) *the degree of exactness m_n of the Newton-Cotes quadrature rule $\mathcal{Q}_n^{NC}$ in (8.4.6), (8.4.7), (8.4.3) is given by*

$$m_n = \begin{cases} n+1, \text{ if } n \text{ is even;} \\[2mm] n, \quad\;\; \text{ if } n \text{ is odd.} \end{cases} \tag{8.5.60}$$

Proof. With the definition

$$Q_n(x) := \begin{cases} x^{n+2}, \text{ if } n \text{ is even;} \\[2mm] x^{n+1}, \text{ if } n \text{ is odd,} \end{cases} \tag{8.5.61}$$

we see that

$$Q_n^{(n+2)}(x) = (n+2)! > 0, \quad x \in \mathbb{R}, \quad \text{if } n \text{ is even;} \tag{8.5.62}$$

$$Q_n^{(n+1)}(x) = (n+1)! > 0, \quad x \in \mathbb{R}, \quad \text{if } n \text{ is odd.} \tag{8.5.63}$$

It follows from (8.5.47) in Theorem 8.5.4, together with (8.5.62) and (8.5.63), that

$$\mathscr{E}_n^{NC}[Q_n] = \begin{cases} -\left(\dfrac{b-a}{n}\right)^{n+3} (n+2)! \, (2\Lambda_{n+3} - \Lambda_{n+2}), \text{ if } n \text{ is even;} \\[4mm] -2\left(\dfrac{b-a}{n}\right)^{n+2} (n+1)! \, \Lambda_{n+2}, \qquad\qquad\;\; \text{ if } n \text{ is odd.} \end{cases} \tag{8.5.64}$$

Also, (8.5.21) and (8.5.22) in Theorem 8.5.3, together with the strict inequalities in (8.5.62) and (8.5.63), imply

$$\mathscr{E}_n^{NC}[Q_n] < 0, \quad n = 0, 1, \dots. \tag{8.5.65}$$

The inequalities in (8.5.59) are now immediate consequences of (8.5.64) and (8.5.65).

Next, we observe from (8.5.64) and (8.5.59) that

$$\mathscr{E}_n^{NC}[Q_n] \neq 0, \quad n = 0, 1, \dots. \tag{8.5.66}$$

Also, the definition (8.5.61) shows that

$$Q_n \in \begin{cases} \pi_{n+2}, & \text{if } n \text{ is even;} \\ \pi_{n+1}, & \text{if } n \text{ is odd.} \end{cases} \tag{8.5.67}$$

It follows from (8.5.66) and (8.5.67) that the degree of exactness m_n of $\mathscr{Q}_n^{NC}$ satisfies

$$m_n \leqslant \begin{cases} n+1, & \text{if } n \text{ is even;} \\ n, & \text{if } n \text{ is odd.} \end{cases} \tag{8.5.68}$$

Suppose now that n is even, and let $P \in \pi_{n+1}$. But then $P^{(n+2)}$ is the zero polynomial, so that we may deduce from (8.5.3) and the first line of (8.5.47) that

$$\int_a^b P(x)dx - \mathscr{Q}_n^{NC}[P] = \mathscr{E}_n^{NC}[P] = 0,$$

that is,

$$\int_a^b P(x)dx = \mathscr{Q}_n^{NC}[P], \quad P \in \pi_{n+1},$$

and thus $m_n \geqslant n+1$, which, together with the first line of (8.5.68), implies the first line of (8.5.60).

Finally, if n is odd, we deduce from (8.4.2) and Theorem 8.1.1 that $m_n \geqslant n$, which, together with the second line of (8.5.68), yields the second line of (8.5.60), and thereby completing our proof. ∎

As our final result of this section, we state, as an immediate consequence of (8.5.47) in Theorem 8.5.4, as well as (8.5.59) in Theorem 8.5.5(a), the following quadrature error bounds for Newton-Cotes quadrature.

Theorem 8.5.6. *For any non-negative integer n, the Newton-Cotes quadrature rule $\mathscr{Q}_n^{NC}$, as given in (8.4.6), (8.4.7), (8.4.3), satisfies the error bounds*

$$\left| \int_a^b f(x)dx - \mathscr{Q}_n^{NC}[f] \right| \leqslant \begin{cases} \left(\dfrac{b-a}{n} \right)^{n+3} (2\Lambda_{n+3} - \Lambda_{n+2}) ||f^{(n+2)}||_\infty, \\ \qquad f \in C^{n+2}[a,b], \text{ if } n \text{ is even;} \\ 2\left(\dfrac{b-a}{n} \right)^{n+2} \Lambda_{n+2} ||f^{(n+1)}||_\infty, \\ \qquad f \in C^{n+1}[a,b], \text{ if } n \text{ is odd,} \end{cases} \tag{8.5.69}$$

with $\{\Lambda_k : k = 0,1,\ldots\}$ denoting the Laplace coefficients as defined in (8.4.37).

Example 8.5.1. For the Newton-Cotes quadrature rules $\mathscr{Q}_n^{NC}, n = 1,\ldots,4$, as formulated in (8.4.56)–(8.4.59) of Example 8.4.1, we apply the result (8.5.69) in Theorem 8.5.6, together with (8.4.54), to obtain the corresponding quadrature error bounds:

$$\left|\int_a^b f(x)dx - \mathscr{Q}_1^{NC}[f]\right| \leqslant \frac{1}{12}(b-a)^3\|f''\|_\infty, \qquad f \in C^2[a,b]; \tag{8.5.70}$$

$$\left|\int_a^b f(x)dx - \mathscr{Q}_2^{NC}[f]\right| \leqslant \frac{1}{90}\left(\frac{b-a}{2}\right)^5 \|f^{(4)}\|_\infty, \qquad f \in C^4[a,b]; \tag{8.5.71}$$

$$\left|\int_a^b f(x)dx - \mathscr{Q}_3^{NC}[f]\right| \leqslant \frac{3}{80}\left(\frac{b-a}{3}\right)^5 \|f^{(4)}\|_\infty, \qquad f \in C^4[a,b]; \tag{8.5.72}$$

$$\left|\int_a^b f(x)dx - \mathscr{Q}_4^{NC}[f]\right| \leqslant \frac{8}{945}\left(\frac{b-a}{4}\right)^7 \|f^{(6)}\|_\infty, \qquad f \in C^6[a,b]. \tag{8.5.73}$$

∎

8.6 Composite Newton-Cotes quadrature

According to Table 8.4.1, the Newton-Cotes quadrature rule sequence $\{\mathscr{Q}_n^{NC} : n = 0,1,\ldots\}$ does not satisfy the non-negative weight condition (8.1.16), since negative weights occur for $n \geqslant 8$. Hence the convergence result

$$\int_a^b f(x)dx - \mathscr{Q}_n^{NC}[f] \to 0, \quad n \to \infty, \quad f \in C[a,b], \tag{8.6.1}$$

is not guaranteed by Theorem 8.1.2. It can in fact be shown, by means of a method beyond the scope of this book, that, for the integrand f chosen as the Runge example (3.1.5) in Example 3.1.3, it holds that

$$\left|\int_{-5}^5 f(x)dx - \mathscr{Q}_n^{NC}[f]\right| \to \infty, \quad n \to \infty. \tag{8.6.2}$$

In order to counteract such divergence phenomena in the Newton-Cotes setting, we present in this section, for any fixed non-negative integer v, the composite Newton-Cotes quadrature rules $\{\mathscr{Q}_{v,n}^{NC} : n = v, 2v, \ldots\}$ for the numerical approximation of the integral $\int_a^b f(x)dx$, according to which, for each $n = v, 2v, \ldots$, the Newton-Cotes quadrature rule $\mathscr{Q}_v^{NC}$ is applied n/v times on successive sub-intervals of $[a,b]$, as formulated more precisely in the following.

Let v and n denote positive integers satisfying the condition

$$\frac{n}{v} \in \mathbb{N}, \tag{8.6.3}$$

and let the point sequence $\{x_{n,j} : j = 0,\ldots,n\}$ be defined by (8.4.3), according to which also (8.4.4) and (8.4.5) are then satisfied. Then, for any $f \in C[a,b]$, we deduce from (8.4.3) and (8.6.3) that

$$\int_a^b f(x)dx = \sum_{j=0}^{\frac{n}{v}-1} \int_{x_{n,jv}}^{x_{n,(j+1)v}} f(x)dx = \sum_{j=0}^{\frac{n}{v}-1} \int_{x_{n,0}}^{x_{n,v}} f\left(x+jv\left(\frac{b-a}{n}\right)\right)dx,$$

that is,

$$\int_a^b f(x)dx = \sum_{j=0}^{\frac{n}{v}-1} \int_a^{x_{n,v}} f_j(x)dx, \qquad (8.6.4)$$

where

$$f_j(x) := f\left(x+jv\left(\frac{b-a}{n}\right)\right), \qquad j=0,\ldots,\frac{n}{v}-1. \qquad (8.6.5)$$

Based on (8.6.4) and (8.6.5), we now consider the numerical approximation

$$\int_a^b f(x)dx \approx \mathscr{Q}_{v,n}^{NC}[f], \qquad f \in C[a,b], \qquad (8.6.6)$$

where the composite Newton-Cotes quadrature rule $\mathscr{Q}_{v,n}^{NC}$ is defined by

$$\mathscr{Q}_{v,n}^{NC}[f] := \sum_{j=0}^{\frac{n}{v}-1} \mathscr{Q}_v^{NC}[f_j], \qquad f \in C[a,b], \qquad (8.6.7)$$

with the functions $\{f_j : j=0,\ldots,\frac{n}{v}-1\} \subset C[a,b]$ defined by (8.6.5), and with $\mathscr{Q}_v^{NC}$ denoting the Newton-Cotes quadrature rule, as given in (8.4.2), with respect to the interpolation points $\{x_{n,0},\ldots,x_{n,v}\}$. Note from (8.6.7) and (8.6.5) that

$$\mathscr{Q}_{v,v}^{NC} = \mathscr{Q}_v^{NC}. \qquad (8.6.8)$$

Observe from (8.6.7), (8.6.5), (8.4.2) and (8.4.3) that, if $n \in \{2v,3v,\ldots\}$, then

$$\mathscr{Q}_{v,n}^{NC}[f] = \int_a^b F_{v,n}^I(x)dx, \qquad f \in C[a,b], \qquad (8.6.9)$$

where, for each $f \in C[a,b]$, $F_{v,n}^I$ is the continuous piecewise polynomial function on $[a,b]$, with polynomial pieces in π_v, and breakpoints at $\{x_{n,jv} : j=1,\ldots,(\frac{n}{v}-1)v\}$, such that $F_{v,n}^I$ interpolates f at the points $\{x_{n,j} : j=0,\ldots,n\}$, that is,

$$F_{v,n}^I(x_{n,j}) = f(x_{n,j}), \qquad j=0,\ldots,n, \qquad f \in C[a,b]. \qquad (8.6.10)$$

For any positive integers v and n satisfying (8.6.3), it follows from (8.4.6), (8.6.5) and (8.4.3), that, for any $j \in \{0,\ldots,\frac{n}{v}-1\}$,

$$\mathscr{Q}_v^{NC}[f_j] = \sum_{k=0}^{v} w_{v,k} f\left(x_{n,k}+jv\left(\frac{b-a}{n}\right)\right) = \sum_{k=0}^{v} w_{v,k} f\left(a+(k+jv)\left(\frac{b-a}{n}\right)\right)$$

$$= \sum_{k=jv}^{(j+1)v} w_{v,k-jv} f(x_{n,k}), \qquad (8.6.11)$$

where, from (8.4.38) in Theorem 8.4.1, with $b - a$ replaced by $x_{n,v} - x_{n,0} = v\left(\dfrac{b-a}{n}\right)$, from (8.4.4), and with n replaced by v, we have

$$w_{v,k} = \frac{b-a}{n}\left[1 - (-1)^k \sum_{\ell=0}^{v} \Lambda_{\ell+1}\left\{\binom{\ell}{k} + (-1)^v \binom{\ell}{v-k}\right\}\right], \quad k = 0,\ldots,v, \quad (8.6.12)$$

with $\{\Lambda_1,\ldots,\Lambda_{v+1}\}$ denoting the Laplace coefficients in (8.4.37). By using also the fact that

$$w_{v,v} = w_{v,0}, \quad (8.6.13)$$

as follows from the symmetry property (8.4.55), we obtain the following formulation of composite Newton-Cotes quadrature as an immediate consequence of (8.6.7), (8.6.11), (8.6.12) and (8.6.13).

Theorem 8.6.1. *For any positive integers v and n satisfying the condition (8.6.3), the composite Newton-Cotes quadrature rule $\mathcal{Q}_{v,n}^{NC}$, as defined by (8.6.7), (8.6.5), satisfies the formulation*

$$\mathcal{Q}_{v,n}^{NC}[f] = \sum_{k=0}^{n} w_{n,k}^{[v]} f(x_{n,k}), \quad f \in C[a,b], \quad (8.6.14)$$

where the weights $\{w_{n,k}^{[v]} : k = 0,\ldots,n\}$ are given by

$$w_{n,jv}^{[v]} = \begin{cases} w_{v,0}, & \text{for } j=0, \text{ or } j=\frac{n}{v}; \\ 2w_{v,0}, & \text{for } j=1,\ldots,\frac{n}{v}-1 \ (\text{if } n \geqslant 2v); \end{cases} \quad (8.6.15)$$

$$w_{n,jv+k}^{[v]} = w_{v,k}, \quad k = 1,\ldots,v-1, \quad \text{for} \quad j = 0,\ldots,\frac{n}{v}-1 \quad (\text{if } v \geqslant 2), \quad (8.6.16)$$

with $\{w_{v,k} : k = 0,\ldots,v\}$ given as in (8.6.12) in terms of the Laplace coefficients $\{\Lambda_1,\ldots,\Lambda_{v+1}\}$ in (8.4.37), and where the points $\{x_{n,k} : k = 0,\ldots,n\}$ are given by (8.4.3).

For the choices $v = 1,\ldots,4$ in Theorem 8.6.1, the corresponding composite Newton-Cotes rules are known as, respectively,
the trapezoidal rule

$$\mathcal{Q}_n^{TR} := \mathcal{Q}_{1,n}^{NC}, \quad n = 1,2,\ldots; \quad (8.6.17)$$

the Simpson rule

$$\mathcal{Q}_n^{SI} := \mathcal{Q}_{2,n}^{NC}, \quad n = 2,4,\ldots; \quad (8.6.18)$$

the 3/8 rule

$$\mathcal{Q}_n^{3/8} := \mathcal{Q}_{3,n}^{NC}, \quad n = 3,6,\ldots; \quad (8.6.19)$$

and the Boole rule

$$\mathscr{Q}_n^{BO} := \mathscr{Q}_{4,n}^{NC}, \quad n = 4, 8, \ldots . \tag{8.6.20}$$

By using (8.6.12)–(8.6.20), and Table 8.4.1, with the notation

$$f_{n,j} := f(x_{n,j}), \quad j = 0, \ldots, n, \tag{8.6.21}$$

for $n = 1, 2, \ldots$, where the points $\{x_{n,j} : j = 0, \ldots, n\}$ are given as in (8.4.3), we obtain, for any $f \in C[a,b]$, the following explicit formulations in composite Newton-Cotes quadrature:

$$\mathscr{Q}_n^{TR}[f] = \frac{b-a}{2n} [f_{n,0} + 2f_{n,1} + 2f_{n,2} + \cdots + 2f_{n,n-1} + f_{n,n}], \quad n = 1, 2, \ldots;$$

$$\tag{8.6.22}$$

$$\mathscr{Q}_n^{SI}[f] = \frac{b-a}{3n} [f_{n,0} + 4f_{n,1} + 2f_{n,2} + 4f_{n,3} + 2f_{n,4} + \cdots + 2f_{n,n-2}$$

$$+ 4f_{n,n-1} + f_{n,n}], \quad n = 2, 4, \ldots; \tag{8.6.23}$$

$$\mathscr{Q}_n^{3/8}[f] = \frac{3(b-a)}{8n} [f_{n,0} + 3f_{n,1} + 3f_{n,2} + 2f_{n,3} + 3f_{n,4} + 3f_{n,5} + 2f_{n,6} + \cdots$$

$$+ 2f_{n,n-3} + 3f_{n,n-2} + 3f_{n,n-1} + f_{n,n}], \quad n = 3, 6, \ldots; \tag{8.6.24}$$

$$\mathscr{Q}_n^{BO}[f] = \frac{2(b-4)}{45n} [7f_{n,0} + 32f_{n,1} + 12f_{n,2} + 32f_{n,3} + 14f_{n,4} + 32f_{n,5} + 12f_{n,6} + 32f_{n,7}$$

$$+ 14f_{n,8} + \cdots + 14f_{n,n-4} + 32f_{n,n-3} + 12f_{n,n-2} + 32f_{n,n-1} + 7f_{n,n}], \quad n = 4, 8, \ldots . \tag{8.6.25}$$

Example 8.6.1. For the numerical approximation of the integral $\int_0^2 f(x)dx$, $f \in C[0,2]$, it follows from (8.6.21)–(8.6.25), together with (8.4.3), that

$$\mathscr{Q}_{12}^{TR}[f] = \frac{1}{12} [f(0) + 2f(\tfrac{1}{6}) + 2f(\tfrac{1}{3}) + 2f(\tfrac{1}{2}) + 2f(\tfrac{2}{3}) + 2f(\tfrac{5}{6}) + 2f(1) + 2f(\tfrac{7}{6})$$

$$+ 2f(\tfrac{4}{3}) + 2f(\tfrac{3}{2}) + 2f(\tfrac{5}{3}) + 2f(\tfrac{11}{6}) + f(2)];$$

$$\mathscr{Q}_{12}^{SI}[f] = \frac{1}{18} [f(0) + 4f(\tfrac{1}{6}) + 2f(\tfrac{1}{3}) + 4f(\tfrac{1}{2}) + 2f(\tfrac{2}{3}) + 4f(\tfrac{5}{6}) + 2f(1) + 4f(\tfrac{7}{6})$$

$$+ 2f(\tfrac{4}{3}) + 4f(\tfrac{3}{2}) + 2f(\tfrac{5}{3}) + 4f(\tfrac{11}{6}) + f(2)];$$

$$\mathscr{Q}_{12}^{3/8}[f] = \frac{1}{16} [f(0) + 3f(\tfrac{1}{6}) + 3f(\tfrac{1}{3}) + 2f(\tfrac{1}{2}) + 3f(\tfrac{2}{3}) + 3f(\tfrac{5}{6}) + 2f(1) + 3f(\tfrac{7}{6})$$

$$+3f(\tfrac{4}{3})+2f(\tfrac{3}{2})+3f(\tfrac{5}{3})+3f(\tfrac{11}{6})+f(2)\big];$$

$$\mathscr{Q}_{12}^{BO}[f] = \frac{1}{135}\big[7f(0)+32f(\tfrac{1}{6})+12f(\tfrac{1}{3})+32f(\tfrac{1}{2})+14f(\tfrac{2}{3})+32f(\tfrac{5}{6})+12f(1)$$

$$+32f(\tfrac{7}{6})+14f(\tfrac{4}{3})+32f(\tfrac{3}{2})+12f(\tfrac{5}{3})+32f(\tfrac{11}{6})+7f(2)\big],$$

which yields, for the case $f(x) = e^x$, the quadrature errors

$$\left| \int_0^2 f(x)dx - \mathscr{Q}_{12}^{TR}[f] \right| \approx 1.478 \times 10^{-2};$$

$$\left| \int_0^2 f(x)dx - \mathscr{Q}_{12}^{SI}[f] \right| \approx 2.730 \times 10^{-5};$$

$$\left| \int_0^2 f(x)dx - \mathscr{Q}_{12}^{3/8}[f] \right| \approx 6.122 \times 10^{-5};$$

$$\left| \int_0^2 f(x)dx - \mathscr{Q}_{12}^{BO}[f] \right| \approx 2.856 \times 10^{-7}.$$

∎

We proceed to analyze the composite Newton-Cotes quadrature error

$$\mathscr{E}_{v,n}^{NC}[f] := \int_a^b f(x)dx - \mathscr{Q}_{v,n}^{NC}[f], \quad f \in C[a,b]. \tag{8.6.26}$$

First, observe from (8.6.26), (8.6.4) and (8.6.7) that, for any positive integers v and n such that (8.6.3) holds, we have

$$\mathscr{E}_{v,n}^{NC}[f] = \sum_{j=0}^{\frac{n}{v}-1}\left\{ \int_{x_{n,0}}^{x_{n,v}} f_j(x)dx - \mathscr{Q}_v^{NC}[f_j]\right\}, \quad f \in C[a,b], \tag{8.6.27}$$

with the functions $\{f_j : j = 0,\dots,n\} \subset C[a,b]$ defined by (8.6.5), and where the points $\{x_{n,j} : j = 0,\dots,n\}$ are given by (8.4.3). It follows from (8.6.27) and (8.6.5), together with (8.5.47) in Theorem 8.5.4, with $b-a$ replaced by $\frac{v}{n}(b-a)$, and with n replaced by v, that there exist points $\{\xi_j : j = 0,\dots,\frac{n}{v}-1\}$ in $[a,b]$ such that

$$\mathscr{E}_{v,n}^{NC}[f] = \begin{cases} -\left(\dfrac{b-a}{n}\right)^{v+3}\{2\Lambda_{v+3}-\Lambda_{v+2}\}\displaystyle\sum_{j=0}^{\frac{n}{v}-1} f^{(v+2)}(\xi_j), \\[2mm] \qquad\qquad\qquad f \in C^{v+2}[a,b], \text{ if } v \text{ is even;} \\[4mm] -2\left(\dfrac{b-a}{n}\right)^{v+2}\Lambda_{v+2}\displaystyle\sum_{j=0}^{\frac{n}{v}-1} f^{(v+1)}(\xi_j), \\[2mm] \qquad\qquad\qquad f \in C^{v+1}[a,b], \text{ if } v \text{ is odd,} \end{cases} \tag{8.6.28}$$

with the Laplace coefficients $\{\Lambda_j : j = 0,1,\dots\}$ defined as in (8.4.37).

Since, for $g \in C[a,b]$, a non-negative integer m, a point sequence $\{\xi_0, \dots, \xi_m\} \subset [a,b]$, and a sequence $\{\alpha_0, \dots, \alpha_m\} \subset \mathbb{R}$, with $\alpha_j > 0$, $j = 0, \dots, m$, we have

$$\min_{a \leqslant x \leqslant b} g(x) \sum_{j=0}^{m} \alpha_j \leqslant \sum_{j=0}^{m} \alpha_j g(\xi_j) \leqslant \max_{a \leqslant x \leqslant b} g(x) \sum_{j=0}^{m} \alpha_j,$$

and thus

$$\min_{a \leqslant x \leqslant b} g(x) \leqslant \frac{\displaystyle\sum_{j=0}^{m} \alpha_j g(\xi_j)}{\displaystyle\sum_{j=0}^{m} \alpha_j} \leqslant \max_{a \leqslant x \leqslant b} g(x),$$

according to which the intermediate value theorem yields

$$\frac{\displaystyle\sum_{j=0}^{m} \alpha_j g(\xi_j)}{\displaystyle\sum_{j=0}^{m} \alpha_j} = g(\xi),$$

for some $\xi \in [a,b]$, that is,

$$\sum_{j=0}^{m} \alpha_j g(\xi_j) = g(\xi) \sum_{j=0}^{m} \alpha_j, \tag{8.6.29}$$

we can now deduce from (8.6.28) and (8.6.29), with $g = f^{(\nu+1)}$, $m = \frac{n}{\nu} - 1$, and $\alpha_j = 1$, $j = 0, \dots, m$, that, for any $f \in C[a,b]$, there exists a point $\xi \in [a,b]$ such that

$$\mathscr{E}_{\nu,n}^{NC}[f] = \begin{cases} -\left(\dfrac{b-a}{n}\right)^{\nu+3} \{2\Lambda_{\nu+3} - \Lambda_{\nu+2}\} \left[\dfrac{n}{\nu} f^{(\nu+2)}(\xi)\right], \\ \qquad\qquad\qquad\qquad\qquad f \in C^{\nu+2}[a,b], \text{ if } \nu \text{ is even;} \\[2mm] -2\left(\dfrac{b-a}{n}\right)^{\nu+2} \Lambda_{\nu+2} \left[\dfrac{n}{\nu} f^{(\nu+1)}(\xi)\right], \\ \qquad\qquad\qquad\qquad\qquad f \in C^{\nu+1}[a,b], \text{ if } \nu \text{ is odd,} \end{cases}$$

that is,

$$\mathscr{E}_{\nu,n}^{NC}[f] = -(b-a) \begin{cases} \left(\dfrac{b-a}{n}\right)^{\nu+2} \left[\dfrac{2\Lambda_{\nu+3} - \Lambda_{\nu+2}}{\nu}\right] f^{(\nu+2)}(\xi), \\ \qquad\qquad\qquad\qquad f \in C^{\nu+2}[a,b], \text{ if } \nu \text{ is even;} \\[2mm] \left(\dfrac{b-a}{n}\right)^{\nu+1} \left[\dfrac{2\Lambda_{\nu+2}}{\nu}\right] f^{(\nu+1)}(\xi), \\ \qquad\qquad\qquad\qquad f \in C^{\nu+1}[a,b], \text{ if } \nu \text{ is odd.} \end{cases} \tag{8.6.30}$$

Since, according to (8.6.26) and (8.6.14), $\mathscr{E}_{\nu,n}^{NC}$ is a linear functional, and since $\mathscr{E}_{\nu,n}^{NC}$ satisfies (8.6.30), an analogous argument to the one which led from (8.5.61) to (8.5.69) now yields the following results.

Theorem 8.6.2. *For any positive integers* v *and* n *satisfying the condition (8.6.3), the composite Newton-Cotes quadrature rule* $\mathcal{Q}_{v,n}^{NC}$, *as defined by (8.6.7), (8.6.5), has degree of exactness* m_v *given by*

$$m_v = \begin{cases} v+1, & \text{if } v \text{ is even;} \\ v, & \text{if } v \text{ is odd,} \end{cases} \tag{8.6.31}$$

and with corresponding quadrature error satisfying

$$\left| \int_a^b f(x)dx - \mathcal{Q}_{v,n}^{NC}[f] \right| \leqslant (b-a) \begin{cases} \left(\dfrac{b-a}{n}\right)^{v+2} \left[\dfrac{2\Lambda_{v+3} - \Lambda_{v+2}}{v}\right] ||f^{(v+2)}||_\infty, \\ \qquad\qquad f \in C^{v+2}[a,b], \text{ if } v \text{ is even;} \\ \left(\dfrac{b-a}{n}\right)^{v+1} \left[\dfrac{2\Lambda_{v+2}}{v}\right] ||f^{(v+1)}||_\infty, \\ \qquad\qquad f \in C^{v+1}[a,b], \text{ if } v \text{ is odd,} \end{cases} \tag{8.6.32}$$

with $\{\Lambda_j : j = 0,1,\ldots\}$ *denoting the Laplace coefficients defined in (8.4.37).*

Observe from (8.4.4) that the factor $(b-a)/n$ appearing in the right hand sides of (8.6.32) represents the spacing between the points $\{x_{n,j} : j = 0,\ldots,n\}$ in (8.4.3), and which halves for each doubling of n.

For the composite Newton-Cotes quadrature rules in (8.6.22)–(8.6.25), we use (8.6.17)–(8.6.20), together with (8.6.31) and (8.6.32) in Theorem 8.6.2, and (8.4.54), to deduce that:

(a) the trapezoidal rule $\mathcal{Q}_n^{TR}$ has degree of exactness $m = 1$, and

$$\left| \int_a^b f(x)dx - \mathcal{Q}_n^{TR}[f] \right| \leqslant \frac{b-a}{12}\left(\frac{b-a}{n}\right)^2 ||f''||_\infty, \ f \in C^2[a,b], \ n = 1,2,\ldots; \tag{8.6.33}$$

(b) the Simpson rule $\mathcal{Q}_n^{SI}$ has degree of exactness $m = 3$, and

$$\left| \int_a^b f(x)dx - \mathcal{Q}_n^{SI}[f] \right| \leqslant \frac{b-a}{180}\left(\frac{b-a}{n}\right)^4 ||f^{(4)}||_\infty, \ f \in C^4[a,b], \ n = 2,4,\ldots; \tag{8.6.34}$$

(c) the 3/8 rule $\mathcal{Q}_n^{3/8}$ has degree of exactness $m = 3$, and

$$\left| \int_a^b f(x)dx - \mathcal{Q}_n^{3/8}[f] \right| \leqslant \frac{b-a}{80}\left(\frac{b-a}{n}\right)^4 ||f^{(4)}||_\infty, \ f \in C^4[a,b], \ n = 3,6,\ldots; \tag{8.6.35}$$

(d) the Boole rule $\mathcal{Q}_n^{BO}$ has degree of exactness $m = 5$, and

$$\left| \int_a^b f(x)dx - \mathcal{Q}_n^{BO}[f] \right| \leqslant \frac{2(b-a)}{945}\left(\frac{b-a}{n}\right)^6 ||f^{(6)}||_\infty, \ f \in C^6[a,b], \ n = 4,8,\ldots. \tag{8.6.36}$$

We proceed to prove the following convergence result for composite Newton-Cotes quadrature.

Theorem 8.6.3. *The sequence* $\{\mathcal{Q}^{NC}_{v,jv} : j = 1,2,\ldots\}$ *of composite Newton-Cotes quadrature rules, as defined in* (8.6.7), (8.6.5), *satisfies the convergence result*

$$\left| \int_a^b f(x)dx - \mathcal{Q}^{NC}_{v,jv}[f] \right| \to 0, \quad j \to \infty, \quad f \in C[a,b]. \tag{8.6.37}$$

Proof. Let $f \in C[a,b]$ and choose $\varepsilon > 0$. Our proof of (8.6.37) will be complete if we can establish the existence of a positive integer $J = J(\varepsilon)$ such that

$$\left| \int_a^b f(x)dx - \mathcal{Q}^{NC}_{v,jv}[f] \right| < \varepsilon, \quad j > J. \tag{8.6.38}$$

To this end, we fix $j \in \mathbb{N}$, and, for the equispaced sequence $\{x_{jv,k} : k = 0,\ldots,jv\}$ defined according to (8.4.3) by

$$x_{jv,k} := a + k\left(\frac{b-a}{jv}\right), \quad k = 0,\ldots,jv, \tag{8.6.39}$$

we let $\{L_{jv,k} : k = 0,\ldots,v\} \subset \pi_v$ denote the corresponding Lagrange fundamental polynomials, as given in (1.2.1) by

$$L_{jv,k}(x) := \prod_{\substack{k \neq i = 0}}^{v} \frac{x - x_{jv,i}}{x_{jv,k} - x_{jv,i}}, \quad k = 0,\ldots,v. \tag{8.6.40}$$

By using (8.6.7), (8.6.5), (8.6.3), (8.6.39), (8.4.2), (8.6.40), as well as (5.1.2) and the Lagrange interpolation formula (1.2.5) in Theorem 1.2.2, together with the identity (1.2.8) in Theorem 1.2.3, we obtain

$$\int_a^b f(x)dx - \mathcal{Q}^{NC}_{v,jv}[f] = \sum_{\ell=0}^{j-1} \int_{x_{jv,\ell v}}^{x_{jv,(\ell+1)v}} f(x)dx - \sum_{\ell=0}^{j-1} \mathcal{Q}^{NC}_v[f_\ell]$$

$$= \sum_{\ell=0}^{j-1} \left[\int_a^{x_{jv,v}} f\left(x + \ell\left(\frac{b-a}{j}\right)\right)dx - \int_a^{x_{jv,v}} \sum_{k=0}^{v} f\left(a + \left(\ell + \frac{k}{v}\right)\left(\frac{b-a}{j}\right)\right) L_{jv,k}(x)dx \right]$$

$$= \sum_{\ell=0}^{j-1} \int_a^{x_{jv,v}} \sum_{k=0}^{v} \left\{ f\left(x + \ell\left(\frac{b-a}{j}\right)\right) - f\left(a + \left(\ell + \frac{k}{v}\right)\left(\frac{b-a}{j}\right)\right) \right\} L_{jv,k}(x)dx,$$

and thus

$$\left| \int_a^b f(x)dx - \mathcal{Q}^{NC}_{v,jv}[f] \right|$$

$$\leq \sum_{\ell=0}^{j-1} \int_a^{x_{jv,v}} \sum_{k=0}^{v} \left| f\left(x + \ell\left(\frac{b-a}{j}\right)\right) - f\left(a + \left(\ell + \frac{k}{v}\right)\left(\frac{b-a}{j}\right)\right) \right| |L_{jv,k}(x)|dx.$$

$$\tag{8.6.41}$$

Now observe from (8.6.39) that, for $x \in [a, x_{jv,v}], k \in \{0, \ldots, v\}$, and $i \in \{0, \ldots, v\} \setminus \{k\}$, we have

$$\left| \frac{x - x_{jv,i}}{x_{jv,k} - x_{jv,i}} \right| \leqslant \frac{x_{jv,v} - a}{(b-a)/jv} = \frac{(b-a)/j}{(b-a)/jv} = v,$$

which, together with (8.6.40), yields the bound

$$|L_{jv,k}(x)| \leqslant v^2, \quad x \in [a, x_{jv,v}], \quad k = 0, \ldots, v. \tag{8.6.42}$$

Hence we may apply the bound (8.6.42) in (8.6.41) to deduce that

$$\left| \int_a^b f(x)\,dx - \mathcal{Q}_{v,jv}^{NC}[f] \right|$$

$$\leqslant v^2 \sum_{\ell=0}^{j-1} \int_a^{x_{jv,v}} \sum_{k=0}^{v} \left| f\left(x + \ell\left(\frac{b-a}{j} \right) \right) - f\left(a + \left(\ell + \frac{k}{v} \right)\left(\frac{b-a}{j} \right) \right) \right| dx. \tag{8.6.43}$$

Since $f \in C[a, b]$, we know from a standard result in calculus that f is uniformly continuous on $[a, b]$, that is, there exists a positive number $\delta = \delta(\varepsilon)$ such that

$$x, y \in [a, b]; |x - y| < \delta \implies |f(x) - f(y)| < \frac{\varepsilon}{v^2(v+1)(b-a)}. \tag{8.6.44}$$

Now observe from (8.6.39) that, for any $x \in [a, x_{jv,v}], k \in \{0, \ldots, v\}$ and $\ell \in \{0, \ldots, j-1\}$, we have

$$-\frac{k}{v}\left(\frac{b-a}{j} \right) \leqslant \left[x + \ell\left(\frac{b-a}{j} \right) \right] - \left[a + \left(\ell + \frac{k}{v} \right)\left(\frac{b-a}{j} \right) \right] \leqslant \frac{v-k}{v}\left(\frac{b-a}{j} \right),$$

and thus

$$\left| \left[x + \ell\left(\frac{b-a}{j} \right) \right] - \left[a + \left(\ell + \frac{k}{v} \right)\left(\frac{b-a}{j} \right) \right] \right| \leqslant \frac{b-a}{j}, \quad x \in [a, x_{jv,v}];$$

$$k = 0, \ldots, v; \ \ell = 0, \ldots, j-1. \tag{8.6.45}$$

Hence, if we define the positive integer $J = J(\varepsilon)$ by

$$J := \lceil (b-a)/\delta \rceil, \tag{8.6.46}$$

according to which

$$0 < \frac{b-a}{J} \leqslant \delta, \tag{8.6.47}$$

it follows from (8.6.45), (8.6.47) and (8.6.44) that, for any $j > J$, it holds that

$$\left| f\left(x + \ell\left(\frac{b-a}{j} \right) \right) - f\left(a + \left(\ell + \frac{k}{v} \right)\left(\frac{b-a}{j} \right) \right) \right| < \frac{\varepsilon}{v^2(v+1)(b-a)},$$

$$x \in [a, x_{jv,v}]; \ k = 0, \ldots, v; \ \ell = 0, \ldots, j-1. \tag{8.6.48}$$

By applying the bound (8.6.48) in (8.6.43), and using (8.6.39), we deduce that, for $j > J$,

$$\left| \int_a^b f(x)dx - \mathcal{Q}_{v,jv}^{NC}[f] \right| < \frac{\varepsilon}{b-a} \sum_{\ell=0}^{j-1} \int_a^{x_{jv,v}} dx = \frac{\varepsilon}{b-a} \sum_{\ell=0}^{j-1} \left[\frac{b-a}{j} \right] = \varepsilon,$$

and thereby showing that the desired result (8.6.38) holds, with the positive integer $J = J(\varepsilon)$ given by (8.6.46). ∎

Finally in this section, we consider two composite quadrature rules based on the interpolation by means of piecewise constants of the integrand $f \in C[a,b]$ in the integral $\int_a^b f(x)dx$. In particular, with the points $\{x_{n,j} : j = 0,\ldots,n\}$ given as in (8.4.3), we define the rectangle rule

$$\mathcal{Q}_n^{RE}[f] := \int_a^b H_n^I(x)dx, \quad f \in C[a,b], \quad n = 1,2,\ldots, \tag{8.6.49}$$

where

$$H_n^I(x) := f(x_{n,j}), \quad x \in [x_{n,j}, x_{n,j+1}), \quad j = 0,\ldots,n-1, \tag{8.6.50}$$

according to which we have the formulation

$$\mathcal{Q}_n^{RE}[f] = \frac{b-a}{n} \sum_{j=0}^{n-1} f(x_{n,j}), \quad f \in C[a,b], \quad n = 1,2,\ldots, \tag{8.6.51}$$

thereby extending the Newton-Cotes quadrature rule $\mathcal{Q}_0^{NC}$, as given by (8.4.9), to the composite setting, and, secondly, the midpoint rule

$$\mathcal{Q}_n^{MI}[f] := \int_a^b \widetilde{H}_n^I(x)dx, \quad f \in C[a,b], \quad n = 1,2,\ldots, \tag{8.6.52}$$

where

$$\widetilde{H}_n^I(x) := f\left(\frac{x_{n,j} + x_{n,j+1}}{2} \right), \quad x \in [x_{n,j}, x_{n,j+1}), \quad j = 0,\ldots,n-1, \tag{8.6.53}$$

which yields the formulation

$$\mathcal{Q}_n^{MI}[f] = \frac{b-a}{n} \sum_{j=0}^{n-1} f\left(\frac{x_{n,j} + x_{n,j+1}}{2} \right), \quad f \in C[a,b], \quad n = 1,2,\ldots. \tag{8.6.54}$$

We proceed to establish the following results with respect to the rectangle and midpoint quadrature rules.

Theorem 8.6.4. *For any integer $n \geqslant 2$, the rectangle rule $\mathcal{Q}_n^{RE}$, as defined by (8.6.49), (8.6.50), has degree of exactness $m = 0$, whereas the midpoint rule $\mathcal{Q}_n^{MI}$, as defined by (8.6.52), (8.6.53), has degree of exactness $m = 1$. Moreover, the corresponding quadrature errors satisfy the bounds*

$$\left| \int_a^b f(x)dx - \mathcal{Q}_n^{RE}[f] \right| \leqslant \frac{b-a}{2}\left(\frac{b-a}{n}\right) \|f'\|_\infty, \quad f \in C^1[a,b], \quad n = 1,2,\ldots; \tag{8.6.55}$$

$$\left| \int_a^b f(x)dx - \mathcal{Q}_n^{MI}[f] \right| \leqslant \frac{b-a}{24}\left(\frac{b-a}{n}\right)^2 \|f''\|_\infty, \quad f \in C^2[a,b], \quad n = 1,2,\ldots. \tag{8.6.56}$$

Proof. For an integer $n \in \mathbb{N}$ and any $f \in C^1[a,b]$, it follows from (8.6.49) and (8.6.50), together with the mean value theorem (8.5.16) for integrals, that, with the points $\{x_{n,j} : j = 0, \ldots, n\}$ as in (8.4.3), so that (8.4.4) holds, we have

$$
\begin{aligned}
\int_a^b f(x)dx - \mathscr{Q}_n^{RE}[f] &= \sum_{j=0}^{n-1} \int_{x_{n,j}}^{x_{n,j+1}} f(x)dx - \sum_{j=0}^{n-1} \int_{x_{n,j}}^{x_{n,j+1}} f(x_{n,j})dx \\
&= \sum_{j=0}^{n-1} \int_{x_{n,j}}^{x_{n,j+1}} [f(x) - f(x_{n,j})]dx \\
&= \sum_{j=0}^{n-1} \int_{x_{n,j}}^{x_{n,j+1}} \left[\int_{x_{n,j}}^{x} f'(t)dt \right] dx \\
&= \sum_{j=0}^{n-1} \int_{x_{n,j}}^{x_{n,j+1}} \left[\int_{t}^{x_{n,j+1}} dx \right] f'(t)dt \\
&= \sum_{j=0}^{n-1} \int_{x_{n,j}}^{x_{n,j+1}} (x_{n,j+1} - t)f'(t)dt \\
&= \sum_{j=0}^{n-1} f'(\xi_j) \int_{x_{n,j}}^{x_{n,j+1}} (x_{n,j+1} - t)dt = \frac{1}{2}\left(\frac{b-a}{n}\right)^2 \left[\sum_{j=0}^{n-1} f'(\xi_j)\right],
\end{aligned}
$$

$$(8.6.57)$$

for some points $\{\xi_j : j = 0, \ldots, n\} \subset [a,b]$. But, as noted before in (8.6.29), an application of the intermediate value theorem yields the existence of a point $\xi \in [a,b]$ such that

$$\sum_{j=0}^{n-1} f'(\xi_j) = nf'(\xi),\tag{8.6.58}$$

which, together with (8.6.57), yields

$$\int_a^b f(x)dx - \mathscr{Q}_n^{RE}[f] = \frac{b-a}{2}\left(\frac{b-a}{n}\right)f'(\xi),\ f \in C^1[a,b],\ n = 1,2,\ldots.\tag{8.6.59}$$

From (8.6.59) we see that the rectangle rule $\mathscr{Q}_n^{RE}$ has degree of exactness $m = 0$, and (8.6.59) also immediately implies the upper bound in (8.6.55).

In order to analyze the midpoint rule $\mathscr{Q}_n^{MI}$, we let $f \in C^2[a,b]$, and first note that for any $x, \alpha \in [a,b]$, we may use integration by parts to obtain

$$
\begin{aligned}
\int_\alpha^x (x-t)f''(t)dt &= \left[(x-t)f'(t)\right]_{t=\alpha}^{t=x} + \int_\alpha^x f'(t)dt \\
&= -(x-\alpha)f'(\alpha) + [f(x) - f(\alpha)],
\end{aligned}
$$

and thus

$$f(x) = f(\alpha) + f'(\alpha)(x - \alpha) + \int_\alpha^x (x - t)f''(t)dt. \tag{8.6.60}$$

It follows from (8.6.52), (8.6.53), as well as (8.6.60), together with the mean value theorem (8.5.16) for integrals, that, for $n \in \mathbb{N}$, and with the points $\{x_{n,j} : j = 0, \ldots, n\}$ as in (8.4.3), so that also (8.4.4) holds, we have (see also Exercise 8.21)

$$\int_a^b f(x) - \mathcal{Q}_n^{MI}[f] = \sum_{j=0}^{n-1} \int_{x_{n,j}}^{x_{n,j+1}} f(x)dx - \sum_{j=0}^{n-1} \int_{x_{n,j}}^{x_{n,j+1}} f\left(\frac{x_{n,j} + x_{n,j+1}}{2}\right) dx$$

$$= \sum_{j=0}^{n-1} \int_{x_{n,j}}^{x_{n,j+1}} \left[f(x) - f\left(\frac{x_{n,j} + x_{n,j+1}}{2}\right)\right] dx$$

$$= \sum_{j=0}^{n-1} \int_{x_{n,j}}^{x_{n,j+1}} \left[f'\left(\frac{x_{n,j} + x_{n,j+1}}{2}\right)\left(x - \frac{x_{n,j} + x_{n,j+1}}{2}\right)\right. $$
$$\left. + \int_{\frac{1}{2}(x_{n,j} + x_{n,j+1})}^x (x - t)f''(t)dt\right] dx$$

$$= \sum_{j=0}^{n-1} \int_{x_{n,j}}^{x_{n,j+1}} \left[\int_{\frac{1}{2}(x_{n,j} + x_{n,j+1})}^x (x - t)f''(t)dt\right] dx$$

$$= \sum_{j=0}^{n-1} \left\{\int_{x_{n,j}}^{\frac{1}{2}(x_{n,j} + x_{n,j+1})} \left[\int_x^{\frac{1}{2}(x_{n,j} + x_{n,j+1})} (t - x)f''(t)dt\right] dx \right.$$
$$\left. + \int_{\frac{1}{2}(x_{n,j} + x_{n,j+1})}^{x_{n,j+1}} \left[\int_{\frac{1}{2}(x_{n,j} + x_{n,j+1})}^x (x - t)f''(t)dt\right]\right\} dx$$

$$= \sum_{j=0}^{n-1} \left\{\int_{x_{n,j}}^{\frac{1}{2}(x_{n,j} + x_{n,j+1})} \left[\int_{x_{n,j}}^t (t - x)dx\right] f''(t)dt \right.$$
$$\left. + \int_{\frac{1}{2}(x_{n,j} + x_{n,j+1})}^{x_{n,j+1}} \left[\int_t^{x_{n,j+1}} (x - t)dx\right] f''(t)dt\right\}$$

$$= \frac{1}{2} \sum_{j=0}^{n-1} \left\{\int_{x_{n,j}}^{\frac{1}{2}(x_{n,j} + x_{n,j+1})} (t - x_{n,j})^2 f''(t)dt \right.$$
$$\left. + \int_{\frac{1}{2}(x_{n,j} + x_{n,j+1})}^{x_{n,j+1}} (x_{n,j+1} - t)^2 f''(t)dt\right\}$$

$$= \frac{1}{2} \sum_{j=0}^{n-1} \left\{f''(\xi_j) \int_{x_{n,j}}^{\frac{1}{2}(x_{n,j} + x_{n,j+1})} (t - x_{n,j})^2 dt \right.$$
$$\left. + f''(\tilde{\xi}_j) \int_{\frac{1}{2}(x_{n,j} + x_{n,j+1})}^{x_{n,j+1}} (x_{n,j+1} - t)^2 dt\right\}$$

$$= \frac{1}{2} \sum_{j=0}^{n-1} \left\{ f''(\xi_j) \frac{1}{3} \left(\frac{b-a}{2n} \right)^3 + f''(\widetilde{\xi}_j) \frac{1}{3} \left(\frac{b-a}{2n} \right)^3 \right\}$$

$$= \frac{1}{48} \left(\frac{b-a}{n} \right)^3 \left\{ \sum_{j=0}^{n-1} f''(\xi_j) + \sum_{j=0}^{n-1} f''(\widetilde{\xi}_j) \right\}, \tag{8.6.61}$$

for some points $\{\xi_j : j = 0, \ldots, n-1\}$, $\{\widetilde{\xi}_j : j = 0, \ldots, n-1\} \subset [a,b]$. But, by applying again (8.6.29), we deduce the existence of points $\overline{\xi}, \xi^* \in [a,b]$ such that

$$\sum_{j=0}^{n-1} f''(\xi_j) + \sum_{j=0}^{n-1} f''(\widetilde{\xi}_j) = nf''(\overline{\xi}) + nf''(\xi^*) = n(f''(\overline{\xi}) + f''(\xi^*))$$

$$= 2nf''(\xi), \tag{8.6.62}$$

for some $\xi \in [a,b]$, after another application of (8.6.29). Hence we may substitute (8.6.62) into (8.6.61) to obtain

$$\int_a^b f(x)dx - \mathscr{Q}_n^{MI}[f] = \frac{b-a}{24} \left(\frac{b-a}{n} \right)^2 f''(\xi), \; f \in C^2[a,b], \; n = 1, 2, \ldots. \tag{8.6.63}$$

It follows from (8.6.63) that the midpoint rule $\mathscr{Q}_n^{MI}$ has order of exactness $m = 1$, whereas (8.6.63) also immediately implies the upper bound in (8.6.56). ∎

According to (8.6.51) and (8.6.54), $\mathscr{Q}_n^{RE}[f]$ and $\mathscr{Q}_n^{MI}[f]$ are both Riemann sums for $f \in C[a,b]$ with respect to the partition points $\{x_{n,j} : j = 0, \ldots, n\}$ of $[a,b]$, as defined by (8.4.3), and therefore satisfying also (8.4.4) and (8.4.5), so that we may state, by virtue of a standard result in calculus for the convergence of a Riemann sum, the following analogue of Theorem 8.6.3.

Theorem 8.6.5. *In Theorem* 8.6.4,

$$\left| \int_a^b f(x)dx - \mathscr{Q}_n^{RE}[f] \right| \to 0, \quad n \to \infty, \; f \in C[a,b]; \tag{8.6.64}$$

$$\left| \int_a^b f(x)dx - \mathscr{Q}_n^{MI}[f] \right| \to 0, \quad n \to \infty, \; f \in C[a,b]. \tag{8.6.65}$$

It is interesting to observe that, although the trapezoidal rule $\mathscr{Q}_n^{TR}$, as given in (8.6.17), is obtained by integrating a (continuous) piecewise linear interpolant of the integrand f, whereas the midpoint rule $\mathscr{Q}_n^{MI}$, as defined by (8.6.52), (8.6.53), is obtained by integrating a (discontinuous) piecewise constant interpolant of f, the quadrature rules $\mathscr{Q}_n^{TR}$ and $\mathscr{Q}_n^{MI}$ have, according to Theorems 8.6.2 and 8.6.4, the same degree of exactness $m = 1$. Moreover, note that the constant $\frac{1}{12}$ in the upper bound (8.6.33) for $\mathscr{Q}_n^{TR}$ is twice as large as the constant $\frac{1}{24}$ appearing in the analogous upper bound in (8.6.56) of Theorem 8.6.4 for $\mathscr{Q}_n^{MI}$.

Example 8.6.2. For the numerical approximation of the integral $\int_0^2 f(x)dx, f \in C[0,2]$, it follows from (8.6.51), (8.6.54) and (8.4.3) that

$$\mathcal{Q}_n^{RE}[f] = \frac{2}{n}\sum_{j=0}^{n-1} f\left(\frac{2j}{n}\right), \quad n = 1,2,\ldots; \tag{8.6.66}$$

$$\mathcal{Q}_n^{MI}[f] = \frac{2}{n}\sum_{j=0}^{n-1} f\left(\frac{2j+1}{n}\right), \quad n = 1,2,\ldots. \tag{8.6.67}$$

For the case $f(x) = e^x$, an application of (8.6.66) and (8.6.67) yields the corresponding quadrature errors

$$\left|\int_0^2 f(x)dx - \mathcal{Q}_8^{RE}[f]\right| \approx 7.654 \times 10^{-1};$$

$$\left|\int_0^2 f(x)dx - \mathcal{Q}_8^{MI}[f]\right| \approx 1.661 \times 10^{-2}.$$

∎

8.7 Exercises

Exercise 8.1 Design the Gauss-Legendre quadrature rule $\mathcal{Q}_2^{GL}$ for the numerical approximation of the integral

$$\mathcal{I}[f] := \int_{-1}^1 f(x)dx,$$

where $f \in C[-1,1]$, and verify that the weights $\{w_{2,0}, w_{2,1}, w_{2,2}\}$ thus obtained satisfy the property (8.1.11), with $w(x) = 1$, $x \in [-1,1]$, as well as the positivity condition (8.2.21) in Theorem 8.2.3. Also, verify that, as given in (8.2.7) of Theorem 8.2.2, $\mathcal{Q}_2^{GL}$ has degree of exactness $m = 5$, that is,

$$\mathcal{E}_2^{GL}[f] := \int_{-1}^1 f(x)dx - \mathcal{Q}_2^{GL}[f] = 0, \quad f \in \pi_5,$$

by explicitly calculating $\mathcal{E}_2^{GL}[f]$ for, respectively,

$$\text{(a) } f(x) = \sum_{j=0}^5 \alpha_j x^j; \qquad \text{(b) } f(x) = x^6,$$

where, in (a), $\{\alpha_0,\ldots,\alpha_5\}$ denotes an arbitrary coefficient sequence in $\mathbb{R}$.

Exercise 8.2 As a continuation of Exercise 8.1, for both of the respective integrands

$$\text{(a) } f(x) = \ln(x+2); \qquad \text{(b) } f(x) = x^6,$$

and after applying integration by parts to obtain

$$\int_{-1}^1 \ln(x+2)dx = 3\ln 3 - 2 \approx 1.296,$$

verify that the corresponding quadrature errors $\mathcal{E}_2^{GL}[f]$ satisfy the upper bound in (8.2.28) of Theorem 8.2.4(b), with $\mathcal{Q}_n^G = \mathcal{Q}_2^{GL}$, and for each $k = 1,2$.

Exercise 8.3 Obtain real numbers A, B, α and β such that the condition

$$\int_0^4 \sqrt{x}f(x)dx = Af(\alpha) + Bf(\beta), \quad f \in \pi_3, \tag{$*$}$$

is satisfied.

[*Hint:* Apply Theorem 8.2.2.]

Exercise 8.4 As a continuation of Exercise 8.3, for any $f \in C^1[0,4]$, apply (8.2.29) in Theorem 8.2.4(b) to obtain an upper bound on the quadrature error

$$\left| \int_0^4 \sqrt{x}f(x)dx - [Af(\alpha) + Bf(\beta)] \right|,$$

and investigate the sharpness of this upper bound for the integrand $f(x) = x^4$.

Exercise 8.5 Repeat Exercise 8.3, with the condition $(*)$ replaced by each of the following:

(a) $\int_{-1}^1 |x|f(x)dx = Af(\alpha) + Bf(\beta), \quad f \in \pi_3$;

(b) $\int_0^1 xf(x)dx = Af(\alpha) + Bf(\beta), \quad f \in \pi_3$;

(c) $\int_{-1}^1 \frac{1}{\sqrt{1-x^2}}f(x)dx = Af(\alpha) + Bf(\beta), \quad f \in \pi_3$.

Exercise 8.6 As a continuation of Exercise 8.5, repeat Exercise 8.4, with $f \in C^1[0,4]$ replaced by, respectively, (a) $f \in C^1[-1,1]$; (b) $f \in C^1[0,1]$; (c) $f \in C^1[-1,1]$.

Exercise 8.7 By explicitly integrating the Chebyshev polynomials in (2.2.2), verify the formula in the first line of (8.3.16) for $k = 0,2,4,6$.

Exercise 8.8 Apply Theorem 8.3.1 to construct the Clenshaw-Curtis quadrature rule $\mathcal{Q}_4^{CC}$ for the numerical approximation of the integral

$$\mathcal{I}[f] := \int_{-5}^5 f(x)dx, \quad f \in C[-5,5],$$

and verify that the weights $\{w_{4,0}, \ldots, w_{4,4}\}$ thus obtained satisfy the property (8.1.11), with $w(x) = 1, x \in [-5,5]$, as well as the positivity condition (8.3.30) in Theorem 8.3.2.

Exercise 8.9 As a continuation of Exercise 8.8, verify that, as given in the first line of (8.3.58) in Theorem 8.3.3, $\mathcal{Q}_4^{CC}$ has degree of exactness $m = 5$, that is,

$$\mathcal{E}_4^{CC}[f] := \int_{-5}^5 f(x)dx - \mathcal{Q}_4^{CC}[f] = 0, \quad f \in \pi_5,$$

by explicitly calculating $\mathcal{E}_4^{CC}[f]$ for, respectively, f chosen as in (a) and (b) of Exercise 8.1.

Exercise 8.10 As a continuation of Exercise 8.9, for each of the integrands

(a) $f(x) = \frac{1}{1+x^2}$ (that is, the Runge example (3.1.5));   (b) $f(x) = x^6$, verify that the quadrature error $\mathscr{E}_4^{CC}[f]$ satisfies the upper bound (8.3.63) of Theorem 8.3.4(b), with $[a,b] = [-5,5]$, $n = 4$, and for each $k = 1, \ldots, 4$.

Exercise 8.11 Prove inductively from (8.4.26) that (8.4.25) holds.

Exercise 8.12 Verify, by means of the recursive formulation (8.4.50) in Theorem 8.4.3, the Laplace coefficient values in (8.4.54), as well as, by applying the formulation (8.4.38) in Theorem 8.4.1, the Newton-Cotes weights in Table 8.4.1.

Exercise 8.13 Extend the quadrature formulas (8.4.56)–(8.4.59) by obtaining explicit formulations for the Newton-Cotes quadrature rules $\mathscr{Q}_5^{NC}$ and $\mathscr{Q}_6^{NC}$.

Exercise 8.14 As a continuation of Exercise 8.13, extend the quadrature error estimates (8.5.70)–(8.5.73) by obtaining analogous quadrature error bounds with respect to $\mathscr{Q}_5^{NC}$ and $\mathscr{Q}_6^{NC}$.

Exercise 8.15 For the function

$$f(x) = x^4, \quad x \in [0,1],$$

find a point $\xi \in [0,1]$, the existence of which is guaranteed by Theorem 8.5.2, such that, as in (8.5.17) with $k = 0$ and $x_0 = 1$,

$$\left[\frac{d}{dx}(f[x,1])\right]\Big|_{x=\frac{1}{2}} = \frac{1}{2}f''(\xi).$$

Exercise 8.16 Verify the integral $\widetilde{G}_2(t)$ in (8.5.35).

Exercise 8.17 Verify the result (8.5.21) of Theorem 8.5.3(a) for the case $n = 2$, $[a,b] = [0,2]$ and

$$f(x) = \frac{1}{(1+x)^2}.$$

[*Hint:* Apply the formula (8.4.57).]

Exercise 8.18 As a continuation of Exercise 8.17, find a point $\xi \in [0,2]$, the existence of which is guaranteed by Theorem 8.5.4, such that the first line of (8.5.47) is satisfied, with $[a,b], n$, and f given as in Exercise 8.17.

[*Hint:* Use (8.4.54).]

Exercise 8.19 Obtain the explicit formulations of the composite Newton-Cotes quadrature rules $\mathscr{Q}_{5,n}^{NC}$ and $\mathscr{Q}_{6,n}^{NC}$, thereby extending the quadrature formulas (8.6.22)–(8.6.25).

Exercise 8.20 As a continuation of Exercise 8.19, obtain quadrature error estimates with respect to $\mathscr{Q}_{5,n}^{NC}$ and $\mathscr{Q}_{6,n}^{NC}$, thereby extending the error estimates (8.6.33) - (8.6.36).

Exercise 8.21 For a bivariate function f that is continuous on a square $[\alpha, \beta] \times [\alpha, \beta]$ in $\mathbb{R}^2$, verify the interchange of integration order result

$$\int_\alpha^\beta \left[\int_x^\beta f(x,t)dt\right] dx = \int_\alpha^\beta \left[\int_\alpha^t f(x,t)dx\right] dt,$$

as used to establish (8.6.61) in the proof of Theorem 8.6.4.

Exercise 8.22 For any positive integer v, let $\{\mathscr{Q}_{v,jv}^{NC} : j = 1,2,\ldots\}$ denote the corresponding sequence of composite Newton-Cotes quadrature rules for the numerical approximation of the integral

$$\int_a^b f(x)dx, \quad f \in C[a,b].$$

Apply the quadrature error estimate (8.6.32) in Theorem 8.6.2 to show that, for a fixed integer $v \in \mathbb{N}$, and any given integrand $f \in C^{\ell_v}[a,b]$, with

$$\ell_v := \begin{cases} v+2, & \text{if } v \text{ is even}; \\ \\ v+1, & \text{if } v \text{ is odd}, \end{cases}$$

it holds that

$$\mathscr{E}_{v,jv}^{NC}[f] := \left| \int_a^b f(x)dx - \mathscr{Q}_{v,jv}^{NC}[f] \right| \leqslant K_j, \quad j \in \mathbb{N},$$

where the upper bounds $\{K_j : j \in \mathbb{N}\}$ satisfy the decay condition

$$K_{j+1} = \frac{1}{(1+\frac{1}{j})^{\ell_v}} K_j, \quad j \in \mathbb{N}. \quad (*)$$

Exercise 8.23 As a continuation of Exercise 8.22, by computing the quadrature error $\mathscr{E}_{v,jv}^{NC}[f]$, with $[a,b] = [0,4]$ and

$$f(x) = \frac{1}{(1+x)^2},$$

and for $j = 1,\ldots,4$; $v = 1,\ldots,4$, investigate numerically whether, analogously to $(*)$ in Exercise 8.22, the decay rate

$$\left| \mathscr{E}_{v,(j+1)v}^{NC}[f] \right| \approx \frac{1}{(1+\frac{1}{j})^{\ell_v}} \left| \mathscr{E}_{v,jv}^{NC}[f] \right|$$

is achieved for this choice of f.

Exercise 8.24 By using the quadrature error estimates (8.6.55) and (8.6.56) in Theorem 8.6.4, and after arguing as in Exercises 8.22 and 8.23, investigate numerically, for the same choices of $[a,b]$ and f as in Exercise 8.23, whether the quadrature errors

$$\mathscr{E}_n^{RE}[f] := \int_0^4 f(x)dx - \mathscr{Q}_n^{RE}[f]; \qquad \mathscr{E}_n^{MI}[f] := \int_0^4 f(x)dx - \mathscr{Q}_n^{MI}[f],$$

satisfy the decay rates

$$\left| \mathscr{E}_{2n}^{RE}[f] \right| \approx \frac{1}{2} \left| \mathscr{E}_n^{RE}[f] \right|; \qquad \left| \mathscr{E}_{2n}^{MI}[f] \right| \approx \frac{1}{4} \left| \mathscr{E}_n^{MI}[f] \right|,$$

for $n = 1,\ldots,4$.

Exercise 8.25 For the function

$$f(x) = e^{-x^2}, \quad x \in [0,1],$$

calculate a numerical approximation $\mathscr{Q}[f]$ of the integral

$$\mathscr{I}[f] := \int_0^1 f(x)dx$$

such that

$$|\mathscr{I}[f] - \mathscr{Q}[f]| < \frac{1}{100},$$

by means of

(a) Gauss-Legendre quadrature;

(b) Clenshaw-Curtis quadrature;

(c) the trapezoidal rule;

(d) the Simpson rule;

(e) the midpoint rule.

Chapter 9

Approximation of Periodic Functions

In the study of approximation of functions in Chapters 1 to 8, the emphasis is on (algebraic) polynomial approximation on the bounded interval $[a,b]$. Since algebraic polynomials are not periodic functions, they are not suitable basis functions for representing and approximating periodic functions on the entire real line $\mathbb{R}$. On the other hand, many natural phenomena can only be represented by periodic functions. It is therefore essential to study approximation of periodic continuous functions $f : \mathbb{R} \to \mathbb{R}$, by the linear span of some elementary periodic functions. This chapter is devoted to the study of this topic by considering basis functions that are formulated in terms of the sine and cosine functions.

9.1 Trigonometric polynomials

Let $\widetilde{f} \in C(\mathbb{R})$, with $C(\mathbb{R})$ denoting the linear space of continuous functions $f : \mathbb{R} \to \mathbb{R}$, and suppose that, moreover, $\widetilde{f}$ is periodic, with period K, that is, K is a positive number such that

$$\widetilde{f}(x+K) = \widetilde{f}(x), \quad x \in \mathbb{R}. \tag{9.1.1}$$

For the function $f \in C(\mathbb{R})$ defined by

$$f(x) := \widetilde{f}\left(\frac{K}{2\pi}x\right), \quad x \in \mathbb{R}, \tag{9.1.2}$$

it then follows from (9.1.2) and (9.1.1) that, for any $x \in \mathbb{R}$,

$$f(x+2\pi) = \widetilde{f}\left(\frac{K}{2\pi}(x+2\pi)\right) = \widetilde{f}\left(\frac{K}{2\pi}x + K\right) = \widetilde{f}\left(\frac{K}{2\pi}x\right),$$

and thus, from (9.1.2),

$$f(x+2\pi) = f(x), \quad x \in \mathbb{R}, \tag{9.1.3}$$

233

that is, f is periodic on $\mathbb{R}$, with period 2π. Hence we shall, without loss of generality, restrict our attention in this chapter to the approximation of functions $f \in C_{2\pi}$, where

$$C_{2\pi} := \{f \in C(\mathbb{R}) : f(x+2\pi) = f(x), \quad x \in \mathbb{R}\}, \tag{9.1.4}$$

since any periodic function $\widetilde{f}$ in $C(\mathbb{R})$ with period $K \neq 2\pi$ can be transformed by means of (9.1.2) into a function $f \in C_{2\pi}$.

For use as approximation set for a given function $f \in C_{2\pi}$, we define, for any $n \in \mathbb{N}$, the linear space

$$\tau_n := \operatorname{span}\{1, \cos x, \ldots, \cos(nx), \sin x, \ldots, \sin(nx)\}, \tag{9.1.5}$$

for which it is then immediately evident that

$$\tau_n \subset C_{2\pi}, \quad n \in \mathbb{N}. \tag{9.1.6}$$

In order to determine the dimension of τ_n, we first observe that $(C_{2\pi}, \langle \cdot, \cdot \rangle_2)$ is an inner product space with respect to the inner product

$$\langle f, g \rangle_2 := \int_{-\pi}^{\pi} f(x)g(x)dx, \quad f, g \in C_{2\pi}, \tag{9.1.7}$$

after having noted in particular that, if $f \in C_{2\pi}$ is such that

$$0 = \langle f, f \rangle_2 = \int_{-\pi}^{\pi} [f(x)]^2 dx,$$

then $f \in C(\mathbb{R})$ implies $f(x) = 0$, $x \in [-\pi, \pi]$, and thus, from (9.1.3), $f(x) = 0, x \in \mathbb{R}$, as required in the condition (4.2.2).

For any inner product space $(X, \langle \cdot, \cdot \rangle)$, we say that a subset Y of X is an orthogonal set in $(X, \langle \cdot, \cdot \rangle)$ if the condition

$$\langle f, g \rangle = 0, \quad f, g \in Y, \quad f \neq g,$$

is satisfied.

The following result then holds.

Theorem 9.1.1. *For any $n \in \mathbb{N}$, the set $\{1, \cos x, \ldots, \cos(nx), \sin x, \ldots, \sin(nx)\}$ is an orthogonal set in the inner product space $(C_{2\pi}, \langle \cdot, \cdot \rangle_2)$, with $\langle \cdot, \cdot \rangle_2$ given as in (9.1.7).*

Proof. For $j, k \in \{0, 1, \ldots, n\}$, with $j \neq k$, we have

$$\int_{-\pi}^{\pi} \cos(jx)\cos(kx)dx = \frac{1}{2} \int_{-\pi}^{\pi} [\cos((j+k)x) + \cos((j-k)x)]dx$$

$$= \frac{1}{2}\left[\frac{\sin((j+k)x)}{j+k} + \frac{\sin((j-k)x)}{j-k}\right]_{-\pi}^{\pi} = 0, \tag{9.1.8}$$

and

$$\int_{-\pi}^{\pi} \sin(jx)\sin(kx)dx = \frac{1}{2}\int_{-\pi}^{\pi}[\cos((j-k)x)-\cos((j+k)x)]\,dx$$

$$= \frac{1}{2}\left[\frac{\sin((j-k)x)}{j-k} - \frac{\sin((j+k)x)}{j-k}\right]_{-\pi}^{\pi} = 0, \qquad (9.1.9)$$

whereas, for $j \in \{0,\dots,n\}$ and $k \in \{1,\dots,n\}$,

$$\int_{-\pi}^{\pi} \cos(jx)\sin(kx)dx = 0, \qquad (9.1.10)$$

since the integrand in (9.1.10) is an odd function on $\mathbb{R}$. It follows from (9.1.8), (9.1.9) and (9.1.10), together with the definition (9.1.7), that $\{1,\cos x,\dots,\cos(nx),\sin x,\dots,\sin(nx)\}$ is an orthogonal set in $(C_{2\pi},\langle\cdot,\cdot\rangle_2)$. ∎

We may now apply Theorem 9.1.1 to deduce from Theorem 7.3.2 that $\{1,\cos x,\dots,\cos(nx),\sin x,\dots,\sin(nx)\}$ is a linearly independent set, so that we have now established the following.

Theorem 9.1.2. *For any $n \in \mathbb{N}$, let the linear space τ_n be defined by (9.1.5). Then:*

(a)

$$\dim(\tau_n) = 2n+1; \qquad (9.1.11)$$

(b)

the set $\{1,\cos x,\dots,\cos(nx),\sin x,\dots,\sin(nx)\}$ is an orthogonal basis for τ_n.

Hence, each $Q \in \tau_n$ has a unique representation of the form

$$Q(x) = \frac{1}{2}a_0 + \sum_{j=1}^{n} a_j\cos(jx) + \sum_{j=1}^{n} b_j\sin(jx), \qquad (9.1.12)$$

for some coefficient sequence $\{a_0,\dots,a_n,b_1,\dots,b_n\} \subset \mathbb{R}$, where the reason for the appearance in (9.1.12) of the factor $\frac{1}{2}$ before the coefficient a_0 will become apparent in Section 9.3.

We proceed to prove the following fundamental properties of the linear space τ_n.

Theorem 9.1.3. *For $n \in \mathbb{N}$, let the linear space τ_n be defined by (9.1.5). Then:*

(a) *For any non-negative integers j and k satisfying $j+k \leqslant n \in \mathbb{N}$, the function Q defined by*

$$Q(x) := (\cos x)^j(\sin x)^k, \qquad (9.1.13)$$

satisfies

$$Q \in \tau_n. \qquad (9.1.14)$$

(b) *For any positive integers m and n, if*

$$Q \in \tau_m \quad ; \quad \tilde{Q} \in \tau_n, \tag{9.1.15}$$

then

$$Q\tilde{Q} \in \tau_{m+n}. \tag{9.1.16}$$

(c) *For any $n \in \mathbb{N}$, if $Q \in \tau_n$, and Q has more than $2n$ distinct zeros in the interval $(-\pi, \pi]$, then Q is the zero function.*

Proof. (a) We use a proof by induction on the integer n. After noting first from (9.1.5) that (9.1.13) trivially implies (9.1.14) if $n = 1$, we suppose next that (9.1.13) implies (9.1.14) for a fixed integer $n \in \mathbb{N}$. For non-negative integers j and k satisfying $j + k \leqslant n + 1$, we shall show that the function Q given by (9.1.13) then satisfies

$$Q \in \tau_{n+1}, \tag{9.1.17}$$

which will complete our inductive proof.

To this end, we apply the inductive hypothesis, and the representation formula (9.1.12), to deduce that, for some coefficient sequence $\{a_0, \ldots, a_n, b_1, \ldots, b_n\} \subset \mathbb{R}$, it holds that, if $k \geqslant 1$, so that $j + (k-1) \leqslant n$,

$$Q(x) = \left[(\cos x)^j (\sin x)^{k-1} \right] \sin x$$

$$= \left[\frac{1}{2} a_0 + \sum_{\ell=1}^{n} a_\ell \cos(\ell x) + \sum_{\ell=1}^{n} b_\ell \sin(\ell x) \right] \sin x$$

$$= \frac{1}{2} a_0 \sin x + \frac{1}{2} \sum_{\ell=1}^{n} a_\ell \left[\sin((\ell+1)x) - \sin((\ell-1)x) \right]$$

$$+ \frac{1}{2} \sum_{\ell=1}^{n} b_\ell \left[\cos((\ell-1)x) - \cos((\ell+1)x) \right], \tag{9.1.18}$$

whereas, if $k = 0$ and $j \geqslant 1$, so that $j - 1 \leqslant n$,

$$Q(x) = (\cos x)^{j-1} \cos x$$

$$= \left[\frac{1}{2} a_0 + \sum_{\ell=1}^{n} a_\ell \cos(\ell x) + \sum_{\ell=1}^{n} b_\ell \sin(\ell x) \right] \cos x$$

$$= \frac{1}{2} a_0 \cos x + \frac{1}{2} \sum_{\ell=1}^{n} a_\ell \left[\cos((\ell+1)x) + \cos((\ell-1)x) \right]$$

$$+ \frac{1}{2} \sum_{\ell=1}^{n} b_\ell \left[\sin((\ell+1)x) + \sin((\ell-1)x) \right], \tag{9.1.19}$$

and, if $j = k = 0$, then (9.1.13) yields

$$Q(x) = 1, \quad x \in \mathbb{R}. \tag{9.1.20}$$

It then follows from (9.1.18), (9.1.19) and (9.1.20), together with the definition (9.1.5), that the desired result (9.1.17) does indeed hold.

(b) Suppose the functions Q and $\widetilde{Q}$ are such that (9.1.15) is satisfied for two given positive integers m and n. The result (9.1.16) is then obtained by formulating Q and $\widetilde{Q}$ by means of (9.1.12), and using the identities

$$\cos(jx)\cos(kx) = \tfrac{1}{2}[\cos((j+k)x) + \cos((j-k)x)], \quad j \in \{0,\dots,m\}, \ k \in \{0,\dots,n\};$$

$$\cos(jx)\sin(kx) = \tfrac{1}{2}[\sin((j+k)x) - \sin((j-k)x)], \quad j \in \{0,\dots,m\}, \ k \in \{1,\dots,n\};$$

$$\sin(jx)\sin(kx) = \tfrac{1}{2}[\cos((j-k)x) - \cos((j+k)x)], \quad j \in \{1,\dots,m\}, \ k \in \{1,\dots,n\},$$

as well as the definition (9.1.5).

(c) For an integer $n \in \mathbb{N}$, let $Q \in \tau_n$ be such that Q has more than $2n$ distinct zeros in the interval $(-\pi, \pi]$.

With the complex number notation $i = \sqrt{-1}$, we now deduce from the representation formula (9.1.12), together with De Moivre's theorem, and $\dfrac{1}{i} = -i$, that, for any $x \in \mathbb{R}$,

$$Q(x) = \frac{1}{2}a_0 + \frac{1}{2}\sum_{j=1}^{n} a_j[\{\cos(jx) + i\sin(jx)\} + \{\cos(-jx) + i\sin(-jx)\}]$$

$$+ \frac{1}{2i}\sum_{j=1}^{n} b_j[\{\cos(jx) + i\sin(jx)\} - \{\cos(-jx) + i\sin(-jx)\}]$$

$$= \frac{1}{2}a_0 + \frac{1}{2}\sum_{j=1}^{n} a_j[(\cos x + i\sin x)^j + (\cos x + i\sin x)^{-j}]$$

$$- \frac{1}{2}i\sum_{j=1}^{n} b_j[(\cos x + i\sin x)^j - (\cos x + i\sin x)^{-j}]$$

$$= \frac{1}{2}a_0 + \frac{1}{2}\sum_{j=1}^{n}(a_j - ib_j)(\cos x + i\sin x)^j + \frac{1}{2}\sum_{j=1}^{n}(a_j + ib_j)(\cos x + i\sin x)^{-j},$$

and thus

$$Q(x) = \sum_{j=-n}^{n} \alpha_j(\cos x + i\sin x)^j, \quad x \in \mathbb{R}, \tag{9.1.21}$$

where

$$\alpha_j := \begin{cases} \tfrac{1}{2}(a_j + ib_j), & j = -n,\dots,-1; \\ \tfrac{1}{2}a_0, & j = 0; \\ \tfrac{1}{2}(a_j - ib_j), & j = 1,\dots,n. \end{cases} \tag{9.1.22}$$

Observe from (9.1.21) that, for any $x \in \mathbb{R}$,

$$Q(x) = \frac{1}{(\cos x + i \sin x)^n} \sum_{j=0}^{2n} \alpha_{j-n} (\cos x + i \sin x)^j$$

$$= \frac{1}{\cos(nx) + i \sin(nx)} \sum_{j=0}^{2n} \alpha_{j-n} (\cos x + i \sin x)^j, \qquad (9.1.23)$$

after another application of De Moivre's theorem.

With the definition

$$P(z) := \sum_{j=0}^{2n} \alpha_{j-n} z^j, \quad z \in \mathbb{C}, \qquad (9.1.24)$$

with $\{\alpha_j : j = -n, \ldots, n\}$ given as in (9.1.22), it follows that P is a polynomial with coefficients in $\mathbb{C}$, with $\mathbb{C}$ denoting the set of complex numbers, and such that $\deg(P) \leqslant 2n$. Since also

$$z = \cos x + i \sin x \Rightarrow |z| = \sqrt{\cos^2 x + \sin^2 x} = 1,$$

we deduce from (9.1.23) and (9.1.24), together with the fact that Q has more than $2n$ distinct zeros on the interval $(-\pi, \pi]$, that the polynomial P has more than $2n$ distinct zeros on the unit circle $|z| = 1$ of the complex plane $\mathbb{C}$, that is,

$$P(z) = \widetilde{P}(z) \prod_{j=1}^{k} (z - z_j), \quad z \in \mathbb{C}, \qquad (9.1.25)$$

with $k \geqslant 2n + 1$, where $\{z_j : j = 1, \ldots, k\}$ are k distinct points in $\mathbb{C}$ such that $|z_j| = 1, j = 1, \ldots, k$, and where $\widetilde{P}$ is a polynomial with coefficients in $\mathbb{C}$. Since, moreover, $\deg(P) \leqslant 2n$, we deduce from (9.1.25) and $k \geqslant 2n + 1$ that $\widetilde{P}$, and therefore also P, must be the zero polynomial. It then follows from (9.1.24) that $\alpha_j = 0, j = -n, \ldots, n$, according to which we deduce from (9.1.21) that Q is the zero function, which completes our proof. ∎

Based on the structural similarities between the linear spaces π_n and τ_n, as is evident from Theorem 9.1.3, we call any function $Q \in \tau_n$, as given by the representation (9.1.12), and with either $a_n \neq 0$, or $b_n \neq 0$, a trigonometric polynomial of degree n.

Recalling also the inclusion (9.1.6), we proceed in the rest of this chapter to analyze the trigonometric polynomial space τ_n as an approximation set for a given function $f \in C_{2\pi}$.

9.2   A Weierstrass result for periodic functions

First in this section, we introduce the normed linear space $(C_{2\pi}, ||\cdot||_\infty)$ with respect to the maximum norm

$$||f||_\infty := \max_{-\pi \leqslant x \leqslant \pi} |f(x)|, \quad f \in C_{2\pi}, \qquad (9.2.1)$$

after having noted in particular that if $f \in C_{2\pi}$ is such that $||f||_\infty = 0$, then (9.2.1) yields $f(x) = 0, x \in [-\pi, \pi]$, and thus, from (9.1.3), $f(x) = 0, x \in \mathbb{R}$, as required in the condition (4.1.2).

We proceed to prove the following analogue for trigonometric polynomials of the Weierstrass theorem, as given in Theorem 3.3.4.

Theorem 9.2.1. *Let $f \in C_{2\pi}$. Then, for each $\varepsilon > 0$, there exists a trigonometric polynomial Q such that*

$$||f - Q||_\infty := \max_{-\pi \leqslant x \leqslant \pi} |f(x) - Q(x)| < \varepsilon. \tag{9.2.2}$$

Proof. First, observe that

$$f = f_1 + f_2, \tag{9.2.3}$$

where

$$f_1(x) := \tfrac{1}{2}[f(x) + f(-x)], \quad x \in \mathbb{R}; \tag{9.2.4}$$

$$f_2(x) := \tfrac{1}{2}[f(x) - f(-x)], \quad x \in \mathbb{R}. \tag{9.2.5}$$

Note from (9.2.4), (9.2.5) and (9.1.3) that, for any $x \in \mathbb{R}$,

$$f_1(x + 2\pi) = \tfrac{1}{2}[f(x + 2\pi) + f(-x + 2\pi)] = \tfrac{1}{2}[f(x) + f(-x)] = f_1(x);$$

$$f_2(x + 2\pi) = \tfrac{1}{2}[f(x + 2\pi) - f(-x + 2\pi)] = \tfrac{1}{2}[f(x) - f(-x)] = f_2(x),$$

from which we deduce that

$$f_1 \in C_{2\pi} \quad ; \quad f_2 \in C_{2\pi}. \tag{9.2.6}$$

Also, (9.2.4) and (9.2.5) yield, for any $x \in \mathbb{R}$,

$$f_1(-x) = \tfrac{1}{2}[f(-x) + f(x)] = f_1(x);$$

$$f_2(-x) = \tfrac{1}{2}[f(-x) - f(x)] = -f_2(x),$$

according to which

$$\left.\begin{array}{l} f_1 \text{ is an even function on } \mathbb{R}; \\ f_2 \text{ is an odd function on } \mathbb{R}. \end{array}\right\} \tag{9.2.7}$$

Let $\varepsilon > 0$. We shall prove that there exist trigonometric polynomials Q_1 and Q_2 such that

$$\left.\begin{array}{l} ||f_1 - Q_1||_\infty < \dfrac{\varepsilon}{4}; \\ ||f_2 - Q_2||_\infty < \dfrac{3\varepsilon}{4}, \end{array}\right\} \tag{9.2.8}$$

which, together with (9.2.3), and the triangle inequality (4.1.4), will then imply that the trigonometric polynomial $Q := Q_1 + Q_2$ satisfies

$$||f - Q||_\infty = ||(f_1 - Q_1) + (f_2 - Q_2)||_\infty$$

$$\leqslant ||f_1 - Q_1||_\infty + ||f_2 - Q_2||_\infty < \frac{\varepsilon}{4} + \frac{3\varepsilon}{4} = \varepsilon,$$

which immediately yields the desired inequality (9.2.2).

To establish the existence of a trigonometric polynomial Q_1 such that the first line of (9.2.8) holds, we first define

$$g_1(t) := f_1(\arccos t), \quad t \in [-1, 1], \tag{9.2.9}$$

according to which, since f_1 is continuous on $\mathbb{R}$, we have $g_1 \in C[-1, 1]$. Hence we may apply the Weierstrass theorem in Theorem 3.3.4 to deduce the existence of an algebraic polynomial P_1 such that

$$|g_1(t) - P_1(t)| < \frac{\varepsilon}{4}, \quad t \in [-1, 1],$$

and thus

$$|g_1(\cos x) - P_1(\cos x)| < \frac{\varepsilon}{4}, \quad x \in [0, \pi]. \tag{9.2.10}$$

Now observe from (9.2.9) that

$$g_1(\cos x) = f_1(x), \quad x \in [0, \pi]. \tag{9.2.11}$$

Also, it follows from Theorem 9.1.3(a) that, since P_1 is an algebraic polynomial, the definition

$$Q_1(x) := P_1(\cos x), \quad x \in \mathbb{R}, \tag{9.2.12}$$

implies that Q_1 is a trigonometric polynomial. Hence we have established the existence of a trigonometric polynomial Q_1 such that, from (9.2.10), (9.2.11) and (9.2.12),

$$|f_1(x) - Q_1(x)| < \frac{\varepsilon}{4}, \quad x \in [0, \pi]. \tag{9.2.13}$$

According to the first line of (9.2.7), as well as (9.2.12), f_1 and Q_1 are both even functions on $\mathbb{R}$, and therefore the difference $f_1 - Q_1$ is an even function on $\mathbb{R}$, so that we may deduce from (9.2.13) that

$$|f_1(x) - Q_1(x)| < \frac{\varepsilon}{4}, \quad x \in [-\pi, \pi],$$

from which, together with the definition (9.2.1), the desired inequality in the first line of (9.2.8) then immediately follows.

It therefore remains to prove the existence of a trigonometric polynomial Q_2 such that the second line of (9.2.8) is satisfied. To this end, we first note from (9.2.5) that

$$f_2(0) = \frac{1}{2}[f(0) - f(0)] = 0, \qquad (9.2.14)$$

and

$$f_2(\pi) = \frac{1}{2}[f(\pi) - f(-\pi)] = 0, \qquad (9.2.15)$$

by virtue of (9.1.3). It follows from (9.2.14) and (9.2.15), together with the continuity on $\mathbb{R}$ of f_2, that there exist points $x_0 \in (0, \frac{\pi}{2}]$ and $x_1 \in [\frac{\pi}{2}, \pi)$, such that x_0 is the largest number in the interval $(0, \frac{\pi}{2}]$ for which it holds that

$$|f_2(x)| \leqslant \frac{\varepsilon}{4}, \quad x \in [0, x_0], \qquad (9.2.16)$$

whereas x_1 is the smallest number in the interval $[\frac{\pi}{2}, \pi)$ for which it holds that

$$|f_2(x)| \leqslant \frac{\varepsilon}{4}, \quad x \in [x_1, \pi]. \qquad (9.2.17)$$

Now define the function $f_3 : \mathbb{R} \to \mathbb{R}$ by

$$f_3(x) := \begin{cases} \dfrac{f_2(x_0)}{\sin x_0}, & x \in [0, x_0]; \\[2mm] \dfrac{f_2(x)}{\sin x}, & x \in (x_0, x_1); \\[2mm] \dfrac{f_2(x_1)}{\sin x_1}, & x \in [x_1, \pi], \end{cases} \qquad (9.2.18)$$

and

$$f_3(x) := \begin{cases} f_3(-x), & x \in [-\pi, 0); \\ f_3(x + 2\pi), & x \in (-\infty, -\pi); \\ f_3(x - 2\pi), & x \in (\pi, \infty), \end{cases} \qquad (9.2.19)$$

from which it follows that $f_3 \in C_{2\pi}$, and f_3 is an even function on $\mathbb{R}$. Hence, as in the proof of the first line of (9.2.8), we may deduce that there exists a trigonometric polynomial Q_3 such that

$$\|f_3 - Q_3\|_\infty < \frac{\varepsilon}{4}, \qquad (9.2.20)$$

and where Q_3 is an even function on $\mathbb{R}$. Now observe that the definition

$$Q_2(x) := Q_3(x)\sin x, \quad x \in \mathbb{R}, \qquad (9.2.21)$$

implies, according to Theorem 9.1.3(a) and (9.1.12), that Q_2 is a trigonometric polynomial. We proceed to prove that Q_2 satisfies the second line of (9.2.8), which will then complete our proof.

First, by using (9.2.18), (9.2.21) and (9.2.20), and keeping in mind also the inclusion $[x_0, x_1] \subset [0, \pi]$, we deduce that

$$|f_2(x) - Q_2(x)| = (\sin x)|f_3(x) - Q_3(x)| \leqslant |f_3(x) - Q_3(x)| < \frac{\varepsilon}{4} < \frac{3\varepsilon}{4}, \quad x \in (x_0, x_1).$$
(9.2.22)

Now observe from (9.2.21), (9.2.16), (9.2.17) and (9.2.20) that, for any $x \in [0, x_0] \cup [x_1, \pi]$,

$$|f_2(x) - Q_2(x)| = |f_2(x) - f_3(x)\sin x + \sin x \{f_3(x) - Q_3(x)\}|$$

$$\leqslant |f_2(x)| + |f_3(x)|\sin x + (\sin x)|f_3(x) - Q_3(x)|$$

$$< \frac{\varepsilon}{4} + |f_3(x)|\sin x + \frac{\varepsilon}{4} = \frac{\varepsilon}{2} + |f_3(x)|\sin x,$$

that is,

$$|f_2(x) - Q_2(x)| < \frac{\varepsilon}{2} + |f_3(x)|\sin x, \quad x \in [0, x_0] \cup [x_1, \pi].$$
(9.2.23)

Moreover, from (9.2.18), (9.2.16) and (9.2.17), we have

$$|f_3(x)|\sin x = \left(\frac{|f_2(x_0)|}{\sin x_0} \right)\sin x \leqslant |f_2(x_0)| < \frac{\varepsilon}{4}, \quad x \in [0, x_0];$$
(9.2.24)

$$|f_3(x)|\sin x = \left(\frac{|f_2(x_1)|}{\sin x_1} \right)\sin x \leqslant |f_2(x_1)| < \frac{\varepsilon}{4}, \quad x \in [x_1, \pi].$$
(9.2.25)

Hence we may now use (9.2.24) and (9.2.25) in (9.2.23) to obtain

$$|f_2(x) - Q_2(x)| < \frac{\varepsilon}{2} + \frac{\varepsilon}{4} = \frac{3\varepsilon}{4}, \quad x \in [0, x_0] \cup [x_1, \pi].$$
(9.2.26)

By combining (9.2.22) and (9.2.26), we deduce that

$$|f_2(x) - Q_2(x)| < \frac{3\varepsilon}{4}, \quad x \in [0, \pi].$$
(9.2.27)

Since Q_3 is an even function on $\mathbb{R}$, it follows from (9.2.21) that Q_2 is an odd function on $\mathbb{R}$, so that, by using also the second line of (9.2.7), the difference $f_2 - Q_2$ is an odd function on $\mathbb{R}$, which, together with (9.2.27), implies

$$|f_2(x) - Q_2(x)| < \frac{3\varepsilon}{4}, \quad x \in [-\pi, \pi].$$
(9.2.28)

The desired inequality in the second line of (9.2.8) is now an immediate consequence of (9.2.28) and the definition (9.2.1). ∎

According to Theorem 9.2.1, any given function $f \in C_{2\pi}$ can be approximated with arbitrary uniform "closeness" by a trigonometric polynomial Q as in (9.1.12).

9.3 The Fourier series operator

According to Theorem 4.2.2, the inner product $\langle \cdot, \cdot \rangle_2$, as defined by (9.1.7), generates the inner product space $(C_{2\pi}, || \cdot ||_2)$, with corresponding norm

$$||f||_2 := \sqrt{\int_{-\pi}^{\pi} [f(x)]^2 dx}, \quad f \in C_{2\pi}. \tag{9.3.1}$$

By recalling (9.1.5), (9.1.6) and Theorem 9.1.2, we deduce from Theorem 4.1.1 and Theorem 4.2.4 that, for any given $f \in C_{2\pi}$, and for each $n \in \mathbb{N}$, there exists precisely one best L^2 approximation Q_n^* from τ_n to f, with respect to the norm $|| \cdot ||_2$ defined in (9.3.1), that is,

$$||f - Q_n^*||_2 < ||f - Q||_2, \quad \text{for} \quad Q \in \tau_n, \quad \text{with} \quad Q \neq Q_n^*. \tag{9.3.2}$$

Hence we may define, for $n \in \mathbb{N}$, the best L^2 trigonometric polynomial approximation operator $\mathcal{T}_n^* : C_{2\pi} \to \tau_n$ by

$$\mathcal{T}_n^* f := Q_n^* \quad , \quad f \in C_{2\pi}, \tag{9.3.3}$$

so that, from (9.3.2), it holds for any $f \in C_{2\pi}$ that

$$||f - \mathcal{T}_n^* f||_2 < ||f - Q||_2, \quad \text{for} \quad Q \in \tau_n, \quad \text{with} \quad Q \neq \mathcal{T}_n^* f. \tag{9.3.4}$$

The approximation operator $\mathcal{T}_n^*$ is called the Fourier series operator. The following properties of $\mathcal{T}_n^*$ are now immediate consequences of Theorem 7.1.3.

Theorem 9.3.1. *For any $n \in \mathbb{N}$, the Fourier series operator $\mathcal{T}_n^* : C_{2\pi} \to \tau_n$, as defined by (9.3.3), satisfies the following properties:*

(a) $\mathcal{T}_n^*$ *is linear;*

(b) $\mathcal{T}_n^*$ *is exact on τ_n, that is,*

$$\mathcal{T}_n^* Q = Q, \quad Q \in \tau_n; \tag{9.3.5}$$

(c) $\mathcal{T}_n^*$ *is bounded with respect to the $|| \cdot ||_2$ norm in (9.3.1), with corresponding operator norm*

$$||\mathcal{T}_n^*||_2 = 1. \tag{9.3.6}$$

Also, according to Theorem 9.1.2(b), the set $\{1, \cos x, \ldots, \cos(nx), \sin x, \ldots, \sin(nx)\}$ is an orthogonal basis for τ_n with respect to the inner product $\langle \cdot, \cdot \rangle_2$ in (9.1.7), so that we may apply the formula (7.3.3) in Theorem 7.3.1 to obtain a formulation of the Fourier series operator $\mathcal{T}_n^*$. Indeed, since

$$\left. \begin{array}{l} \displaystyle\int_{-\pi}^{\pi} [\cos(jx)]^2 dx = \int_{-\pi}^{\pi} \frac{1 + \cos(2jx)}{2} dx = \pi, \\[3mm] \displaystyle\int_{-\pi}^{\pi} [\sin(jx)]^2 dx = \int_{-\pi}^{\pi} \frac{1 - \cos(2jx)}{2} dx = \pi, \end{array} \right\} \quad j = 1, 2, \ldots, \tag{9.3.7}$$

whereas

$$\int_{-\pi}^{\pi} 1^2 dx = 2\pi, \tag{9.3.8}$$

the following formulation follows immediately from (7.3.3) in Theorem 7.3.1, together with (9.3.7) and (9.3.8).

Theorem 9.3.2. *For any $n \in \mathbb{N}$, the Fourier series operator $\mathscr{T}_n^* : C_{2\pi} \to \tau_n$, as defined by* (9.3.3), *satisfies the formulation*

$$(\mathscr{T}_n^* f)(x) = \frac{1}{2}a_0^* + \sum_{j=1}^{n} a_j^* \cos(jx) + \sum_{j=1}^{n} b_j^* \sin(jx), \quad f \in C_{2\pi}, \tag{9.3.9}$$

where

$$\left.\begin{array}{l} a_j^* := \dfrac{1}{\pi} \displaystyle\int_{-\pi}^{\pi} f(x)\cos(jx)dx, \ j = 0,\ldots,n, \\[2mm] b_j^* := \dfrac{1}{\pi} \displaystyle\int_{-\pi}^{\pi} f(x)\sin(jx)dx, \ j = 1,\ldots,n, \end{array}\right\} \quad f \in C_{2\pi}. \tag{9.3.10}$$

Observe that the inclusion of the factor $\frac{1}{2}$ before a_0 in the general representation formula (9.1.12) has now been justified, since the first line of (9.3.10) holds for each $j = 0, 1, \ldots$.

For those cases where a quadrature rule is required for the computation of the integrals in (9.3.10) of Theorem 9.3.2, we proceed to establish the fact that specifically the trapezoidal rule is remarkably accurate when applied to numerically approximate the integral of a smooth periodic function over its full period. To this end, we first prove the Euler-Maclaurin formula, which gives a relationship between the quadrature error for the trapezoidal rule and the endpoint derivatives of the integrand.

In our proof below, we shall rely on Taylor's theorem with explicit integral form of the remainder, according to which, for any bounded interval $[a,b]$ and non-negative integer k, it holds for any $c \in [a,b]$ that

$$g(x) = \sum_{j=0}^{k} \frac{g^{(j)}(c)}{j!}(x-c)^j + \frac{1}{k!}\int_c^x (x-t)^k g^{(k+1)}(t)dt, \ x \in [a,b], \ g \in C^{k+1}[a,b]. \tag{9.3.11}$$

For an inductive proof of (9.3.11), we recall first from (3.5.3), (3.5.2) and (8.6.49) that (9.3.11) holds for $k = 0$ and $k = 1$, whereas, if (9.3.11) holds for a fixed integer $k \in \mathbb{N}$, then integration by parts, together with the inductive hypothesis (9.3.11), yields, for any $g \in C^{k+2}[a,b]$ and $x \in [a,b]$,

$$\frac{1}{(k+1)!}\int_c^x (x-t)^{k+1} g^{(k+2)}(t)dt = \left[\frac{(x-t)^{k+1} g^{(k+1)}(t)}{(k+1)!}\right]_{t=c}^{t=x} + \frac{1}{k!}\int_c^x (x-t)^k g^{(k+1)}(t)dt$$

$$= -\frac{g^{(k+1)}(c)}{(k+1)!}(x-c)^{k+1} + \left[g(x) - \sum_{j=0}^{k} \frac{g^{(j)}(c)}{j!}(x-c)^j\right],$$

which shows that (9.3.11) holds with k replaced by $k+1$.

The trapezoidal rule quadrature error has the following representation, in which the convention $\displaystyle\sum_{j=j_0}^{j_1} \alpha_j := 0$, if $j_1 < j_0$, is adopted once again.

Theorem 9.3.3 (Euler-Maclaurin formula). *For an arbitrary bounded interval $[a,b]$, and any positive integer n, the quadrature error corresponding to the trapezoidal rule $\mathscr{Q}_n^{TR}$, as given by (8.6.22), (8.6.21), (8.4.3), satisfies, for $m = 0,1,\ldots$, the formula*

$$\int_a^b f(x)dx - \mathscr{Q}_n^{TR}[f] = -\sum_{j=1}^m \frac{B_{2j}}{(2j)!}\left[f^{(2j-1)}(b) - f^{(2j-1)}(a)\right]\left(\frac{b-a}{n}\right)^{2j}$$

$$-\frac{[(b-a)/n]^{2m+2}}{(2m+2)!}\int_a^b K_{2m}\left(\frac{x-a}{(b-a)/n}\right)f^{(2m+2)}(x)dx,$$

$$f \in C^{2m+2}[a,b], \qquad (9.3.12)$$

where the real numbers $\{B_{2j} : j = 1,2,\ldots\}$ are defined recursively by

$$B_{2j} := \frac{1}{2j+1}\left[j - \frac{1}{2} - \sum_{k=1}^{j-1}\binom{2j+1}{2k}B_{2k}\right], \quad j = 1,2,\ldots, \qquad (9.3.13)$$

and where $K_{2m} \in C(\mathbb{R})$ is a periodic function on $\mathbb{R}$, with period 1, that is,

$$K_{2m}(t+1) = K_{2m}(t), \quad t \in \mathbb{R}, \qquad (9.3.14)$$

with

$$K_{2m}(t) := \frac{(-1)^{m-1}}{2}\sum_{j=0}^m (-1)^j(2m-2j)!\binom{2m+2}{2j+2}\Gamma_{2m-2j}[(1-t)^{2j+2} + t^{2j+2} - 1],$$

$$t \in [0,1], \qquad (9.3.15)$$

where the real numbers $\{\Gamma_{2j} : j = 0,1,\ldots,\}$ are defined recursively by

$$\left.\begin{array}{l}\Gamma_0 := 1; \\[2mm] \Gamma_{2j} := \displaystyle\sum_{k=1}^j \frac{(-1)^{k-1}}{(2k+1)!}\Gamma_{2j-2k}, \quad j = 1,2,\ldots.\end{array}\right\} \qquad (9.3.16)$$

Proof. For any non-negative integer m, let $\ell \in \{0,\ldots,m\}$, and suppose $f \in C^{2m+2}[a,b]$, so that $f^{(2\ell)} \in C^{2m-2\ell+2}[a,b]$. Also, for any positive integer n, define

$$x_j := a + j\left(\frac{b-a}{n}\right), \quad j = 0,\ldots,n, \qquad (9.3.17)$$

according to which

$$x_{j+1} - x_j = \frac{b-a}{n}, \quad j = 0,\ldots,n-1. \qquad (9.3.18)$$

Let $j \in \{0,\dots,n-1\}$ be fixed. By applying Taylor's theorem (9.3.11) with $k = 2m - 2\ell + 1$, $g = f^{(2\ell)}$, and, respectively, $c = x_j$ and $c = x_{j+1}$, and using (9.3.18), we obtain

$$\int_{x_j}^{x_{j+1}} f^{(2\ell)}(x)dx = \frac{1}{2}\int_{x_j}^{x_{j+1}} [f^{(2\ell)}(x) + f^{(2\ell)}(x)]dx$$

$$= \frac{1}{2}\left[\sum_{k=0}^{2m-2\ell+1} \frac{f^{(2\ell+k)}(x_j)}{k!}\int_{x_j}^{x_{j+1}}(x-x_j)^k dx + \sum_{k=0}^{2m-2\ell+1} \frac{f^{(2\ell+k)}(x_{j+1})}{k!}\int_{x_j}^{x_{j+1}}(x-x_{j+1})^k dx\right]$$

$$+ \frac{1}{2(2m-2\ell+1)!}\left[\int_{x_j}^{x_{j+1}}\left\{\int_{x_j}^x (x-t)^{2m-2\ell+1} f^{(2m+2)}(t)dt\right\}dx\right.$$

$$\left. + \int_{x_j}^{x_{j+1}}\left\{\int_x^{x_{j+1}} (t-x)^{2m-2\ell+1} f^{(2m+2)}(t)dt\right\}dx\right]$$

$$= \frac{1}{2}\sum_{k=0}^{2m-2\ell+1} \frac{[(b-a)/n]^{k+1}}{(k+1)!}\left[f^{(2\ell+k)}(x_j) + (-1)^k f^{(2\ell+k)}(x_{j+1})\right]$$

$$+ \frac{1}{2(2m-2\ell+1)!}\left[\int_{x_j}^{x_{j+1}}\left\{\int_t^{x_{j+1}}(x-t)^{2m-2\ell+1}dx\right\}f^{(2m+2)}(t)dt\right.$$

$$\left. + \int_{x_j}^{x_{j+1}}\left\{\int_{x_j}^t (t-x)^{2m-2\ell+1}dx\right\}f^{(2m+2)}(t)dt\right]$$

$$= \frac{1}{2}\left\{\sum_{k=0}^{m-\ell} \frac{[(b-a)/n]^{2k+1}}{(2k+1)!}\left[f^{(2\ell+2k)}(x_j) + f^{(2\ell+2k)}(x_{j+1})\right]\right.$$

$$\left. + \sum_{k=1}^{m-\ell+1} \frac{[(b-a)/n]^{2k}}{(2k)!}\left[f^{(2\ell+2k-1)}(x_j) - f^{(2\ell+2k-1)}(x_{j+1})\right]\right\}$$

$$+ \frac{1}{2(2m-2\ell+2)!}\int_{x_j}^{x_{j+1}}\left[(x_{j+1}-t)^{2m-2\ell+2} + (t-x_j)^{2m-2\ell+2}\right]f^{(2m+2)}(t)dt$$

$$= \frac{b-a}{2n}\left[f^{(2\ell)}(x_j) + f^{(2\ell)}(x_{j+1})\right] + \sum_{k=1}^{m-\ell}\alpha_{2k}\left\{\frac{b-a}{2n}\left[f^{(2\ell+2k)}(x_j) + f^{(2\ell+2k)}(x_{j+1})\right]\right.$$

$$\left. - \left(k+\frac{1}{2}\right)\int_{x_j}^{x_{j+1}} f^{(2\ell+2k)}(x)dx\right\} + \frac{1}{2(2m-2\ell+2)!}$$

$$\times \int_{x_j}^{x_{j+1}}\left[(x_{j+1}-x)^{2m-2\ell+2} + (x-x_j)^{2m-2\ell+2} - \left(\frac{b-a}{n}\right)^{2m-2\ell+2}\right]f^{(2m+2)}(x)dx,$$

$$(9.3.19)$$

where

$$\alpha_{2k} := \frac{[(b-a)/n]^{2k}}{(2k+1)!}, \quad k = 0, \ldots, m. \tag{9.3.20}$$

By using (9.3.17), (9.3.18), (9.3.19), (8.6.17), (8.6.7), (8.6.5) and (8.4.56), and introducing the notation

$$e_{2\ell} := \int_a^b f^{(2\ell)}(x)dx - \mathcal{Q}_n^{TR}[f^{(2\ell)}], \quad \ell = 0, \ldots, m, \tag{9.3.21}$$

we deduce that

$$e_{2\ell} + \sum_{k=1}^{m-\ell} \alpha_{2k} e_{2\ell+2k} = g_{2\ell} + h_{2\ell}, \quad \ell = 0, \ldots, m, \tag{9.3.22}$$

where

$$g_{2\ell} := \sum_{k=1}^{m-\ell} \alpha_{2k}(\tfrac{1}{2} - k) \left[f^{(2\ell+2k-1)}(b) - f^{(2\ell+2k-1)}(a) \right], \quad \ell = 0, \ldots, m; \tag{9.3.23}$$

$$h_{2\ell} := \frac{1}{2(2m - 2\ell + 2)!} \sum_{j=0}^{n-1} \int_{x_j}^{x_{j+1}} \left[(x_{j+1} - x)^{2m-2\ell+2} + (x - x_j)^{2m-2\ell+2} \right.$$

$$\left. - \left(\frac{b-a}{n} \right)^{2m-2\ell+2} \right] f^{(2m+2)}(x)dx,$$

$$\ell = 0, \ldots, m. \tag{9.3.24}$$

Now observe that the linear system (9.3.22) is equivalent to the matrix-vector equation

$$Ae = g + h, \tag{9.3.25}$$

where A is the $(m+1) \times (m+1)$ upper triangular matrix

$$A := \begin{bmatrix} 1 & \alpha_2 & \alpha_4 & \cdots & \alpha_{2m} \\ 0 & 1 & \alpha_2 & \cdots & \alpha_{2m-2} \\ \vdots & \vdots & \vdots & & \vdots \\ 0 & 0 & 0 & \cdots & 1 \end{bmatrix}, \tag{9.3.26}$$

and with $e, g, h \in \mathbb{R}^{m+1}$ denoting the column vectors

$$e := [e_0, e_2, \ldots, e_{2m}]^T; \quad g := [g_0, g_2, \ldots, g_{2m}]^T; \quad h := [h_0, h_2, \ldots, h_{2m}]^T. \tag{9.3.27}$$

Since (9.3.26) implies $\det(A) = 1 \neq 0$, it follows that A is an invertible matrix. By using the fact that the inverse matrix A^{-1} satisfies $A^{-1}A = I$, the identity matrix, we may now use (9.3.26) to deduce that the first row $[\beta_0, \beta_2, \ldots, \beta_{2m}]$ of A^{-1} satisfies the conditions

$$\beta_{2k} + \sum_{\ell=0}^{k-1} \alpha_{2k-2\ell} \beta_{2\ell} = \delta_k, \quad k = 0, \ldots, m, \tag{9.3.28}$$

with $\{\delta_k : k = 0,\ldots,m\}$ denoting the Kronecker delta sequence as in (1.2.2), and thus

$$
\left.\begin{aligned}
\beta_0 &= 1;\\
\beta_{2k} &= -\sum_{\ell=0}^{k-1}\alpha_{2k-2\ell}\beta_{2\ell}, \quad k = 1,\ldots,m.
\end{aligned}\right\} \tag{9.3.29}
$$

We claim that the sequence $\{\beta_0,\ldots,\beta_{2m}\}$ given by (9.3.29) satisfies the formulation

$$
\beta_{2k} = (-1)^k \Gamma_{2k}\left(\frac{b-a}{n}\right)^{2k}, \quad k = 0,\ldots,m, \tag{9.3.30}
$$

where the real numbers $\{\Gamma_0,\Gamma_2,\ldots,\Gamma_{2m}\}$ are obtained recursively from (9.3.16). To prove (9.3.30), we first note from the first lines of (9.3.29) and (9.3.16) that (9.3.30) holds for $k = 0$, whereas (9.3.29), (9.3.20) and (9.3.16) give

$$
\beta_2 = -\alpha_2\beta_0 = -\frac{1}{6}\left(\frac{b-a}{n}\right)^2;
$$

$$
\Gamma_2 = \frac{1}{6}\Gamma_0 = \frac{1}{6},
$$

from which it follows that (9.3.30) is also satisfied for $k = 1$. Suppose next that (9.3.30) holds for a fixed integer $k \in \{0,\ldots,m-1\}$. But then, from the second line of (9.3.29), together with (9.3.20), and the second line of (9.3.16), we obtain

$$
\begin{aligned}
\beta_{2k+2} &= -\sum_{\ell=0}^{k}\left\{\frac{[(b-a)/n]^{2k-2\ell+2}}{(2k-2\ell+3)!}\right\}\left\{(-1)^\ell\Gamma_{2\ell}\left(\frac{b-a}{n}\right)^{2\ell}\right\}\\
&= -\left(\frac{b-a}{n}\right)^{2k+2}\sum_{\ell=0}^{k}\left[\frac{(-1)^\ell}{(2k-2\ell+3)!}\Gamma_{2\ell}\right]\\
&= (-1)^{k+1}\left(\frac{b-a}{n}\right)^{2k+2}\sum_{\ell=1}^{k+1}\left[\frac{(-1)^{\ell-1}}{(2\ell+1)!}\Gamma_{2k+2-2\ell}\right] = (-1)^{k+1}\Gamma_{2k+2}\left(\frac{b-a}{n}\right)^{2k+2},
\end{aligned}
$$

according to which (9.3.30) holds with k replaced by $k+1$, and thereby completing our inductive proof of (9.3.30).

Since $[\beta_0,\beta_2,\ldots,\beta_{2m}]$ is the first row of the inverse matrix A^{-1}, and since (9.3.25) implies

$$
\mathbf{e} = A^{-1}(\mathbf{g}+\mathbf{h}), \tag{9.3.31}
$$

we may now apply (9.3.21), (9.3.27) and (9.3.31) to obtain

$$
\int_a^b f(x)dx - \mathscr{Q}_n^{TR}[f] = e_0 = \sum_{k=0}^{m}\beta_{2k}g_{2k} + \sum_{k=0}^{m}\beta_{2k}h_{2k}. \tag{9.3.32}
$$

After noting from (9.3.23) that $g_{2m} = 0$, we next apply (9.3.23), (9.3.30) and (9.3.20) to deduce that

$$
\sum_{k=0}^{m}\beta_{2k}g_{2k} = \sum_{k=0}^{m-1}\beta_{2k}\sum_{j=k}^{m-1}\alpha_{2j-2k+2}(-\tfrac{1}{2}-j+k)\left[f^{(2j+1)}(b) - f^{(2j+1)}(a)\right]
$$

$$= \sum_{j=0}^{m-1} \left[\sum_{k=0}^{j} \alpha_{2j-2k+2}(-\tfrac{1}{2}-j+k)\beta_{2k} \right] \left[f^{(2j+1)}(b) - f^{(2j+1)}(a) \right]$$

$$= \sum_{j=0}^{m-1} \left[\sum_{k=0}^{j} \frac{[(b-a)/n]^{2j-2k+2}}{(2j-2k+3)!}(-\tfrac{1}{2}-j+k)(-1)^k \Gamma_{2k} \left(\frac{b-a}{n} \right)^{2k} \right]$$

$$\times [f^{(2j+1)}(b) - f^{(2j+1)}(a)]$$

$$= \sum_{j=1}^{m} \left[\sum_{k=0}^{j-1} (-1)^k \frac{(\tfrac{1}{2}-j+k)}{(2j-2k+1)!} \Gamma_{2k} \right] \left[f^{(2j-1)}(b) - f^{(2j-1)}(a) \right] \left(\frac{b-a}{n} \right)^{2j}$$

$$= \sum_{j=1}^{m} (-1)^j \left[\sum_{k=1}^{j} (-1)^{k-1} \frac{(k-\tfrac{1}{2})}{(2k+1)!} \Gamma_{2j-2k} \right] \left[f^{(2j-1)}(b) - f^{(2j-1)}(a) \right] \left(\frac{b-a}{n} \right)^{2j}$$

$$= -\sum_{j=1}^{m} \frac{B_{2j}}{(2j)!} \left[f^{(2j-1)}(b) - f^{(2j-1)}(a) \right] \left(\frac{b-a}{n} \right)^{2j}, \tag{9.3.33}$$

where

$$B_{2j} := (-1)^{j-1}(2j)! \sum_{k=1}^{j} \frac{(-1)^{k-1}(k-\tfrac{1}{2})}{(2k+1)!} \Gamma_{2j-2k}, \quad j=1,\dots,m. \tag{9.3.34}$$

Observe from (9.3.34) and (9.3.16) that, for $j=1,\dots,m$,

$$B_{2j} = \frac{j-\tfrac{1}{2}}{2j+1} + (-1)^{j-1}(2j)! \sum_{k=1}^{j-1} \frac{(-1)^{k-1}(k-\tfrac{1}{2})}{(2k+1)!} \Gamma_{2j-2k}$$

$$= \frac{j-\tfrac{1}{2}}{2j+1} + (-1)^{j-1}(2j)! \sum_{k=1}^{j-1} \frac{(-1)^{k-1}(k-\tfrac{1}{2})}{(2k+1)!} \sum_{\ell=1}^{j-k} \frac{(-1)^{\ell-1}}{(2\ell+1)!} \Gamma_{2j-2k-2\ell}$$

$$= \frac{j-\tfrac{1}{2}}{2j+1} + (-1)^{j-1}(2j)! \sum_{\ell=1}^{j-1} \frac{(-1)^{\ell-1}}{(2\ell+1)!} \sum_{k=1}^{j-\ell} \frac{(-1)^{k-1}(k-\tfrac{1}{2})}{(2k+1)!} \Gamma_{2j-2\ell-2k}$$

$$= \frac{j-\tfrac{1}{2}}{2j+1} + (-1)^{j-1}(2j)! \sum_{\ell=1}^{j-1} \frac{(-1)^{\ell-1}}{(2\ell+1)!} \left[\frac{(-1)^{j-\ell-1}}{(2j-2\ell)!} B_{2j-2\ell} \right]$$

$$= \frac{j-\tfrac{1}{2}}{2j+1} - \frac{1}{2j+1} \sum_{\ell=1}^{j-1} \binom{2j+1}{2\ell+1} B_{2j-2\ell} = \frac{1}{2j+1} \left[j - \frac{1}{2} - \sum_{\ell=1}^{j-1} \binom{2j+1}{2j-2\ell+1} B_{2\ell} \right],$$

from which it then follows that the sequence $\{B_0, B_2, \dots, B_{2m}\}$ satisfies the recursion formulation (9.3.13). Finally, we use (9.3.24), (9.3.30), (9.3.17) and (9.3.18) to obtain

$$\sum_{k=0}^{m} B_{2k} h_{2k} = \sum_{j=0}^{n-1} \int_{x_j}^{x_{j+1}} \sum_{k=0}^{m} \frac{\beta_{2k}}{2(2m-2k+2)!} \left[(x_{j+1}-x)^{2m-2k+2} + (x-x_j)^{2m-2k+2} \right.$$

$$-\left(\frac{b-a}{n}\right)^{2m-2k+2}\Bigg]f^{(2m+2)}(x)dx$$

$$=-\frac{[(b-a)/n]^{2m+2}}{(2m+2)!}\sum_{j=0}^{n-1}\int_{x_j}^{x_{j+1}}P_{2m}\left(\frac{x-a}{(b-a)/n}-j\right)f^{(2m+2)}(x)dx,\qquad(9.3.35)$$

where

$$P_{2m}(t):=\frac{(-1)^{m-1}}{2}\sum_{k=0}^{m}(-1)^{k}(2m-2k)!\binom{2m+2}{2k+2}\Gamma_{2m-2k}[(1-t)^{2k+2}+t^{2k+2}-1].$$
$$(9.3.36)$$

Let $K_{2m}\in C(\mathbb{R})$ be defined by (9.3.15) and (9.3.14), according to which K_{2m} is a periodic function on $\mathbb{R}$, with period 1, and thus also, by noting that the right hand sides of (9.3.15) and (9.3.36) are equivalent, we have

$$P_{2m}(t-j)=K_{2m}(t),\quad t\in[j,j+1],\quad j=0,\dots,n-1,$$

so that, by using also (9.3.17) and (9.3.18),

$$P_{2m}\left(\frac{x-a}{(b-a)/n}-j\right)=K_{2m}\left(\frac{x-a}{(b-a)/n}\right),\quad x\in[x_j,x_{j+1}],\quad j=0,\dots,n-1.\quad(9.3.37)$$

Hence we may substitute (9.3.37) into (9.3.35) to deduce that

$$\sum_{k=0}^{m}\beta_{2k}h_{2k}=-\frac{[(b-a)/n]^{2m+2}}{(2m+2)!}\int_{a}^{b}K_{2m}\left(\frac{x-a}{(b-a)/n}\right)f^{(2m+2)}(x)dx.\qquad(9.3.38)$$

The Euler-Maclaurin formula (9.3.12) is now an immediate consequence of (9.3.32), (9.3.33) and (9.3.38). ∎

The well-known Bernoulli numbers $\{B_j:j=0,1,\dots\}$ are defined recursively by

$$B_0:=1;\quad B_j:=-\frac{1}{j+1}\sum_{k=0}^{j-1}\binom{j+1}{k}B_k,\quad j=1,2,\dots,\qquad(9.3.39)$$

by means of which we calculate that

$$B_1=-\frac{1}{2}.\qquad(9.3.40)$$

Also, it is known (see Exercise 9.5) that

$$B_{2j+1}=0,\quad j=1,2,\dots.\qquad(9.3.41)$$

By using (9.3.39), (9.3.40) and (9.3.41), we obtain, for $j=1,2,\dots$,

$$B_{2j}=-\frac{1}{2j+1}\sum_{k=0}^{2j-1}\binom{2j+1}{k}B_k$$

$$=-\frac{1}{2j+1}\left[1-(j+\tfrac{1}{2})+\sum_{k=2}^{2j-1}\binom{2j+1}{k}B_k\right]=\frac{1}{2j+1}\left[j-\frac{1}{2}-\sum_{k=1}^{j-1}\binom{2j+1}{2k}B_{2k}\right],$$

which agrees with (9.3.13). Hence the real numbers $\{B_0, B_2, \ldots, B_{2m}\}$ in Theorem 9.3.3 are precisely the Bernoulli numbers with even indices.

In order to be able to apply the mean value theorem (8.5.16) for integrals to the integral in the right hand side of the Euler-Maclaurin formula (9.3.12), we proceed to first prove the following properties of the polynomial P_{2m} in (9.3.36), which agrees with K_{2m} on $[0, 1]$, as in (9.3.15).

Theorem 9.3.4. *In Theorem 9.3.3, the polynomials* $\{P_{2m} : m = 0, 1, \ldots\}$, *as defined by* (9.3.36), *and the real numbers* $\{B_{2m} : m = 1, 2, \ldots\}$, *as given by* (9.3.13), *satisfy the following:*

(a)

$$(-1)^m P_{2m}(t) > 0, \quad t \in (0, 1), \quad m = 0, 1, \ldots; \tag{9.3.42}$$

(b)

$$\int_0^1 P_{2m}(t)dt = B_{2m+2}, \quad m = 0, 1, \ldots; \tag{9.3.43}$$

(c)

$$(-1)^m B_{2m} > 0, \quad m = 1, 2, \ldots. \tag{9.3.44}$$

Proof. (a) Since (9.3.16) yields

$$\Gamma_0 = 1; \quad \Gamma_2 = \frac{1}{6}, \tag{9.3.45}$$

we calculate from (9.3.36) that

$$P_0(t) = t(1-t); \quad P_2(t) = -t^2(1-t)^2, \tag{9.3.46}$$

and thus also

$$P_0'(t) = 1 - 2t; \quad P_2'(t) = 2t(1-t)(2t-1). \tag{9.3.47}$$

Observe from (9.3.46) that (9.3.42) is satisfied for $m = 0$ and $m = 1$.

Let $m \in \mathbb{N}$, and note from (9.3.36) that

$$P_{2m}'(t) = \frac{(-1)^{m-1}}{2} \sum_{j=0}^{m} (-1)^j (2m-2j)! \binom{2m+2}{2j+2} \Gamma_{2m-2j}(2j+2)[-(1-t)^{2j+1} + t^{2j+1}], \tag{9.3.48}$$

according to which

$$P_{2m}'(0) = P_{2m}'(\tfrac{1}{2}) = P_{2m}'(1) = 0. \tag{9.3.49}$$

We proceed to prove inductively that

$$(-1)^m P'_{2m}(t) \begin{cases} > 0, t \in (0, \tfrac{1}{2}); \\ < 0, t \in (\tfrac{1}{2}, 1), \end{cases} \tag{9.3.50}$$

which, together with the fact that (9.3.36) yields

$$P_{2m}(0) = P_{2m}(1) = 0, \tag{9.3.51}$$

will then imply the desired result (9.3.42).

Since, according to the second equation in (9.3.47), the inequalities (9.3.50) are satisfied for $m = 1$, suppose next that (9.3.50) holds for a fixed integer $m \in \mathbb{N}$. From the fact that (9.3.36) gives

$$P_{2m+2}(t) = \frac{(-1)^m}{2} \sum_{j=0}^{m+1} (-1)^j (2m+2-2j)! \binom{2m+4}{2j+2} \Gamma_{2m+2-2j}[(1-t)^{2j+2} + t^{2j+2} - 1],$$

we deduce that

$$P'''_{2m+2}(t) = \frac{(-1)^m}{2} \sum_{j=1}^{m+1} (-1)^j \frac{(2m+4)!}{(2j+2)!} \Gamma_{2m+2-2j}(2j+2)(2j+1)(2j)$$

$$\times [-(1-t)^{2j-1} + t^{2j-1}]$$

$$= \frac{(-1)^m}{2} \sum_{j=1}^{m+1} (-1)^j \frac{(2m+4)!}{(2j)!} \Gamma_{2m+2-2j}(2j)[-(1-t)^{2j-1} + t^{2j-1}]$$

$$= \frac{(-1)^{m-1}}{2}(2m+4)(2m+3) \sum_{j=0}^{m} (-1)^j (2m-2j)! \binom{2m+2}{2j+2}$$

$$\times \Gamma_{2m-2j}(2j+2)[-(1-t)^{2j+1} + t^{2j+1}]$$

$$= (2m+4)(2m+3)P'_{2m}(t),$$

by virtue of (9.3.48), and thus, from the inductive hypothesis (9.3.50),

$$(-1)^m P'''_{2m+2}(t) \begin{cases} > 0, t \in (0, \tfrac{1}{2}); \\ < 0, t \in (\tfrac{1}{2}, 1). \end{cases} \tag{9.3.52}$$

Since also

$$P'_{2m+2}(0) = P'_{2m+2}(\tfrac{1}{2}) = P'_{2m+2}(1) = 0,$$

from the fact that (9.3.49) is satisfied for each $m \in \mathbb{N}$, whereas $P'''_{2m+2} = (P'_{2m+2})''$ satisfies (9.3.52), it follows that (9.3.50) is satisfied with m replaced by $m+1$, and thereby completing our inductive proof of (9.3.50).

(b) For any non-negative integer m, we deduce from (9.3.36) and (9.3.34) that

$$\int_0^1 P_{2m}(t)dt = \frac{(-1)^{m-1}(2m+2)!}{2} \sum_{j=0}^{m} \frac{(-1)^j \Gamma_{2m-2j}}{(2j+2)!} \int_0^1 [(1-t)^{2j+2} + t^{2j+2} - 1]dt$$

$$= \frac{(-1)^m(2m+2)!}{2} \sum_{j=0}^{m} \frac{(-1)^j(2j+1)}{(2j+3)!} \Gamma_{2m-2j}$$

$$= (-1)^m(2m+2)! \sum_{j=1}^{m+1} \frac{(-1)^{j-1}(j-\frac{1}{2})}{(2j+1)!} \Gamma_{2m+2-2j} = B_{2m+2},$$

which proves (9.3.43).

(c) The inequality (9.3.44) is an immediate consequence of (9.3.43) and (9.3.42). ∎

By using Theorem 9.3.4, we can now prove the following consequence of Theorem 9.3.3.

Theorem 9.3.5. *In Theorem 9.3.3, the Euler-Maclaurin formula (9.3.12) has the alternative formulation*

$$\int_a^b f(x)dx - \mathscr{Q}_n^{TR}[f] = -\sum_{j=1}^{m} \frac{B_{2j}}{(2j)!}[f^{(2j-1)}(b) - f^{(2j-1)}(a)]\left(\frac{b-a}{n}\right)^{2j}$$

$$- \frac{(b-a)B_{2m+2}}{(2m+2)!} f^{(2m+2)}(\xi)\left(\frac{b-a}{n}\right)^{2m+2}, \quad f \in C^{2m+2}[a,b],$$

$$(9.3.53)$$

for some point $\xi \in [a,b]$.

Proof. According to (9.3.15), (9.3.14) and (9.3.36), the function K_{2m} satisfies the property

$$K_{2m}(t-j) = K_{2m}(t) = P_{2m}(t), \quad t \in [0,1], \quad j = 0,\ldots,n-1. \tag{9.3.54}$$

Also, from (9.3.14) and (9.3.15), together with (9.3.42) in Theorem 9.3.4(a), the function K_{2m} does not change sign on the interval $[0,n]$, so that we may apply the mean value theorem (8.5.16) for integrals, as well as (9.3.54) and (9.3.43), to deduce that there exists a point $\xi \in [a,b]$ such that

$$\int_a^b K_{2m}\left(\frac{x-a}{(b-a)/n}\right) f^{(2m+2)}(x)dx = f^{(2m+2)}(\xi)\int_a^b K_{2m}\left(\frac{x-a}{(b-a)/n}\right)dx$$

$$= f^{(2m+2)}(\xi)\frac{b-a}{n}\int_0^n K_{2m}(t)dt$$

$$= f^{(2m+2)}(\xi)\frac{b-a}{n}\sum_{j=0}^{n-1}\left[\int_j^{j+1} K_{2m}(t)dt\right]$$

$$= f^{(2m+2)}(\xi)\frac{b-a}{n}\sum_{j=0}^{n-1}\left[\int_0^1 K_{2m}(t-j)dt\right]$$

$$= f^{(2m+2)}(\xi)\frac{b-a}{n}\sum_{j=0}^{n-1}\left[\int_0^1 P_{2m}(t)dt\right]$$

$$= (b-a)B_{2m+2}f^{(2m+2)}(\xi). \qquad (9.3.55)$$

The desired result (9.3.53) then follows immediately by substituting (9.3.55) into the Euler-Maclaurin formula (9.3.12). ∎

Calculating by means of (9.3.13), we obtain the values

$$B_2 = \frac{1}{6}; \quad B_4 = -\frac{1}{30}; \quad B_6 = \frac{1}{42}; \quad B_8 = -\frac{1}{30}; \quad B_{10} = \frac{5}{66}. \qquad (9.3.56)$$

By setting $m = 0$ in (9.3.53), and using the value $B_2 = \frac{1}{6}$, we obtain

$$\int_a^b f(x) - \mathscr{Q}_n^{TR}[f] = -\frac{b-a}{12}\left(\frac{b-a}{n}\right)^2 f''(\xi), \quad f \in C^2[a,b], \qquad (9.3.57)$$

which, since $\Lambda_3 = \frac{1}{24}$ from (8.4.54), is consistent with the case $v = 1$ in the second line of (8.6.30), together with (8.6.26) and (8.6.17).

As an immediate consequence of Theorem 9.3.5, we have the following result.

Theorem 9.3.6. *For a bounded interval $[a,b]$ and any positive integer m, suppose $f \in C^{2m+2}[a,b]$, with, moreover,*

$$f^{(2j-1)}(a) = f^{(2j-1)}(b), \quad j = 1,\ldots,m. \qquad (9.3.58)$$

Then, for any positive integer n, the quadrature error corresponding to the trapezoidal rule $\mathscr{Q}_n^{TR}$, as given by (8.6.22), (8.6.21), (8.4.3), satisfies

$$\left|\int_a^b f(x)dx - \mathscr{Q}_n^{TR}[f]\right| \leqslant \frac{(b-a)|B_{2m+2}|}{(2m+2)!}\left(\frac{b-a}{n}\right)^{2m+2}||f^{(2m+2)}||_\infty, \qquad (9.3.59)$$

with the real number B_{2m+2} given as in (9.3.13).

Returning to the setting of periodic continuous functions in $C_{2\pi}$, as given in (9.1.4), we now define, for any non-negative integer k, the function space

$$C_{2\pi}^k := C_{2\pi} \cap C^k(\mathbb{R}), \qquad (9.3.60)$$

with $C^k(\mathbb{R})$ denoting the linear space of functions $f : \mathbb{R} \to \mathbb{R}$ such that $f^{(\ell)} \in C(\mathbb{R})$, $\ell = 0,\ldots,k$, and thus $C^0(\mathbb{R}) = C(\mathbb{R})$, according to which then also $C_{2\pi}^0 = C_{2\pi}$. Observe from (9.3.60) that, if $f \in C_{2\pi}^k$ for a positive integer k, then, for any $x \in \mathbb{R}$,

$$f'(x+2\pi) = \lim_{h\to 0}\frac{f(x+2\pi+h) - f(x+2\pi)}{h} = \lim_{h\to 0}\frac{f(x+h) - f(x)}{h} = f'(x),$$

and thus, from repeated use of this argument, we deduce that, for any $k \in \mathbb{N}$,

$$f^{(j)}(x+2\pi) = f^{(j)}(x), \quad x \in \mathbb{R}, \quad j = 1,\ldots,k, \quad f \in C_{2\pi}^k, \tag{9.3.61}$$

or equivalently,

$$f \in C_{2\pi}^k \Rightarrow f^{(j)} \in C_{2\pi}, \quad j = 0,\ldots,k. \tag{9.3.62}$$

In particular, by setting $x = 0$ in (9.3.61), we deduce that, for any $k \in \mathbb{N}$,

$$f^{(j)}(2\pi) = f^{(j)}(0), \quad j = 1,\ldots,k, \quad f \in C_{2\pi}^k. \tag{9.3.63}$$

Next, for $f \in C_{2\pi}$, we use (9.1.3) to deduce that, for any $\alpha \in \mathbb{R}$,

$$\int_{-\pi+\alpha}^{\pi+\alpha} f(x)dx = \int_{-\pi+\alpha}^{0} f(x)dx + \int_{0}^{\pi+\alpha} f(x)dx$$

$$= \int_{-\pi+\alpha}^{0} f(x+2\pi)dx + \int_{0}^{\pi+\alpha} f(x)dx$$

$$= \int_{\pi+\alpha}^{2\pi} f(x)dx + \int_{0}^{\pi+\alpha} f(x)dx = \int_{0}^{2\pi} f(x)dx,$$

that is

$$\int_{-\pi+\alpha}^{\pi+\alpha} f(x)dx = \int_{0}^{2\pi} f(x)dx, \quad \alpha \in \mathbb{R}, \quad f \in C_{2\pi}. \tag{9.3.64}$$

In particular, by setting $\alpha = 0$ in (9.3.64), we obtain

$$\int_{-\pi}^{\pi} f(x)dx = \int_{0}^{2\pi} f(x)dx, \quad f \in C_{2\pi}. \tag{9.3.65}$$

It follows from (9.3.65) that the formulas (9.3.10) in Theorem 9.3.2 can equivalently be formulated as

$$\left.\begin{aligned} a_j^* &:= \frac{1}{\pi}\int_{0}^{2\pi} f(x)\cos(jx)dx, \ j = 0,\ldots,n, \\ b_j^* &:= \frac{1}{\pi}\int_{0}^{2\pi} f(x)\sin(jx)dx, \ j = 1,\ldots,n, \end{aligned}\right\} f \in C_{2\pi}. \tag{9.3.66}$$

For any positive integer N, we now apply the trapezoidal rule $\mathscr{Q}_N^{TR}$, as obtained from (8.6.22), (8.6.21), (8.4.3), to the integrals in the right hand side of (9.3.66) to obtain the numerical approximations

$$\left.\begin{aligned} a_j^* &\approx \tilde{a}_{N,j} := \frac{2}{N}\sum_{k=0}^{N-1} f\left(\frac{2k}{N}\pi\right)\cos\left(\frac{2kj}{N}\pi\right), \ j = 0,\ldots,n, \\ b_j^* &\approx \tilde{b}_{N,j} := \frac{2}{N}\sum_{k=0}^{N-1} f\left(\frac{2k}{N}\pi\right)\sin\left(\frac{2kj}{N}\pi\right), \ j = 1,\ldots,n, \end{aligned}\right\} f \in C_{2\pi}, \tag{9.3.67}$$

after having used also the fact that (9.1.3) implies $f(0) = f(2\pi)$. The resulting approximation operator $\widetilde{\mathscr{T}}_{n,N} : C_{2\pi} \to \tau_n$, as defined by

$$(\widetilde{\mathscr{T}}_{n,N}f)(x) := \frac{1}{2}\tilde{a}_{N,0} + \sum_{j=1}^{n} \tilde{a}_{N,j}\cos(jx) + \sum_{j=1}^{n} \tilde{b}_{N,j}\sin(jx), \ f \in C_{2\pi}, \tag{9.3.68}$$

is called the discrete Fourier series operator.

The remarkable accuracy of the trapezoidal rule numerical approximation in (9.3.67), for sufficiently differentiable functions $f \in C_{2\pi}$, can now be derived by means of Theorem 9.3.6, as follows.

Theorem 9.3.7. *For any non-negative integer* m, *let* $f \in C_{2\pi}^{2m+2}$, *and, for* $n \in \mathbb{N}$, *let the Fourier series approximation* $\mathscr{T}_n^* f$ *be defined by* (9.3.9), (9.3.10). *Then, for any* $N \in \mathbb{N}$, *the trapezoidal rule quadrature error in the numerical approximation* (9.3.67) *satisfies*

$$\left.\begin{aligned}\left|a_j^* - \widetilde{a}_{N,j}\right| &\leqslant \frac{2\pi|B_{2m+2}|}{(2m+2)!}\left(\frac{2\pi}{N}\right)^{2m+2}\|u_j^{(2m+2)}\|_\infty, \ j=0,\ldots,n; \\ \left|b_j^* - \widetilde{b}_{N,j}\right| &\leqslant \frac{2\pi|B_{2m+2}|}{(2m+2)!}\left(\frac{2\pi}{N}\right)^{2m+2}\|v_j^{(2m+2)}\|_\infty, \ j=1,\ldots,n,\end{aligned}\right\} \tag{9.3.69}$$

where

$$\left.\begin{aligned}u_j(x) &:= \frac{1}{\pi}f(x)\cos(jx), \ j=0,\ldots,n; \\ v_j(x) &:= \frac{1}{\pi}f(x)\sin(jx), \ j=1,\ldots,n,\end{aligned}\right\} \tag{9.3.70}$$

and with B_{2m+2} *denoting the real number given as in* (9.3.13).

Proof. Since $f \in C_{2\pi}^{2m+2}$, it follows from (9.3.70) that $u_j \in C_{2\pi}^{2m+2}, j=0,\ldots,n$, and $v_j \in C_{2\pi}^{2m+2}, j=1,\ldots,n$, according to which, from (9.3.63), we have

$$\left.\begin{aligned}u_j^{(2k-1)}(2\pi) &= u_j^{(2k-1)}(0), \ k=1,\ldots,m; \ j=0,\ldots,n; \\ v_j^{(2k-1)}(2\pi) &= v_j^{(2k-1)}(0), \ k=1,\ldots,m; \ j=1,\ldots,n.\end{aligned}\right\} \tag{9.3.71}$$

Now recall from (9.3.67) that

$$\left.\begin{aligned}\widetilde{a}_{N,j} &= \mathscr{Q}_N^{TR}[u_j], \ j=0,\ldots,n; \\ \widetilde{b}_{N,j} &= \mathscr{Q}_N^{TR}[v_j], \ j=1,\ldots,n,\end{aligned}\right\} \tag{9.3.72}$$

with $\mathscr{Q}_N^{TR}$ denoting the trapezoidal rule as given by (8.6.22), (8.6.21), (8.4.3).

It follows from (9.3.71) and (9.3.72) that we may apply (9.3.59) in Theorem 9.3.6 to deduce that the desired upper bounds in (9.3.69) are satisfied. ∎

Observe that if, for some positive integer M, we choose $N=2M$ and $n=M$ in (9.3.67), then

$$\left.\begin{aligned}\widetilde{a}_{2M,j} &= A_j + B_j, \\ \widetilde{a}_{2M,M-j} &= A_j - B_j,\end{aligned}\right\} \ j=0,\ldots,M, \tag{9.3.73}$$

where

$$
\left. \begin{aligned}
A_j &:= \frac{1}{M} \sum_{k=0}^{M-1} f\left(\frac{2k}{M}\pi\right) \cos\left(\frac{2kj}{M}\pi\right), \\
B_j &:= \frac{1}{M} \sum_{k=0}^{M-1} f\left(\frac{2k+1}{M}\pi\right) \cos\left(\frac{(2k+1)j}{M}\pi\right),
\end{aligned} \right\} \quad j = 0, \ldots, M,
\tag{9.3.74}
$$

and similarly,

$$
\left. \begin{aligned}
\widetilde{b}_{2M,j} &= C_j + D_j, \\
\widetilde{b}_{2M,M-j} &= -C_j + D_j,
\end{aligned} \right\} \quad j = 1, \ldots, M,
\tag{9.3.75}
$$

where

$$
\left. \begin{aligned}
C_j &:= \frac{1}{M} \sum_{k=0}^{M-1} f\left(\frac{2k}{M}\pi\right) \sin\left(\frac{2kj}{M}\pi\right), \\
D_j &:= \frac{1}{M} \sum_{k=0}^{M-1} f\left(\frac{2k+1}{M}\pi\right) \sin\left(\frac{(2k+1)j}{M}\pi\right),
\end{aligned} \right\} \quad j = 1, \ldots, M.
\tag{9.3.76}
$$

Hence, if N is even and $n = \frac{N}{2}$, the separate computation of the sums A_j and B_j in (9.3.74) yields, according to (9.3.73), both $\widetilde{a}_{N,j}$ and $\widetilde{a}_{N,\frac{N}{2}-j}$, whereas, similarly, the separate computation of the sums C_j and D_j in (9.3.76) yields, according to (9.3.75), both $\widetilde{b}_{N,j}$ and $\widetilde{b}_{N,\frac{N}{2}-j}$. The computational efficiency thus gained is significant; indeed, (9.3.73) - (9.3.76) form the basis of the widely used Fast Fourier Transform (FFT), a detailed presentation of which is beyond the scope of this book.

9.4 Fourier series

Our principal aim in the rest of this chapter is to study the convergence properties of the Fourier series operator $\mathscr{T}_n^*$. Our first step in this direction is to show, analogously to the weighted L^2 case in Theorem 6.4.2, how the Weierstrass result of Theorem 9.2.1 can be used to prove the following convergence result.

Theorem 9.4.1. *The sequence* $\{\mathscr{T}_n^* : n = 1, 2, \ldots\}$ *of Fourier series operators, as defined by (9.3.3), satisfies the convergence result*

$$
\|f - \mathscr{T}_n^* f\|_2 \to 0, \quad n \to \infty, \quad f \in C_{2\pi},
\tag{9.4.1}
$$

with the norm $\|\cdot\|_2$ *given by (9.3.1).*

Proof. Suppose $f \in C_{2\pi}$, and let $\varepsilon > 0$. Our proof of (9.4.1) will be complete if we can show that there exists a positive integer $N = N(\varepsilon)$ such that

$$
\|f - \mathscr{T}_n^* f\|_2 < \varepsilon, \quad n > N.
\tag{9.4.2}
$$

To this end, we first apply Theorem 9.2.1 to deduce that there exist a positive integer $N = N(\varepsilon)$ and a trigonometric polynomial $Q \in \tau_N$ such that

$$||f - Q||_\infty < \frac{\varepsilon}{\sqrt{2\pi}}, \tag{9.4.3}$$

with the maximum norm $|| \cdot ||_\infty$ defined as in (9.2.1). Now observe, analogously to (5.3.4), that (9.3.1) and (9.2.1) imply

$$||f - Q||_2 = \sqrt{\int_{-\pi}^{\pi} [f(x) - Q(x)]^2 dx} \leqslant \sqrt{2\pi}\, ||f - Q||_\infty. \tag{9.4.4}$$

It follows from (9.3.4), (9.4.4) and (9.4.3) that

$$||f - \mathscr{T}_N^* f||_2 \leqslant ||f - Q||_2 < \sqrt{2\pi} \left(\frac{\varepsilon}{\sqrt{2\pi}} \right) = \varepsilon,$$

that is,

$$||f - \mathscr{T}_N^* f||_2 < \varepsilon. \tag{9.4.5}$$

Next, we note from (9.1.5) that the nesting property

$$\tau_n \subset \tau_{n+1}, \quad n = 1, 2, \ldots, \tag{9.4.6}$$

holds. Together, (9.3.4) and (9.4.6) imply

$$||f - \mathscr{T}_n^* f||_2 \leqslant ||f - \mathscr{T}_N^* f||_2, \quad n > N. \tag{9.4.7}$$

The desired result (9.4.2) is then an immediate consequence of (9.4.7) and (9.4.5). ∎

Observe from (9.3.9) in Theorem 9.3.2, together with (9.3.1), that the convergence result (9.4.1) can be formulated as

$$\lim_{n \to \infty} \sqrt{\int_{-\pi}^{\pi} \left[f(x) - \left\{ \frac{1}{2}a_0^* + \sum_{j=1}^{n} a_j^* \cos(jx) + \sum_{j=1}^{n} b_j^* \sin(jx) \right\} \right]^2 dx} = 0, \ f \in C_{2\pi}, \tag{9.4.8}$$

with the coefficient sequences $\{a_j^* : j = 0, \ldots, n\}$ and $\{b_j^* : j = 1, \ldots, n\}$ given by (9.3.10). For any given $f \in C_{2\pi}$, the infinite series

$$\frac{1}{2}a_0^* + \sum_{j=1}^{\infty} a_j^* \cos(jx) + \sum_{j=1}^{\infty} b_j^* \sin(jx), \quad x \in \mathbb{R}, \tag{9.4.9}$$

where

$$\left. \begin{aligned} a_j^* &:= \frac{1}{\pi} \int_{-\pi}^{\pi} f(x) \cos(jx) dx, \ j = 0, 1, \ldots; \\ b_j^* &:= \frac{1}{\pi} \int_{-\pi}^{\pi} f(x) \sin(jx) dx, \ j = 1, 2, \ldots, \end{aligned} \right\} \tag{9.4.10}$$

is known as the Fourier series of f, and the coefficients $\{a_j^* : j = 0, 1, \ldots\}$ and $\{b_j^* : j = 1, 2, \ldots\}$ in (9.4.10) are called the Fourier coefficients of f. The result (9.4.8) shows that the Fourier series (9.4.9) of each $f \in C_{2\pi}$ converges to f with respect to the $||\cdot||_2$ norm, as given by (9.3.1).

We proceed to show how the results of Theorem 7.3.4 and Theorem 9.3.1 can be applied to yield the following useful identity satisfied by the Fourier coefficients of a function $f \in C_{2\pi}$.

Theorem 9.4.2. (Parseval identity) *For any $f \in C_{2\pi}$, let the Fourier coefficients $\{a_j^* : 0, 1, \ldots\}$ and $\{b_j^* : j = 1, 2, \ldots\}$ of f be given as in (9.4.10). Then*

$$\frac{1}{2}(a_0^*)^2 + \sum_{j=1}^{\infty} [(a_j^*)^2 + (b_j^*)^2] = \frac{1}{\pi} \int_{-\pi}^{\pi} [f(x)]^2 dx. \tag{9.4.11}$$

Proof. By applying Theorem 9.1.2(b), and (7.3.31) in Theorem 7.3.4, together with (9.3.7), (9.3.8), and using the definitions (9.1.7) and (9.3.1), we deduce that, for any $n \in \mathbb{N}$,

$$\frac{1}{2\pi}\left[\int_{-\pi}^{\pi} f(x)dx\right]^2 + \frac{1}{\pi}\sum_{j=1}^{n}\left\{\left[\int_{-\pi}^{\pi} f(x)\cos(jx)dx\right]^2 + \left[\int_{-\pi}^{\pi} f(x)\sin(jx)dx\right]^2\right\}$$

$$= \int_{-\pi}^{\pi} [f(x)]^2 dx - (||f - \mathscr{T}_n^* f||_2)^2. \tag{9.4.12}$$

Now observe from (9.4.10) that

$$\left. \begin{array}{l} \int_{-\pi}^{\pi} f(x)\cos(jx)dx = \pi a_j^*, \ j = 0, \ldots, n; \\ \int_{-\pi}^{\pi} f(x)\sin(jx)dx = \pi b_j^*, \ j = 1, \ldots, n. \end{array} \right\} \tag{9.4.13}$$

Substitution of (9.4.13) into (9.4.12) yields

$$\frac{1}{2}(a_0^*)^2 + \sum_{j=1}^{n} [(a_j^*)^2 + (b_j^*)^2] = \frac{1}{\pi} \int_{-\pi}^{\pi} [f(x)]^2 dx - \frac{1}{\pi}(||f - \mathscr{T}_n^* f||_2)^2, n = 1, 2, \ldots. \tag{9.4.14}$$

The Parseval identity (9.4.11) now follows by an application in (9.4.14) of the convergence result (9.4.1) of Theorem 9.4.1. ∎

A standard result in the theory of infinite series states that, if $\sum_{j=0}^{\infty} \alpha_j$ is a convergent series, then we must have $\alpha_j \to 0$, $j \to \infty$. Hence we have the following immediate consequence of Theorem 9.4.2.

Theorem 9.4.3. *For any $f \in C_{2\pi}$, the Fourier coefficients $\{a_j^* : j = 0, 1, \ldots\}$ and $\{b_j^* : j = 1, 2, \ldots\}$ of f, as given in (9.4.10), satisfy*

$$a_j^* \to 0, \quad j \to \infty; \qquad b_j^* \to 0, \quad j \to \infty. \tag{9.4.15}$$

The Parseval identity (9.4.11) in Theorem 9.4.2 is useful in the calculation of the sum of certain infinite series, as illustrated by the following example.

Example 9.4.1. Let $f \in C_{2\pi}$ be the "sawtooth" function defined on $[-\pi, \pi]$ by

$$f(x) := |x|, \quad x \in [-\pi, \pi]. \tag{9.4.16}$$

Then the first line of (9.4.10) gives, for $j \in \mathbb{N}$,

$$a_j^* = \frac{1}{\pi} \int_{-\pi}^{\pi} |x| \cos(jx)dx = \frac{2}{\pi} \int_0^{\pi} x \cos(jx)dx$$

$$= \frac{2}{\pi} \left\{ \left[\frac{x \sin(jx)}{j} \right]_0^{\pi} - \frac{1}{j} \int_0^{\pi} \sin(jx)dx \right\}$$

$$= -\frac{2}{\pi j^2} \left[-\cos(jx) \right]_0^{\pi} = \frac{2}{\pi j^2} \left[(-1)^j - 1 \right], \tag{9.4.17}$$

and, for $j = 0$,

$$a_0^* = \frac{1}{\pi} \int_{-\pi}^{\pi} |x| dx = \frac{2}{\pi} \int_0^{\pi} x dx = \pi, \tag{9.4.18}$$

whereas the second line of (9.4.10) yields, for $j = 1, 2, \ldots,$

$$b_j^* = \frac{1}{\pi} \int_{-\pi}^{\pi} |x| \sin(jx)dx = 0, \tag{9.4.19}$$

since the integrand is an odd function on $\mathbb{R}$. Moreover,

$$\frac{1}{\pi} \int_{-\pi}^{\pi} [f(x)]^2 dx = \frac{2}{\pi} \int_0^{\pi} x^2 dx = \frac{2\pi^2}{3}. \tag{9.4.20}$$

Observe from (9.4.17) and (9.4.19) that the limits (9.4.15) in Theorem 9.4.3 are satisfied. By substituting (9.4.17) - (9.4.20) into the Parseval identity (9.4.11), we obtain

$$\frac{\pi^2}{2} + \frac{4}{\pi^2} \sum_{j=1}^{\infty} \frac{[(-1)^j - 1]^2}{j^4} = \frac{2\pi^2}{3},$$

that is,

$$\frac{\pi^2}{2} + \frac{16}{\pi^2} \sum_{j=1}^{\infty} \frac{1}{(2j-1)^4} = \frac{2\pi^2}{3},$$

which is equivalent to

$$\sum_{j=1}^{\infty} \frac{1}{(2j-1)^4} = \frac{\pi^4}{96}.$$

Also, according to (9.4.9), (9.4.17), (9.4.18) and (9.4.19), the Fourier series of f is given by

$$\frac{\pi}{2} - \frac{4}{\pi} \sum_{j=1}^{\infty} \frac{1}{(2j-1)^2} \cos((2j-1)x), \quad x \in \mathbb{R}, \tag{9.4.21}$$

for which (9.4.1) in Theorem 9.4.1 yields the convergence result

$$\lim_{n \to \infty} \sqrt{\int_{-\pi}^{\pi} \left[|x| - \left\{ \frac{\pi}{2} - \frac{4}{\pi} \sum_{j=1}^{n} \frac{1}{(2j-1)^2} \cos((2j-1)x) \right\} \right]^2 dx} = 0. \tag{9.4.22}$$

∎

We have therefore established that the Fourier series (9.4.9) of any $f \in C_{2\pi}$ converges to f in the sense of (9.4.8). An interesting question is whether the Fourier series of any $f \in C_{2\pi}$ converges pointwise to f on $\mathbb{R}$, that is,

$$f(x) = \frac{1}{2}a_0^* + \sum_{j=1}^{\infty} a_j^* \cos(jx) + \sum_{j=1}^{\infty} b_j^* \sin(jx), \quad x \in \mathbb{R}. \tag{9.4.23}$$

The answer is negative, since, as will be explained after the proof of Theorem 9.5.3, there exists a function $f \in C_{2\pi}$ such that

$$\|f - \mathscr{T}_n^* f\|_\infty \to \infty, \quad n \to \infty, \tag{9.4.24}$$

with $\{\mathscr{T}_n^* : n = 1,2,\ldots\}$ denoting the Fourier series operators as defined by (9.3.3), and where $\|\cdot\|_\infty$ is defined by (9.2.1), which implies that the pointwise convergence result (9.4.23) does not hold for this particular function f.

Hence we proceed, in the rest of this chapter, to identify a subspace M of $C_{2\pi}$ for which the uniform convergence result

$$\|f - \mathscr{T}_n^* f\|_\infty \to 0, \quad n \to \infty, \quad f \in M, \tag{9.4.25}$$

is satisfied, and from which, together with (9.3.9), (9.3.10) and (9.2.1), it will then follow that the Fourier series (9.4.9) of each $f \in M$ is indeed pointwise convergent to f on $\mathbb{R}$.

According to (9.1.6) and Theorem 9.1.2(a), it holds for any $n \in \mathbb{N}$ that τ_n is a finite-dimensional subspace of $C_{2\pi}$, so that we may apply Theorem 4.1.1 to deduce that there exists a best approximation from τ_n to each $f \in C_{2\pi}$ with respect to the maximum norm $\|\cdot\|_\infty$ in (9.2.1), and with respect to which we introduce the best approximation error functional $\mathscr{E}_n^* : C_{2\pi} \to \mathbb{R}$ defined by

$$\mathscr{E}_n^*[f] := \min_{Q \in \tau_n} \|f - Q\|_\infty, \quad f \in C_{2\pi}, \tag{9.4.26}$$

for $n = 1,2,\ldots$.

The following result then holds.

Theorem 9.4.4. *The best approximation error functionals* $\{\mathscr{E}_n^* : n = 1,2,\ldots\}$, *as defined in* (9.4.26), *satisfy*

$$\mathscr{E}_n^*[f] \to 0, \quad n \to \infty, \quad f \in C_{2\pi}. \tag{9.4.27}$$

Proof. Suppose $f \in C_{2\pi}$, and let $\varepsilon > 0$. According to Theorem 9.2.1, there exists a positive integer $N = N(\varepsilon)$ and a trigonometric polynomial $Q \in \tau_N$ such that $\|f - Q\|_\infty < \varepsilon$, and thus, since (9.4.26) implies $\mathscr{E}_N^*[f] \leq \|f - Q\|_\infty$, we deduce that

$$\mathscr{E}_N^*[f] < \varepsilon. \tag{9.4.28}$$

Moreover, since the nesting property (9.4.6) is satisfied, we have, from (9.4.26), and analogously to (9.4.7),

$$\mathscr{E}_n^*[f] \leqslant \mathscr{E}_N^*[f], \quad n > N,$$

which, together with (9.4.28), yields

$$\mathscr{E}_n^*[f] < \varepsilon, \quad n > N. \tag{9.4.29}$$

The desired result (9.4.27) is equivalent to (9.4.29).   ∎

We shall show that if, in (9.4.26), the function $f \in C_{2\pi}$ is such that the convergence of the sequence $\{\mathscr{E}_n^*[f] : n = 1, 2, \ldots\}$ to zero, as given in (9.4.27) of Theorem 9.4.4, occurs at a sufficiently rapid rate, then the uniform convergence result (9.4.25) is obtained, after which we shall then identify a subspace M of $C_{2\pi}$ such that, for any $f \in M$, this required convergence rate is indeed achieved.

To this purpose, we proceed in the next section to investigate the boundedness of the Fourier series operator $\mathscr{T}_n^*$ with respect to the maximum norm $\|\cdot\|_\infty$.

9.5 The Lebesgue constant in the maximum norm

Recalling from Theorem 9.3.1(c) that, for $n \in \mathbb{N}$, the Fourier series operator $\mathscr{T}_n^*$ is bounded with respect to the $\|\cdot\|_2$ norm, with corresponding Lebesgue constant $\|\mathscr{T}_n^*\|_2 = 1$, we proceed here to show that $\mathscr{T}_n^*$ is also bounded with respect to the maximum norm $\|\cdot\|_\infty$, and we investigate the corresponding Lebesgue constant $\|\mathscr{T}_n^*\|_\infty$ as a function of n.

First, we prove the following integral representation formula for $\mathscr{T}_n^*$.

Theorem 9.5.1. *For any $f \in C_{2\pi}$, and $n \in \mathbb{N}$, the Fourier series operator $\mathscr{T}_n^* : C_{2\pi} \to \tau_n$, as defined by (9.3.3), satisfies the formulation*

$$(\mathscr{T}_n^* f)(x) = \frac{1}{\pi} \int_{-\pi}^{\pi} K_n^*(\theta) f(x + \theta) \, d\theta, \quad x \in \mathbb{R}, \tag{9.5.1}$$

where the function K_n^* is defined by

$$K_n^*(\theta) := \frac{1}{2} + \sum_{j=1}^{n} \cos(j\theta), \quad \theta \in [-\pi, \pi], \tag{9.5.2}$$

and where also

$$K_n^*(\theta) = \begin{cases} \dfrac{\sin((n + \frac{1}{2})\theta)}{2\sin(\frac{1}{2}\theta)}, & \theta \in [-\pi, 0) \cup (0, \pi]; \\ n + \frac{1}{2}, & \theta = 0. \end{cases} \tag{9.5.3}$$

Proof. Let $x \in \mathbb{R}$ be fixed. By substituting the formulas (9.3.10) into (9.3.9), we obtain

$$(\mathcal{T}_n^* f)(x) = \frac{1}{2\pi} \int_{-\pi}^{\pi} f(\theta)d\theta + \frac{1}{\pi} \sum_{j=1}^{n} \left[\int_{-\pi}^{\pi} f(\theta)\cos(j\theta)d\theta \right] \cos(jx)$$

$$+ \frac{1}{\pi} \sum_{j=1}^{n} \left[\int_{-\pi}^{\pi} f(\theta)\sin(j\theta)d\theta \right] \sin(jx)$$

$$= \frac{1}{\pi} \int_{-\pi}^{\pi} \left[\frac{1}{2} + \sum_{j=1}^{n} \{\cos(j\theta)\cos(jx) + \sin(j\theta)\sin(jx)\} \right] f(\theta)d\theta$$

$$= \frac{1}{\pi} \int_{-\pi}^{\pi} \left[\frac{1}{2} + \sum_{j=1}^{n} \cos(j(\theta - x)) \right] f(\theta)d\theta. \tag{9.5.4}$$

Now observe that the function

$$g(\theta) := \frac{1}{2} + \sum_{j=1}^{n} \cos(j(\theta - x)), \quad \theta \in \mathbb{R}, \tag{9.5.5}$$

satisfies $g(\theta + 2\pi) = g(\theta), \theta \in \mathbb{R}$, and thus $g \in C_{2\pi}$, so that we may apply (9.3.64), (9.3.65) to deduce that

$$\int_{-\pi}^{\pi} g(\theta)d\theta = \int_{-\pi+x}^{\pi+x} g(\theta)d\theta = \int_{-\pi}^{\pi} g(\theta + x)dx. \tag{9.5.6}$$

It follows from (9.5.4), (9.5.5) and (9.5.6) that the formulation (9.5.1), (9.5.2) is satisfied. After noting from (9.5.2) that the second line of (9.5.3) is satisfied, we observe next, for any $\theta \in [-\pi, 0) \cup (0, \pi]$, that

$$\sin(\tfrac{1}{2}\theta) \sum_{j=1}^{n} \cos(j\theta) = \sum_{j=1}^{n} \left[\cos(j\theta)\sin(\tfrac{1}{2}\theta) \right]$$

$$= \tfrac{1}{2} \sum_{j=1}^{n} \left[\sin((j+\tfrac{1}{2})\theta) - \sin((j-\tfrac{1}{2})\theta) \right]$$

$$= \tfrac{1}{2} \left\{ \left[\sin(\tfrac{3}{2}\theta) - \sin(\tfrac{1}{2}\theta) \right] + \left[\sin(\tfrac{5}{2}\theta) - \sin(\tfrac{3}{2}\theta) \right] \right.$$

$$\left. + \cdots + \left[\sin((n+\tfrac{1}{2})\theta) - \sin((n-\tfrac{1}{2})\theta) \right] \right\}$$

$$= \tfrac{1}{2} \left[\sin((n+\tfrac{1}{2})\theta) - \sin(\tfrac{1}{2}\theta) \right],$$

and thus

$$\frac{1}{2} + \sum_{j=1}^{n} \cos(j\theta) = \frac{\sin((n+\tfrac{1}{2})\theta)}{2\sin(\tfrac{1}{2}\theta)}, \quad \theta \in [-\pi, 0) \cup (0, \pi]. \tag{9.5.7}$$

The first line of (9.5.3) follows immediately from (9.5.2) and (9.5.7). ∎

The formulas in Theorem 9.5.1 enable us to prove the boundedness with respect to the maximum norm $\| \cdot \|_\infty$ of the Fourier series operator $\mathcal{T}_n^*$, as follows.

Theorem 9.5.2. *For* $n \in \mathbb{N}$, *the Fourier series operator* $\mathcal{T}_n^*$, *as defined by* (9.3.3), *is bounded with respect to the maximum norm* $|| \cdot ||_\infty$, *as given by* (9.2.1), *with corresponding Lebesgue constant* $||\mathcal{T}_n^*||_\infty$ *bounded above by*

$$||\mathcal{T}_n^*||_\infty \leqslant 1 + \ln(2n+1). \tag{9.5.8}$$

Proof. Let $n \in \mathbb{N}$ be fixed. From (9.5.1) in Theorem 9.5.1, and by using also the fact that, according to (9.5.2), K_n^* is an even function on $[-\pi, \pi]$, we obtain, for any $f \in C_{2\pi}$ and $x \in [-\pi, \pi]$,

$$|(\mathcal{T}_n^* f)(x)| \leqslant \frac{1}{\pi} \int_{-\pi}^{\pi} |K_n^*(\theta)| \, |f(x+\theta)| d\theta$$

$$\leqslant \left[\frac{1}{\pi} \int_{-\pi}^{\pi} |K_n^*(\theta)| d\theta \right] ||f||_\infty = \left[\frac{2}{\pi} \int_0^{\pi} |K_n^*(\theta)| d\theta \right] ||f||_\infty,$$

and thus

$$||\mathcal{T}_n^* f||_\infty \leqslant \left[\frac{2}{\pi} \int_0^{\pi} |K_n^*(\theta)| d\theta \right] ||f||_\infty,$$

from which it then follows that

$$\frac{||\mathcal{T}_n^* f||_\infty}{||f||_\infty} \leqslant \frac{2}{\pi} \int_0^{\pi} |K_n^*(\theta)| d\theta, \quad \text{for} \quad f \in C_{2\pi}, \text{ with } \ f \neq 0. \tag{9.5.9}$$

It follows from (9.5.9) that $\{ ||\mathcal{T}_n^* f|| / ||f|| : f \in C_{2\pi}; f \neq 0 \}$ is a bounded set, and thus the approximation operator $\mathcal{T}_n^*$ is indeed bounded with respect to the maximum norm $|| \cdot ||_\infty$, with, according to the definition (5.2.2), corresponding Lebesgue constant $||\mathcal{T}_n^*||_\infty$ satisfying the upper bound

$$||\mathcal{T}_n^*||_\infty \leqslant \frac{2}{\pi} \int_0^{\pi} |K_n^*(\theta)| d\theta. \tag{9.5.10}$$

The desired result (9.5.8) will therefore follow if we can prove the inequality

$$\int_0^{\pi} |K_n^*(\theta)| d\theta \leqslant \frac{\pi}{2} [1 + \ln(2n+1)]. \tag{9.5.11}$$

To prove (9.5.11), we let μ denote an arbitrary number such that $0 < \mu < \pi$, and note from (9.5.2) that

$$|K_n^*(\theta)| = \left| \frac{1}{2} + \sum_{j=1}^{n} \cos(j\theta) \right| \leqslant \frac{1}{2} + n, \quad \theta \in [0, \mu]. \tag{9.5.12}$$

Next, we observe that the inequality (6.5.25) is equivalent to

$$\sin(\tfrac{1}{2}\theta) \geqslant \frac{\theta}{\pi}, \quad \theta \in [0, \pi], \tag{9.5.13}$$

which, together with the first line of (9.5.3), yields

$$|K_n^*(\theta)| \leqslant \frac{1}{2|\sin(\tfrac{1}{2}\theta)|} \leqslant \frac{\pi}{2\theta}, \quad \theta \in [\mu, \pi]. \tag{9.5.14}$$

It follows from (9.5.12) and (9.5.14) that

$$\int_0^\pi |K_n^*(\theta)|\,d\theta = \int_0^\mu |K_n^*(\theta)|\,d\theta + \int_\mu^\pi |K_n^*(\theta)|\,d\theta$$

$$\leqslant \int_0^\mu \left(\frac{1}{2}+n\right)d\theta + \int_\mu^\pi \frac{\pi}{2\theta}\,d\theta$$

$$= \left(\frac{1}{2}+n\right)\mu + \frac{\pi}{2}\left[\ln\pi - \ln\mu\right]. \tag{9.5.15}$$

Recalling that μ is an arbitrary number in the interval $(0,\pi)$, we may now choose $\mu = \dfrac{\pi}{2n+1}$ in (9.5.15) to obtain

$$\int_0^\pi |K_n^*(\theta)|\,d\theta \leqslant \frac{2n+1}{2}\left(\frac{\pi}{2n+1}\right) + \frac{\pi}{2}\left[\ln\pi - \{\ln\pi - \ln(2n+1)\}\right]$$

$$= \frac{\pi}{2}[1 + \ln(2n+1)],$$

and thereby yielding (9.5.11).   ∎

Note that the upper bound in (9.5.8) satisfies

$$1 + \ln(2n+1) \to \infty, \quad n \to \infty. \tag{9.5.16}$$

We proceed to show that it in fact holds that

$$\|\mathscr{T}_n^*\|_\infty \to \infty, \quad n \to \infty, \tag{9.5.17}$$

as follows from our next result.

Theorem 9.5.3. *The Lebesgue constant* $\|\mathscr{T}_n^*\|_\infty$ *in Theorem 9.5.2 satisfies*

$$\|\mathscr{T}_n^*\|_\infty = \frac{2}{\pi}\int_0^\pi |K_n^*(\theta)|\,d\theta > \frac{4}{\pi^2}\ln(n+1), \quad n \in \mathbb{N}. \tag{9.5.18}$$

Proof. Our first step is to show that

$$\|\mathscr{T}_n^*\|_\infty \geqslant \frac{2}{\pi}\int_0^\pi |K_n^*(\theta)|\,d\theta, \tag{9.5.19}$$

which, together with (9.5.10), will then yield the formula in (9.5.18) for $\|\mathscr{T}_n^*\|_\infty$.

To this end, we fix $n \in \mathbb{N}$, and observe from (9.5.3) in Theorem 9.5.1 that, with the definition

$$\theta_k := \begin{cases} \dfrac{2k\pi}{2n+1}, & k = 0,\ldots,n; \\ \pi, & k = n+1, \end{cases} \tag{9.5.20}$$

we have (see Exercise 9.18)

$$K_n^*(\theta) > 0, \theta \in [\theta_0, \theta_1); \tag{9.5.21}$$

$$(-1)^k K_n^*(\theta) > 0, \theta \in (\theta_k, \theta_{k+1}), \quad k = 1,\ldots,n-1 \ (\text{if } n \geqslant 2); \tag{9.5.22}$$

$$(-1)^n K_n^*(\theta) > 0, \theta \in (\theta_n, \theta_{n+1}], \tag{9.5.23}$$

with, moreover,

$$K_n^*(\theta_k) = 0, \quad k = 1, \ldots, n. \tag{9.5.24}$$

For any $j \in \mathbb{N}$, define the points

$$\left.\begin{array}{l} \psi_{j,k} := \theta_k - \dfrac{\pi}{(2n+1)(j+2)}, \, k = 1, \ldots, n; \\[3mm] \widetilde{\psi}_{j,k} := \theta_k + \dfrac{\pi}{(2n+1)(j+2)}, \, k = 1, \ldots, n, \end{array}\right\} \tag{9.5.25}$$

with $\{\theta_1, \ldots, \theta_n\}$ as in (9.5.20), and according to which, for each $j \in \mathbb{N}$, we have (see Exercise 9.19) the ordering

$$0 = \theta_0 < \psi_{j,1} < \theta_1 < \widetilde{\psi}_{j,1} < \psi_{j,2} < \theta_2 < \widetilde{\psi}_{j,2} < \cdots < \psi_{j,n} < \theta_n < \widetilde{\psi}_{j,n} < \theta_{n+1} = \pi. \tag{9.5.26}$$

Also, for $j \in \mathbb{N}$, let $g_j \in C_{2\pi}$ be the even function on $\mathbb{R}$, as defined on $[0, \pi]$ by

$$g_j(x) := \begin{cases} 1, & x \in [\theta_0, \psi_{j,1}); \\ (-1)^k, & x \in [\widetilde{\psi}_{j,k}, \psi_{j,k+1}), \quad k = 1, \ldots, n-1; \\ (-1)^n, & x \in [\widetilde{\psi}_n, \theta_{n+1}], \end{cases} \tag{9.5.27}$$

and

$$g_j(x) := (-1)^k \frac{(2n+1)(j+2)}{\pi} (x - \theta_k), \quad x \in [\psi_{j,k}, \widetilde{\psi}_{j,k}], \, k = 1, \ldots, n, \tag{9.5.28}$$

according to which, together with (9.5.21) - (9.5.26), we have (see Exercise 9.19)

$$K_n^*(\theta) g_j(\theta) \geqslant 0, \quad \theta \in [0, \pi], \quad j \in \mathbb{N}, \tag{9.5.29}$$

as well as

$$\|g_j\|_\infty = 1, \quad j \in \mathbb{N}. \tag{9.5.30}$$

Next, by using (9.5.26) and (9.5.29), as well as (9.5.27), we deduce that, for any $j \in \mathbb{N}$,

$$\int_0^\pi K_n^*(\theta) g_j(\theta) d\theta = \int_0^\pi |K_n^*(\theta) g_j(\theta)| d\theta$$

$$= \int_0^{\psi_{j,1}} |K_n^*(\theta)| d\theta + \sum_{k=1}^{n-1} \int_{\widetilde{\psi}_{j,k}}^{\psi_{j,k+1}} |K_n^*(\theta)| d\theta$$

$$+ \int_{\widetilde{\psi}_{j,n}}^\pi |K_n^*(\theta)| d\theta + \sum_{k=1}^n \int_{\psi_{j,k}}^{\widetilde{\psi}_{j,k}} |K_n^*(\theta) g_j(\theta)| d\theta. \tag{9.5.31}$$

Since (9.5.2) implies

$$|K_n^*(\theta)| \leqslant \tfrac{1}{2} + n, \quad \theta \in [-\pi, \pi], \tag{9.5.32}$$

we may now apply (9.5.32), (9.5.30) and (9.5.25) to obtain, for any $j \in \mathbb{N}$,

$$\sum_{k=1}^{n} \int_{\psi_{j,k}}^{\tilde{\psi}_{j,k}} |K_n^*(\theta) g_j(\theta)| d\theta \leqslant (n + \tfrac{1}{2})(\tilde{\psi}_{j,k} - \psi_{j,k})n = \frac{n\pi}{j+2},$$

and thus

$$\lim_{j \to \infty} \sum_{k=1}^{n} \int_{\psi_{j,k}}^{\tilde{\psi}_{j,k}} |K_n^*(\theta) g_j(\theta)| d\theta = 0. \tag{9.5.33}$$

By using the implication of the fundamental theorem of calculus that, for any $f \in C[0, \pi]$, the integrals $\int_0^x f(t)dt$ and $\int_\pi^x f(t)dt$ are both continuous for $x \in [0, \pi]$, and since (9.5.25) yields, for any $k \in \{1, \ldots, n\}$, the limits

$$\lim_{j \to \infty} \psi_{j,k} = \theta_k; \quad \lim_{j \to \infty} \tilde{\psi}_{j,k} = \theta_k, \tag{9.5.34}$$

we further deduce that

$$\lim_{j \to \infty} \left[\int_0^{\psi_{j,1}} |K_n^*(\theta)| d\theta + \sum_{k=1}^{n-1} \int_{\tilde{\psi}_{j,k}}^{\psi_{j,k+1}} |K_n^*(\theta)| d\theta + \int_{\tilde{\psi}_{j,n}}^{\pi} |K_n^*(\theta)| d\theta \right]$$

$$= \lim_{j \to \infty} \left[\int_0^{\psi_{j,1}} |K_n^*(\theta)| d\theta + \sum_{k=1}^{n-1} \left\{ \int_0^{\psi_{j,k+1}} |K_n^*(\theta)| d\theta - \int_0^{\tilde{\psi}_{j,k}} |K_n^*(\theta)| d\theta \right\} + \int_{\tilde{\psi}_{j,n}}^{\pi} |K_n^*(\theta)| d\theta \right]$$

$$= \int_0^{\theta_1} |K_n^*(\theta)| d\theta + \sum_{k=1}^{n-1} \left\{ \int_0^{\theta_{k+1}} |K_n^*(\theta)| d\theta - \int_0^{\theta_k} |K_n^*(\theta)| d\theta \right\} + \int_{\theta_n}^{\pi} |K_n^*(\theta)| d\theta$$

$$= \sum_{k=0}^{n} \int_{\theta_k}^{\theta_{k+1}} |K_n^*(\theta)| d\theta = \int_0^{\pi} |K_n^*(\theta)| d\theta, \tag{9.5.35}$$

by virtue of (9.5.20).

By combining (9.5.31), (9.5.35) and (9.5.33), we obtain the limit

$$\lim_{j \to \infty} \left[\int_0^{\pi} K_n^*(\theta) g_j(\theta) d\theta \right] = \int_0^{\pi} |K_n^*(\theta)| d\theta. \tag{9.5.36}$$

Now let $x \in [0, \pi]$ be fixed, and define, for $j = 1, 2, \ldots$,

$$f_j(\theta) := g_j(\theta - x), \quad \theta \in \mathbb{R}, \tag{9.5.37}$$

with g_j denoting the function in $C_{2\pi}$, as defined by (9.5.27), (9.5.28), and the condition that g_j is even on $\mathbb{R}$. It follows from (9.5.37) that $f_j \in C_{2\pi}$, $j \in \mathbb{N}$, with, moreover, from (9.5.30),

$$\|f_j\|_\infty = 1, \quad j \in \mathbb{N}. \tag{9.5.38}$$

Hence we may apply the formula (9.5.1) in Theorem 9.5.1, together with (9.5.37), to obtain, for any $j \in \mathbb{N}$,

$$(\mathscr{T}_n^* f_j)(x) = \frac{1}{\pi} \int_{-\pi}^{\pi} K_n^*(\theta) f_j(x + \theta) d\theta = \frac{1}{\pi} \int_{-\pi}^{\pi} K_n^*(\theta) g_j(\theta) d\theta = \frac{2}{\pi} \int_0^{\pi} K_n^*(\theta) g_j(\theta) d\theta, \tag{9.5.39}$$

since (9.5.2) implies that K_n^* is an even function on $\mathbb{R}$, whereas g_j is by definition an even function on $\mathbb{R}$.

According to Theorem 9.3.1(a) and Theorem 9.5.2, the Fourier series operator $\mathscr{T}_n^*$ is linear, and bounded with respect to the maximum norm $||\cdot||_\infty$. Hence we may apply the formula (5.2.10) in Theorem 5.2.3, together with (9.5.38), and (9.5.39), to obtain, for any $j \in \mathbb{N}$,

$$||\mathscr{T}_n^*||_\infty = \sup\{||\mathscr{T}_n^* f||_\infty : f \in C_{2\pi};\ ||f||_\infty = 1\}$$

$$\geqslant ||\mathscr{T}_n^* f_j||_\infty = \max_{-1 \leqslant t \leqslant 1} |(\mathscr{T}_n^* f_j)(t)| \geqslant (\mathscr{T}_n^* f_j)(x) = \frac{2}{\pi} \int_0^\pi K_n^*(\theta) g_j(\theta) d\theta,$$

and thus, by using also (9.5.36),

$$||\mathscr{T}_n^*||_\infty \geqslant \frac{2}{\pi} \int_0^\pi K_n^*(\theta) g_j(\theta) \to \frac{2}{\pi} \int_0^\pi |K_n^*(\theta)| d\theta, \quad j \to \infty,$$

which immediately yields the desired inequality (9.5.19), and thereby completing the proof of the formula in (9.5.18) for the Lebesgue constant $||\mathscr{T}_n^*||_\infty$.

Next, to prove the inequality in (9.5.18), we first use (9.5.19), (9.5.20) and (9.5.3) to obtain

$$||\mathscr{T}_n^*||_\infty > \frac{2}{\pi} \int_0^{\theta_n} |K_n^*(\theta)| d\theta = \frac{2}{\pi} \sum_{j=0}^{n-1} \int_{\theta_j}^{\theta_{j+1}} \frac{|\sin((n+\frac{1}{2})\theta)|}{2\sin(\frac{1}{2}\theta)} d\theta. \qquad (9.5.40)$$

Now observe that the function

$$u(\theta) := \theta - \sin\theta$$

satisfies

$$u(0) = 0; \qquad u'(\theta) = 1 - \cos\theta \geqslant 0, \quad \theta \in \mathbb{R},$$

and thus

$$u(\theta) \geqslant 0, \quad \theta \in \mathbb{R},$$

from which we then obtain the inequality

$$\sin(\tfrac{1}{2}\theta) \leqslant \tfrac{1}{2}\theta, \quad \theta \geqslant 0. \qquad (9.5.41)$$

Hence, from (9.5.41) and (9.5.20),

$$\sum_{j=0}^{n-1} \int_{\theta_j}^{\theta_{j+1}} \frac{|\sin((n+\frac{1}{2})\theta)|}{2\sin(\frac{1}{2}\theta)} d\theta \geqslant \sum_{j=0}^{n-1} \int_{\theta_j}^{\theta_{j+1}} \frac{|\sin((n+\frac{1}{2})\theta)|}{\theta} d\theta$$

$$\geqslant \sum_{j=0}^{n-1} \frac{1}{\theta_{j+1}} \int_{\theta_j}^{\theta_{j+1}} |\sin((n+\tfrac{1}{2})\theta)| d\theta$$

$$= \sum_{j=0}^{n-1} \frac{(-1)^j}{\theta_{j+1}} \int_{\theta_j}^{\theta_{j+1}} \sin((n+\tfrac{1}{2})\theta) d\theta$$

$$= \sum_{j=0}^{n-1} \frac{(-1)^j}{\theta_{j+1}} \frac{2}{2n+1} \left[\cos\left((n+\tfrac{1}{2})\theta_j\right) - \cos\left((n+\tfrac{1}{2})\theta_{j+1}\right) \right]$$

$$= \frac{1}{\pi} \sum_{j=0}^{n-1} \frac{(-1)^j}{j+1} \left[(-1)^j - (-1)^{j+1} \right]$$

$$= \frac{2}{\pi} \sum_{j=0}^{n-1} \frac{1}{j+1} = \frac{2}{\pi} \sum_{j=1}^{n} \frac{1}{j}. \tag{9.5.42}$$

Finally, we prove the inequality

$$\sum_{j=1}^{n} \frac{1}{j} > \ln(n+1), \quad n \in \mathbb{N}, \tag{9.5.43}$$

which, together with (9.5.40) and (9.5.42), will then complete our proof of the inequality in (9.5.18).

To this end, we define, for any $n \in \mathbb{N}$, the piecewise constant function

$$h(x) := \begin{cases} \dfrac{1}{j}, & x \in [j, j+1), \quad j = 1, \dots, n; \\ \dfrac{1}{n+1}, & x = n+1, \end{cases} \tag{9.5.44}$$

for which it then follows that

$$\frac{1}{x} \leqslant h(x), \quad x \in [1, n+1]. \tag{9.5.45}$$

By using (9.5.45) and (9.5.44), we deduce that

$$\ln(n+1) = \int_1^{n+1} \frac{1}{x} dx \leqslant \int_1^{n+1} h(x) dx = \sum_{j=1}^{n} \frac{1}{j},$$

and thereby proving the desired inequality (9.5.43). ∎

Since (9.5.17) holds, as follows immediately from (9.5.18) in Theorem 9.5.3, we may apply the Banach-Steinhaus theorem, that is, the principle of uniform boundedness, which is a standard result in functional analysis, and the proof of which is beyond the scope of this book, to deduce the existence of a function $f \in C_{2\pi}$ such that

$$\|\mathscr{T}_n^* f\|_\infty \to \infty, \quad n \to \infty,$$

and thus, from (4.1.3) and (4.1.5),

$$\|f - \mathscr{T}_n^* f\|_\infty = \|\mathscr{T}_n^* f - f\|_\infty \geqslant \big| \|\mathscr{T}_n^* f\|_\infty - \|f\|_\infty \big| \geqslant \|\mathscr{T}_n^* f\|_\infty - \|f\|_\infty \to \infty, \ n \to \infty,$$

and thereby justifying the statement in Section 9.4 that there exists a function $f \in C_{2\pi}$ such that the divergence result (9.4.24) holds.

Finally in this section, we prove the following explicit formulation of the Lebesgue constant $\|\mathscr{T}_n^*\|_\infty$.

Theorem 9.5.4. *For any $n \in \mathbb{N}$, the Lebesgue constant $||\mathscr{T}_n^*||_\infty$ in Theorem 9.5.2 is given explicitly by the formula*

$$||\mathscr{T}_n^*||_\infty = \frac{1}{2n+1} + \frac{2}{\pi}\sum_{k=1}^{n}\frac{1}{k}\tan\left(\frac{k}{2n+1}\pi\right). \tag{9.5.46}$$

Proof. Let the points $\{\theta_0,\ldots,\theta_{n+1}\}$ be defined as in (9.5.20). It then follows from the formulation of the Lebesgue constant $||\mathscr{T}_n^*||_\infty$ in (9.5.18) of Theorem 9.5.3, together with the properties (9.5.21) - (9.5.24) of the function K_n^*, as well as (9.5.2) in Theorem 9.5.1, that

$$||\mathscr{T}_n^*||_\infty = \frac{2}{\pi}\sum_{j=0}^{n}\int_{\theta_j}^{\theta_{j+1}}|K_n^*(\theta)|d\theta$$

$$= \frac{2}{\pi}\sum_{j=0}^{n}(-1)^j\int_{\theta_j}^{\theta_{j+1}}\left[\frac{1}{2}+\sum_{k=1}^{n}\cos(k\theta)\right]d\theta$$

$$= \frac{2}{\pi}\sum_{j=0}^{n}(-1)^j\left[\frac{1}{2}(\theta_{j+1}-\theta_j)+\sum_{k=1}^{n}\frac{1}{k}\{\sin(k\theta_{j+1})-\sin(k\theta_j)\}\right]$$

$$= \frac{1}{\pi}\sum_{j=0}^{n}(-1)^j(\theta_{j+1}-\theta_j)+\frac{2}{\pi}\sum_{k=1}^{n}\frac{1}{k}\sum_{j=0}^{n}(-1)^j[\sin(k\theta_{j+1})-\sin(k\theta_j)]. \tag{9.5.47}$$

By using (9.5.20), we deduce that

$$\sum_{j=0}^{n}(-1)^j(\theta_{j+1}-\theta_j) = \sum_{j=0}^{n-1}(-1)^j(\theta_{j+1}-\theta_j)+(-1)^n\frac{\pi}{2n+1}$$

$$= \sum_{j=0}^{n-1}(-1)^j\frac{2\pi}{2n+1}+(-1)^n\frac{\pi}{2n+1}$$

$$= \frac{\pi}{2n+1}\left[\sum_{j=0}^{n-1}2(-1)^j+(-1)^n\right]. \tag{9.5.48}$$

Now observe that

$$\sum_{j=0}^{n-1}2(-1)^j+(-1)^n = \begin{cases} 2-1=1, & \text{if } n \text{ is odd;} \\ 0+1=1, & \text{if } n \text{ is even,} \end{cases}$$

and thus, from (9.5.48),

$$\sum_{j=0}^{n}(-1)^j(\theta_{j+1}-\theta_j) = \frac{\pi}{2n+1}. \tag{9.5.49}$$

Next, note from (9.5.20) that, for any $k \in \{1,\ldots,n\}$,

$$\sum_{j=0}^{n}(-1)^j[\sin(k\theta_{j+1})-\sin(k\theta_j)] = [\sin(k\theta_1)-0]-[\sin(k\theta_2)-\sin(k\theta_1)]$$

$$+ [\sin(k\theta_3) - \sin(k\theta_2)] + \cdots + (-1)^n[\sin(k\pi) - \sin(k\theta_n)]$$

$$= 2\sum_{j=1}^{n}(-1)^{j-1}\sin(k\theta_j) = 2\sum_{j=1}^{n}(-1)^{j-1}\sin\left(\frac{2kj}{2n+1}\pi\right).$$

$$(9.5.50)$$

But, for any $k \in \{1,\ldots,n\}$,

$$\cos\left(\frac{k}{2n+1}\pi\right)\sum_{j=1}^{n}(-1)^{j-1}\sin\left(\frac{2kj}{2n+1}\pi\right)$$

$$= \sum_{j=1}^{n}(-1)^{j-1}\sin\left(\frac{2kj}{2n+1}\pi\right)\cos\left(\frac{k}{2n+1}\pi\right)$$

$$= \frac{1}{2}\sum_{j=1}^{n}(-1)^{j-1}\left[\sin\left(\frac{(2j+1)k}{2n+1}\pi\right) + \sin\left(\frac{(2j-1)k}{2n+1}\pi\right)\right]$$

$$= \frac{1}{2}\left\{\left[\sin\left(\frac{3k}{2n+1}\pi\right) + \sin\left(\frac{k}{2n+1}\pi\right)\right] - \left[\sin\left(\frac{5k}{2n+1}\pi\right) + \sin\left(\frac{3k}{2n+1}\pi\right)\right]\right.$$

$$\left. + \cdots + (-1)^{n-1}\left[\sin(k\pi) + \sin\left(\frac{(2n-1)k}{2n+1}\pi\right)\right]\right\}$$

$$= \frac{1}{2}\sin\left(\frac{k}{2n+1}\pi\right),$$

and thus

$$\sum_{j=1}^{n}(-1)^{j-1}\sin\left(\frac{2kj}{2n+1}\pi\right) = \frac{1}{2}\tan\left(\frac{k}{2n+1}\pi\right), \quad k = 1,\ldots,n. \qquad (9.5.51)$$

The formula (9.5.46) is an immediate consequence of (9.5.47), (9.5.49), (9.5.50) and (9.5.51). ∎

Calculating by means of (9.5.46), we obtain the values of $||\mathscr{T}_n^*||_\infty$ for $n = 1,\ldots,10$, as given in Table 9.5.1.

It can in fact be shown that, for any $n \in \mathbb{N}$, if $\mathscr{A} : C_{2\pi} \to \tau_n$ is a linear approximation operator satisfying the exactness condition

$$\mathscr{A}f = f, \quad f \in \tau_n,$$

and $\mathscr{A}$ is bounded with respect to the maximum norm $||\cdot||_\infty$, then

$$||\mathscr{A}||_\infty \geqslant ||\mathscr{T}_n^*||_\infty.$$

It follows from Theorem 9.3.1(a) and (b), as well as Theorem 9.5.2, that we may apply the Lebesgue inequality in Theorem 5.3.2 to immediately deduce the following upper bound

Table 9.5.1   The Lebesgue constant $||\mathscr{T}_n^*||_\infty$ for $n = 1, \ldots, 10$.

n	$||\mathscr{T}_n^*||_\infty$	n	$||\mathscr{T}_n^*||_\infty$
1	1.436	6	2.029
2	1.642	7	2.087
3	1.778	8	2.138
4	1.880	9	2.183
5	1.961	10	2.223

on the approximation error with respect to the maximum norm $||\cdot||_\infty$ for the Fourier series operator.

Theorem 9.5.5. *For any $n \in \mathbb{N}$, the uniform approximation error for the Fourier series operator $\mathscr{T}_n^* : C_{2\pi} \to \tau_n$, as defined by (9.3.3), satisfies the Lebesgue inequality*

$$||f - \mathscr{T}_n^* f||_\infty \leqslant [2 + \ln(2n+1)]\mathscr{E}_n^*[f], \quad f \in C_{2\pi}, \tag{9.5.52}$$

with $\mathscr{E}_n^* : C_{2\pi} \to \mathbb{R}$ denoting the best approximation error functional, as defined by (9.4.26).

After recalling also (9.4.27) in Theorem 9.4.4, as well as (9.5.17), together with the Lebesgue inequality (5.3.7) in Theorem 5.3.2 for the approximation operator $\mathscr{T}_n^*$, we deduce from (9.5.52) in Theorem 9.5.5 the following sufficient condition on a function $f \in C_{2\pi}$ for the uniform convergence result (9.4.25) to be satisfied.

Theorem 9.5.6. *In Theorem 9.5.5, suppose $f \in C_{2\pi}$ satisfies the condition*

$$[\ln(2n+1)]\mathscr{E}_n^*[f] \to 0, \quad n \to \infty. \tag{9.5.53}$$

Then the uniform convergence result

$$||f - \mathscr{T}_n^* f||_\infty \to 0, \quad n \to \infty, \tag{9.5.54}$$

holds.

In the next section, we establish a subspace M of $C_{2\pi}$ for which any $f \in M$ satisfies the condition (9.5.53), and therefore also, according to Theorem 9.5.6, the uniform convergence result (9.4.25).

9.6 Sufficient condition for uniform convergence

In this section, in order to establish a subspace M of $C_{2\pi}$ such that the sufficient condition (9.5.53) in Theorem 9.5.6 holds for each $f \in M$, we investigate the convergence rate with respect to the result (9.4.27) in Theorem 9.4.4, for specific classes of functions f.

We shall rely on the following trigonometric polynomial interpolation result.

Theorem 9.6.1. *For the function*

$$f(x) := x, \tag{9.6.1}$$

and any $n \in \mathbb{N}$, there exists precisely one trigonometric polynomial

$$Q_n \in \mathrm{span}\{\sin x, \sin(2x), \ldots, \sin(nx)\} \tag{9.6.2}$$

satisfying the interpolation conditions

$$Q_n\left(\frac{j\pi}{n+1}\right) = f\left(\frac{j\pi}{n+1}\right), \quad j = 1, \ldots, n, \tag{9.6.3}$$

with, moreover,

$$\int_0^\pi |f(x) - Q_n(x)|\, dx = \int_0^\pi |x - Q_n(x)|\, dx = \frac{\pi^2}{2(n+1)}. \tag{9.6.4}$$

Proof. With the notation

$$x_{n,j} := \frac{j\pi}{n+1}, \quad j = 1, \ldots, n, \tag{9.6.5}$$

the existence and uniqueness statement of the theorem will follow if we can show that the matrix

$$A_n := \begin{bmatrix} \sin x_{n,1} & \sin(2x_{n,1}) & \cdots & \sin(nx_{n,1}) \\ \sin x_{n,2} & \sin(2x_{n,2}) & \cdots & \sin(nx_{n,2}) \\ \vdots & \vdots & \vdots & \vdots \\ \sin x_{n,n} & \sin(2x_{n,n}) & \cdots & \sin(nx_{n,n}) \end{bmatrix} \tag{9.6.6}$$

is invertible, or equivalently, from a standard result in linear algebra, if we can show that the homogeneous linear system $A_n \mathbf{c} = \mathbf{0}$ has only the trivial solution $\mathbf{c} = \mathbf{0} \in \mathbb{R}^n$.

To this end, suppose $\mathbf{c} = (c_1, \ldots, c_n)^T \subset \mathbb{R}^n$ is such that $A_n \mathbf{c} = \mathbf{0}$. It follows from (9.6.6) that the trigonometric polynomial

$$Q(x) := \sum_{k=1}^n c_k \sin(kx) \tag{9.6.7}$$

then satisfies

$$Q(x_{n,j}) = 0, \quad j = 1, \ldots, n. \tag{9.6.8}$$

According to (9.6.7), Q is an odd function on $\mathbb{R}$, and thus (9.6.8) implies

$$Q(-x_{n,j}) = 0, \quad j = 1, \ldots, n. \tag{9.6.9}$$

By noting from (9.6.5) that

$$\{x_{n,1}, \ldots, x_{n,n}\} \subset (0, \pi); \quad \{-x_{n,1}, \ldots, -x_{n,n}\} \subset (-\pi, 0),$$

we deduce from (9.6.8) and (9.6.9), together with $Q(0) = 0$, as follows from (9.6.7), that Q has at least $2n+1$ distinct zeros in $(-\pi, \pi)$. Moreover, (9.6.7) and (9.1.5) imply that $Q \in \tau_n$. Hence we may apply Theorem 9.1.3(c) to deduce that Q must be the zero function, and thus, from (9.6.7), $\mathbf{c} = (c_1, \ldots, c_n) = \mathbf{0} \in \mathbb{R}^n$, from which it then follows that the matrix A_n in (9.6.6) is indeed invertible, thereby yielding the existence of precisely one trigonometric polynomial Q_n as in (9.6.2), and satisfying the interpolation condition (9.6.3) with respect to the function f given by (9.6.1).

For the error function

$$E_n(x) := x - Q_n(x), \tag{9.6.10}$$

it follows from (9.6.1), (9.6.3) and (9.6.5) that

$$E_n(x_{n,j}) = 0, \quad j = 1, \ldots, n. \tag{9.6.11}$$

We claim that $\{x_{n,j} : j = 1, \ldots, n\}$ are the only zeros of E_n in $(0, \pi)$, and are all simple zeros of E_n, in the sense that

$$E_n'(x_{n,j}) \neq 0, \quad j = 1, \ldots, n, \tag{9.6.12}$$

according to which E_n has sign changes at all of its zeros $\{x_{n,j} : j = 1, \ldots, n\}$ in $(0, \pi)$. To prove this statement, suppose there exists a point $\tilde{x} \in (0, \pi) \setminus \{x_{n,1}, \ldots, x_{n,n}\}$ such that $E_n(\tilde{x}) = 0$. Since also, from (9.6.10) and (9.6.2), $E_n(0) = 0$, it follows that E_n has at least $n+2$ distinct zeros in $[0, \pi)$. An application of Rolle's theorem then shows that the derivative E_n' has at least $n+1$ distinct zeros in $(0, \pi)$. But (9.6.10) gives

$$E_n'(x) = 1 - Q_n'(x), \tag{9.6.13}$$

which, together with (9.6.2), shows that E_n' is an even function on $\mathbb{R}$, and thus E_n' also has at least $n+1$ distinct zeros in $(-\pi, 0)$, so that E_n' has at least $2n+2$ distinct zeros in $(-\pi, \pi)$. Moreover, from (9.6.13), (9.6.2) and (9.1.5), we have $E_n' \in \tau_n$. Hence we may apply Theorem 9.1.3(c) to deduce that E_n' is the zero function, that is, from (9.6.13), $Q_n'(x) = 1, x \in \mathbb{R}$, which, together with (9.6.3), (9.6.1), yields $Q_n(x) = x, x \in \mathbb{R}$, and thus, by using also (9.6.2), we obtain the contradiction $0 = Q_n(\pi) = \pi$. It follows that $\{x_{n,1}, \ldots, x_{n,n}\}$ are the only zeros of E_n in $(0, \pi)$.

To prove (9.6.12), suppose $E_n'(x_{n,k}) = 0$ for some integer $k \in \{1,\ldots,n\}$. From (9.6.11) and $E_n(0) = 0$, an application of Rolle's theorem shows that E_n' has at least n distinct zeros in $(0,\pi)$, all of which are different from $x_{n,k}$. Hence E_n' has at least $n+1$ distinct zeros in $(0,\pi)$, which has already been shown above to lead to a contradiction, and thereby proving (9.6.12).

By virtue of the fact that $\{x_{n,j} : j = 0,\ldots,n\}$ are the only zeros of E_n in $(0,\pi)$, and are all simple zeros, it follows from (9.6.10) and (9.6.5) that

$$\int_0^\pi |x - Q_n(x)| dx = \sum_{j=0}^n \int_{\frac{j\pi}{n+1}}^{\frac{(j+1)\pi}{n+1}} |x - Q_n(x)| dx$$

$$= \left| \sum_{j=0}^n \int_{\frac{j\pi}{n+1}}^{\frac{(j+1)\pi}{n+1}} (-1)^j [x - Q_n(x)] dx \right|$$

$$= \left| \sum_{j=0}^n (-1)^j \int_{\frac{j\pi}{n+1}}^{\frac{(j+1)\pi}{n+1}} x\, dx - \sum_{j=0}^n (-1)^j \int_{\frac{j\pi}{n+1}}^{\frac{(j+1)\pi}{n+1}} Q_n(x) dx \right|. \qquad (9.6.14)$$

We claim that

$$\sum_{j=0}^n (-1)^j \int_{\frac{j\pi}{n+1}}^{\frac{(j+1)\pi}{n+1}} Q_n(x) dx = 0, \qquad (9.6.15)$$

which, according to (9.6.2), would follow if we can show that

$$I_k := \sum_{j=0}^n (-1)^j \int_{\frac{j\pi}{n+1}}^{\frac{(j+1)\pi}{n+1}} \sin(kx) dx = 0, \quad k = 1,\ldots,n. \qquad (9.6.16)$$

To prove (9.6.16), let the piecewise constant function $\sigma_n : \mathbb{R} \to \mathbb{R}$ be defined by

$$\sigma_n(x) := \begin{cases} (-1)^j, \; x \in \left[\frac{j\pi}{n+1}, \frac{(j+1)\pi}{n+1} \right), & j = 0,\ldots,n; \\ (-1)^n, \; x = \pi; \end{cases} \qquad (9.6.17)$$

$$\sigma_n(x) := -\sigma_n(-x), \quad x \in (-\pi, 0); \qquad (9.6.18)$$

$$\sigma_n(x) := \sigma_n(x + 2\pi), \quad x \in (-\infty, -\pi]; \qquad (9.6.19)$$

$$\sigma_n(x) := \sigma_n(x - 2\pi), \quad x \in (\pi, \infty). \qquad (9.6.20)$$

It follows from (9.6.17)–(9.6.20) that

$$\sigma_n(x + 2\pi) = \sigma_n(x), \quad x \in \mathbb{R}; \qquad (9.6.21)$$

$$\sigma_n\left(x + \frac{\pi}{n+1} \right) = -\sigma_n(x), \quad x \in [-\pi, \pi] \setminus \left\{ -\frac{\pi}{n+1}, \frac{n\pi}{n+1} \right\}. \qquad (9.6.22)$$

Since, as in the argument leading to (9.3.64) and (9.3.65), we can show that, for any $\alpha \in \mathbb{R}$, we have

$$\int_{-\pi+\alpha}^{\pi+\alpha} g(x)dx = \int_{-\pi}^{\pi} g(x)dx, \tag{9.6.23}$$

for any piecewise constant function g satisfying $g(x+2\pi) = g(x), x \in \mathbb{R}$, it now follows from (9.6.16), (9.6.17), (9.6.21), (9.6.22) and (9.6.23) that, for any $k \in \{1, \ldots, n\}$,

$$I_k = \sum_{j=0}^{n} \int_{\frac{j\pi}{n+1}}^{\frac{(j+1)\pi}{n+1}} \sigma_n(x) \sin(kx) dx$$

$$= \int_0^{\pi} \sigma_n(x) \sin(kx) dx$$

$$= \frac{1}{2} \int_{-\pi}^{\pi} \sigma_n(x) \sin(kx) dx$$

$$= \frac{1}{2} \int_{-\pi-\frac{\pi}{n+1}}^{\pi-\frac{\pi}{n+1}} \sigma_n\left(x+\frac{\pi}{n+1}\right) \sin\left(k\left(x+\frac{\pi}{n+1}\right)\right) dx$$

$$= \frac{1}{2} \int_{-\pi}^{\pi} \sigma_n\left(x+\frac{\pi}{n+1}\right) \sin\left(k\left(x+\frac{\pi}{n+1}\right)\right) dx$$

$$= -\frac{1}{2} \int_{-\pi}^{\pi} \sigma_n(x) \sin\left(kx+\frac{k\pi}{n+1}\right) dx$$

$$= -\frac{1}{2}\cos\left(\frac{k\pi}{n+1}\right) \int_{-\pi}^{\pi} \sigma_n(x) \sin(kx) dx - \frac{1}{2}\sin\left(\frac{k\pi}{n+1}\right) \int_{-\pi}^{\pi} \sigma_n(x)\cos(kx) dx$$

$$= -\cos\left(\frac{k\pi}{n+1}\right) I_k - \frac{1}{2}\left[\sin\left(\frac{k\pi}{n+1}\right)\right](0),$$

and thus

$$\left[1+\cos\left(\frac{k\pi}{n+1}\right)\right] I_k = 0,$$

from which, since

$$1+\cos\left(\frac{k\pi}{n+1}\right) \neq 0, \quad k=1,\ldots,n,$$

the result (9.6.16), and therefore also (9.6.15), then follow.

By substituting (9.6.15) into (9.6.14), we obtain

$$\int_0^{\pi} |x - Q_n(x)| dx = \left| \sum_{j=0}^{n} (-1)^j \int_{\frac{j\pi}{n+1}}^{\frac{(j+1)\pi}{n+1}} x dx \right| = \frac{\pi^2}{2(n+1)^2} \left| \sum_{j=0}^{n} (-1)^j [(j+1)^2 - j^2] \right|$$

$$= \frac{\pi^2}{2(n+1)^2} \left| \sum_{j=0}^{n} (-1)^j (2j+1) \right|. \tag{9.6.24}$$

We claim that

$$\sum_{j=0}^{n}(-1)^{j}(2j+1) = (-1)^{n}(n+1), \quad n \in \mathbb{N}, \tag{9.6.25}$$

which, together with (9.6.24), will yield the desired result (9.6.4).

To prove (9.6.25), we note first that, if $n = 1$, (9.6.25) holds with both sides equal to -2.

To advance the inductive hypothesis from n to $n+1$, we use (9.6.25) to obtain

$$\sum_{j=0}^{n+1}(-1)^{j}(2j+1) = (-1)^{n}(n+1)+(-1)^{n+1}(2n+3)$$

$$= (-1)^{n+1}(-n-1+2n+3) = (-1)^{n+1}((n+1)+1),$$

which shows that (9.6.25) holds with n replaced by $n+1$, and thereby completing our proof. ∎

The result (9.6.4) of Theorem 9.6.1 is instrumental in the proof of the following "Jackson's first theorem", which specifies a convergence rate with respect to the result (9.4.27) in Theorem 9.4.4 for the error functional $\mathscr{E}_{n}^{*}$ corresponding to the Fourier series operator $\mathscr{T}_{n}^{*}$, and for functions $f \in C_{2\pi}$ that are continuously differentiable on $\mathbb{R}$.

Theorem 9.6.2 (Jackson I). *For any positive integer n, the error functional $\mathscr{E}_{n}^{*} : C_{2\pi} \to \mathbb{R}$, as defined by (9.4.26), satisfies*

$$\mathscr{E}_{n}^{*}[f] \leqslant \frac{\pi}{2(n+1)}||f'||_{\infty}, \quad f \in C_{2\pi}^{1}. \tag{9.6.26}$$

Proof. Let $f \in C_{2\pi}^{1}$. By using also (9.3.64), (9.3.65), as well as (9.1.3), we deduce by means of integration by parts that, for any $x \in \mathbb{R}$,

$$\int_{-\pi}^{\pi}\theta f'(\theta+x+\pi)d\theta = \left[\theta f(\theta+x+\pi)\right]_{\theta=-\pi}^{\theta=\pi} - \int_{-\pi}^{\pi}f(\theta+x+\pi)d\theta$$

$$= \pi[f(x+2\pi)+f(x)] - \int_{-\pi+(x+\pi)}^{\pi+(x+\pi)}f(\theta)d\theta$$

$$= 2\pi f(x) - \int_{-\pi}^{\pi}f(\theta)d\theta,$$

and thus

$$f(x) - \frac{1}{2\pi}\int_{-\pi}^{\pi}f(\theta)d\theta + \frac{1}{2\pi}\int_{-\pi}^{\pi}\theta f'(\theta+x+\pi)d\theta, \quad x \in \mathbb{R}. \tag{9.6.27}$$

It follows from (9.4.26), (9.2.1) and (9.6.27) that

$$\mathscr{E}_{n}^{*}[f] = \min_{Q \in \tau_{n}} \max_{-\pi \leqslant x \leqslant \pi} \left| \frac{1}{2\pi}\int_{-\pi}^{\pi}\theta f'(\theta+x+\pi)d\theta - \left\{Q(x) - \frac{1}{2\pi}\int_{-\pi}^{\pi}f(\theta)d\theta\right\} \right|$$

$$= \min_{Q \in \tau_{n}} \max_{-\pi \leqslant x \leqslant \pi} \left| \frac{1}{2\pi}\int_{-\pi}^{\pi}\theta f'(\theta+x+\pi)d\theta - Q(x) \right|, \tag{9.6.28}$$

by virtue of the fact that, according to (9.1.5), the function $Q(x) = 1, x \in \mathbb{R}$, belongs to the set τ_n.

Our next step is to show that, for any given $g \in C_{2\pi}$, and $Q \in \tau_n$, the function

$$\psi(x) := \int_{-\pi}^{\pi} Q(\theta)g(\theta + x)d\theta, \quad x \in \mathbb{R}, \tag{9.6.29}$$

satisfies

$$\psi \in \tau_n. \tag{9.6.30}$$

To this end, we first use the fact that, for $g \in C_{2\pi}$ and $Q \in \tau_n$, it holds from (9.1.6) that, for each fixed $x \in \mathbb{R}$, the product function $Q(\cdot - x)g$ belongs to $C_{2\pi}$, so that we use (9.3.64), (9.3.65) to deduce from (9.6.29) that

$$\psi(x) = \int_{-\pi+x}^{\pi+x} Q(\theta - x)g(\theta)d\theta = \int_{-\pi}^{\pi} Q(\theta - x)g(\theta)d\theta, \quad x \in \mathbb{R}. \tag{9.6.31}$$

Since $Q \in \tau_n$, it follows from the representation formula (9.1.12) that there exist coefficient sequences $\{a_j : j = 0, \ldots, n\}$ and $\{b_j : j = 1, \ldots, n\}$ such that, for $\theta, x \in \mathbb{R}$,

$$Q(\theta - x) = \frac{1}{2}a_0 + \sum_{j=1}^{n} a_j \cos(j(\theta - x)) + \sum_{j=1}^{n} b_j \sin(j(\theta - x))$$

$$= \frac{1}{2}a_0 + \sum_{j=1}^{n} [a_j \cos(j\theta) + b_j \sin(j\theta)] \cos(jx)$$

$$+ \sum_{j=1}^{n} [a_j \sin(j\theta) - b_j \cos(j\theta)] \sin(jx),$$

which, together with (9.6.31), yields, for any $x \in \mathbb{R}$,

$$\psi(x) = \left[\frac{1}{2}a_0 \int_{-\pi}^{\pi} g(\theta)d\theta \right] + \sum_{j=1}^{n} \left\{ \int_{-\pi}^{\pi} [a_j \cos(j\theta) + b_j \sin(j\theta)]g(\theta)d\theta \right\} \cos(jx)$$

$$+ \sum_{j=1}^{n} \left\{ \int_{-\pi}^{\pi} [a_j \sin(j\theta) - b_j \cos(j\theta)]g(\theta)d\theta \right\} \sin(jx). \tag{9.6.32}$$

It follows from (9.6.32) and (9.1.5) that $\psi \in \tau_n$, as required.

Since $f \in C_{2\pi}^1$, it follows from (9.3.62) that the function

$$h(x) := f'(x + \pi) \tag{9.6.33}$$

satisfies $h \in C_{2\pi}$. Moreover, note from (9.6.2) and (9.1.5) that the trigonometric polynomial Q_n in Theorem 9.6.1 satisfies $Q_n \in \tau_n$. Hence we may choose $g = \frac{1}{2\pi}h$ and $Q = Q_n$ in (9.6.29) to deduce from (9.6.30), (9.6.33), as well as (9.6.4) in Theorem 9.6.1, that

$$\min_{Q \in \tau_n} \max_{-\pi \leqslant x \leqslant \pi} \left| \frac{1}{2\pi} \int_{-\pi}^{\pi} \theta f'(\theta + x + \pi)d\theta - Q(x) \right|$$

$$\leqslant \max_{-\pi \leqslant x \leqslant \pi} \left| \frac{1}{2\pi} \int_{-\pi}^{\pi} \theta f'(\theta + x + \pi) d\theta - \frac{1}{2\pi} \int_{-\pi}^{\pi} Q_n(\theta) f'(\theta + x + \pi) d\theta \right|$$

$$= \frac{1}{2\pi} \max_{-\pi \leqslant x \leqslant \pi} \left| \int_{-\pi}^{\pi} [\theta - Q_n(\theta)] f'(\theta + x + \pi) d\theta \right|$$

$$\leqslant \frac{||f'||_\infty}{2\pi} \int_{-\pi}^{\pi} |\theta - Q_n(\theta)| d\theta$$

$$= \frac{||f'||_\infty}{\pi} \int_{0}^{\pi} |\theta - Q_n(\theta)| d\theta = \frac{\pi}{2(n+1)} ||f'||_\infty, \qquad (9.6.34)$$

since the integrand is an even function on $\mathbb{R}$, by virtue of the fact that, from (9.6.2), Q is an odd function on $\mathbb{R}$. The desired result (9.6.26) is now an immediate consequence of (9.6.28) and (9.6.34).   ∎

A function $f : \mathbb{R} \to \mathbb{R}$ is called a Lipschitz-continuous function on $\mathbb{R}$ if there exists a non-negative constant K_f such that

$$|f(x) - f(y)| \leqslant K_f |x - y|, \quad x, y \in \mathbb{R}. \qquad (9.6.35)$$

The constant K_f in (9.6.35) is called a Lipschitz constant for f on $\mathbb{R}$. The class of all functions f that are periodic on $\mathbb{R}$ in the sense of (9.1.3), and Lipschitz-continuous on $\mathbb{R}$, will be denoted by $C_{2\pi}^{\mathrm{Lip}}$. It follows from (9.6.35) that $C_{2\pi}^{\mathrm{Lip}}$ is a linear space. In fact, $C_{2\pi}^{\mathrm{Lip}}$ is situated between the two linear spaces $C_{2\pi}$ and $C_{2\pi}^1$, as follows.

Theorem 9.6.3. *The linear space $C_{2\pi}^{\mathrm{Lip}}$ of Lipschitz-continuous functions on $\mathbb{R}$, as characterised by the conditions (9.1.3) and (9.6.35), satisfies*

$$C_{2\pi}^1 \subset C_{2\pi}^{\mathrm{Lip}} \subset C_{2\pi}. \qquad (9.6.36)$$

Proof. The second inclusion in (9.6.36) is an immediate consequence of (9.1.3), together with the fact that (9.6.35) yields

$$\lim_{y \to x} |f(x) - f(y)| = 0, \quad x \in \mathbb{R}, \quad f \in C_{2\pi}^{\mathrm{Lip}}.$$

To prove the first inclusion in (9.6.36), suppose $f \in C_{2\pi}^1$, and let $x, y \in \mathbb{R}$. But then the mean value theorem implies the existence of a point ξ in the interval joining x and y such that

$$f(x) - f(y) = f'(\xi)(x - y),$$

and thus

$$|f(x) - f(y)| = |f'(\xi)| \, |x - y| \leqslant ||f'||_\infty |x - y|, \qquad (9.6.37)$$

after having used also the fact that $f' \in C_{2\pi}$, from (9.3.62). It follows from (9.6.37) that f is Lipschitz-continuous on $\mathbb{R}$ as in (9.6.35), with Lipschitz constant $K_f = ||f'||_\infty$. Since f

also satisfies (9.1.3), we conclude that $f \in C_{2\pi}^{\text{Lip}}$, and thereby proving the first inclusion in (9.6.36). ∎

Example 9.6.1. In order to obtain a function $f \in C_{2\pi}^{\text{Lip}} \setminus C_{2\pi}^1$, let $f \in C_{2\pi}$ be chosen as the "sawtooth" function, as defined on $[-\pi, \pi]$ in (9.4.16) of Example 9.4.1, for which we immediately observe that $f \notin C_{2\pi}^1$.

Now let $x, y \in \mathbb{R}$. It follows from (9.4.16), together with the facts that $f \in C_{2\pi}$, and f is an even function on $\mathbb{R}$, that there exist points $\tilde{x}, \tilde{y} \in [0, \pi]$, with $|\tilde{x} - \tilde{y}| \leq |x - y|$, such that $f(\tilde{x}) = f(x)$; $f(\tilde{y}) = f(y)$ and thus

$$|f(x) - f(y)| = |f(\tilde{x}) - f(\tilde{y})| = |\tilde{x} - \tilde{y}| \leq |x - y|,$$

according to which f satisfies the Lipschitz condition (9.6.35), with Lipschitz constant $K_f = 1$. We have therefore shown that it indeed holds that $f \in C_{2\pi}^{\text{Lip}} \setminus C_{2\pi}^1$. ∎

By using Theorem 9.6.2, we proceed to prove "Jackson's second theorem" for Lipschitz-continuous periodic functions, as follows.

Theorem 9.6.4 (Jackson II). *For any positive integer n, the error functional $\mathscr{E}_n^* : C_{2\pi} \to \mathbb{R}$, as defined by (9.4.26), satisfies*

$$\mathscr{E}_n^*[f] \leq \frac{\pi}{2(n+1)} K_f, \quad f \in C_{2\pi}^{\text{Lip}}, \tag{9.6.38}$$

with K_f denoting a Lipschitz constant of f, as in (9.6.35).

Proof. Let $f \in C_{2\pi}^{\text{Lip}}$. Our method of proof consists of approximating f with arbitrary uniform "closeness" by a function $\tilde{f} \in C_{2\pi}^1$, followed by an application of Theorem 9.6.2 to $\tilde{f}$.

Let δ be an arbitrary positive number, and define the function

$$\tilde{f}(x) := \frac{1}{2\delta} \int_{x-\delta}^{x+\delta} f(\theta) d\theta, \quad x \in \mathbb{R}. \tag{9.6.39}$$

Since (9.6.39) implies

$$\tilde{f}(x) = \frac{1}{2\delta} \left[\int_0^{x+\delta} f(\theta) d\theta - \int_0^{x-\delta} f(\theta) d\theta \right], \quad x \in \mathbb{R},$$

and since f is continuous on $\mathbb{R}$ by virtue of the second inclusion in (9.6.36) of Theorem 9.6.3, we may apply the fundamental theorem of calculus to deduce that $\tilde{f} \in C^1(\mathbb{R})$, with

$$\tilde{f}'(x) = \frac{1}{2\delta} [f(x+\delta) - f(x-\delta)], \quad x \in \mathbb{R}. \tag{9.6.40}$$

Moreover, since f satisfies the periodicity condition (9.1.3), we may use (9.3.64), (9.3.65) to deduce from (9.6.39) that

$$\tilde{f}(x+2\pi) = \tilde{f}(x), \quad x \in \mathbb{R},$$

and thus $\tilde{f} \in C_{2\pi}^1$.

Hence we may apply (9.6.26) in Theorem 9.6.2 to obtain the inequality

$$\mathscr{E}_n^*[\tilde{f}] \leqslant \frac{\pi}{2(n+1)}||\tilde{f}'||_\infty. \tag{9.6.41}$$

The fact that f is Lipschitz-continuous on $\mathbb{R}$ implies the existence of a constant K_f such that the Lipschitz condition (9.6.35) holds, which, together with (9.6.40), yields

$$|\tilde{f}'(x)| \leqslant \frac{1}{2\delta}K_f|(x+\delta) - (x-\delta)| = K_f, \quad x \in \mathbb{R},$$

and thus

$$||\tilde{f}'||_\infty \leqslant K_f. \tag{9.6.42}$$

It follows from (9.6.41) and (9.6.42) that

$$\mathscr{E}_n^*[\tilde{f}] \leqslant \frac{\pi}{2(n+1)}K_f. \tag{9.6.43}$$

Next, we use (9.6.39) and (9.6.35) to obtain, for any $x \in \mathbb{R}$,

$$|f(x) - \tilde{f}(x)| = \frac{1}{2\delta}\left|\int_{x-\delta}^{x+\delta} [f(x) - f(\theta)]d\theta\right|$$

$$\leqslant \frac{1}{2\delta}\int_{x-\delta}^{x+\delta} |f(x) - f(\theta)|d\theta \leqslant \frac{K_f}{2\delta}\int_{x-\delta}^{x+\delta} |x - \theta|d\theta$$

$$= \frac{K_f}{2\delta}\int_{-\delta}^{\delta} |\theta|d\theta = \frac{K_f}{\delta}\int_0^\delta \theta d\theta = \frac{1}{2}K_f\delta,$$

and thus

$$||f - \tilde{f}||_\infty \leqslant \frac{1}{2}K_f\delta. \tag{9.6.44}$$

Let $Q^* \in \tau_n$ be defined as in (9.4.26) with f replaced by $\tilde{f}$, that is,

$$||\tilde{f} - Q^*||_\infty = \mathscr{E}_n^*[\tilde{f}]. \tag{9.6.45}$$

It follows from (9.4.26), the triangle inequality (4.1.4), together with (9.6.44), (9.6.45) and (9.6.43), that

$$\mathscr{E}_n^*[f] \leqslant ||f - Q^*||_\infty$$

$$= ||(f - \tilde{f}) + (\tilde{f} - Q^*)||_\infty \leqslant ||f - \tilde{f}||_\infty + ||\tilde{f} \quad Q^*||_\infty \leqslant \frac{1}{2}K_f\delta + \frac{\pi}{2(n+1)}K_f,$$

and thus

$$\mathscr{E}_n^*[f] - \frac{\pi}{2(n+1)}K_f \leqslant \frac{1}{2}K_f\delta, \quad \text{for each} \quad \delta > 0,$$

from which the desired inequality (9.6.38) then immediately follows. ∎

By combining Theorem 9.5.6 and Theorem 9.6.4, we can now prove the following uniform convergence result for Fourier series.

Theorem 9.6.5 (Dini-Lipschitz). *The sequence $\{\mathscr{T}_n^* : n = 1,2,\ldots\}$ of Fourier series operators, as defined by (9.3.3), satisfies the uniform convergence result*

$$||f - \mathscr{T}_n^* f||_\infty \to 0, \quad n \to \infty, \quad f \in C_{2\pi}^{\text{Lip}}, \tag{9.6.46}$$

according to which the Fourier series of any $f \in C_{2\pi}^{\text{Lip}}$ converges pointwise to f on $\mathbb{R}$, that is,

$$\frac{1}{2}a_0^* + \sum_{j=1}^\infty a_j^* \cos(jx) + \sum_{j=1}^\infty b_j^* \sin(jx) = f(x), \quad x \in \mathbb{R}, \quad f \in C_{2\pi}^{\text{Lip}}, \tag{9.6.47}$$

with $\{a_j^* : j = 0,1,\ldots\}$ and $\{b_j^* : j = 1,2,\ldots\}$ denoting the Fourier coefficients of f, as given in (9.4.10).

Proof. Let $f \in C_{2\pi}^{\text{Lip}}$. It follows from (9.6.38) in Theorem 9.6.4 that

$$[\ln(2n+1)]\mathscr{E}_n^*[f] \leqslant \frac{\pi K_f}{2} \frac{\ln(2n+1)}{n+1}, \quad n = 1,2,\ldots, \tag{9.6.48}$$

with K_f denoting the Lipschitz constant of f on $\mathbb{R}$, as in (9.6.35). Now observe that an application of L'Hospital's rule yields the limit

$$\lim_{x \to \infty} \frac{\ln(2x+1)}{x+1} = \lim_{x \to \infty} \left[\frac{\frac{2}{2x+1}}{1}\right] = \lim_{x \to \infty} \left[\frac{2}{2x+1}\right] = 0,$$

and thus also

$$\lim_{n \to \infty} \frac{\ln(2n+1)}{n+1} = 0. \tag{9.6.49}$$

It follows from (9.6.48) and (9.6.49) that

$$[\ln(2n+1)]\mathscr{E}_n^*[f] \to 0, \quad n \to \infty,$$

which, together with (9.5.54) in Theorem 9.5.6, yields the uniform convergence result (9.6.46). The pointwise convergence result (9.6.47) is then an immediate consequence of (9.6.46), (9.3.9), (9.3.10), together with the definition (9.2.1) of the maximum norm $||\cdot||_\infty$. ∎

Similar to the Parseval identity (9.4.11) in Theorem 9.4.2, the pointwise convergence result (9.6.47) in Theorem 9.6.5 is also useful for obtaining the sum of certain convergent infinite series, as illustrated in the following example.

Example 9.6.2. Let f denote the "sawtooth" function, as in Examples 9.4.1 and 9.6.1. According to Example 9.6.1, we have that $f \in C_{2\pi}^{\text{Lip}}$, and thus, according to (9.6.47) in Theorem 9.6.5, the Fourier series (9.4.21) of f converges pointwise to f on $\mathbb{R}$, that is,

$$\frac{\pi}{2} - \frac{4}{\pi} \sum_{j=1}^\infty \frac{1}{(2j-1)^2} \cos((2j-1)x) = f(x), \quad x \in \mathbb{R}, \tag{9.6.50}$$

or, in particular, from (9.4.16),

$$\frac{\pi}{2} - \frac{4}{\pi} \sum_{j=1}^{\infty} \frac{1}{(2j-1)^2} \cos((2j-1)x) = |x|, \quad x \in [-\pi, \pi]. \qquad (9.6.51)$$

For example, we may set $x = 0$ in (9.6.51) to obtain

$$\frac{\pi}{2} - \frac{4}{\pi} \sum_{j=1}^{\infty} \frac{1}{(2j-1)^2} = 0,$$

that is,

$$\sum_{j=1}^{\infty} \frac{1}{(2j-1)^2} = \frac{\pi^2}{8}.$$

∎

9.7   Exercises

Exercise 9.1 In Theorem 9.3.3, by applying the recursive formulation (9.3.16), compute the numbers $\{\Gamma_{2j} : j = 1, \ldots, 4\}$.

Exercise 9.2 Verify the Euler-Maclaurin formula for the case $f(x) = x^6$, $[a, b] = [0, 1]$, $n = m = 2$, by explicitly calculating both sides of equation (9.3.12) in Theorem 9.3.3.
[*Hint:* Use (8.6.22), (8.6.21), (8.4.3), (9.3.56) and (9.3.15), as well as Exercise 9.1.]

Exercise 9.3 Verify the validity of Theorem 9.3.4 (b) for the case $m = 2$ by explicitly calculating the integral in the left hand side of equation (9.3.43).

Exercise 9.4 Compute the Bernoulli numbers B_{12} and B_{14} by means of (9.3.13) and (9.3.56), and verify that the condition (9.3.44) in Theorem 9.3.4(c) is satisfied for $m = 6$ and $m = 7$.

Exercise 9.5 By applying the recursive formulation (9.3.39), prove that the odd-indexed Bernoulli numbers $\{B_{2j+1} : j = 1, 2, \ldots\}$ are all zero, as in (9.3.41).

Exercise 9.6 Let $f \in C^{2m+2}[a, b]$ for some non-negative integer m, and suppose f satisfies the condition (9.3.58) of Theorem 9.3.6. Show that the upper bounds $\{C_n : n \in \mathbb{N}\}$ in the trapezoidal rule error estimate

$$\left| \mathcal{E}_n^{TR}[f] \right| := \left| \int_a^b f(x) dx - \mathcal{Q}_n^{TR}[f] \right| \leq C_n,$$

as obtained from (9.3.59), satisfy the decay condition

$$C_{2n} = \left(\frac{1}{2} \right)^{2m+2} C_n, \quad n \in \mathbb{N}. \qquad (*)$$

Exercise 9.7 Show that the function

$$f(x) = \frac{1}{2 + \sin x}, \quad x \in \mathbb{R},$$

satisfies the condition (9.3.58) of Theorem 9.3.6, with $[a,b] = [-\pi, \pi]$, and for each $m \in \mathbb{N}$.
[*Hint:* Apply (9.3.62).]

Exercise 9.8 As a continuation of Exercise 9.7, and analogously to Exercises 8.23 and 8.24, by computing the quadrature error $\mathscr{E}_n^{TR}[f]$ for $n = 1, \ldots, 4$, investigate numerically whether, analogously to $(*)$ in Exercise 9.6, the decay rate

$$|\mathscr{E}_{2n}^{TR}[f]| \approx \left(\frac{1}{2}\right)^{2m+2} |\mathscr{E}_n^{TR}[f]|$$

is achieved for this choice of f, and where the non-negative number m may be chosen arbitrarily.
[*Hint:* Show first that

$$\int_{-\pi}^{\pi} \frac{1}{2 + \sin x} dx = \frac{2\pi}{\sqrt{3}} \approx 3.62759873,$$

by setting $x = 2 \arctan t$.]

Exercise 9.9 For any positive integers n and m, the Euler-Maclaurin quadrature rule $\mathscr{Q}_{n,m}^{EM}$ for the numerical approximation of the integral

$$\int_a^b f(x)dx, \quad f \in C^m[a,b],$$

is defined by

$$\mathscr{Q}_{n,m}^{EM}[f] := \mathscr{Q}_n^{TR}[f] - \sum_{j=1}^m \frac{B_{2j}}{(2j)!}[f^{(2j-1)}(b) - f^{(2j-1)}(a)]\left(\frac{b-a}{n}\right)^{2j}, \quad f \in C^m[a,b],$$

with $\mathscr{Q}_n^{TR}$ denoting the trapezoidal rule given by (8.6.22), (8.6.21), (8.4.3), and where $\{B_{2j} : j = 1, 2, \ldots\}$ are the even-indexed Bernoulli numbers as defined recursively in (9.3.13). Prove that the quadrature rule $\mathscr{Q}_{n,m}^{EM}$ has degree of exactness equal to $2m+1$, and satisfies the quadrature error estimate

$$\left|\int_a^b f(x)dx - \mathscr{Q}_{n,m}^{EM}[f]\right| \leqslant (b-a)\frac{|B_{2m+2}|}{(2m+2)!}\left(\frac{b-a}{n}\right)^{2m+2}||f^{(2m+2)}||_\infty,$$

$$f \in C^{2m+2}[a,b]. \qquad (**)$$

[*Hint:* Apply Theorem 9.3.5.]

Exercise 9.10 As a continuation of Exercise 9.9, and as an extension of Exercise 8.25, calculate the value of $\mathscr{Q}_{2,m}^{EM}[f]$, with $[a,b] = [0,1]$, and

$$f(x) = e^{-x^2}, \quad x \in [0,1],$$

and where m is the smallest value for which it holds, according to the quadrature error estimate $(**)$ in Exercise 9.9, that

$$\left|\int_0^1 f(x)dx - \mathscr{Q}_{2,m}^{EM}[f]\right| < \frac{1}{100}.$$

Exercise 9.11 For the function
$$f(x) = \frac{1}{2 + \sin x}, \quad x \in \mathbb{R},$$
apply the method described in (9.3.73) - (9.3.76) to obtain the trigonometric polyno-mial $\widetilde{\mathscr{F}}_{2,4} f \in \tau_2$, with $\widetilde{\mathscr{F}}_{2,4}$ denoting the discrete Fourier series operator given in (9.3.68), (9.3.67).

Exerice 9.12 As a continuation of Exercise 9.11, consider the trigonometric polynomial $\mathscr{F}_2^* f \in \tau_2$, with $\mathscr{F}_2^*$ denoting the Fourier series operator given in (9.3.9), (9.3.10), and calculate the upper bounds in the Fourier coefficient error estimates (9.3.69).

Exercise 9.13 As a continuation of Exercise 9.12, apply the bounds obtained there to obtain an upper bound on the quantity
$$\| \mathscr{F}_2^* f - \widetilde{\mathscr{F}}_{2,4} f \|_\infty := \max_{x \in [-\pi, \pi]} \left| (\mathscr{F}_2^* f)(x) - (\widetilde{\mathscr{F}}_{2,4} f)(x) \right|.$$

Exercise 9.14 In each of the following two cases, apply the formulations in (9.4.10) to calculate the Fourier coefficients $\{a_j^* : j = 0, 1, \ldots\}$ and $\{b_j^* : j = 1, 2, \ldots\}$ for the function $f \in C_{2\pi}$, as given on the interval $(-\pi, \pi]$ by:

$$\text{(a)} \ f(x) = x^2, \quad x \in (-\pi, \pi];$$

$$\left. \begin{array}{l} \text{(b)} \ f(x) = 1 - \dfrac{4}{\pi^2} \left(x - \dfrac{\pi}{2} \right)^2, \quad x \in [0, \pi]; \\[2mm] \qquad f(x) = -f(-x), \qquad\qquad x \in (0, \pi). \end{array} \right\}$$

Exercise 9.15 As a continuation of Exercise 9.14, verify for each of the cases (a) and (b) that the Fourier coefficients $\{a_j^* : j = 0, 1, \ldots\}$ and $\{b_j^* : j = 1, 2, \ldots\}$ satisfy the property (9.4.15) in Theorem 9.4.3.

Exercise 9.16 As a further continuation of Exercise 9.14, write down the Fourier series of the functions f in (a) and (b).

Exercise 9.17 As yet another continuation of Exercise 9.14, apply the Parseval identity (9.4.11) in Theorem 9.4.2 to each of the functions f in (a) and (b) of Exercise 9.14 to obtain, respectively, the sum of the infinite series
$$\text{(a)} \ 1 + \frac{1}{2^4} + \frac{1}{3^4} + \frac{1}{4^4} + \cdots; \qquad \text{(b)} \ 1 + \frac{1}{3^6} + \frac{1}{5^6} + \frac{1}{7^6} + \cdots.$$

Exercise 9.18 In the proof of Theorem 9.5.3, provide the details in the derivations of (9.5.21) - (9.5.24).

Exercise 9.19 In the proof of Theorem 9.5.3, provide the details in the derivations of (9.5.26), (9.5.29), (9.5.30).

Exercise 9.20 According to (9.5.8) in Theorem 9.5.2 and (9.5.18) in Theorem 9.5.3, it holds that
$$\frac{4}{\pi^2} \ln(n+1) < \| \mathscr{F}_n^* \|_\infty \leqslant 1 + \ln(2n+1), \ n \in \mathbb{N}.$$

Verify these inequalities for $n = 1, \ldots, 10$, by using the computed values of $\|\mathscr{T}_n^*\|_\infty$ in Table 9.5.1.

Exercise 9.21 For the case $n = 2$ of Theorem 9.6.1, use the matrix formulation (9.6.6) in the proof to obtain the inverse matrix A_2^{-1}, and write down an explicit formula for the trigonometric polynomial Q_2 of the theorem.

Exercise 9.22 As a continuation of Exercise 9.21, show by means of explicit integration that (9.6.4) holds for $n = 2$, that is,

$$\int_0^\pi |x - Q_2(x)| dx = \frac{\pi^2}{6}.$$

Exercise 9.23 For the function f as in Exercise 9.11, show that

$$\|f'\|_\infty := \max_{x \in [-\pi, \pi]} |f'(x)| = \frac{1}{4},$$

and then apply this result, together with Jackson's first theorem, as formulated in (9.6.26) of Theorem 9.6.2, as well as the Lebesgue inequality (9.5.52) in Theorem 9.5.5, to establish the error estimate

$$\|f - \mathscr{T}_n^* f\|_\infty \leqslant \frac{\pi}{8} \left[\frac{2 + \ln(2n + 1)}{n + 1} \right], \quad n \in \mathbb{N}. \tag{$*$}$$

Exercise 9.24 As a continuation of Exercise 9.23, find the smallest value of n for which we are guaranteed, according to the error estimate $(*)$ of Exercise 9.23, that

$$\|f - \mathscr{T}_n^* f\|_\infty < \frac{1}{10}.$$

Exercise 9.25 Prove that the function $f \in C_{2\pi}$, as defined on $(-\pi, \pi]$ by

$$f(x) = \begin{cases} x \sin \frac{1}{x}, & x \in (0, \pi]; \\ 0, & x = 0; \\ f(-x), & x \in (-\pi, 0), \end{cases}$$

satisfies

$$f \in C_{2\pi} \setminus C_{2\pi}^{\text{Lip}}.$$

[*Hint:* To prove that $f \notin C_{2\pi}^{\text{Lip}}$, show that the definition

$$x_0 := 0 \quad ; \quad x_n := \frac{1}{n}, \, n \in \mathbb{N},$$

yields

$$\frac{f(x_0) - f(x_n)}{x_0 - x_n} \to \infty, \quad n \to \infty.]$$

Exercise 9.26 For each of the two functions $f \in C_{2\pi}$ of (a) and (b) in Exercise 9.14, show that

$$f \in C_{2\pi}^{\text{Lip}} \setminus C_{2\pi}^1,$$

and give the corresponding Lipschitz constants K_f.

[*Hint:* Apply the method used in Example 9.6.1.]

Exercise 9.27 As a continuation of Exercise 9.26, and analogously to Exercises 9.23 and 9.24, apply Jackson's second theorem, as formulated in (9.6.38) of Theorem 9.6.4, together with the Lebesgue inequality (9.5.52) in Theorem 9.5.5, to find the smallest value of n for which we are guaranteed that

$$\|f - \mathscr{T}_n^* f\|_\infty < \frac{1}{10},$$

for each of the functions f in (a) and (b) of Exercise 9.14.

Exercise 9.28 By applying the Dini-Lipschitz theorem, as formulated by (9.6.46) in Theorem 9.6.5, and using Exercise 9.26, prove the identities

(a) $x^2 = \frac{1}{2}a_0^* + \sum_{j=1}^{\infty} a_j^* \cos(jx) + \sum_{j=1}^{\infty} b_j^* \sin(jx), \quad x \in [-\pi, \pi];$

(b) $1 - \frac{4}{\pi^2}(x - \frac{\pi}{2})^2 = \frac{1}{2}a_0^* + \sum_{j=1}^{\infty} a_j^* \cos(jx) + \sum_{j=1}^{\infty} b_j^* \sin(jx), \quad x \in [0, \pi],$

with $\{a_j^* : j = 0, 1, \ldots\}$ and $\{b_j^* : j = 1, 2, \ldots\}$ denoting the Fourier coefficients calculated in Exercise 9.14 for the functions f in, respectively, (a) and (b) of Exercise 9.14.

Exercise 9.29 Apply the identity in Exercise 9.28(a) to obtain the sum of each of the following two infinite series:

(i) $1 + \frac{1}{2^2} + \frac{1}{3^2} + \frac{1}{4^2} + \cdots;$ (ii) $1 - \frac{1}{2^2} + \frac{1}{3^2} - \frac{1}{4^2} + \cdots.$

Exercise 9.30 Apply the identity in Exercise 9.28(b) to obtain the sum of the infinite series

$$1 - \frac{1}{3^3} + \frac{1}{5^3} - \frac{1}{7^3} + \cdots.$$

Chapter 10

Spline Approximation

The main focus in the previous chapters has been approximation of functions by algebraic (or trigonometric) polynomials, while achieving arbitrarily desirable approximation accuracy by increasing the polynomial degrees, at the expense of increasing computational complexity and undesirable features, such as increase in oscillation of the polynomial approximant. To avoid the increase of polynomial degrees, this chapter is devoted to the study of approximation by piecewise polynomials with fixed degrees, while achieving arbitrarily desirable approximation accuracy by allowing the decrease in spacing of the break-points of the polynomial pieces. To meet the need of smooth piecewise polynomial approximants for the representation of functions $f \in C[a,b]$, certain smoothing conditions are imposed on the adjacent polynomial pieces. The resulting basis functions are called B-splines and the breakpoints of the polynomial pieces that constitute the B-splines are called knots.

10.1 Spline spaces

For integers $m \geqslant 0$ and $r \geqslant 1$, and any sequence $\{\tau_1, \ldots, \tau_r\} \subset \mathbb{R}$ satisfying

$$\tau_1 < \tau_2 < \cdots < \tau_r, \tag{10.1.1}$$

the spline space $\sigma_m(\tau_1, \ldots, \tau_r)$ is defined as the linear space of all piecewise polynomials S such that, for some polynomial sequence

$$\{P_0, \ldots, P_r\} \subset \pi_m, \tag{10.1.2}$$

the function $S : \mathbb{R} \to \mathbb{R}$ is given by

$$S(x) := \begin{cases} P_0(x) \, , x \in (-\infty, \tau_1); \\ P_j(x) \, , x \in [\tau_j, \tau_{j+1}), \quad j = 1, \ldots, r-1; \\ P_r(x) \, , x \in [\tau_r, \infty), \end{cases} \tag{10.1.3}$$

289

and with S satisfying the continuity condition

$$S \in C^{m-1}(\mathbb{R}), \quad \text{if} \quad m \in \mathbb{N}. \tag{10.1.4}$$

The points $\{\tau_1, \ldots, \tau_r\}$ are called the knots of the spline space $\sigma_m(\tau_1, \ldots, \tau_r)$, and a piece-wise polynomial $S \in \sigma_m(\tau_1, \ldots, \tau_r)$ is called a spline. Observe that the inclusion

$$\pi_m \subset \sigma_m(\tau_1, \ldots, \tau_r) \tag{10.1.5}$$

is satisfied, since, if $P \in \pi_m$, we may choose, in (10.1.2),

$$P_j = P, \quad j = 0, \ldots, r, \tag{10.1.6}$$

according to which (10.1.3) yields $S = P$, so that (10.1.4) is also satisfied, and thus $P \in \sigma_m(\tau_1, \ldots, \tau_r)$.

To motivate the choice (10.1.4) for the continuity degree of splines in $\sigma_m(\tau_1, \ldots, \tau_r)$, suppose $\widetilde{\sigma}_m(\tau_1, \ldots, \tau_r)$ is the linear space of piecewise polynomials as in (10.1.2), (10.1.3), but with the continuity condition (10.1.4) replaced by

$$S \in C^m(\mathbb{R}). \tag{10.1.7}$$

Let $S \in \widetilde{\sigma}_m(\tau_1, \ldots, \tau_r)$. For any fixed $j \in \{1, \ldots, r\}$, it then follows from (10.1.3) and (10.1.7) that

$$P_{j-1}^{(k)}(\tau_j) = P_j^{(k)}(\tau_j), \quad k = 0, \ldots, m, \tag{10.1.8}$$

according to which the polynomial

$$\widetilde{P}_j := P_j - P_{j-1} \in \pi_m \tag{10.1.9}$$

satisfies

$$\widetilde{P}_j^{(k)}(\tau_j) = 0, \quad k = 0, \ldots, m. \tag{10.1.10}$$

Now recall the Taylor expansion polynomial identity

$$P(x) = \sum_{k=0}^{m} \frac{P^{(k)}(c)}{k!}(x-c)^k, \quad x \in \mathbb{R}, \quad P \in \pi_m, \tag{10.1.11}$$

for any $c \in \mathbb{R}$, as follows immediately from (9.3.11). By applying the identity (10.1.11), with the choice $c = \tau_j$, to the polynomial $\widetilde{P}_j$ in (10.1.9), and using (10.1.10), we deduce that $\widetilde{P}_j$ is the zero polynomial, and it follows from (10.1.9) that (10.1.6) is satisfied for some polynomial $P \in \pi_m$, that is, $S \in \pi_m$. Hence we have shown that $\widetilde{\sigma}_m(\tau_1, \ldots, \tau_r) \subset \pi_m$, and since also, analogously to (10.1.5), we have $\pi_m \subset \widetilde{\sigma}_m(\tau_1, \ldots, \tau_r)$, it follows that

$$\widetilde{\sigma}_m(\tau_1, \ldots, \tau_r) = \pi_m. \tag{10.1.12}$$

By recalling also the continuity requirement (10.1.7) for $\tilde{\sigma}_m(\tau_1, \ldots, \tau_r)$, it follows from (10.1.12) that (10.1.4) is the optimal continuity condition for which the corresponding linear space of piecewise polynomials S as in (10.1.3), (10.1.2) have the potential to provide an extension of π_m, in the sense that (10.1.5) is a proper inclusion.

For any non-negative integer m, we define the truncated power

$$x_+^m := \begin{cases} x^m, & x \geq 0; \\ 0, & x < 0, \end{cases} \qquad (10.1.13)$$

where $0^0 := 1$. It follows from (10.1.13) (see Exercise 10.1) that

$$(\cdot)_+^m \in C^{m-1}(\mathbb{R}), \quad m \in \mathbb{N}. \qquad (10.1.14)$$

For any spline space $\sigma_m(\tau_1, \ldots, \tau_r)$, we note from (10.1.13) that

$$(x - \tau_j)_+^m = \left. \begin{cases} 0, & x < \tau_j, \\ (x - \tau_j)^m, & x \geq \tau_j, \end{cases} \right\} \quad j = 1, \ldots r, \qquad (10.1.15)$$

from which, together with (10.1.14), we deduce that

$$(\cdot - \tau_j)_+^m \in \sigma_m(\tau_1, \ldots, \tau_r), \quad j = 1, \ldots, r. \qquad (10.1.16)$$

Also, (10.1.15) shows that

$$(\cdot - \tau_j)_+^m \notin \pi_m, \quad j = 1, \ldots, r, \qquad (10.1.17)$$

and it follows that π_m is indeed a proper subspace of $\sigma_m(\tau_1, \ldots, \tau_r)$.

The result (10.1.16) enables us to extend the standard basis $\{1, x, \ldots, x^m\}$ of π_m to obtain a basis for the spline space $\sigma_m(\tau_1, \ldots, \tau_r)$, as follows.

Theorem 10.1.1. *The spline space $\sigma_m(\tau_1, \ldots, \tau_r)$ is finite-dimensional, with dimension*

$$\dim \sigma_m(\tau_1, \ldots, \tau_r) = m + 1 + r, \qquad (10.1.18)$$

and the set

$$X := \{1, x, \ldots, x^m, (x - \tau_1)_+^m, (x - \tau_2)_+^m, \ldots, (x - \tau_r)_+^m\} \qquad (10.1.19)$$

is a basis for $\sigma_m(\tau_1, \ldots, \tau_r)$.

Proof. Since, according to (10.1.5) and (10.1.16), the set X in (10.1.19) satisfies $X \subset \sigma_m(\tau_1, \ldots, \tau_r)$, with X containing precisely $m + 1 + r$ elements, our result will be proved if we can show that

$$\sigma_m(\tau_1, \ldots, \tau_r) = \text{span } X, \qquad (10.1.20)$$

and

$$X \text{ is a linearly independent set.} \tag{10.1.21}$$

For $S \in \operatorname{span} X$, it follows from (10.1.19), (10.1.5) and (10.1.16) that $S \in \sigma_m(\tau_1, \ldots, \tau_r)$, and thus

$$\operatorname{span} X \subset \sigma_m(\tau_1, \ldots, \tau_r). \tag{10.1.22}$$

Next, suppose $S \in \sigma_m(\tau_1, \ldots, \tau_r)$, and denote by $\{P_0, \ldots, P_r\}$ the polynomial sequence in π_m for which (10.1.3) is satisfied. Let $j \in \{1, \ldots, r\}$ be fixed. Since S satisfies the continuity condition (10.1.4), it follows from (10.1.3) that

$$P_{j-1}^{(k)}(\tau_j) = P_j^{(k)}(\tau_j), \quad k = 0, \ldots, m-1, \tag{10.1.23}$$

and thus the polynomial

$$\widetilde{P}_j := P_j - P_{j-1} \in \pi_m \tag{10.1.24}$$

satisfies

$$\widetilde{P}_j^{(k)}(\tau_j) = 0, \quad k = 0, \ldots, m-1. \tag{10.1.25}$$

Moreover, since $\widetilde{P}_j \in \pi_m$, we may apply the Taylor expansion polynomial identity (10.1.11), with $c = \tau_j$, together with (10.1.25), to obtain

$$\widetilde{P}_j(x) = d_j(x - \tau_j)_+^m, \quad x \in \mathbb{R}, \tag{10.1.26}$$

where

$$d_j := \frac{\widetilde{P}^{(m)}(\tau_j)}{m!}. \tag{10.1.27}$$

Note from (10.1.24) and (10.1.26) that

$$P_j(x) = P_{j-1}(x) + d_j(x - \tau_j)_+^m, \quad x \in \mathbb{R}, \quad j = 1, \ldots, r. \tag{10.1.28}$$

We claim that (10.1.28) implies the formula

$$S(x) = P_0(x) + \sum_{j=1}^{r} d_j(x - \tau_j)_+^m, \quad x \in \mathbb{R}. \tag{10.1.29}$$

To prove (10.1.29), we first note from the first lines of (10.1.3) and (10.1.15) that (10.1.29) holds for $x \in (-\infty, \tau_1)$. Next, for $x \in [\tau_1, \tau_2)$, it follows from the second line of (10.1.3), together with (10.1.28), and the second line of (10.1.15), that

$$S(x) = P_1(x) = P_0(x) + d_1(x - \tau_1)^m, \tag{10.1.30}$$

which, according to (10.1.15), agrees with (10.1.29). Similarly, for $x \in [\tau_2, \tau_3)$, we get

$$S(x) = P_2(x) = P_1(x) + d_2(x - \tau_2)^m = P_0(x) + \sum_{j=1}^{2} d_j(x - \tau_j)^m,$$

from (10.1.30), and again yielding (10.1.29). By repeating this argument for the successive intervals $[\tau_3, \tau_4), \ldots, [\tau_{r-1}, \tau_r), [\tau_r, \infty)$, the formula (10.1.29) is proved. Since, moreover, $P_0 \in \pi_m = \mathrm{span}\{1, x, \ldots, x^m\}$, it follows from (10.1.29) and (10.1.19) that $S \in \mathrm{span}\, X$. Hence we have shown that $\sigma_m(\tau_1, \ldots, \tau_r) \subset \mathrm{span}\, X$, which, together with (10.1.22), then proves (10.1.20).

To prove (10.1.21), suppose the coefficient sequence $\{c_0, \ldots, c_m, d_1, \ldots, d_r\} \subset \mathbb{R}$ satisfies

$$\sum_{j=0}^{m} c_j x^j + \sum_{j=1}^{r} d_j(x - \tau_j)_+^m = 0, \quad x \in \mathbb{R}. \tag{10.1.31}$$

It follows from (10.1.31) and the first line of (10.1.15) that

$$\sum_{j=0}^{m} c_j x^j = 0, \quad x \in (-\infty, \tau_1),$$

and thus

$$c_j = 0, \quad j = 0, \ldots, m, \tag{10.1.32}$$

which, together with (10.1.31), gives

$$\sum_{j=1}^{r} d_j(x - \tau_j)_+^m = 0, \quad x \in \mathbb{R}. \tag{10.1.33}$$

By using (10.1.15), we deduce from (10.1.33) that

$$d_1(x - \tau_1)^m = 0, \quad x \in [\tau_1, \tau_2),$$

and thus $d_1 = 0$, which, together with (10.1.33), implies

$$\sum_{j=2}^{r} d_j(x - \tau_j)_+^m = 0, \quad x \in \mathbb{R}. \tag{10.1.34}$$

By repeating the same argument for the successive intervals $[\tau_2, \tau_3), \ldots, [\tau_{r-1}, \tau_r), [\tau_r, \infty)$, we obtain $d_2 = \cdots = d_r = 0$, and thus

$$d_1 = \cdots = d_r = 0. \tag{10.1.35}$$

According to (10.1.32) and (10.1.35), we have now proved the desired linear independence result (10.1.21). ∎

In Section 10.2 we shall construct a spline sequence that provides a more efficient basis for $\sigma_m(\tau_1, \ldots, \tau_r)$ than the set X in (10.1.19).

We shall rely on two further properties of spline spaces, as formulated in the following two theorems.

Theorem 10.1.2. *Suppose $S \in \sigma_m(\tau_1, \ldots, \tau_r)$, where $m \geqslant 2$. Then the derivatives of S satisfy*

$$S^{(k)} \in \sigma_{m-k}(\tau_1, \ldots, \tau_r), \quad k = 1, \ldots, m-1. \tag{10.1.36}$$

Proof. Let $k \in \{1, \ldots, m-1\}$ be fixed. Since (10.1.4) is satisfied, we deduce that

$$S^{(k)} \in C^{m-1-k}(\mathbb{R}). \tag{10.1.37}$$

Also, with $\{P_0, \ldots, P_r\} \subset \pi_m$ denoting the polynomial sequence satisfying (10.1.3), it follows from (10.1.3) and (10.1.37) that

$$S^{(k)}(x) = \begin{cases} P_0^{(k)}(x) \,, x \in (-\infty, \tau_1); \\ P_j^{(k)}(x) \,, x \in [\tau_j, \tau_{j+1}), \quad j = 1, \ldots, r-1; \\ P_r^{(k)}(x) \,, x \in [\tau_r, \infty). \end{cases} \tag{10.1.38}$$

Moreover, (10.1.2) implies

$$\{P_0^{(k)}, \ldots, P_r^{(k)}\} \subset \pi_{m-k}. \tag{10.1.39}$$

The desired result (10.1.36) is now an immediate consequence of (10.1.39), (10.1.38) and (10.1.37). ∎

Our next result shows that, if knots are removed from a spline space $\sigma_m(\tau_1, \ldots, \tau_r)$, then the new spline space thus obtained is a subspace of $\sigma_m(\tau_1, \ldots, \tau_r)$.

Theorem 10.1.3. *For any integers r and ρ such that $r \geqslant 2$ and $1 \leqslant \rho \leqslant r - 1$, suppose the integer set $\{j_1, \ldots, j_\rho\}$ satisfies*

$$j_1 < j_2 < \cdots < j_\rho \tag{10.1.40}$$

and

$$\{j_1, \ldots, j_\rho\} \subset \{1, \ldots, r\}. \tag{10.1.41}$$

Then

$$\sigma_m(\tau_{j_1}, \ldots, \tau_{j_\rho}) \subset \sigma_m(\tau_1, \ldots, \tau_r). \tag{10.1.42}$$

Proof. Let $S \in \sigma_m(\tau_{j_1}, \ldots, \tau_{j_\rho})$. As in (10.1.2), (10.1.3), it then holds, for some polynomial sequence $\{\widetilde{P}_0, \ldots, \widetilde{P}_\rho\} \subset \pi_m$, that

$$S(x) = \begin{cases} \widetilde{P}_0(x), \, x \in (-\infty, \tau_{j_1}); \\ \widetilde{P}_\ell(x), \, x \in [\tau_{j_\ell}, \tau_{j_{\ell+1}}), \quad \ell = 1, \ldots, \rho - 1; \\ \widetilde{P}_\rho(x), \, x \in [\tau_{j_\rho}, \infty). \end{cases} \tag{10.1.43}$$

It then follows from (10.1.40), (10.1.41) and (10.1.43) that the polynomial sequence $\{P_0, \ldots, P_r\} \subset \pi_m$ defined by

$$P_j := \begin{cases} \widetilde{P}_0, \, j = 0, \ldots, j_1 - 1; \\ \widetilde{P}_1, \, j = j_\ell, \ldots, j_{\ell+1} - 1, \quad \ell = 1, \ldots, \rho - 1; \\ \widetilde{P}_\rho, \, j = j_\rho, \ldots, r, \end{cases} \tag{10.1.44}$$

satisfies (10.1.3), from which we then deduce that $S \in \sigma_m(\tau_1, \ldots, \tau_r)$, and thereby completing our proof of the inclusion (10.1.42). ∎

10.2   B-splines

If a function $f : \mathbb{R} \to \mathbb{R}$ satisfies

$$f(x) = 0, \quad x \in \mathbb{R} \setminus [\alpha, \beta], \tag{10.2.1}$$

for some bounded interval $[\alpha, \beta]$, we say that f is a finitely supported function.

For any spline space $\sigma_m(\tau_1, \ldots, \tau_r)$, we define the subspace

$$\sigma_{m,0}(\tau_1, \ldots, \tau_r) := \{ S \in \sigma_m(\tau_1, \ldots, \tau_r) : S \text{ is finitely supported} \}. \tag{10.2.2}$$

The following result then holds.

Theorem 10.2.1. *The spline subspace $\sigma_{m,0}(\tau_1, \ldots, \tau_r)$, as defined by* (10.2.2), *satisfies:*
(a) *If $S \in \sigma_{m,0}(\tau_1, \ldots, \tau_r)$, then*

(i)

$$S(x) = 0, \quad x \in \mathbb{R} \setminus [\tau_1, \tau_r); \tag{10.2.3}$$

(ii)

$$S \in \operatorname{span}\{ (\cdot - \tau_1)_+^m, \ldots, (\cdot - \tau_r)_+^m \}. \tag{10.2.4}$$

(b) *The subspace $\sigma_{m,0}(\tau_1, \ldots, \tau_r)$ is non-trivial, that is*

$$\sigma_{m,0}(\tau_1, \ldots, \tau_r) \neq \{0\}, \tag{10.2.5}$$

if and only if

$$r \geqslant m + 2. \tag{10.2.6}$$

(c) *For $r = m + 2$, it holds that*

$$S \in \sigma_{m,0}(\tau_1, \ldots, \tau_{m+2}) \tag{10.2.7}$$

if and only if

$$S(x) = c \sum_{j=1}^{m+2} \left[\frac{1}{\displaystyle\prod_{\substack{j \neq k=1}}^{m+2} (\tau_k - \tau_j)} \right] (x - \tau_j)_+^m, \quad x \in \mathbb{R}, \tag{10.2.8}$$

with c denoting an arbitrary real constant.

Proof. (a) Let $S \in \sigma_{m,0}(\tau_1,\ldots,\tau_r)$.

(i) According to the definition (10.2.2), the polynomials P_0 and P_r in the formulation (10.1.3) must both be the zero polynomial, which proves (10.2.3).

(ii) Since $\sigma_{m,0}(\tau_1,\ldots,\tau_r)$ is a subspace of $\sigma_m(\tau_1,\ldots,\tau_r)$, it follows from Theorem 10.1.1 that $S = P + \widetilde{S}$, where $P \in \pi_m$ and $\widetilde{S} \in \text{span}\{(\cdot - \tau_1)_+^m,\ldots,(\cdot - \tau_r)_+^m\}$, and thus, from (10.2.3) and the first line of (10.1.15),

$$0 = S(x) = P(x), \quad x \in (-\infty, \tau_1),$$

according to which P is the zero polynomial, and therefore $S = \widetilde{S}$, which proves (10.2.4).

(b) Let $S \in \sigma_{m,0}(\tau_1,\ldots,\tau_r)$. It follows from (10.2.4) that

$$S(x) = \sum_{j=1}^{r} d_j(x - \tau_j)_+^m, \quad x \in \mathbb{R}, \tag{10.2.9}$$

for some coefficient sequence $\{d_1,\ldots,d_r\} \subset \mathbb{R}$. Hence, from (10.2.9), together with (10.2.3), as well as the second line of (10.1.15), we obtain

$$\sum_{j=1}^{r} d_j(x - \tau_j)^m = 0, \quad x \in [\tau_r, \infty),$$

or equivalently,

$$\sum_{j=1}^{r} d_j(x - \tau_j)^m = 0, \quad x \in \mathbb{R}. \tag{10.2.10}$$

Since, for any $x \in \mathbb{R}$, we have

$$\sum_{j=1}^{r} d_j(x - \tau_j)^m = \sum_{j=1}^{r} d_j \sum_{k=0}^{m} \binom{m}{k} x^{m-k}(-1)^k \tau_j^k = \sum_{k=0}^{m} (-1)^k \binom{m}{k}\left[\sum_{j=1}^{r} \tau_j^k d_j\right] x^{m-k},$$

it follows that (10.2.10) is equivalent to

$$\sum_{k=0}^{m} (-1)^k \binom{m}{k}\left[\sum_{j=1}^{r} \tau_j^k d_j\right] x^{m-k} = 0, \quad x \in \mathbb{R},$$

which holds if and only if

$$\sum_{j=1}^{r} \tau_j^k d_j = 0, \quad k = 0,\ldots,m. \tag{10.2.11}$$

Hence we have shown that $S \in \sigma_{m,0}(\tau_1,\ldots,\tau_r)$ if and only if S is given by the formulation (10.2.9), where $\{d_1,\ldots,d_r\}$ is a real coefficient sequence satisfying the homogeneous linear system (10.2.11).

Now observe that (10.2.11) has the matrix-vector formulation

$$A\mathbf{d} = \mathbf{0}, \tag{10.2.12}$$

where A is the $(m+1) \times r$ matrix

$$A := \begin{bmatrix} 1 & 1 & \cdots & 1 \\ \tau_1 & \tau_2 & \cdots & \tau_r \\ \tau_1^2 & \tau_2^2 & \cdots & \tau_r^2 \\ \vdots & \vdots & & \vdots \\ \tau_1^m & \tau_2^m & \cdots & \tau_r^m \end{bmatrix}, \tag{10.2.13}$$

and $\mathbf{d} \in \mathbb{R}^r$ is the (column) vector

$$\mathbf{d} := [d_1, \ldots, d_r]^T. \tag{10.2.14}$$

Suppose first $r \leqslant m+1$. Then A contains the square $r \times r$ submatrix

$$\tilde{A} := \begin{bmatrix} 1 & 1 & \cdots & 1 \\ \tau_1 & \tau_2 & \cdots & \tau_r \\ \tau_1^2 & \tau_2^2 & \cdots & \tau_r^2 \\ \vdots & \vdots & & \vdots \\ \tau_1^{r-1} & \tau_2^{r-1} & \cdots & \tau_r^{r-1} \end{bmatrix}$$

(where $\tilde{A} = A$ if $r = m+1$), with transpose

$$\tilde{A}^T = \begin{bmatrix} 1 & \tau_1 & \tau_1^2 & \cdots & \tau_1^{r-1} \\ 1 & \tau_2 & \tau_2^2 & \cdots & \tau_2^{r-1} \\ \vdots & \vdots & \vdots & & \vdots \\ 1 & \tau_r & \tau_r^2 & \cdots & \tau_r^{r-1} \end{bmatrix},$$

which is a Vandermonde matrix as in (1.1.7), so that, since (10.1.1) holds, we may apply Theorem 1.1.2 to deduce that $\tilde{A}^T$ is invertible, and hence $\tilde{A}$ is invertible. Since A is a matrix with number of rows at least equal to its number of columns, and A contains an invertible submatrix, we deduce that a vector $\mathbf{d} \in \mathbb{R}^r$ satisfying (10.2.11) must be the zero vector, and thus, from (10.2.9), if $S \in \sigma_{m,0}(\tau_1, \ldots, \tau_r)$, with $r \leqslant m+1$, then S is the zero function, and thereby completing our proof of the statement that (10.2.5) implies (10.2.6).

Conversely, if the inequality (10.2.6) is satisfied, then the homogeneous linear system (10.2.11) has more unknowns than equations, and it follows from a standard result in linear algebra that there exists a non-trivial solution $\{d_1, \ldots, d_r\} \in \mathbb{R}^r$ of (10.2.10), which, together with (10.2.9), then yields a spline $S \in \sigma_{m,0}(\tau_1, \ldots, \tau_r)$ which is not the zero function, and thereby completing our proof of the fact that (10.2.6) implies (10.2.5).

(c) To prove the equivalence of (10.2.7) and (10.2.8), we consider the case $r = m+2$ of the matrix A in (10.2.13), to obtain its transpose

$$A^T = \begin{bmatrix} 1 & \tau_1 & \tau_1^2 & \cdots & \tau_1^m \\ 1 & \tau_2 & \tau_2^2 & \cdots & \tau_2^m \\ \vdots & \vdots & \vdots & & \vdots \\ 1 & \tau_{m+2} & \tau_{m+2}^2 & \cdots & \tau_{m+2}^m \end{bmatrix}, \tag{10.2.15}$$

which contains as submatrix the invertible $(m+1) \times (m+1)$ Vandermonde matrix

$$A^* := \begin{bmatrix} 1 & \tau_1 & \tau_1^2 & \cdots & \tau_1^m \\ 1 & \tau_2 & \tau_2^2 & \cdots & \tau_2^m \\ \vdots & \vdots & \vdots & & \vdots \\ 1 & \tau_{m+1} & \tau_{m+1}^2 & \cdots & \tau_{m+1}^m \end{bmatrix}. \tag{10.2.16}$$

Since the matrix A^T in (10.2.15) has $m+1$ columns, its rank satisfies

$$\operatorname{rank}(A^T) \leqslant m+1. \tag{10.2.17}$$

Moreover, since A^* is invertible, we know from standard linear algebra theory that

$$\dim(\text{row space of } A^*) = m+1,$$

and thus, since the rows of A^* are precisely the first $m+1$ rows of A^T, we have

$$\operatorname{rank}(A^T) := \dim(\text{row space of} A^T) \geqslant m+1,$$

which, together with (10.2.17), yields $\operatorname{rank}(A^T) = m+1$, and thus, since also $\operatorname{rank}(A^T) = \operatorname{rank}(A)$, we deduce that

$$\operatorname{rank}(A) = m+1. \tag{10.2.18}$$

By applying the dimension theorem for matrices, we deduce from (10.2.18) that

$$\dim(\text{nullspace of} A) = (\text{number of columns of} A) - \operatorname{rank}(A)$$
$$= (m+2) - (m+1) = 1, \tag{10.2.19}$$

according to which, together with the case $r = m+2$ of (10.2.12), (10.2.11) and (10.2.9), we deduce that (10.2.7) is satisfied by a spline S if and only if

$$S(x) = c \sum_{j=1}^{m+2} d_j (x - \tau_j)_+^m, \quad x \in \mathbb{R}, \tag{10.2.20}$$

with c denoting an arbitrary real constant, and where $\{d_1,\ldots,d_{m+2}\}$ is any non-trivial solution of the $(m+1) \times (m+2)$ homogeneous linear system

$$\sum_{j=1}^{m+2} \tau_j^k d_j = 0, \quad k = 0,\ldots,m. \tag{10.2.21}$$

To obtain an explicit non-trivial solution of (10.2.21), we first use the polynomial identity (1.2.7) in Theorem 1.2.3, together with the definition (1.2.1) of Lagrange fundamental polynomials, to deduce that

$$(-1)^{m+1} \sum_{j=1}^{m+2} \tau_j^k \left[\frac{\displaystyle\prod_{j\neq\ell=1}^{m+2} (x - \tau_\ell)}{\displaystyle\prod_{j\neq\ell=1}^{m+2} (\tau_\ell - \tau_j)} \right] = x^k, \quad x \in \mathbb{R}, \quad k = 0,\ldots,m+1, \tag{10.2.22}$$

according to which

$$(-1)^{m+1} \left\{ \sum_{j=1}^{m+2} \tau_j^k \left[\frac{1}{\displaystyle\prod_{j\neq\ell=1}^{m+2} (\tau_\ell - \tau_j)} \right] \right\} x^{m+1} + P(x) = x^k, \quad x \in \mathbb{R}, \ k = 0,\ldots,m+1, \tag{10.2.23}$$

for some polynomial $P \in \pi_m$. It follows from (10.2.23) that

$$\left\{ \sum_{j=1}^{m+2} \tau_j^k \left[\frac{1}{\displaystyle\prod_{j\neq\ell=1}^{m+2} (\tau_\ell - \tau_j)} \right] \right\} x^{m+1} = \widetilde{P}(x), \quad x \in \mathbb{R}, \quad k = 0,\ldots,m, \tag{10.2.24}$$

for some polynomial $\widetilde{P} \in \pi_m$, from which we deduce that

$$\sum_{j=1}^{m+2} \tau_j^k \left[\frac{1}{\displaystyle\prod_{j\neq\ell=1}^{m+2} (\tau_\ell - \tau_j)} \right] = 0, \quad k = 0,\ldots,m. \tag{10.2.25}$$

According to (10.2.25), a non-trivial solution of (10.2.21) is given by

$$d_j := \frac{1}{\displaystyle\prod_{j\neq\ell=1}^{m+2} (\tau_\ell - \tau_j)}, \quad j = 1,\ldots,m+2,$$

which, together with (10.2.20), then completes our proof of the statement that (10.2.7) and (10.2.8) are equivalent. ∎

For a spline space $\sigma_m(\tau_1,\ldots,\tau_r)$ and a bounded interval $[a,b]$ satisfying the condition

$$[\tau_1,\tau_r] \subset (a,b), \tag{10.2.26}$$

we write $\sigma_m([a,b];\tau_1,\ldots,\tau_r)$ for the linear space of splines in $\sigma_m(\tau_1,\ldots,\tau_r)$ with domains restricted to $[a,b]$. Since (10.2.26) is satisfied, we see from (10.1.3) that Theorem 10.1.1 also holds with $\sigma_m(\tau_1,\ldots,\tau_r)$ replaced by $\sigma_m([a,b];\tau_1,\ldots,\tau_r)$, that is:

$$\dim \sigma_m([a,b];\tau_1,\ldots,\tau_r) = m+1+r, \tag{10.2.27}$$

and

the set X in (10.1.19) is a basis for $\sigma_m([a,b]; \tau_1, \ldots, \tau_r)$. $\qquad$ (10.2.28)

Moreover, (10.1.4) implies that

$$S \in C^{m-1}[a,b] \quad \text{for} \quad S \in \sigma_m([a,b]; \tau_1, \ldots, \tau_r), \quad m \in \mathbb{N}. \qquad (10.2.29)$$

We proceed to show how Theorem 10.2.1 can be used to construct a basis for $\sigma_m([a,b]; \tau_1, \ldots, \tau_r)$ that is more efficient than the one in (10.2.28). To this end, for integers μ and ν satisfying

$$\mu \leqslant -m; \quad \nu \geqslant r+m+1, \qquad (10.2.30)$$

we let $\{\tau_\mu, \ldots, \tau_\nu\}$ be any extension of the knot sequence $\{\tau_1, \ldots, \tau_r\}$, such that

$$\tau_\mu < \cdots < \tau_0 := a < \tau_1 < \cdots < \tau_r < b =: \tau_{r+1} < \cdots < \tau_\nu, \qquad (10.2.31)$$

where we have kept also in mind the conditions (10.1.1) and (10.2.26). The functions

$$N_{m,j}(x) := (\tau_{j+m+1} - \tau_j) \sum_{k=j}^{j+m+1} \left[\frac{1}{\prod_{\substack{k \neq \ell = j}}^{j+m+1} (\tau_\ell - \tau_k)} \right] (x - \tau_k)_+^m,$$

$$j = \mu, \ldots, \nu - m - 1, \qquad (10.2.32)$$

as based on the choice $c = (-1)^{m+1}(\tau_{j+m+1} - \tau_j)$ in (10.2.8) of Theorem 10.2.1, are then called the B-splines of degree m with respect to the knot sequence $\{\tau_\mu, \ldots, \tau_\nu\}$, and for which we proceed to prove the following result.

Theorem 10.2.2. *The B-splines $\{N_{m,j} : j = \mu, \ldots, \nu - m - 1\}$, as defined by (10.2.32) in terms of an extended knot sequence $\{\tau_\mu, \ldots, \tau_\nu\}$ as in (10.2.31), (10.2.30), and with $[a,b]$ denoting any bounded interval, satisfy:*

(a)

$$\left.\begin{array}{l} N_{m,j} \in \sigma_{m,0}(\tau_j, \ldots, \tau_{j+m+1}), \\[2mm] N_{m,j}(x) = 0, \quad x \in \mathbb{R} \setminus [\tau_j, \tau_{j+m+1}), \end{array}\right\} \quad j = \mu, \ldots, \nu - m - 1; \qquad (10.2.33)$$

with

(b)

$$\{N_{m,j} : j = \mu, \ldots, \nu - m - 1\} \subset \sigma_{m,0}(\tau_\mu, \ldots, \tau_\nu); \qquad (10.2.34)$$

(c)

$$N_{m,j}\Big|_{[a,b]} \in \sigma_m([a,b]; \tau_1, \ldots, \tau_r), \quad j = -m, \ldots, r; \qquad (10.2.35)$$

(d) *the set*

$$\left\{ N_{m,j}\Big|_{[a,b]} : j = -m, \ldots, r \right\} \qquad (10.2.36)$$

is a basis for $\sigma_m([a,b]; \tau_1, \ldots, \tau_r)$.

Proof. (a) The result (10.2.33) follows from (10.2.32), together with the equivalence of (10.2.7) and (10.2.8) in Theorem 10.2.1(c), as well as (10.2.3) in Theorem 10.2.1(a)(i).

(b) The inclusion (10.2.34) is a direct consequence of the first line of (10.2.33), together with (10.2.31) and Theorem 10.1.3.

(c) For any $j \in \{-m, \ldots, r\}$, it follows from (10.2.33) and (10.2.31) that the restriction to $[a,b]$ of the B-spline $N_{m,j}$ satisfies (10.2.35).

(d) Since (10.2.35) and (10.2.27) hold, and since the set (10.2.36) contains precisely $m + 1 + r$ elements, it follows from a standard result in linear algebra that it will suffice to prove that the set (10.2.36) is linearly independent on $[a,b]$. Suppose therefore that the sequence $\{c_{-m}, \ldots, c_r\} \subset \mathbb{R}$ satisfies the condition

$$\sum_{j=-m}^{r} c_j N_{m,j}(x) = 0, \quad x \in [a,b]. \tag{10.2.37}$$

Our proof will be complete if we can show that

$$c_{-m} = \cdots = c_r = 0. \tag{10.2.38}$$

To this end, we define the function

$$S(x) := \sum_{j=-m}^{r} c_j N_{m,j}(x), \quad x \in \mathbb{R}, \tag{10.2.39}$$

for which it follows from (10.2.34), with $\mu = -m$ and $\nu = r+m+1$, that

$$S \in \sigma_{m,0}(\tau_{-m}, \ldots, \tau_{r+m+1}). \tag{10.2.40}$$

Morever, by applying the second line of (10.2.33), as well as (10.2.37), we deduce from (10.2.39), together with the fact that (10.2.31) gives $a = \tau_0$, that

$$S(x) = 0, \quad x \in \mathbb{R} \setminus [\tau_{-m}, \tau_0). \tag{10.2.41}$$

Since (10.2.40) holds, it follows from (10.2.4) in Theorem 10.2.1(a)(ii) that

$$S(x) = \sum_{j=-m}^{r+m+1} d_j (x - \tau_j)_+^m, \quad x \in \mathbb{R}, \tag{10.2.42}$$

for some coefficient sequence $\{d_{-m}, \ldots, d_{r+m+1}\} \subset \mathbb{R}$. Now observe from (10.2.41), (10.2.42) and (10.1.15) that

$$0 = S(x) = \sum_{j=-m}^{0} d_j (x - \tau_j)^m, \quad x \in [\tau_0, \tau_1),$$

and thus

$$\sum_{j=-m}^{0} d_j (x - \tau_j)^m = 0, \quad x \in \mathbb{R},$$

which, together with (10.2.42) and (10.1.15), yields

$$S(x) = \sum_{j=1}^{r+m+1} d_j(x - \tau_j)_+^m = 0, \quad x \in [\tau_1, \infty). \tag{10.2.43}$$

It follows from (10.2.43), and (10.1.15), by choosing successively $x \in [\tau_1, \tau_2), \ldots, [\tau_{r+m}, \tau_{r+m+1}), [\tau_{r+m+1}, \infty)$, that

$$d_1 = \cdots = d_{r+m+1} = 0, \tag{10.2.44}$$

which can now be substituted into (10.2.42) to yield

$$S(x) = \sum_{j=-m}^{0} d_j(x - \tau_j)_+^m, \quad x \in \mathbb{R}. \tag{10.2.45}$$

By also noting from (10.1.13) and (10.1.14) that, analogously to (10.1.16), it holds that

$$(\cdot - \tau_j)_+^m \in \sigma_m(\tau_{-m}, \ldots, \tau_0), \quad j = -m, \ldots, 0,$$

we deduce from (10.2.45) and (10.2.41) that

$$S \in \sigma_{m,0}(\tau_{-m}, \ldots, \tau_0). \tag{10.2.46}$$

Since the space $\sigma_{m,0}(\tau_{-m}, \ldots, \tau_0)$ has precisely $m + 1$ knots, it follows from (10.2.46), together with the fact that (10.2.5) implies (10.2.6) in Theorem 10.2.1(b), that S is the zero function on $\mathbb{R}$, and thus, from (10.2.39),

$$\sum_{j=-m}^{r} c_j N_{m,j}(x) = 0, \quad x \in [\tau_{-m}, \tau_{r+1}]. \tag{10.2.47}$$

Suppose (10.2.38) is not satisfied, and denote by λ the smallest integer in the set $\{-m, \ldots, r\}$ for which it holds that $c_\lambda \neq 0$, according to which, by using also (10.2.47), together with the definition (10.2.32), (10.2.31), as well as (10.1.15), we have, for any $x \in (\tau_\lambda, \tau_{\lambda+1})$,

$$0 = c_\lambda N_{m,\lambda}(x) = c_\lambda(\tau_{\lambda+m+1} - \tau_\lambda) \frac{(x - \tau_\lambda)^m}{\prod\limits_{\ell=\lambda+1}^{\lambda+m+1} (\tau_\ell - \tau_\lambda)} \neq 0,$$

from (10.2.31), and thereby yielding a contradiction. Hence (10.2.38) is satisfied, and our proof is complete. ∎

Example 10.2.1. By setting $m = 0$ and $m = 1$ in (10.2.32), and using (10.1.15), we obtain (see Exercise 10.6) the B-spline formulations

$$N_{0,j}(x) = \left\{ \begin{array}{l} 1, x \in [\tau_j, \tau_{j+1}), \\ 0, x \in \mathbb{R} \setminus [\tau_j, \tau_{j+1}), \end{array} \right\} \quad j = \mu, \ldots, \nu - 1; \tag{10.2.48}$$

and

$$N_{1,j}(x) = \begin{cases} \dfrac{x - \tau_j}{\tau_{j+1} - \tau_j} & , x \in [\tau_j, \tau_{j+1}), \\[2mm] \dfrac{\tau_{j+2} - x}{\tau_{j+2} - \tau_{j+1}} & , x \in [\tau_{j+1}, \tau_{j+2}), \\[2mm] 0 & , x \in \mathbb{R} \setminus [\tau_j, \tau_{j+2}), \end{cases} \qquad j = \mu, \ldots, \nu - 2, \qquad (10.2.49)$$

from which we also note in particular that

$$N_{1,j}(\tau_{j+1}) = \delta_j, \quad j = \mu, \ldots, \nu - 2. \qquad (10.2.50)$$

∎

Observe that the B-spline formulation (10.2.32) is in terms of the truncated powers in the basis (10.1.19) of the spline space $\sigma_m(\tau_\mu, \ldots, \tau_\nu)$. Our statement that the B-spline basis (10.2.36) is more efficient than the basis (10.1.19) for $\sigma_m([a,b]; \tau_1, \ldots, \tau_r)$, is based on the fact, to be established below, that B-splines can be computed recursively with respect to the spline degree m.

Our first step in this direction is the following result, which gives a formulation for B-splines in terms of the divided difference of a truncated power, and which is an immediate consequence of (10.2.32), (1.3.3) and (1.3.2).

Theorem 10.2.3. *For any* $j \in \{\mu, \ldots, \nu - m - 1\}$, *the B-spline $N_{m,j}$, as defined by* (10.2.32), (10.2.31), *satisfies the formulation*

$$N_{m,j}(x) = (-1)^{m+1} (\tau_{j+m+1} - \tau_j) \left\{ (x - \cdot)_+^m [\tau_j, \ldots, \tau_{j+m+1}] \right\}, \quad x \in \mathbb{R}, \qquad (10.2.51)$$

that is, for any fixed $x \in \mathbb{R}$, $N_{m,j}(x)$ *is given by* $(-1)^{m+1}(\tau_{j+m+1} - \tau_j)$ *times the divided difference (in terms of the t variable), with respect to the points* $\{\tau_j, \ldots, \tau_{j+m+1}\}$, *of the function* $(x - t)_+^m, t \in \mathbb{R}$.

The recursion formula for B-splines in Theorem 10.2.5 below will be derived from (10.2.51) in Theorem 10.2.3, together with the following formula for the divided difference of the product of two functions.

Theorem 10.2.4. *For any non-negative integer n, the divided difference of the product of two functions f and g with respect to any sequence of $n+1$ distinct points $\{x_0, \ldots, x_n\} \subset \mathbb{R}$ is given by the formula*

$$(fg)[x_0, \ldots, x_n] = \sum_{j=0}^{n} f[x_0, \ldots, x_j] g[x_j, \ldots, x_n]. \qquad (10.2.52)$$

Proof. First, observe from the first equation in (1.3.4) that (10.2.52) is satisfied for $n = 0$. Proceeding inductively, we suppose next that (10.2.52) holds for a fixed non-negative integer n. Now let $\{x_0, \ldots, x_{n+1}\}$ denote any sequence of $n+2$ distinct points in $\mathbb{R}$. Our inductive proof will be complete if we can show that then

$$(fg)[x_0, \ldots, x_{n+1}] = \sum_{j=0}^{n+1} f[x_0, \ldots, x_j] g[x_j, \ldots, x_{n+1}]. \tag{10.2.53}$$

To prove (10.2.53), we first apply the recursion formula (1.3.21) in Theorem 1.3.4 to obtain

$$(fg)[x_0, \ldots, x_{n+1}] = \frac{(fg)[x_1, \ldots, x_{n+1}] - (fg)[x_0, \ldots, x_n]}{x_{n+1} - x_0}. \tag{10.2.54}$$

Next, we deduce from the inductive hypothesis (10.2.52), together with (1.3.21), that

$$(fg)[x_1, \ldots, x_{n+1}] - (fg)[x_0, \ldots, x_n]$$

$$= \sum_{j=1}^{n+1} f[x_1, \ldots, x_j] g[x_j, \ldots, x_{n+1}] - \sum_{j=0}^{n} f[x_0, \ldots, x_j] g[x_j, \ldots, x_n]$$

$$= \sum_{j=0}^{n} f[x_1, \ldots, x_{j+1}] g[x_{j+1}, \ldots, x_{n+1}] - \sum_{j=0}^{n} f[x_0, \ldots, x_j] g[x_j, \ldots, x_n]$$

$$= \sum_{j=0}^{n} \{ f[x_1, \ldots, x_{j+1}] - f[x_0, \ldots, x_j] \} g[x_{j+1}, \ldots, x_{n+1}]$$

$$+ \sum_{j=0}^{n} f[x_0, \ldots, x_j] \{ g[x_{j+1}, \ldots, x_{n+1}] - g[x_j, \ldots, x_n] \}$$

$$= \sum_{j=0}^{n} (x_{j+1} - x_0) f[x_0, \ldots, x_{j+1}] g[x_{j+1}, \ldots, x_{n+1}]$$

$$+ \sum_{j=0}^{n} (x_{n+1} - x_j) f[x_0, \ldots, x_j] g[x_j, \ldots, x_{n+1}]$$

$$= \sum_{j=0}^{n+1} (x_j - x_0) f[x_0, \ldots, x_j] g[x_j, \ldots, x_{n+1}]$$

$$+ \sum_{j=0}^{n+1} (x_{n+1} - x_j) f[x_0, \ldots, x_j] g[x_j, \ldots, x_{n+1}]$$

$$= (x_{n+1} - x_0) \sum_{j=0}^{n+1} f[x_0, \ldots, x_j] g[x_j, \ldots, x_{n+1}],$$

which, together with (10.2.54), yields the desired result (10.2.53). ∎

By using Theorems 10.2.3 and 10.2.4, as well as the fact that

$$x_+^m = x \left(x_+^{m-1} \right), \quad x \in \mathbb{R}, \quad m \in \mathbb{N}, \tag{10.2.55}$$

as is immediately evident from the definition (10.1.13), we can now establish the following recursion formula satisfied by B-splines.

Theorem 10.2.5. *For* $m \in \mathbb{N}$, *the* B-*spline sequence* $\{N_{m,j} : j = \mu, \ldots, \nu - m - 1\}$, *as defined by* (10.2.32), (10.2.31), *satisfies, for any* $x \in \mathbb{R}$, *the recursion formulation given by*

$$N_{m,j}(x) = \frac{x - \tau_j}{\tau_{j+m} - \tau_j} N_{m-1,j}(x) + \frac{\tau_{j+m+1} - x}{\tau_{j+m+1} - \tau_{j+1}} N_{m-1,j+1}(x),$$

$$j = \mu, \ldots, \nu - m - 1, \qquad (10.2.56)$$

together with (10.2.48).

Proof. Let $j \in \{\mu, \ldots, \nu - m - 1\}$ and $x \in \mathbb{R}$ be fixed. It follows from (10.2.51) in Theorem 10.2.3, as well as (10.2.55), and (10.2.52) in Theorem 10.2.4, that

$$N_{m,j}(x) = (-1)^{m+1} (\tau_{j+m+1} - \tau_j) \left\{ ((x - \cdot)(x - \cdot)_+^{m-1}) [\tau_j, \ldots, \tau_{j+m+1}] \right\}$$

$$= (-1)^{m+1} (\tau_{j+m+1} - \tau_j) \sum_{k=j}^{j+m+1} \left\{ (x - \cdot)[\tau_j, \ldots, \tau_k] \right\} \left\{ (x - \cdot)_+^{m-1} [\tau_k, \ldots, \tau_{j+m+1}] \right\}.$$

$$(10.2.57)$$

Now observe from (1.3.4) that

$$(x - \cdot)[\tau_j] = x - \tau_j; \qquad (10.2.58)$$

$$(x - \cdot)[\tau_j, \tau_{j+1}] = \frac{(x - \tau_{j+1}) - (x - \tau_j)}{\tau_{j+1} - \tau_j} = -1, \qquad (10.2.59)$$

whereas, since $(x - \cdot) \in \pi_1$, we deduce from (2.1.10) in Theorem 2.1.2 that

$$(x - \cdot)[\tau_j, \ldots, \tau_k] = 0, \quad \text{if} \quad k \geqslant j + 2. \qquad (10.2.60)$$

By substituting (10.2.58), (10.2.59) and (10.2.60) into (10.2.57), and using the recursion formula (1.3.21) in Theorem 1.3.4, we obtain

$$N_{m,j}(x) = (-1)^{m+1} (\tau_{j+m+1} - \tau_j) \Big[(x - \tau_j) \left\{ (x - \cdot)_+^{m-1} [\tau_j, \ldots, \tau_{j+m+1}] \right\}$$

$$- (x - \cdot)_+^{m-1} [\tau_{j+1}, \ldots, \tau_{j+m+1}] \Big]$$

$$= (-1)^{m+1} (\tau_{j+m+1} - \tau_j) \Bigg\{ (x - \tau_j) \frac{(x - \cdot)_+^{m-1} [\tau_{j+1}, \ldots, \tau_{j+m+1}] - (x - \cdot)_+^{m-1} [\tau_j, \ldots, \tau_{j+m}]}{\tau_{j+m+1} - \tau_j}$$

$$- (x - \cdot)_+^{m-1} [\tau_{j+1}, \ldots, \tau_{j+m+1}] \Bigg\}$$

$$= (-1)^{m+1}(x-\tau_j)\left\{(x-\cdot)_+^{m-1}[\tau_{j+1},\ldots,\tau_{j+m+1}] - (x-\cdot)_+^{m-1}[\tau_j,\ldots,\tau_{j+m}]\right\}$$

$$+(-1)^m(\tau_{j+m+1}-\tau_j)\left\{(x-\cdot)_+^{m-1}[\tau_{j+1},\ldots,\tau_{j+m+1}]\right\}$$

$$= (-1)^m\left[(x-\tau_j)\left\{(x-\cdot)_+^{m-1}[\tau_j,\ldots,\tau_{j+m}]\right\}\right.$$

$$\left.+(\tau_{j+m+1}-x)\left\{(x-\cdot)_+^{m-1}[\tau_{j+1},\ldots,\tau_{j+1+m}]\right\}\right]$$

$$= \frac{x-\tau_j}{\tau_{j+m}-\tau_j}N_{m-1,j}(x) + \frac{\tau_{j+m+1}-x}{\tau_{j+m+1}-\tau_{j+1}}N_{m-1,j+1}(x),$$

from (10.2.51) in Theorem 10.2.3, and thereby completing the proof of (10.2.56). ∎

The formula (10.2.56), together with (10.2.55) and (10.2.48), can be used for the recursive evaluation of the B-spline value $N_{m,j}(x)$, for any $x \in (\tau_j, \tau_{j+m+1})$, as illustrated in Fig.10.2.1.

$$N_{0,j}(x) \qquad N_{1,j}(x) \qquad N_{2,j}(x) \qquad \cdots \qquad N_{m-1,j}(x) \qquad N_{m,j}(x)$$

$$N_{0,j+1}(x) \qquad N_{1,j+1}(x) \qquad N_{2,j+1}(x) \qquad \cdots \qquad N_{m-1,j+1}(x)$$

$$\vdots$$

$$N_{0,j+2}(x) \qquad \vdots$$

$$N_{2,j+m-2}(x)$$

$$\vdots \qquad N_{1,j+m-1}(x)$$

$$N_{0,j+m}(x)$$

Fig. 10.2.1 *B-spline evaluation based on the recursion formula* (10.2.56), *together with* (10.2.48).

Note from (10.2.48) and (10.2.55) that, for any given $x \in (\tau_j, \tau_{j+m+1})$, and with k denoting the (unique) integer in $\{j,\ldots,j+m\}$ such that $x \in [\tau_k, \tau_{k+1})$, the first column in Fig 10.2.1 satisfies

$$N_{0,\ell}(x) = \delta_{k-\ell}, \quad \ell = j,\ldots,j+m. \tag{10.2.61}$$

Example 10.2.2. For $m = 3$ and $\{\tau_0,\ldots,\tau_4\} = \{0,1,2,4,5\}$, we have $N_{3,0} \in \sigma_{3,0}(\tau_0,\ldots,\tau_4)$, and the B-spline value $N_{3,0}(3)$ can be computed recursively by means of (10.2.56) and (10.2.61), to obtain Table 10.2.1.

Table 10.2.1   Recursive computation of $N_{3,0}(3)$ in Example 10.2.2.

j	$N_{0,j}(3)$	$N_{1,j}(3)$	$N_{2,j}(3)$	$N_{3,j}(3)$
0	0	0	$\frac{1}{6}$	$\frac{25}{72}$
1	0	$\frac{1}{2}$	$\frac{2}{3}$	
2	1	$\frac{1}{2}$		
3	0			

The values in Table 10.2.1 are obtained by the following calculations by means of (10.2.56), and by using the second line of (10.2.33) in Theorem 10.2.2 whenever applicable:

$$N_{1,0}(3) = 0;$$

$$N_{1,1}(3) = \frac{3-\tau_1}{\tau_2-\tau_1}N_{0,1}(3) + \frac{\tau_3-3}{\tau_3-\tau_2}N_{0,2}(3) = \frac{3-1}{2-1}(0) + \frac{4-3}{4-2}(1) = \frac{1}{2};$$

$$N_{1,2}(3) = \frac{3-\tau_2}{\tau_3-\tau_2}N_{0,2}(3) + \frac{\tau_4-3}{\tau_4-\tau_3}N_{0,3}(3) = \frac{3-2}{4-2}(1) + \frac{5-3}{5-4}(0) = \frac{1}{2};$$

$$N_{2,0}(3) = \frac{3-\tau_0}{\tau_2-\tau_0}N_{1,0}(3) + \frac{\tau_3-3}{\tau_3-\tau_1}N_{1,1}(3) = \frac{3-0}{2-0}(0) + \frac{4-3}{4-1}\left(\frac{1}{2}\right) = \frac{1}{6};$$

$$N_{2,1}(3) = \frac{3-\tau_1}{\tau_3-\tau_1}N_{1,1}(3) + \frac{\tau_4-3}{\tau_4-\tau_2}N_{1,2}(3) = \frac{3-1}{4-1}\left(\frac{1}{2}\right) + \frac{5-3}{5-2}\left(\frac{1}{2}\right) = \frac{1}{3} + \frac{1}{3} = \frac{2}{3};$$

$$N_{3,0}(3) = \frac{3-\tau_0}{\tau_3-\tau_0}N_{2,0}(3) + \frac{\tau_4-3}{\tau_4-\tau_1}N_{2,1}(3) = \frac{3-0}{4-0}\left(\frac{1}{6}\right) + \frac{4-3}{4-1}\left(\frac{2}{3}\right) = \frac{1}{8} + \frac{2}{9} = \frac{25}{72}.$$ ∎

The following two further properties of B-splines, both of which will be required later in this chapter, can now be proved by means of the recursion formula (10.2.56) in Theorem 10.2.5.

Theorem 10.2.6. *For any non-negative integer m, the B-spline sequence* $\{N_{m,j} : j = -m,\ldots,r\}$, *as obtained from the formulation* (10.2.32), (10.2.31), *satisfies:*

(a) *For any* $j \in \{-m,\ldots,r\}$, *it holds that*

$$N_{m,j}(x) > 0, \quad x \in (\tau_j, \tau_{j+m+1}); \tag{10.2.62}$$

(b)

$$\sum_{j=-m}^{r} N_{m,j}(x) = 1, \quad x \in [a,b]. \tag{10.2.63}$$

Proof. (a) Let m denote any non-negative integer. Now observe from (10.2.48), with $\mu = -m$ and $v = r+m+1$, that

$$N_{0,j}(x) > 0, \; x \in [\tau_j, \tau_{j+1}), \quad j = -m, \ldots, r+m, \tag{10.2.64}$$

which proves (10.2.62) for $m = 0$.

We proceed to prove inductively that

$$N_{k,j}(x) > 0, \; x \in (\tau_j, \tau_{j+k+1}), \; j = -m, \ldots, r+m-k, \tag{10.2.65}$$

in which we may then set $k = m$ to deduce the desired result (10.2.62).

After first observing from (10.2.64) that (10.2.65) is satisfied for $m = 0$, we suppose next that (10.2.65) holds for a fixed non-negative integer k. It follows from the recursion formula (10.2.56) in Theorem 10.2.5 that

$$N_{k+1,j}(x) = \frac{x-\tau_j}{\tau_{j+k+1} - \tau_j} N_{k,j}(x) + \frac{\tau_{j+k+2} - x}{\tau_{j+k+2} - \tau_{j+1}} N_{k,j+1}(x), \; x \in (\tau_j, \tau_{j+k+2}),$$

$$j = -m, \ldots, r+m-k-1. \tag{10.2.66}$$

After also noting from the second line of (10.2.33) in Theorem 10.2.2(a) that

$$\left. \begin{array}{l} N_{k,j}(x) \;\; = 0, x \in [\tau_{j+k+1}, \tau_{j+k+2}), \\[4pt] N_{k,j+1}(x) = 0, x \in (\tau_j, \tau_{j+1}), \end{array} \right\} \; j = -m, \ldots, r+m-k-1,$$

we deduce from (10.2.66) and (10.2.65) that

$$N_{k+1,j}(x) > 0, \quad x \in (\tau_j, \tau_{j+k+2}), \quad j = -m, \ldots, r+m-k-1,$$

which then completes our inductive proof of (10.2.65).

(b) Our proof is once again by induction on m. First, observe from (10.2.48) that (10.2.63) is satisfied for $m = 0$. Next, suppose that (10.2.63) holds for a fixed non-negative integer m, and consider a knot sequence as in (10.2.31), with $\mu = -m-1$ and $v = r+m+2$. It follows from (10.2.56), together with the second line of (10.2.33), and the inductive hypothesis (10.2.63), that, for any $x \in [a,b] = [\tau_0, \tau_{r+1}]$,

$$\sum_{j=-m-1}^{r} N_{m+1,j}(x) = \sum_{j=-m-1}^{r} \frac{x-\tau_j}{\tau_{j+m+1} - \tau_j} N_{m,j}(x) + \sum_{j=-m-1}^{r} \frac{\tau_{j+m+2} - x}{\tau_{j+m+2} - \tau_{j+1}} N_{m,j+1}(x)$$

$$= \sum_{j=-m-1}^{r} \frac{x-\tau_j}{\tau_{j+m+1} - \tau_j} N_{m,j}(x) + \sum_{j=-m}^{r+1} \frac{\tau_{j+m+1} - x}{\tau_{j+m+1} - \tau_j} N_{m,j}(x)$$

$$= \sum_{j=-m}^{r} \frac{x-\tau_j}{\tau_{j+m+1} - \tau_j} N_{m,j}(x) + \sum_{j=-m}^{r} \frac{\tau_{j+m+1} - x}{\tau_{j+m+1} - \tau_j} N_{m,j}(x)$$

$$= \sum_{j=-m}^{r} \frac{(x-\tau_j)+(\tau_{j+m+1}-x)}{\tau_{j+m+1}-\tau_j} N_{m,j}(x) = \sum_{j=-m}^{r} N_{m,j}(x) = 1,$$

which completes our inductive proof of (10.2.63).   ∎

Finally in this section, we consider the important special case where the knots $\{\tau_\mu, \ldots, \tau_\nu\}$ in (10.2.31), with $[a,b] = [0, r+1]$, are chosen as the integers, that is,

$$\tau_j = j, \quad j = \mu, \ldots, \nu, \tag{10.2.67}$$

for any integers μ and ν satisfying (10.2.30), and for which we proceed to prove the following result.

Theorem 10.2.7. *For any $j \in \{\mu, \ldots, \nu - m - 1\}$, where μ and ν are any integers satisfying* (10.2.30), *let $N_{m,j}$ denote the B-spline defined by* (10.2.32), (10.2.31), *where $[a,b] = [0, r+1]$, and with integer knot sequence as in* (10.2.67). *Then*

$$N_{m,j}(x) = N_m(x-j), \quad x \in \mathbb{R}, \tag{10.2.68}$$

with

$$N_m(x) := N_{m,0}(x) = \frac{1}{m!} \sum_{k=0}^{m+1} (-1)^k \binom{m+1}{k} (x-k)_+^m, \quad x \in \mathbb{R}, \tag{10.2.69}$$

and where the spline N_m satisfies:

(a)

$$\left.\begin{array}{c} N_m \in \sigma_{m,0}(0, \ldots, m+1), \\[4pt] N_m(x) = 0, \quad x \in \mathbb{R} \setminus [0, m+1); \end{array}\right\} \tag{10.2.70}$$

with

(b)

$$N_m(x) > 0, \quad x \in (0, m+1); \tag{10.2.71}$$

(c)

$$\sum_{j=-m}^{r} N_m(x-j) = 1, \quad x \in [a, b]; \tag{10.2.72}$$

(d) *the recursive formulation*

$$\left.\begin{array}{l} N_0(x) = \begin{cases} 1, x \in [0, 1); \\[4pt] 0, x \in \mathbb{R} \setminus [0, 1); \end{cases} \\[12pt] N_m(x) = \dfrac{x}{m} N_{m-1}(x) + \dfrac{m+1-x}{m} N_{m-1}(x-1), \quad x \in \mathbb{R}, \quad m = 1, 2, \ldots; \end{array}\right\} \tag{10.2.73}$$

(e) *for $m \in \mathbb{N}$, the symmetry condition*

$$N_m(m+1-x) = N_m(x), \quad x \in \mathbb{R},$$

or equivalently,

$$N_m\left(\frac{m+1}{2}-x\right) = N_m\left(\frac{m+1}{2}+x\right), \quad x \in \mathbb{R}; \qquad (10.2.74)$$

(f)

$$N_m(k) = \frac{1}{m!}\sum_{j=0}^{k-1}(-1)^j\binom{m+1}{j}(k-j)^m, \quad k=1,\ldots,m. \qquad (10.2.75)$$

Proof. Let $j \in \{\mu,\ldots,v-m-1\}$ be fixed. By using (10.2.51) in Theorem 10.2.3, together with (10.2.67), as well as (3.4.2) in Theorem 3.4.1, we obtain, for any $x \in \mathbb{R}$,

$$N_{m,j}(x) = (-1)^{m+1}(m+1)\left\{(x-\cdot)_+^m[j,\ldots,j+m+1]\right\}$$

$$= (-1)^{m+1}(m+1)\left\{\frac{(-1)^{m+1}}{(m+1)!}\sum_{k=0}^{m+1}(-1)^k\binom{m+1}{k}(x-(j+k))_+^m\right\}$$

$$= \frac{1}{m!}\sum_{k=0}^{m+1}(-1)^k\binom{m+1}{k}((x-j)-k)_+^m,$$

which proves (10.2.68), (10.2.69).

The properties (10.2.70), (10.2.71) and (10.2.72) of N_m are immediate consequences of the definition $N_m := N_{m,0}$ in (10.2.69), together with, respectively, (10.2.33) in Theorem 10.2.2(a), (10.2.62) in Theorem 10.2.6(a), and (10.2.63) in Theorem 10.2.6(b), whereas the recursive formulation (10.2.73) follows likewise from (10.2.48), as well as (10.2.56) in Theorem 10.2.5, together with the case $j=1$ of (10.2.68).

To prove the symmetry condition (10.2.74) for $m \in \mathbb{N}$, we fix $x \in [0,m+1)$, and let ℓ denote the (unique) integer in the integer set $\{0,\ldots,m\}$ for which $x \in [\ell,\ell+1)$. It then follows from the formula in (10.2.69), together with (10.1.15), that

$$N_m(x) = \frac{1}{m!}\sum_{k=0}^{\ell}(-1)^k\binom{m+1}{k}(x-k)^m, \qquad (10.2.76)$$

and similarly, since $m+1-x \in (m-\ell,m-\ell+1]$, and using also the fact that $N_m(0)=0$ if $m \in \mathbb{N}$, as can be seen from (10.2.69) and (10.1.15), we have

$$N_m(m+1-x) = \frac{1}{m!}\sum_{k=0}^{m-\ell}(-1)^k\binom{m+1}{k}(m+1-x-k)^m$$

$$= \frac{1}{m!}\sum_{k=0}^{m-\ell}(-1)^{k+m}\binom{m+1}{k}[x-(m+1-k)]^m$$

$$= \frac{1}{m!}\sum_{k=\ell+1}^{m+1}(-1)^{k+1}\binom{m+1}{m+1-k}(x-k)^m$$

$$= -\frac{1}{m!} \sum_{k=\ell+1}^{m+1} (-1)^k \binom{m+1}{k} (x-k)^m,$$

which, together with (10.2.76), yields

$$N_m(x) - N_m(m+1-x) = \frac{1}{m!} \sum_{k=0}^{m+1} (-1)^k \binom{m+1}{k} (x-k)^m. \qquad (10.2.77)$$

Now observe from the second line of (10.2.70), together with (10.2.69), as well as (10.1.15), that

$$0 = N_m(x) = \frac{1}{m!} \sum_{k=0}^{m+1} (-1)^k \binom{m+1}{k} (x-k)^m, \quad x \geqslant m+1,$$

and thus also

$$\frac{1}{m!} \sum_{k=0}^{m+1} (-1)^k \binom{m+1}{k} (x-k)^m = 0, \quad x \in \mathbb{R}. \qquad (10.2.78)$$

The symmetry result in the first line of (10.2.74) now follows from (10.2.77) and (10.2.78) for $x \in [0, m+1)$, whereas it follows from the second line in (10.2.70), together with $N_m(0) = 0$, for $x \in \mathbb{R} \setminus [0, m+1)$.

Finally, observe that (10.2.75) is an immediate consequence of the formula in (10.2.69), together with (10.1.15). ∎

The spline N_m defined by (10.2.69) in Theorem 10.2.7 is called the cardinal B-spline of degree m.

Example 10.2.3. By setting $m = 1, m = 2$ and $m = 3$ in the formula (10.2.69), and using (10.1.15), we obtain (see Exercise 10.10) the explicit cardinal B-spline formulations

$$N_1(x) = \left\{ \begin{array}{ll} x, & x \in [0,1); \\ 2-x, & x \in [1,2); \\ 0, & x \in \mathbb{R} \setminus [0,2); \end{array} \right\} \qquad (10.2.79)$$

$$N_2(x) = \left\{ \begin{array}{ll} \frac{1}{2}x^2, & x \in [0,1); \\ -x^2 + 3x - \frac{3}{2}, & x \in [1,2); \\ \frac{1}{2}x^2 - 3x + \frac{9}{2}, & x \in [2,3); \\ 0, & x \in \mathbb{R} \setminus [0,3); \end{array} \right\} \qquad (10.2.80)$$

$$N_3(x) = \left\{ \begin{array}{ll} \frac{1}{6}x^3, & x \in [0,1); \\ -\frac{1}{2}x^3 + 2x^2 - 2x + \frac{2}{3}, & x \in [1,2); \\ \frac{1}{2}x^3 - 4x^2 + 10x - \frac{22}{3}, & x \in [2,3); \\ -\frac{1}{6}x^3 + 2x^2 - 8x + \frac{32}{3}, & x \in [3,4); \\ 0, & x \in \mathbb{R} \setminus [0,4). \end{array} \right\} \qquad (10.2.81)$$

Observe that (10.2.79) corresponds precisely with the integer knot case (10.2.67) of (10.2.49) in Example 10.2.1.

Also, by using either the formula (10.2.75), or directly the explicit formulations (10.2.80), (10.2.81), we calculate the values

$$N_2(1) = \frac{1}{2} \quad ; \quad N_2(2) = \frac{1}{2}; \tag{10.2.82}$$

$$N_3(1) = \frac{1}{6}; \quad N_3(2) = \frac{2}{3}; \quad N_3(3) = \frac{1}{6}. \tag{10.2.83}$$

■

10.3   Spline interpolation

For a bounded interval $[a,b]$, let the function $f : [a,b] \to \mathbb{R}$ be given, and, for a positive integer n, let $\{x_0,\ldots,x_n\}$ denote a sequence of $n+1$ distinct points in $[a,b]$, with

$$a \leqslant x_0 < \cdots < x_n \leqslant b. \tag{10.3.1}$$

In this section, we investigate the existence of a spline $S \in \sigma_m([a,b]; \tau_1,\ldots,\tau_r)$ which interpolates f at the points $\{x_0,\ldots,x_n\}$, that is,

$$S(x_j) = f(x_j), \quad j = 0,\ldots,n. \tag{10.3.2}$$

By observing that (10.3.2) consists of precisely $n+1$ interpolation conditions, and recalling from (10.2.27) that the spline space $\sigma_m([a,b]; \tau_1,\ldots,\tau_r)$ has dimension $m+1+r$, we impose the condition

$$m+1+r = n+1. \tag{10.3.3}$$

Based on (10.3.3), as well as (10.2.31), (10.2.30), for integers n and m satisfying $n > m \geqslant 0$, and an extended knot sequence $\{\tau_\mu,\ldots,\tau_\nu\}$ satisfying

$$\tau_\mu < \cdots < \tau_0 := a < \tau_1 < \cdots < \tau_{n-m} < b =: \tau_{n-m+1} < \cdots < \tau_\nu, \tag{10.3.4}$$

where

$$\mu \leqslant -m; \quad \nu \geqslant n+1, \tag{10.3.5}$$

we shall therefore seek to obtain a spline

$$S \in \sigma_m([a,b]; \tau_1,\ldots,\tau_{n-m}) \tag{10.3.6}$$

satisfying the interpolation conditions (10.3.2).

By applying Theorem 10.2.2(d), we deduce that there exists a spline S satisfying (10.3.6) and (10.3.2) if and only if

$$S(x) := \sum_{j=-m}^{n-m} c_j N_{m,j}(x), \quad x \in [a,b], \tag{10.3.7}$$

where $\{c_{-m}, \ldots, c_{n-m}\} \subset \mathbb{R}$ satisfies the linear system

$$\sum_{j=-m}^{n-m} c_j N_{m,j}(x_k) = f(x_k), \quad k = 0, \ldots, n, \tag{10.3.8}$$

or equivalently, in matrix-vector formulation, where the (column) vector $\mathbf{c} = (c_{-m}, \ldots, c_{n-m})^T$
$\in \mathbb{R}^{n+1}$ is a solution of the equation

$$A_{m,n}\mathbf{c} = \mathbf{f}_n, \tag{10.3.9}$$

with $A_{m,n}$ denoting the $(n+1) \times (n+1)$ matrix

$$A_{m,n} := \begin{bmatrix} N_{m,-m}(x_0) & N_{m,-m+1}(x_0) & \cdots & N_{m,n-m}(x_0) \\ N_{m,-m}(x_1) & N_{m,-m+1}(x_1) & \cdots & N_{m,n-m}(x_1) \\ \vdots & \vdots & & \vdots \\ N_{m,-m}(x_n) & N_{m,-m+1}(x_n) & \cdots & N_{m,n-m}(x_n) \end{bmatrix}, \tag{10.3.10}$$

and where $\mathbf{f}_n \in \mathbb{R}^{n+1}$ is the (column) vector

$$\mathbf{f}_n := (f(x_0), \ldots, f(x_n))^T. \tag{10.3.11}$$

Hence we proceed to establish, in terms of the sequences $\{x_0, \ldots, x_n\}$ and $\{\tau_{-m}, \ldots, \tau_{n+1}\}$, a necessary and sufficient condition for the invertibility of the matrix $A_{m,n}$.
To this end, for $m \in \mathbb{N}$, and integers κ and λ such that

$$\lambda - \kappa \geqslant m + 1; \quad \mu \leqslant \kappa < \lambda \leqslant \nu; \tag{10.3.12}$$

let S denote any finitely supported spline, with

$$S \in \sigma_{m,0}(\tau_\kappa, \ldots, \tau_\lambda), \tag{10.3.13}$$

according to which, by applying (10.2.3) in Theorem 10.2.1, together with the fact that, from (10.1.4), S is continuous at τ_κ, we also have

$$S(x) = 0, \quad x \in \mathbb{R} \setminus (\tau_\kappa, \tau_\lambda). \tag{10.3.14}$$

We shall say that S has a sign change in $(\tau_\kappa, \tau_\lambda)$ if either:

(i) S changes sign at a point $t \in (\tau_\kappa, \tau_\lambda)$, in the sense that

$$\left. \begin{aligned} S(t) &= 0; \\ S(t-\varepsilon)S(t+\varepsilon) &< 0 \text{ for sufficiently small } \varepsilon > 0, \end{aligned} \right\} \tag{10.3.15}$$

or

(ii) S changes sign with respect to an interval $[\tau_j, \tau_k] \subset (\tau_\kappa, \tau_\lambda)$, in the sense that

$$\left. \begin{aligned} S(x) &= 0, \quad x \in [\tau_j, \tau_k]; \\ S(\tau_j-\varepsilon)S(\tau_k+\varepsilon) &< 0 \text{ for sufficiently small } \varepsilon > 0. \end{aligned} \right\} \tag{10.3.16}$$

For $m \geqslant 2$, we observe, if $S(\xi_1) = S(\xi_2) = 0$, where $\tau_\kappa \leqslant \xi_1 < \xi_2 \leqslant \tau_\lambda$, that we may apply Rolle's theorem to deduce that S' has at least one sign change, of either one of the types (10.3.15) or (10.3.16) above, in the interval (ξ_1, ξ_2).

In our investigation of the invertibility of the matrix $A_{m,n}$ in (10.3.10), we shall require the following result on the zeros of a finitely supported spline.

Theorem 10.3.1. *For $m \in \mathbb{N}$, let S denote a finitely supported spline as in (10.3.13), (10.3.12), and suppose that, moreover, S has a finite number of zeros in $(\tau_\kappa, \tau_\lambda)$. Then the non-negative integer ρ defined by*

$$\rho := \text{number of distinct zeros of } S \text{ in } (\tau_\kappa, \tau_\lambda), \tag{10.3.17}$$

satisfies the inequality

$$\rho + m + 1 \leqslant \lambda - \kappa. \tag{10.3.18}$$

Proof. Observe that, by applying (10.1.36) in Theorem 10.1.2, as well as (10.3.14), we have

$$S^{(k)} \in \sigma_{m-k,0}(\tau_\kappa, \ldots, \tau_\lambda), \quad k = 0, \ldots, m-1, \tag{10.3.19}$$

and thus

$$S^{(k)}(\tau_\kappa) = S^{(k)}(\tau_\lambda) = 0, \quad k = 0, \ldots, m-1. \tag{10.3.20}$$

Let $m = 1$, for which we see from (10.3.19) that $S(\tau_k) = S(\tau_\lambda) = 0$. Also, since (10.3.13) holds, we note from (10.1.2), (10.1.3) and (10.1.4) that S is a continuous linear piecewise polynomial with breakpoints at $\{\tau_\kappa, \ldots, \tau_\lambda\}$, and thus S can have no zeros in $(\tau_\kappa, \tau_{\kappa+1}]$ and $[\tau_{\lambda-1}, \tau_\lambda)$, and at most one zero in each of the $\lambda - \kappa - 2$ successive intervals $[\tau_{\kappa+1}, \tau_{\kappa+2}], \ldots, [\tau_{\lambda-2}, \tau_{\lambda-1}]$. Hence $\rho \leqslant \lambda - \kappa - 2$, which proves the inequality (10.3.18) for $m = 1$.

Suppose next $m \geqslant 2$, for which (10.1.4) gives $S \in C^1(\mathbb{R})$. Since (10.3.19) implies $S(\tau_\kappa) = S(\tau_\lambda) = 0$, it follows from the definition (10.3.17) that S has $\rho + 2$ distinct zeros in $[\tau_\kappa, \tau_\lambda]$, so that an application of Rolle's theorem yields

$$\text{number of sign changes of } S' \text{ in } (\tau_\kappa, \tau_\lambda) \geqslant \rho + 1, \quad \text{if } m \geqslant 2. \qquad (10.3.21)$$

If $m = 2$, it follows from (10.3.19) that $S' \in \sigma_{1,0}(\tau_\kappa, \ldots, \tau_\lambda)$, with also, from (10.3.20), $S'(\tau_\kappa) = S'(\tau_\lambda) = 0$. Now observe that

$$\widetilde{S} \in \sigma_{1,0}(\tau_\kappa, \ldots, \tau_\lambda) \Rightarrow \text{number of sign changes of } \widetilde{S} \text{ in } (\tau_\kappa, \tau_\lambda) \leqslant \lambda - \kappa - 2, \quad (10.3.22)$$

since any linear spline $\widetilde{S} \in \sigma_{1,0}(\tau_\kappa, \ldots, \tau_\lambda)$ satisfies $\widetilde{S}(\tau_\kappa) = \widetilde{S}(\tau_\lambda) = 0$, so that $\widetilde{S}$ can have no sign changes in either of the intervals $(\tau_\kappa, \tau_{\kappa+1})$ or $[\tau_{\lambda-1}, \tau_\lambda)$, whereas $\widetilde{S}$ can have at most one sign change in each of the $\lambda - \kappa - 2$ successive intervals $(\tau_{\kappa+1}, \tau_{\kappa+2}], \ldots, (\tau_{\lambda-2}, \tau_{\lambda-1}]$. By applying (10.3.22), with $\widetilde{S} = S'$, as well as (10.3.21), we obtain $\rho + 1 \leqslant \lambda - \kappa - 2$, which is equivalent to the inequality (10.3.18) for $m = 2$.

Next, let $m \geqslant 3$, for which (10.1.4) gives $S' \in C^1(\mathbb{R})$. Since (10.3.21) then holds, as well as, from (10.3.20), $S'(\tau_\kappa) = S'(\tau_\lambda) = 0$, it follows that S' has at least $\rho + 3$ distinct zeros in $(\tau_\kappa, \tau_\lambda)$, so that an application of Rolle's theorem yields

$$\text{number of sign changes of } S'' \text{ in } (\tau_\kappa, \tau_\lambda) \geqslant \rho + 2, \text{ if } m \geqslant 3. \qquad (10.3.23)$$

Moreover, (10.3.19) gives $S'' \in \sigma_{1,0}(\tau_\kappa, \ldots, \tau_\lambda)$, with, from (10.3.20), $S''(\tau_\kappa) = S''(\tau_\lambda) = 0$, according to which we may apply (10.3.22), with $\widetilde{S} = S''$, together with (10.3.23), to obtain $\rho + 2 \leqslant \lambda - \kappa - 2$, which is equivalent to the inequality (10.3.18) for $m = 3$.

For $m \geqslant 4$, we continue in this fashion, for each m showing that $S^{(m-1)} \in \sigma_{1,0}(\tau_\kappa, \ldots, \tau_\lambda)$ possesses at least $\rho + m - 1$ sign changes in $(\tau_\kappa, \tau_\lambda)$, and employing (10.3.22), with $\widetilde{S} = S^{(m-1)}$, to deduce that $\rho + m - 1 \leqslant \lambda - \kappa - 2$, which is equivalent to the desired inequality (10.3.18). ∎

By using Theorem 10.3.1, we can now prove the following necessary and sufficient condition for spline interpolation.

Theorem 10.3.2 (Schoenberg-Whitney). *For any bounded interval $[a, b]$, and integers n and m satisfying $n > m \geqslant 0$, let $\{x_0, \ldots, x_n\}$ denote a sequence of $n + 1$ distinct points as in (10.3.1), and let $\{\tau_\mu, \ldots, \tau_\nu\}$ be a knot sequence as in (10.3.4), (10.3.5). Then the matrix $A_{m,n}$ in (10.3.10) is invertible if and only if*

$$N_{m, j-m}(x_j) \neq 0, \quad j = 0, \ldots, n, \qquad (10.3.24)$$

with the B-splines $\{N_{m,-m},\ldots,N_{m,n-m}\}$ *defined as in* (10.2.32), *in which case there exists precisely one spline* $S_{m,n}^I$ *in the space* $\sigma_m([a,b];\tau_1,\ldots,\tau_{n-m})$ *satisfying the interpolation conditions* (10.3.2), *where* $S_{m,n}^I$ *is given by the formula*

$$S_{m,n}^I(x) := \sum_{j=-m}^{n-m} (A_{m,n}^{-1}\mathbf{f}_n)_j N_{m,j}(x), \quad x \in [a,b], \tag{10.3.25}$$

with $\mathbf{f}_n \in \mathbb{R}^{n+1}$ *defined by* (10.3.11).

Proof. First, we show that if (10.3.24) is not satisfied, then the matrix $A_{m,n}$ is not invertible, which will then prove that (10.3.24) is a necessary condition for the invertibility of $A_{m,n}$. Suppose therefore that (10.3.24) does not hold, and denote by ℓ the smallest integer in the set $\{0,\ldots,n\}$ such that

$$N_{m,\ell-m}(x_\ell) = 0. \tag{10.3.26}$$

By applying the second line of (10.2.33) in Theorem 10.2.2(a), we deduce from (10.3.26) that either:

(i) $x_\ell < \tau_{\ell-m}$, in which case (10.3.1) and (10.3.4) imply $\ell \geqslant m+1$, and the last $n+1-\ell$ columns of $A_{m,n}$ in (10.3.10) each contains at most $n-\ell$ non-zero entries, all of which occur in the last $n-\ell$ positions, according to which these columns form a linearly dependent set in $\mathbb{R}^{n+1}$, and thus the columns of $A_{m,n}$ are linearly dependent; or

(ii) $x_\ell \geqslant \tau_{\ell+1}$, in which case (10.3.1) and (10.3.4) imply $\ell \leqslant n-m-1$, and the last $n+1-\ell$ rows of $A_{m,n}$ in (10.3.10) each contains at most $n-\ell$ non-zero entries, all of which occur in the last $n-\ell$ positions, according to which these rows form a linearly dependent set in $\mathbb{R}^{n+1}$, and thus the rows of $A_{m,n}$ are linearly dependent.

Hence, either columns or rows of the matrix $A_{m,n}$ are linearly dependent, so that we may deduce from a standard result in linear algebra that $A_{m,n}$ is not invertible.

Conversely, suppose that the condition (10.3.24) is satisfied, and assume that the matrix $A_{m,n}$ is not invertible. We shall now proceed to derive a contradiction, which will then prove that (10.3.24) is a sufficient condition for the invertibility of $A_{m,n}$.

Since $A_{m,n}$ is not invertible, there exists a non-trivial solution $\mathbf{c} = (c_{-m},\ldots,c_{n-m})^T$ of the homogeneous linear system

$$A_{m,n}\mathbf{c} = \mathbf{0}. \tag{10.3.27}$$

It then follows from (10.3.27) and (10.3.10) that the spline $\widetilde{S}$ defined by

$$\widetilde{S}(x) := \sum_{j=-m}^{n-m} c_j N_{m,j}(x), \quad x \in \mathbb{R}, \tag{10.3.28}$$

satisfies

$$\widetilde{S}(x_k) = 0, \quad k = 0,\ldots,n. \tag{10.3.29}$$

Note from (10.3.28), together with (10.2.34) in Theorem 10.2.2(b), that

$$\widetilde{S} \in \sigma_{m,0}(\tau_{-m},\ldots,\tau_{n+1}). \tag{10.3.30}$$

Also, since $\mathbf{c} = (c_{-m},\ldots,c_{n-m})^T$ is not the zero vector, we deduce from (10.3.28), together with the second line of (10.2.33) in Theorem 10.2.2(a), as well as (10.2.62) in Theorem 10.2.6(a), that $\widetilde{S}$ is not the zero function. Hence, by keeping in mind also (10.1.2), (10.1.3), and recalling Theorem 10.2.1(b), we deduce from (10.3.29) that there exist integers κ and λ, with

$$\lambda - \kappa \geqslant m+1; \quad -m \leqslant \kappa < \lambda \leqslant n+1, \tag{10.3.31}$$

such that the spline S defined by

$$S(x) := \begin{cases} \widetilde{S}(x), & x \in [\tau_\kappa, \tau_\lambda); \\ 0, & x \in \mathbb{R} \setminus [\tau_\kappa, \tau_\lambda), \end{cases} \tag{10.3.32}$$

is not identically zero on any of the successive intervals $[\tau_\kappa, \tau_{\kappa+1}),\ldots,[\tau_{\lambda-1},\tau_\lambda)$, according to which S has at most a finite number of zeros in $(\tau_\kappa, \tau_\lambda)$. Observe from (10.3.32) and (10.3.28) that

$$S \in \sigma_{m,0}(\tau_\kappa,\ldots,\tau_\lambda). \tag{10.3.33}$$

Now observe from (10.3.24), together with the second line of (10.2.33) in Theorem 10.2.2(a), that

$$x_j \in (\tau_j, \tau_{j+m+1}), \quad j = \kappa,\ldots,\lambda - m - 1, \tag{10.3.34}$$

and thus, from (10.3.4),

$$\{x_\kappa,\ldots,x_{\lambda-m-1}\} \subset (\tau_\kappa, \tau_\lambda). \tag{10.3.35}$$

Also, note from (10.3.32) and (10.3.29) that

$$S(x_j) = 0, \quad j = \kappa,\ldots,\lambda - m - 1. \tag{10.3.36}$$

If $m = 0$, we deduce from (10.3.36), (10.3.34) and (10.3.33) that S is identically zero on at least one of the intervals $[\tau_\kappa, \tau_{\kappa+1}),\ldots,[\tau_{\lambda-1},\tau_\lambda)$, which is a contradiction.

Suppose next $m \in \mathbb{N}$. But then we may apply Theorem 10.3.1 to deduce that, with the non-negative number ρ defined as in (10.3.17), the inequality (10.3.18) is satisfied, that is,

$$\rho \leqslant \lambda - \kappa - m - 1. \tag{10.3.37}$$

But, according to (10.3.36) and (10.3.35), as well as (10.3.1),

$$\rho \geqslant (\lambda - m - 1) - \kappa + 1 = \lambda - \kappa - m,$$

which contradicts (10.3.37), and thereby completing our proof.   ■

According to Theorem 10.3.2, we may define the spline interpolation operator $\mathscr{S}^l_{m,n}$:
$C[a,b] \to \sigma_m([a,b]; \tau_1, \ldots, \tau_{n-m})$ by

$$\mathscr{S}^l_{m,n} f := S^l_{m,n}, \quad f \in C[a,b], \tag{10.3.38}$$

with the spline $S^l_{m,n}$ defined by (10.3.25), and for which the following properties can now
be proved.

Theorem 10.3.3. *The spline interpolation operator* $\mathscr{S}^l_{m,n} : C[a,b] \to \sigma_m([a,b]; \tau_1, \ldots,$
$\tau_{n-m})$, *as defined by* (10.3.38), *with* $S^l_{m,n}$ *as in Theorem* 10.3.2, *satisfies the following:*
(a) $\mathscr{S}^l_{m,n}$ *is linear;*
(b) $\mathscr{S}^l_{m,n}$ *is exact on* $\sigma_m([a,b]; \tau_1, \ldots, \tau_{n-m})$, *that is,*

$$\mathscr{S}^l_{m,n} f = f, \quad f \in \sigma_m([a,b]; \tau_1, \ldots, \tau_{n-m}). \tag{10.3.39}$$

Proof. (a) For $f, g \in C[a,b]$, and $\lambda, \mu \in \mathbb{R}$, it follows from (10.3.38) and (10.3.25) that,
with the definition

$$\mathbf{g}_n := (g(x_0), \ldots, g(x_n))^T,$$

$$\mathscr{S}^l_{m,n}(\lambda f + \mu g) = \sum_{j=-m}^{n-m} \left(A^{-1}_{m,n}(\lambda \mathbf{f}_n + \mu \mathbf{g}_n) \right)_j N_{m,j}$$

$$= \lambda \left[\sum_{j=-m}^{n-m} \left(A^{-1}_{m,n} \mathbf{f}_n \right)_j N_{m,j} \right] + \mu \left[\sum_{j=-m}^{n-m} \left(A^{-1}_{m,n} \mathbf{g}_n \right)_j N_{m,j} \right]$$

$$= \lambda (\mathscr{S}^l_{m,n} f) + \mu (\mathscr{S}^l_{m,n} g),$$

and thus $\mathscr{S}^l_{m,n}$ is linear.

(b) Let $S \in \sigma_m([a,b]; \tau_1, \ldots, \tau_{n-m})$. Then S trivially satisfies the interpolation conditions
(10.3.2) for the choice $f = S$, and thus, by using also the uniqueness statement in Theo-
rem 10.3.2, as well as the definition (10.3.38), we deduce that the exactness result (10.3.39)
does indeed hold.   ■

Example 10.3.1. For any function $f \in C[0,5]$, consider the problem of obtaining a spline

$$S \in \sigma_2([0,5]; 1, 2, 3, 4) \tag{10.3.40}$$

satisfying the interpolation conditions

$$S(x_j) = f(x_j), \quad j = 0, \ldots, 6, \tag{10.3.41}$$

where

$$\{x_0, \ldots, x_6\} = \{0, \tfrac{1}{2}, \tfrac{3}{2}, \tfrac{5}{2}, \tfrac{7}{2}, \tfrac{9}{2}, 5\}. \tag{10.3.42}$$

With $n = 6$ and $m = 2$, and with the extended knot sequence $\{\tau_{-2}, \tau_{-1}, \ldots, \tau_7\}$ given by

$$\tau_j = j, \quad j = -2, \ldots, 7, \tag{10.3.43}$$

so that (10.3.4) is satisfied, with

$$[a, b] = [0, 5], \quad \text{and} \quad \{\mu, \nu\} = \{-2, 7\},$$

it can now be verified by means of (10.3.42), together with (10.2.68) and (10.2.71) in Theorem 10.2.7, that the condition (10.3.24) of Theorem 10.3.2 is satisfied. It follows from Theorem 10.3.2, together with (10.3.38) and (10.3.25), that there exists precisely one spline $\mathscr{S}_{2,6}^I f = S_{2,6}^I$ in $\sigma_2([0,5]; 1, 2, 3, 4)$ such that

$$(\mathscr{S}_{2,6}^I f)(x_j) = f(x_j), \quad j = 0, \ldots, 6, \tag{10.3.44}$$

and where, by using also (10.2.68) in Theorem 10.2.7,

$$(\mathscr{S}_{2,6}^I f)(x) = \sum_{j=-2}^{4} \left(A_{2,6}^{-1} \mathbf{f}_6 \right)_j N_2(x - j), \quad x \in [0, 5], \tag{10.3.45}$$

with N_2 denoting the quadratic cardinal B-spline in (10.2.80), where $A_{2,6}$ is the invertible 7×7 matrix in (10.3.10), and $\mathbf{f}_6 \in \mathbb{R}^7$ the vector in (10.3.11).

To compute the matrix $A_{2,6}$, we first use the formulas in (10.2.80) to obtain the values

$$N_2(\tfrac{1}{2}) = \tfrac{1}{8}; \quad N_2(\tfrac{3}{2}) = \tfrac{3}{4}; \quad N_2(\tfrac{5}{2}) = \tfrac{1}{8}. \tag{10.3.46}$$

By using (10.3.46), as well as the values (10.2.82) obtained in Example 10.2.3, it now follows from (10.3.10) and (10.3.42), together with (10.2.68), that

$$A_{2,6} = \begin{bmatrix} \tfrac{1}{2} & \tfrac{1}{2} & 0 & 0 & 0 & 0 & 0 \\ \tfrac{1}{8} & \tfrac{3}{4} & \tfrac{1}{8} & 0 & 0 & 0 & 0 \\ 0 & \tfrac{1}{8} & \tfrac{3}{4} & \tfrac{1}{8} & 0 & 0 & 0 \\ 0 & 0 & \tfrac{1}{8} & \tfrac{3}{4} & \tfrac{1}{8} & 0 & 0 \\ 0 & 0 & 0 & \tfrac{1}{8} & \tfrac{3}{4} & \tfrac{1}{8} & 0 \\ 0 & 0 & 0 & 0 & \tfrac{1}{8} & \tfrac{3}{4} & \tfrac{1}{8} \\ 0 & 0 & 0 & 0 & 0 & \tfrac{1}{2} & \tfrac{1}{2} \end{bmatrix}, \tag{10.3.47}$$

the inverse of which is given by

$$A_{2,6}^{-1} = \begin{bmatrix} \frac{5741}{2378} & -\frac{1970}{1189} & \frac{338}{1189} & -\frac{2}{41} & \frac{10}{1189} & -\frac{2}{1189} & \frac{1}{2378} \\[4pt] -\frac{985}{2378} & \frac{1970}{1189} & -\frac{338}{1189} & \frac{2}{41} & -\frac{10}{1189} & \frac{2}{1189} & -\frac{1}{2378} \\[4pt] \frac{169}{2378} & -\frac{338}{1189} & \frac{1690}{1189} & -\frac{10}{41} & \frac{50}{1189} & -\frac{10}{1189} & \frac{5}{2378} \\[4pt] -\frac{1}{82} & \frac{2}{41} & -\frac{10}{41} & \frac{58}{41} & -\frac{10}{41} & \frac{2}{41} & -\frac{1}{82} \\[4pt] \frac{5}{2378} & -\frac{10}{1189} & \frac{50}{1189} & -\frac{10}{41} & \frac{1690}{1189} & -\frac{338}{1189} & \frac{169}{2378} \\[4pt] -\frac{1}{2378} & \frac{2}{1189} & -\frac{10}{1189} & \frac{58}{1189} & -\frac{338}{1189} & \frac{1970}{1189} & -\frac{985}{2378} \\[4pt] \frac{1}{2378} & -\frac{2}{1189} & \frac{10}{1189} & -\frac{58}{1189} & \frac{338}{1189} & -\frac{1970}{1189} & \frac{5741}{2378} \end{bmatrix}. \tag{10.3.48}$$

Since, moreover, (10.3.11) and (10.3.42) give

$$\mathbf{f}_6 = (f(0), f(\tfrac{1}{2}), f(\tfrac{3}{2}), f(\tfrac{5}{2}), f(\tfrac{7}{2}), f(\tfrac{9}{2}), f(5))^T, \tag{10.3.49}$$

it follows from (10.3.45), (10.3.48) and (10.3.49) that the spline interpolant $\mathscr{S}_{2,6}^I f$ is given explicitly, for $x \in [0,5]$, by the formula

$$\begin{aligned}
(\mathscr{S}_{2,6}^I f)(x) =\; & \big[\tfrac{5741}{2378} f(0) - \tfrac{1970}{1189} f(\tfrac{1}{2}) + \tfrac{338}{1189} f(\tfrac{3}{2}) - \tfrac{2}{41} f(\tfrac{5}{2}) \\
& + \tfrac{10}{1189} f(\tfrac{7}{2}) - \tfrac{2}{1189} f(\tfrac{9}{2}) + \tfrac{1}{2378} f(5)\big] N_2(x+2) \\
& + \big[-\tfrac{985}{2378} f(0) + \tfrac{1970}{1189} f(\tfrac{1}{2}) - \tfrac{338}{1189} f(\tfrac{3}{2}) + \tfrac{2}{41} f(\tfrac{5}{2}) \\
& - \tfrac{10}{1189} f(\tfrac{7}{2}) + \tfrac{2}{1189} f(\tfrac{9}{2}) - \tfrac{1}{2378} f(5)\big] N_2(x+1) \\
& + \big[\tfrac{169}{2378} f(0) - \tfrac{338}{1189} f(\tfrac{1}{2}) + \tfrac{1690}{1189} f(\tfrac{3}{2}) - \tfrac{10}{41} f(\tfrac{5}{2}) \\
& + \tfrac{50}{1189} f(\tfrac{7}{2}) - \tfrac{10}{1189} f(\tfrac{9}{2}) + \tfrac{5}{2378} f(5)\big] N_2(x) \\
& + \big[-\tfrac{1}{82} f(0) + \tfrac{2}{41} f(\tfrac{1}{2}) - \tfrac{10}{41} f(\tfrac{3}{2}) + \tfrac{58}{41} f(\tfrac{5}{2}) \\
& - \tfrac{10}{41} f(\tfrac{7}{2}) + \tfrac{2}{41} f(\tfrac{9}{2}) - \tfrac{1}{82} f(5)\big] N_2(x-1) \\
& + \big[\tfrac{5}{2378} f(0) - \tfrac{10}{1189} f(\tfrac{1}{2}) + \tfrac{50}{1189} f(\tfrac{3}{2}) - \tfrac{10}{41} f(\tfrac{5}{2}) \\
& + \tfrac{1690}{1189} f(\tfrac{7}{2}) - \tfrac{338}{1189} f(\tfrac{9}{2}) + \tfrac{169}{2378} f(5)\big] N_2(x-2) \\
& + \big[-\tfrac{1}{2378} f(0) + \tfrac{2}{1189} f(\tfrac{1}{2}) - \tfrac{10}{1189} f(\tfrac{3}{2}) + \tfrac{58}{1189} f(\tfrac{5}{2}) \\
& - \tfrac{338}{1189} f(\tfrac{7}{2}) + \tfrac{1970}{1189} f(\tfrac{9}{2}) - \tfrac{985}{2378} f(5)\big] N_2(x-3) \\
& + \big[\tfrac{1}{2378} f(0) - \tfrac{2}{1189} f(\tfrac{1}{2}) + \tfrac{10}{1189} f(\tfrac{3}{2}) - \tfrac{58}{1189} f(\tfrac{5}{2}) \\
& + \tfrac{338}{1189} f(\tfrac{7}{2}) - \tfrac{1970}{1189} f(\tfrac{9}{2}) + \tfrac{5741}{2378} f(5)\big] N_2(x-4). \tag{10.3.50}
\end{aligned}$$

■

According to (10.3.38) and (10.3.25), the computation of $(\mathscr{S}_{m,n}^I f)(x)$ depends on the inversion of the matrix $A_{m,n}$ in (10.3.10), which, as follows from the second line of (10.2.33) in Theorem 10.2.2(a), and as illustrated by (10.3.47) in Example 10.3.1, is a banded matrix. Moreover, the inverse matrix $A_{m,n}^{-1}$ is, in general, and as illustrated by (10.3.48) in Example 10.3.1, not banded, but a full matrix, with the result that $\mathscr{S}_{m,n}^I$ is not a local approximation operator, in the sense that, for any $x \in [a,b]$, the value of $(\mathscr{S}_{m,n}^I f)(x)$ depends on all, or most, of the function values $\{f(x_0),\ldots,f(x_n)\}$, as illustrated by the formula (10.3.50) in Example 10.3.1. In Section 10.4, we shall construct a class of local spline approximation operators with explicit formulations.

Finally in this section, we show how Theorem 10.3.1 and its proof may be used to prove the following result, according to which B-splines are "bell-shaped".

Theorem 10.3.4. *For any integer $m \geqslant 2$ and $j \in \{\mu,\ldots,\nu - m - 1\}$, let $N_{m,j}$ denote the B-spline as defined in (10.2.32), (10.2.31). Then, for $k = 1,\ldots,m-1$, the k-th derivative $N_{m,j}^{(k)}$ has precisely k distinct zeros in the interval (τ_j, τ_{j+m+1}), each of which is a sign change of the type (10.3.15).*

Proof. In Theorem 10.3.1, let $S = N_{m,j}$, $\kappa = j$ and $\lambda = j + m + 1$, so that, from (10.2.62) in Theorem 10.2.6, we have $\rho = 0$. It follows as in the final paragraph in the proof of Theorem 10.3.1 that

$$m - 1 = 0 + (m-1) \leqslant \text{number of sign changes of } N_{m,j}^{(m-1)} \text{ in } (\tau_j, \tau_{j+m+1})$$
$$\leqslant (m+1) - 2 = m - 1,$$

and thus $N_{m,j}^{(m-1)}$ has precisely $m - 1$ sign changes in the interval (τ_j, τ_{j+m+1}). Since also, from (10.2.33) in Theorem 10.2.2(a), $N_{m,j}^{(m-1)}$ vanishes identically on $(-\infty, \tau_j] \cup [\tau_{j+m+1},\infty)$, and since, moreover, from (10.1.36) in Theorem 10.1.2, we have $N_{m,j}^{(m-1)} \in \sigma_{1,0}(\tau_j,\ldots,\tau_{j+m+1})$, that is, $N_{m,j}^{(m-1)}$ is a continuous linear piecewise polynomial on $[\tau_j, \tau_{j+m+1}]$, with breakpoints at $\{\tau_{j+1},\ldots,\tau_{j+m}\}$, we deduce that the $m - 1$ sign changes in (τ_j, τ_{j+m+1}) of $N_{m,j}^{(m-1)}$ are the only zeros of $N_{m,j}^{(m-1)}$ in (τ_j, τ_{j+m+1}), and are all of the type (10.3.15), with, moreover, these zeros occurring in the $m - 1$ successive intervals $(\tau_{j+1},\tau_{j+2}),\ldots,(\tau_{j+m-1},\tau_{j+m})$. Hence we have now established the theorem for $k = m-1$. Next, observe from the proof of Theorem 10.3.1 that

$$\text{number of sign changes of } N_{m,j}^{(m-2)} \text{ in } (\tau_j,\tau_{j+m+1}) \geqslant 0 + (m-2) = m-2. \quad (10.3.51)$$

Furthermore, observe that $N_{m,j}^{(m-2)}$ is not identically zero on any of the intervals $[\tau_j, \tau_{j+1}]$, $\ldots,[\tau_{j+1},\tau_{j+m+1}]$, for if it did vanish identically on such an interval, then $(N_{m,j}^{(m-2)})' =$

$N_{m,j}^{(m-1)}$ would also be identically zero on that interval, and thereby contradicting the fact that $N_{m,j}^{(m-1)}$ has a finite number $(= m-1)$ of zeros in (τ_j, τ_{j+m+1}). Hence the sign changes of $N_{m,j}^{(m-2)}$ in (τ_j, τ_{j+m+1}) are all of the type (10.3.15).

Suppose now that

$$\text{number of distinct zeros of } N_{m,j}^{(m-2)} \text{ in } (\tau_j, \tau_{j+m+1}) \geqslant m-1. \tag{10.3.52}$$

Since also, from (10.3.20) in the proof of Theorem 10.3.1,

$$N_{m,j}^{(m-2)}(\tau_j) = N_{m,j}^{(m-2)}(\tau_{j+m+1}) = 0,$$

so that $N_{m,j}^{(m-2)}$ has at least $m+1$ distinct zeros in $[\tau_j, \tau_{j+m+1}]$, it follows that $(N_{m,j}^{(m-2)})' = N_{m,j}^{(m-1)}$ has at least m sign changes in (τ_j, τ_{j+m+1}), which contradicts the fact that $N_{m,j}^{(m-1)}$ has precisely $m-1$ sign changes in (τ_j, τ_{j+m+1}). Hence (10.3.52) is not true, that is,

$$\text{number of distinct zeros of } N_{m,j}^{(m-2)} \text{ in } (\tau_j, \tau_{j+m+1}) \leqslant m-2. \tag{10.3.53}$$

It follows from (10.3.51) and (10.3.53) that $N_{m,j}^{(m-2)}$ has precisely $m-2$ distinct zeros in (τ_j, τ_{j+m+1}), all of which are sign changes. Hence we have established the theorem for $k = m-2$.

By proceeding inductively as above, the cases $k = m-3, \ldots, 1$, are proved successively.

∎

10.4   Local quasi-interpolation

In this section, we shall establish an explicitly formulated spline approximation operator $\mathscr{S} : C[a,b] \to \sigma_m([a,b]; \tau_1, \ldots, \tau_r)$ such that $\mathscr{S}$ is exact on π_m, that is,

$$(\mathscr{S}f)(x) = f(x), \quad x \in [a,b], \quad f \in \pi_m, \tag{10.4.1}$$

in which case $\mathscr{S}$ is called a quasi-interpolation operator, and, moreover, such that $\mathscr{S}$ is a local approximation operator, in the sense that, for any $x \in [a,b]$, the value $(\mathscr{S}f)(x)$ is independent of the values $\{f(x) : x \in [a,b] \setminus [\alpha, \beta]\}$, for some subinterval $[\alpha, \beta]$ of $[a,b]$, the size of which depends only on the spline degree m.

We shall rely on the following identity for B-splines, in which we employ, as before, the convention $\prod_{j=j_0}^{j_1} a_j := 1$, if $j_1 < j_0$.

Theorem 10.4.1. (Marsden identity) *For any non-negative integer m, and an extended knot sequence $\{\tau_{-m}, \ldots, \tau_{r+m+1}\}$ as in (10.2.31), (10.2.30), the B-splines $\{N_{m,-m}, \ldots, N_{m,r}\}$, as defined in (10.2.32), satisfy the identity*

$$(t-x)^m = \sum_{j=-m}^{r} g_{m,j}(t) N_{m,j}(x), \quad x \in [a,b], \quad t \in \mathbb{R}, \tag{10.4.2}$$

where the polynomial sequence $\{g_{m,j} : j = -m, \ldots, r\} \subset \pi_m$ *is defined by*

$$g_{m,j}(t) := \prod_{k=1}^{m}(t - \tau_{j+k}), \quad j = -m, \ldots, r. \tag{10.4.3}$$

Proof. Let $t \in \mathbb{R}$ be fixed. Since the inclusion (10.1.5) holds, and $(t - \cdot)^m \in \pi_m$, we deduce from Theorem 10.2.2(d) that there exists a (unique) coefficient sequence $\{g_{m,j}(t) : j = -m, \ldots, r\} \subset \mathbb{R}$ such that the identity (10.4.2) is satisfied.

Next, to prove (10.4.2), (10.4.3), we first note from (10.2.63) in Theorem 10.2.6 that, if $m = 0$, then (10.4.2) holds, with

$$g_{0,j}(t) := 1, \quad j = 0, \ldots, r,$$

that is, (10.4.2), (10.4.3) are satisfied for $m = 0$.

Proceeding inductively, we suppose next that (10.4.2), (10.4.3) hold for a fixed non-negative integer m, and, for an extended knot sequence $\{\tau_{-m-1}, \ldots, \tau_{r+m+2}\}$ as in (10.2.31), (10.2.30), denote by $\{N_{m+1,-m-1}, \ldots, N_{m+1,r}\}$ the B-splines as obtained from (10.2.32). With $\{g_{m+1,j}(t) : j = -m-1, \ldots, r\} \subset \mathbb{R}$ denoting the coefficient sequence obtained from (10.4.3), we use the recursive formulation (10.2.56) in Theorem 10.2.5, as well as (10.2.31) and the second line of (10.2.33) in Theorem 10.2.2(a), together with (10.4.3), and eventually (10.4.2), to obtain, for any $x \in [a, b]$,

$$\sum_{j=-m-1}^{r} g_{m+1,j}(t) N_{m+1,j}(x)$$

$$= \sum_{j=-m-1}^{r} g_{m+1,j}(t) \left[\frac{x - \tau_j}{\tau_{j+m+1} - \tau_j} \right] N_{m,j}(x) + \sum_{j=-m-1}^{r} g_{m+1,j}(t) \left[\frac{\tau_{j+m+2} - x}{\tau_{j+m+2} - \tau_{j+1}} \right] N_{m,j+1}(x)$$

$$= \sum_{j=-m-1}^{r} g_{m+1,j}(t) \left[\frac{x - \tau_j}{\tau_{j+m+1} - \tau_j} \right] N_{m,j}(x) + \sum_{j=-m}^{r+1} g_{m+1,j-1}(t) \left[\frac{\tau_{j+m+1} - x}{\tau_{j+m+1} - \tau_j} \right] N_{m,j}(x)$$

$$= \sum_{j=-m}^{r} \frac{(x - \tau_j) g_{m+1,j}(t) + (\tau_{j+m+1} - x) g_{m+1,j-1}(t)}{\tau_{j+m+1} - \tau_j} N_{m,j}(x)$$

$$= \sum_{j=-m}^{r} \frac{[(x - \tau_j)(t - \tau_{j+m+1}) + (\tau_{j+m+1} - x)(t - \tau_j)] g_{m,j}(t)}{\tau_{j+m+1} - \tau_j} N_{m,j}(x)$$

$$= \sum_{j=-m}^{r} \frac{(t - x)(\tau_{j+m+1} - \tau_j) g_{m,j}(t)}{\tau_{j+m+1} - \tau_j} N_{m,j}(x) = (t - x) \sum_{j=-m}^{r} g_{m,j}(t) N_{m,j}(x)$$

$$= (t - x)(t - x)^m = (t - x)^{m+1},$$

that is, (10.4.2), (10.4.3) also hold with m replaced by $m+1$, and thereby completing our inductive proof of the formula (10.4.2). ∎

For $m \in \mathbb{N}$, and any integer $\ell \in \{1,\ldots,m\}$, we may now differentiate the identity (10.4.2) ℓ times with respect to t to obtain

$$(t-x)^{m-\ell} = \frac{(m-\ell)!}{m!} \sum_{j=-m}^{r} g_{m,j}^{(\ell)}(t) N_{m,j}(x), \quad x \in [a,b], \quad t \in \mathbb{R},$$

or equivalently,

$$(t-x)^{\ell} = \frac{\ell!}{m!} \sum_{j=-m}^{r} g_{m,j}^{(m-\ell)}(t) N_{m,j}(x), \quad x \in [a,b], \quad t \in \mathbb{R},$$

in which we may now set $t = 0$ to obtain the following result.

Theorem 10.4.2. *The B-splines* $\{N_{m,-m},\ldots,N_{m,r}\}$ *in Theorem 10.4.1 satisfy the identity*

$$x^{\ell} = (-1)^{\ell} \frac{\ell!}{m!} \sum_{j=-m}^{r} g_{m,j}^{(m-\ell)}(0) N_{m,j}(x), \quad x \in [a,b], \quad \ell = 0,\ldots,m, \tag{10.4.4}$$

with the polynomial sequence $\{g_{m,j} : j = -m,\ldots,r\} \subset \pi_m$ *defined as in* (10.4.3).

We shall also require the following explicit formulation in terms of Lagrange fundamental polynomials of the inverse of a Vandermonde matrix.

Theorem 10.4.3. *For any non-negative integer* n, *the inverse of the Vandermonde matrix* V_n *in Theorem 1.1.2 is given by*

$$V_n^{-1} = \begin{bmatrix} L_{n,0}(0) & L_{n,1}(0) & \cdots & L_{n,n}(0) \\ L_{n,0}'(0) & L_{n,1}'(0) & \cdots & L_{n,n}'(0) \\ \dfrac{L_{n,0}''(0)}{2!} & \dfrac{L_{n,1}''(0)}{2!} & \cdots & \dfrac{L_{n,n}''(0)}{2!} \\ \vdots & \vdots & & \vdots \\ \dfrac{L_{n,0}^{(n)}(0)}{n!} & \dfrac{L_{n,1}^{(n)}(0)}{n!} & \cdots & \dfrac{L_{n,n}^{(n)}(0)}{n!} \end{bmatrix}, \tag{10.4.5}$$

with $\{L_{n,0},\ldots,L_{n,n}\} \subset \pi_n$ *denoting the Lagrange fundamental polynomials, as defined in* (1.2.1).

Proof. First, observe from the interpolation formulas given in (1.1.16) of Theorem 1.1.2, and in (1.2.5) of Theorem 1.2.2, that

$$\sum_{j=0}^{n} (V_n^{-1}\mathbf{f})_j x^j = P_n^I(x) = \sum_{j=0}^{n} f(x_j) L_{n,j}(x), \quad x \in \mathbb{R}, \tag{10.4.6}$$

where $\mathbf{f} \in \mathbb{R}^{n+1}$ is defined by (1.1.8). Since $P_n^I \in \pi_n$, we may now apply the Taylor expansion polynomial identity (10.1.11), with $P = P_n^I$ and $c = 0$, to deduce from the second equation in (10.4.6) that

$$\sum_{j=0}^{n} f(x_j) L_{n,j}(x) = \sum_{j=0}^{n} \left[\frac{\sum_{k=0}^{n} f(x_k) L_{n,k}^{(j)}(0)}{j!} \right] x^j = \sum_{j=0}^{n} \left[\frac{1}{j!} \sum_{k=0}^{n} L_{n,k}^{(j)}(0) f(x_k) \right] x^j, \quad x \in \mathbb{R},$$

(10.4.7)

which, together with (10.4.6), yields

$$(V_n^{-1} \mathbf{f})_j = \frac{1}{j!} \sum_{k=0}^{n} L_{n,k}^{(j)}(0) f(x_k), \quad j = 0, \dots, n.$$

(10.4.8)

According to (1.1.8), we have

$$(V_n^{-1} \mathbf{f})_j = \sum_{k=0}^{n} (V_n^{-1})_{jk} f(x_k), \quad j = 0, \dots, n,$$

which, together with (10.4.8), yields

$$\sum_{k=0}^{n} \left[(V_n^{-1})_{jk} - \frac{1}{j!} L_{n,k}^{(j)}(0) \right] f(x_k) = 0, \quad j = 0, \dots, n.$$

(10.4.9)

Since the function f is arbitrary, we may now, for any fixed $\ell \in \{0, \dots, n\}$, choose f such that

$$f(x_k) = \delta_{\ell-k}, \quad k = 0, \dots, n,$$

with the Kronecker delta sequence $\{\delta_j : j \in \mathbb{Z}\}$ as in (1.2.2), to deduce from (10.4.9) that

$$(V_n^{-1})_{j\ell} = \frac{L_{n,\ell}^{(j)}(0)}{j!}, \quad j, \ell = 0, \dots, n,$$

which is equivalent to the desired result (10.4.5). ∎

In order to obtain an approximation operator $\mathscr{S} : C[a,b] \to \sigma_m([a,b]; \tau_1, \dots, \tau_r)$ satisfying the polynomial exactness condition (10.4.1), we let $\{\tau_{-m}, \dots, \tau_{r+m+1}\}$ denote a knot sequence as in (10.2.31), (10.2.30), for integers $m \geqslant 0$ and $r \geqslant 1$ satisfying

$$m \leqslant r+1,$$

(10.4.10)

and define, for any coefficient sequence $\{\alpha_{j,k} : k = j, \dots, j+m; j = -m, \dots, r\}$, the spline approximation operator $\mathscr{S}$ by

$$(\mathscr{S}f)(x) := \sum_{j=-m}^{r} \left[\sum_{k=j}^{j+m} \alpha_{j,k} f(\tau_k) \right] N_{m,j}(x), \quad x \in [a,b], \quad f \in C[a,b],$$

(10.4.11)

where the values of f at the knots are extended from $\{\tau_0, \dots, \tau_{r+1}\}$ to $\{\tau_{-m}, \dots, \tau_{r+m}\}$ by means of the polynomial extrapolation method

$$f(\tau_k) := \begin{cases} (\mathscr{P}_m^I f)(\tau_k), \; k = -m, \dots, -1 \; (\text{if } m \geqslant 1); \\ (\widetilde{\mathscr{P}}_m^I f)(\tau_k), \; k = r+2, \dots, r+m \; (\text{if } m \geqslant 2), \end{cases}$$

(10.4.12)

with $\mathscr{P}_m^l f$ and $\widetilde{\mathscr{P}}_m^l f$ denoting, as in (5.1.2), the polynomials in π_m interpolating f at, respectively, the point sequences $\{\tau_0,\ldots,\tau_m\}$ and $\{\tau_{r+1-m},\ldots,\tau_{r+1}\}$, that is,

with

$$\left.\begin{array}{l} \mathscr{P}_m^l f \in \pi_m \quad ; \quad \widetilde{\mathscr{P}}_m^l f \in \pi_m, \\[4pt] (\mathscr{P}_m^l f)(\tau_k) = f(\tau_k), \quad k = 0,\ldots,m; \\[4pt] (\widetilde{\mathscr{P}}_m^l f)(\tau_k) = f(\tau_k), \quad k = r+1-m,\ldots,r+1. \end{array}\right\} \quad (10.4.13)$$

Note from (10.4.11) and (10.4.12), together with Theorem 5.1.1(a), that $\mathscr{S}$ is a linear approximation operator.

By interchanging the order of summation in the defining formula (10.4.11), we obtain the formulation

$$(\mathscr{S}f)(x) = \sum_{k=-m}^{r+m} f(\tau_k) U_{m,k}(x), \quad x \in [a,b], \quad f \in C[a,b], \quad (10.4.14)$$

where

$$U_{m,k}(x) := \sum_{j=\max\{k-m,-m\}}^{\min\{k,r\}} \alpha_{j,k} N_{m,j}(x), \quad x \in [a,b], \quad k = -m,\ldots,r+m, \quad (10.4.15)$$

according to which, together with (10.2.34) in Theorem 10.2.2(b), we have

$$\{U_{m,k} : k = -m,\ldots,r+m\} \subset \sigma_m([a,b];\tau_1,\ldots,\tau_r), \quad (10.4.16)$$

with also, from (10.4.15), together with the second line of (10.2.33) in Theorem 10.2.2(a),

$$U_{m,k}(x) = 0, \quad x \in \left\{\begin{array}{ll} [\tau_{k+m+1},b], & k=-m,\ldots,m \text{ (if } r \geqslant 2m+1); \\[4pt] [a,\tau_{k-m}] \cup [\tau_{k+m+1},b], & k=m+1,\ldots,r-m-1 \text{ (if } r \geqslant 2m+2); \\[4pt] [a,\tau_{k-m}], & k=r-m,\ldots,r+m. \end{array}\right.$$
$$(10.4.17)$$

Observe from (10.4.14) and (10.4.17), together with (10.4.12), (10.4.13), that $\mathscr{S}$ is a local approximation operator, in the sense that, for any fixed $x \in [a,b]$, the value of $(\mathscr{S}f)(x)$ is independent of the values $\{f(x) : x \in [a,b] \setminus [\tau_\kappa,\tau_\lambda]\}$, for integers κ and λ satisfying $\lambda - \kappa \leqslant 2m+1$.

We proceed to show that there exists a unique sequence $\{\alpha_{j,k} : k = j,\ldots,j+m; j = -m,\ldots,r\} \subset \mathbb{R}$ such that the operator $\mathscr{S}$ in (10.4.11), (10.4.12) satisfies the polynomial exactness condition (10.4.1), from which it will then follow that $\mathscr{S}$ is an optimally local quasi-interpolation operator.

To this end, we first observe that the polynomial exactness condition (10.4.1) has the equivalent formulation

$$(\mathscr{S}f_\ell)(x) = f_\ell(x), \quad x \in [a,b], \quad \ell = 0,\ldots,m, \quad (10.4.18)$$

where

$$f_\ell(x) := x^\ell, \quad x \in [a,b], \quad \ell = 0, \ldots, m. \tag{10.4.19}$$

It follows from (10.4.11), (10.4.12), (10.4.13), together with Theorem 5.1.1(b), that

$$(\mathscr{S} f_\ell)(x) = \sum_{j=-m}^{r} \left[\sum_{k=j}^{j+m} \alpha_{j,k} \tau_k^\ell \right] N_{m,j}(x), \quad x \in [a,b], \quad \ell = 0, \ldots, m. \tag{10.4.20}$$

By applying (10.4.4) in Theorem 10.4.2, we deduce from (10.4.20) and (10.4.19) that the condition (10.4.18) is equivalent to

$$\sum_{j=-m}^{r} \left[\sum_{k=j}^{j+m} \alpha_{j,k} \tau_k^\ell - (-1)^\ell \frac{\ell!}{m!} g_{m,j}^{(m-\ell)}(0) \right] N_{m,j}(x) = 0, \; x \in [a,b], \; \ell = 0, \ldots, m, \tag{10.4.21}$$

with $\{g_{m,j} : j = -m, \ldots, r\} \subset \pi_m$ denoting the polynomial sequence defined by (10.4.3). By using Theorem 10.2.2(d), it furthermore follows that (10.4.21) holds if and only if the sequence $\{\alpha_{j,k} : k = j, \ldots, j+m; j = -m, \ldots, r\}$ satisfies the linear system

$$\sum_{k=j}^{j+m} \tau_k^\ell \alpha_{j,k} = (-1)^\ell \frac{\ell!}{m!} g_{m,j}^{(m-\ell)}(0), \; \ell = 0, \ldots, m, \quad j = -m, \ldots, r, \tag{10.4.22}$$

or equivalently, in matrix-vector notation,

$$B_{m,j} \alpha_j = (-1)^\ell \frac{\ell!}{m!} g_{m,j}, \quad j = -m, \ldots, r, \tag{10.4.23}$$

where the $(m+1) \times (m+1)$ matrix sequence $\{B_{m,j} : j = -m, \ldots, r\}$ is given by

$$B_{m,j} := \begin{bmatrix} 1 & 1 & \cdots & 1 \\ \tau_j & \tau_{j+1} & \cdots & \tau_{j+m} \\ \tau_j^2 & \tau_{j+1}^2 & \cdots & \tau_{j+m}^2 \\ \vdots & \vdots & & \vdots \\ \tau_j^m & \tau_{j+1}^m & \cdots & \tau_{j+m}^m \end{bmatrix}, \quad j = -m, \ldots, r, \tag{10.4.24}$$

and with $\{\alpha_j : j = -m, \ldots, r\}$ and $\{g_{m,j} : j = -m, \ldots, r\}$ denoting the (column) vector sequences in $\mathbb{R}^{m+1}$ defined by

$$\left. \begin{aligned} \alpha_j &:= (\alpha_{j,j}, \ldots, \alpha_{j,j+m})^T, \\ g_{m,j} &:= (g_{m,j}^{(m)}(0), g_{m,j}^{(m-1)}(0), \ldots, g_{m,j}'(0), g_{m,j}(0))^T, \end{aligned} \right\} j = -m, \ldots, r. \tag{10.4.25}$$

We have therefore established that the approximation operator $\mathscr{S}$ in (10.4.11), (10.4.12) satisfies the polynomial exactness condition (10.4.1) if and only if the coefficient sequence $\{\alpha_{j,k} : k = j, \ldots, j+m; j = -m, \ldots, r\}$ satisfies the linear system (10.4.23), (10.4.24),

(10.4.25). To solve the equation (10.4.23), we fix $j \in \{-m,\ldots,r\}$, and observe that the transpose of the matrix $B_{m,j}$ in (10.4.24) is given by

$$B_{m,j}^T := \begin{bmatrix} 1 & \tau_j & \tau_j^2 & \cdots & \tau_j^m \\ 1 & \tau_{j+1} & \tau_{j+1}^2 & \cdots & \tau_{j+1}^m \\ \vdots & \vdots & \vdots & & \vdots \\ 1 & \tau_{j+m} & \tau_{j+m}^2 & \cdots & \tau_{j+m}^m \end{bmatrix}, \tag{10.4.26}$$

which, since also (10.2.31) is satisfied, is a Vandermonde matrix as in Theorem 1.1.2. It follows from Theorem 1.1.2 that $B_{m,j}^T$ is an invertible matrix, with, from (10.4.5) in Theorem 10.4.3, inverse given by

$$(B_{m,j}^T)^{-1} = \begin{bmatrix} L_{m,j}(0) & L_{m,j+1}(0) & \cdots & L_{m,j+m}(0) \\ L_{m,j}'(0) & L_{m,j+1}'(0) & \cdots & L_{m,j+m}'(0) \\ \dfrac{L_{m,j}''(0)}{2!} & \dfrac{L_{m,j+1}''(0)}{2!} & & \dfrac{L_{m,j+m}''(0)}{2!} \\ \vdots & \vdots & & \vdots \\ \dfrac{L_{m,j}^{(m)}(0)}{m!} & \dfrac{L_{m,j+1}^{(m)}(0)}{m!} & \cdots & \dfrac{L_{m,j+m}^{(m)}(0)}{m!} \end{bmatrix}, \tag{10.4.27}$$

where $\{L_{m,j},\ldots,L_{m,j+m}\} \subset \pi_m$ are the Lagrange fundamental polynomials defined, according to (1.2.1), by

$$L_{m,k}(x) := \prod_{\substack{k \neq \ell = j}}^{j+m} \frac{x - \tau_\ell}{\tau_k - \tau_\ell}, \quad k = j,\ldots,j+m. \tag{10.4.28}$$

Since the Vandermonde matrix $B_{m,j}^T$ is invertible, a standard result in linear algebra guarantees that the matrix $B_{m,j}$ is invertible, with inverse given, according to (10.4.27), by

$$B_{m,j}^{-1} = \left((B_{m,j}^T)^T\right)^{-1} = \left((B_{m,j}^T)^{-1}\right)^T$$

$$= \begin{bmatrix} L_{m,j}(0) & L_{m,j}'(0) & \dfrac{L_{m,j}''(0)}{2!} & \cdots & \dfrac{L_{m,j}^{(m)}(0)}{m!} \\ L_{m,j+1}(0) & L_{m,j+1}'(0) & \dfrac{L_{m,j+1}''(0)}{2!} & \cdots & \dfrac{L_{m,j+1}^{(m)}(0)}{m!} \\ \vdots & \vdots & \vdots & & \vdots \\ L_{m,j+m}(0) & L_{m,j+m}'(0) & \dfrac{L_{m,j+m}''(0)}{2!} & \cdots & \dfrac{L_{m,j+m}^{(m)}(0)}{m!} \end{bmatrix}, \tag{10.4.29}$$

and where the unique solution $\alpha_j = \alpha_{m,j}$ of the equation (10.4.23) is therefore given by

$$\alpha_{m,j} = (-1)^\ell \frac{\ell!}{m!} B_{m,j}^{-1} \mathbf{g}_{m,j}. \tag{10.4.30}$$

Hence, according to (10.4.30), (10.4.29) and (10.3.25), and by recalling also (10.4.14)-(10.4.17), we have established the following result.

Theorem 10.4.4. *For integers $m \geqslant 0$ and $r \geqslant 1$ satisfying $m \leqslant r+1$, and a knot sequence $\{\tau_{-m}, \ldots, \tau_{r+m+1}\}$ as in (10.2.31), (10.2.30), let the spline approximation operator $\mathscr{S}_{m,r}^{QI}$: $C[a,b] \to \sigma_m([a,b]; \tau_1, \ldots, \tau_r)$ be defined by*

$$(\mathscr{S}_{m,r}^{QI} f)(x) := \sum_{j=-m}^{r} \left[\sum_{k=j}^{j+m} \alpha_{m,j,k} f(\tau_k) \right] N_{m,j}(x), \quad x \in [a,b], \quad f \in C[a,b], \quad (10.4.31)$$

where the values of f at the knots are extended from $\{\tau_0, \ldots, \tau_{r+1}\}$ to $\{\tau_{-m}, \ldots, \tau_{r+m}\}$ by means of the polynomial extrapolation method (10.4.12), (10.4.13), and where the coefficient sequence $\{\alpha_{m,j,k} : k = j, \ldots, j+m; j = -m, \ldots, r\} \subset \mathbb{R}$ is given by

$$\alpha_{m,j,k} := \frac{1}{m!} \sum_{\ell=0}^{m} (-1)^\ell L_{m,k}^{(\ell)}(0) g_{m,j}^{(m-\ell)}(0), \ k = j, \ldots, j+m; \ j = -m, \ldots, r, \quad (10.4.32)$$

with the polynomial sequences $\{L_{m,k} : k = j, \ldots, j+m; j = -m, \ldots, r\} \subset \pi_m$ and $\{g_{m,j} : j = -m, \ldots, r\} \subset \pi_m$ defined as in, respectively, (10.4.28) and (10.4.3), and with the B-splines $\{N_{m,j} : j = -m, \ldots, r\}$ given as in (10.2.32). Then $\mathscr{S}_{m,r}^{QI}$ is linear, and $\mathscr{S}_{m,r}^{QI}$ is an optimally local quasi-interpolation operator satisfying the polynomial exactness condition (10.4.1). Moreover, $\mathscr{S}_{m,r}^{QI}$ satisfies the formulation

$$(\mathscr{S}_{m,r}^{QI} f)(x) = \sum_{k=-m}^{r+m} f(\tau_k) U_{m,k}(x), \quad x \in [a,b], \quad f \in C[a,b], \quad (10.4.33)$$

where the splines $\{U_{m,k} : k = -m, \ldots, r+m\}$ are given by

$$U_{m,k}(x) := \sum_{j=\max\{k-m,-m\}}^{\min\{k,r\}} \alpha_{m,j,k} N_{m,j}(x), \ x \in [a,b], \ k = -m, \ldots, r+m, \quad (10.4.34)$$

and satisfy the properties (10.4.16) and (10.4.17).

Our next step is to obtain an explicit formulation in terms of the knot sequence $\{\tau_{-m}, \ldots, \tau_{m+r+1}\}$ for the expression in the right hand side of (10.4.32), for the purpose of which we first introduce the following notation.

For any integer sequence $\{\ell_1, \ldots, \ell_n\}$, we define, for $k \in \mathbb{N}$, with $k \leqslant n$,

$$\text{per}_k\{\ell_1, \ldots, \ell_n\} := \text{the set of all permutations } \{j_1, \ldots, j_k\} \text{ of } \{\ell_1, \ldots, \ell_n\}; \quad (10.4.35)$$

$$\text{com}_k\{\ell_1, \ldots, \ell_n\} := \text{the set of all combinations } \{j_1, \ldots, j_k\} \text{ of } \{\ell_1, \ldots, \ell_n\}; \quad (10.4.36)$$

$$\text{per}\{\ell_1, \ldots, \ell_n\} := \text{per}_n\{\ell_1, \ldots, \ell_n\}. \quad (10.4.37)$$

Note from (10.4.35) and (10.4.36) that:

number of sequences in $\text{per}_k\{\ell_1,\ldots,\ell_n\} = k!(\text{number of sequences in com}_k\{\ell_1,\ldots,\ell_n\})$.

$$(10.4.38)$$

Also, observe from (10.4.36) that, for any coefficient sequence $\{a_1,\ldots,a_n\} \subset \mathbb{R}$, it holds that

$$\prod_{j=1}^{n}(x-a_j) = \sum_{k=0}^{n}(-1)^k h_k(a_1,\ldots,a_n)x^{n-k}, \quad x \in \mathbb{R}, \qquad (10.4.39)$$

where, for $k \in \{0,\ldots,n\}$, the function $h_k : \mathbb{R}^n \to \mathbb{R}$ is the classical symmetric function defined by

$$h_k(a_1,\ldots,a_n) := \begin{cases} \displaystyle\sum_{\{j_1,\ldots,j_k\}\in\text{com}_k\{1,\ldots,n\}} \prod_{\ell=1}^{k} a_{j_\ell}, & \text{if } k = 1,\ldots,n; \\ 1, & \text{if } k = 0. \end{cases} \qquad (10.4.40)$$

We shall rely on the following result.

Theorem 10.4.5. *For $n \in \mathbb{N}$, and any real sequences $\{\beta_1,\ldots,\beta_n\},\{\gamma_1,\ldots,\gamma_n\}$, let the polynomials $P,Q \in \pi_n$ be defined by*

$$P(x) := \prod_{j=1}^{n}(x-\beta_j); \quad Q(x) := \prod_{j=1}^{n}(x-\gamma_j). \qquad (10.4.41)$$

Then

$$\sum_{\ell=0}^{n}(-1)^\ell P^{(\ell)}(0)Q^{(n-\ell)}(0) = \sum_{\{j_1,\ldots,j_n\}\in\text{per}\{1,\ldots,n\}} \prod_{k=1}^{n}(\gamma_{j_k} - \beta_k), \qquad (10.4.42)$$

with the set $\text{per}\{1,\ldots,n\}$ *defined as in* (10.4.37), (10.4.35).

Proof. First, observe from (10.4.41) and (10.4.39) that

$$\left.\begin{aligned} P^{(\ell)}(0) &= (-1)^{n-\ell}\ell! h_{n-\ell}(\beta_1,\ldots,\beta_n), \\ Q^{(n-\ell)}(0) &= (-1)^\ell(n-\ell)! h_\ell(\gamma_1,\ldots,\gamma_n), \end{aligned}\right\} \quad \ell = 0,\ldots,n,$$

and thus

$$\sum_{\ell=0}^{n}(-1)^\ell P^{(\ell)}(0)Q^{(n-\ell)}(0) = (-1)^n \sum_{\ell=0}^{n}(-1)^\ell \ell!(n-\ell)! h_{n-\ell}(\beta_1,\ldots,\beta_n)h_\ell(\gamma_1,\ldots,\gamma_n),$$

$$(10.4.43)$$

with the sequence $\{h_\ell : \ell = 0,\ldots,n\}$ defined as in (10.4.40).

Now let the multivariate polynomial $F : \mathbb{R}^n \to \mathbb{R}$ be defined by

$$F(\mathbf{x}) := \sum_{\{j_1,\ldots,j_n\}\in\text{per}\{1,\ldots,n\}} \prod_{k=1}^{n}(\gamma_{j_k} - x_k), \quad \mathbf{x} = (x_1,\ldots,x_n) \in \mathbb{R}^n. \qquad (10.4.44)$$

It then follows from (10.4.44), (10.4.37) and (10.4.35) that

$$F(\mathbf{x}) = \sum_{j_1=1}^{n} (\gamma_{j_1} - x_1) \sum_{\substack{j_2=1 \\ j_2 \neq j_1}}^{n} (\gamma_{j_2} - x_2) \cdots \sum_{\substack{j_n=1 \\ j_n \neq j_1,\ldots,j_{n-1}}}^{n} (\gamma_{j_n} - x_n),$$

$$\mathbf{x} = (x_1,\ldots,x_n) \in \mathbb{R}^n. \quad (10.4.45)$$

It follows from the multivariate polynomial structure of (10.4.45), together with the multivariate extension of the Taylor expansion polynomial identity (10.1.11), as well as (10.4.35), that

$$F(\mathbf{x}) = \sum_{\ell=0}^{n} \frac{1}{\ell!} \sum_{\{v_1,\ldots,v_\ell\} \in \text{per}_\ell\{1,\ldots,n\}} \frac{\partial^\ell F}{\partial x_{v_1} \ldots \partial x_{v_\ell}}(\mathbf{0}) \prod_{k=1}^{\ell} x_{v_k},$$

$$\mathbf{x} = (x_1,\ldots,x_n) \in \mathbb{R}^n. \quad (10.4.46)$$

By using (10.4.44),(10.4.37), (10.4.35) and (10.4.38), as well as the first line of (10.4.40), we obtain

$$F(\mathbf{0}) = \sum_{\{j_1,\ldots,j_n\} \in \text{per}\{1,\ldots,n\}} \prod_{k=1}^{n} \gamma_{j_k}$$

$$= n! \sum_{\{j_1,\ldots,j_n\} \in \text{com}_n\{1,\ldots,n\}} \prod_{k=1}^{n} \gamma_{j_k} = n! h_n(\gamma_1,\ldots,\gamma_n). \quad (10.4.47)$$

Next, for $\ell \in \{0,\ldots,n\}$, and any distinct integer sequence $\{v_1,\ldots,v_\ell\} \subset \{1,\ldots,n\}$, by noting also that the order in which the components of $\mathbf{x} = (x_1,\ldots,x_n)$ appear in the definition (10.4.44) can be permuted arbitrarily without changing $F(\mathbf{x})$, we deduce from (10.4.45), (10.4.35), (10.4.36), (10.4.38) and (10.4.40) that

$$\frac{\partial^\ell F}{\partial x_{v_1} \ldots \partial x_{v_\ell}}(\mathbf{0})$$

$$= (-1)^\ell \sum_{v_1=1}^{n} \cdots \sum_{\substack{v_\ell=1; v_\ell \neq v_1,\ldots,v_{\ell-1}}}^{n} \left[\sum_{\substack{v_{\ell+1}=1; v_{\ell+1} \neq v_1,\ldots,v_\ell}}^{n} \gamma_{v_{\ell+1}} \cdots \sum_{\substack{v_n=1; v_n \neq v_1,\ldots,v_{n-1}}}^{n} \gamma_{v_n} \right]$$

$$= (-1)^\ell \ell! \left[\sum_{\substack{v_{\ell+1}=1; v_{\ell+1} \neq v_1,\ldots,v_\ell}}^{n} \gamma_{v_{\ell+1}} \cdots \sum_{\substack{v_n=1; v_n \neq v_1,\ldots,v_{n-1}}}^{n} \gamma_{v_n} \right]$$

$$= (-1)^\ell \ell! \sum_{\{v_1,\ldots,v_{n-\ell}\} \in \text{per}_{n-\ell}\{1,\ldots,n\}} \prod_{k=1}^{n-\ell} \gamma_{v_k}$$

$$= (-1)^\ell \ell! (n-\ell)! \sum_{\{v_1,\ldots,v_{n-\ell}\} \in \text{com}_{n-\ell}\{1,\ldots,n\}} \prod_{k=1}^{n-\ell} \gamma_{v_k}$$

$$= (-1)^\ell \ell! (n-\ell)! h_{n-\ell}(\gamma_1,\ldots,\gamma_n), \quad (10.4.48)$$

whereas, for any non-distinct integer sequence $\{v_1, \ldots, v_\ell\} \subset \{1, \ldots, n\}$, with $2 \leqslant \ell \leqslant n$, it follows from (10.4.45) that

$$\frac{\partial^\ell F}{\partial x_{v_1} \ldots \partial x_{v_\ell}}(\mathbf{x}) = 0, \quad \mathbf{x} = (x_1, \ldots, x_n) \in \mathbb{R}^n. \tag{10.4.49}$$

By using (10.4.47), (10.4.48) and (10.4.49) in (10.4.46), and setting $\mathbf{x} = \beta = (\beta_1, \ldots, \beta_n)$, we deduce from (10.4.35), (10.4.36), (10.4.38) and (10.4.40) that

$$
\begin{aligned}
F(\beta) &= \sum_{\ell=0}^{n}(-1)^\ell(n-\ell)! h_{n-\ell}(\gamma_1, \ldots, \gamma_n) \sum_{\{v_1, \ldots, v_\ell\}\in\mathrm{per}_\ell\{1, \ldots, n\}} \prod_{k=1}^{\ell} \beta_{v_k} \\
&= \sum_{\ell=0}^{n}(-1)^\ell(n-\ell)! h_{n-\ell}(\gamma_1, \ldots, \gamma_n) \left[\ell! \sum_{\{v_1, \ldots, v_\ell\}\in\mathrm{com}_\ell\{1, \ldots, n\}} \prod_{k=1}^{\ell} \beta_{v_k}\right] \\
&= \sum_{\ell=0}^{n}(-1)^\ell(n-\ell)! \ell! h_{n-\ell}(\gamma_1, \ldots, \gamma_n) h_\ell(\beta_1, \ldots, \beta_n) \\
&= (-1)^n \sum_{\ell=0}^{n}(-1)^\ell \ell! (n-\ell)! h_\ell(\gamma_1, \ldots, \gamma_n) h_{n-\ell}(\beta_1, \ldots, \beta_n). \tag{10.4.50}
\end{aligned}
$$

The desired result (10.4.42) is now an immediate consequence of (10.4.43), (10.4.44) and (10.4.50).   ∎

For $j \in \{-m, \ldots, r\}$ and $k \in \{j, \ldots, j+m\}$, we now apply the formula (10.4.42) in Theorem 10.4.5, with $n = m$, and

$$\left.\begin{aligned}
\{\beta_1, \ldots, \beta_m\} &= \{\tau_j, \ldots, \tau_{j+m}\} \setminus \{\tau_k\}; \\
\{\gamma_1, \ldots, \gamma_m\} &= \{\tau_{j+1}, \ldots, \tau_{j+m}\},
\end{aligned}\right\} \tag{10.4.51}$$

to deduce from (10.4.32), (10.4.28) and (10.4.3) the following explicit formulation of the coefficient sequence in Theorem 10.4.4.

Theorem 10.4.6. *In Theorem* 10.4.4, *the coefficient sequence* $\{\alpha_{m,j,k} : k = j, \ldots, j+m; j = -m, \ldots, r\}$ *satisfies the explicit formulation*

$$\alpha_{m,j,j+k} = \frac{1}{m!} \frac{\displaystyle\sum_{\{v_0, \ldots, v_m\}\setminus\{v_k\}\in\mathrm{per}\{1, \ldots, m\}} \prod_{\substack{k\neq\ell=0}}^{m} (\tau_{j+v_\ell} - \tau_{j+\ell})}{\displaystyle\prod_{\substack{k\neq\ell=0}}^{m}(\tau_{j+k} - \tau_{j+\ell})},$$

$$k = 0, \ldots, m; \quad j = -m, \ldots, r. \tag{10.4.52}$$

Example 10.4.1. (a) For $m = 1$, we calculate by means of (10.4.52) that, for $j = -1, \ldots, r$,

$$\alpha_{1,j,j} = 0 \quad ; \quad \alpha_{1,j,j+1} = 1, \tag{10.4.53}$$

which, together with (10.4.31), yields the approximation operator

$$(\mathscr{S}^{QI}_{1,r}f)(x) = \sum_{j=0}^{r+1} f(\tau_j)N_{1,j-1}(x), \quad x \in [a,b], \quad f \in C[a,b]. \tag{10.4.54}$$

Also, by inserting the coefficient values (10.4.53) into (10.4.34), we find that $U_{1,-1}$ is the zero function, whereas

$$U_{1,k}(x) = N_{1,k-1}(x), \quad k = 0,\ldots,r+1, \tag{10.4.55}$$

which, together with (10.4.33), and (10.4.12), (10.4.13), is consistent with (10.4.54). Observe from (10.4.54) and (10.2.50) that

$$(\mathscr{S}^{QI}_{1,r}f)(\tau_k) = f(\tau_k), \quad k = 0,\ldots,r+1, \tag{10.4.56}$$

that is, $\mathscr{S}^{QI}_{1,r}f$ is the piecewise linear interpolant of f with respect to the interpolation points $\{\tau_0,\ldots,\tau_{r+1}\}$, and thus

$$\mathscr{S}^{QI}_{1,r}f = f, \quad f \in \pi_1, \tag{10.4.57}$$

as also guaranteed by Theorem 10.4.4.

(b) For $m = 2$, an application of (10.4.52) yields (see Exercise 10.42) the coefficients

$$\left.\begin{aligned}
\alpha_{2,j,j} &= -\frac{1}{2}\frac{(\tau_{j+2}-\tau_{j+1})^2}{(\tau_{j+1}-\tau_j)(\tau_{j+2}-\tau_j)}, \\
\alpha_{2,j,j+1} &= \frac{1}{2}\frac{\tau_{j+2}-\tau_j}{\tau_{j+1}-\tau_j}, \\
\alpha_{2,j,j+2} &= \frac{1}{2}\frac{\tau_{j+1}-\tau_j}{\tau_{j+2}-\tau_j},
\end{aligned}\right\} \quad j = -2,\ldots,r, \tag{10.4.58}$$

from which, together with (10.4.34) in Theorem 10.4.4, we obtain, for $x \in [a,b]$, the formulas

$$\left.\begin{aligned}
U_{2,-2}(x) &= -\frac{1}{2}\frac{(\tau_0-\tau_{-1})^2}{(\tau_{-1}-\tau_{-2})(\tau_0-\tau_{-2})}N_{2,-2}(x); \\
U_{2,-1}(x) &= \frac{1}{2}\frac{\tau_0-\tau_{-2}}{\tau_{-1}-\tau_{-2}}N_{2,-2}(x) - \frac{1}{2}\frac{(\tau_1-\tau_0)^2}{(\tau_0-\tau_{-1})(\tau_1-\tau_{-1})}N_{2,-1}(x); \\
U_{2,k}(x) &= \frac{1}{2}\frac{\tau_{k-1}-\tau_{k-2}}{\tau_k-\tau_{k-2}}N_{2,k-2}(x) + \frac{1}{2}\frac{\tau_{k+1}-\tau_{k-1}}{\tau_k-\tau_{k-1}}N_{2,k-1}(x) \\
&\quad -\frac{1}{2}\frac{(\tau_{k+2}-\tau_{k+1})^2}{(\tau_{k+1}-\tau_k)(\tau_{k+2}-\tau_k)}N_{2,k}(x), \quad k = 0,\ldots,r; \\
U_{2,r+1}(x) &= \frac{1}{2}\frac{\tau_r-\tau_{r-1}}{\tau_{r+1}-\tau_{r-1}}N_{2,r-1}(x) + \frac{1}{2}\frac{\tau_{r+2}-\tau_r}{\tau_{r+1}-\tau_r}N_{2,r}(x); \\
U_{2,r+2}(x) &= \frac{1}{2}\frac{\tau_{r+1}-\tau_r}{\tau_{r+2}-\tau_r}N_{2,r}(x).
\end{aligned}\right\} \tag{10.4.59}$$

According to (10.4.33) in Theorem 10.4.4, as well as (10.4.12), (10.4.13), together with (5.1.2) and the Lagrange interpolation formula (1.2.5) in Theorem 1.2.2, and the definition

(1.2.1) of the Lagrange fundamental polynomials, we obtain, for any $f \in C[a,b]$, and $x \in [a,b]$, the approximation operator formulation

$$(\mathscr{S}_{2,r}^{QI} f)(x) = \sum_{k=-2}^{-1} \left[\sum_{j=0}^{2} \left\{ \prod_{j \neq \ell=0}^{2} \frac{\tau_k - \tau_\ell}{\tau_j - \tau_\ell} \right\} f(\tau_j) \right] U_{2,k}(x) + \sum_{k=0}^{r+1} f(\tau_k) U_{2,k}(x)$$

$$+ \left[\sum_{j=r-1}^{r+1} \left\{ \prod_{j \neq \ell=r-1}^{r+1} \frac{\tau_{r+2} - \tau_\ell}{\tau_j - \tau_\ell} \right\} f(\tau_j) \right] U_{2,r+2}(x), \quad (10.4.60)$$

with the splines $\{U_{m,k} : k = -2, \ldots, r+2\}$ defined as in (10.4.59), and for which, according to Theorem 10.4.4, it holds that

$$\mathscr{S}_{2,r}^{QI} f = f, \quad f \in \pi_2. \tag{10.4.61}$$

For the case of the integer knot sequence as in (10.2.31), where $[a,b] = [0, r+1]$, and satisfying (10.2.67), that is,

$$\tau_j = j, \quad j = -2, \ldots, r+3, \tag{10.4.62}$$

it follows from (10.4.59), together with (10.2.68) in Theorem 10.2.7, and with N_2 denoting the quadratic cardinal B-spline as defined in (10.2.69), that (see Exercise 10.42), for any $x \in [0, r+1]$,

$$\left.\begin{aligned}
U_{2,-2}(x) &= -\tfrac{1}{4} N_2(x+2); \\
U_{2,-1}(x) &= N_2(x+2) - \tfrac{1}{4} N_2(x+1); \\
U_{2,k}(x) &= \tfrac{1}{4} N_2(x+2-k) + N_2(x+1-k) - \tfrac{1}{4} N_2(x-k), \quad k = 0, \ldots, r; \\
U_{2,r+1}(x) &= \tfrac{1}{4} N_2(x-r+1) + N_2(x-r); \\
U_{2,r+2}(x) &= \tfrac{1}{4} N_2(x-r),
\end{aligned}\right\} \quad (10.4.63)$$

and, from (10.4.60), for any $f \in C[0, r+1]$, and $x \in [0, r+1]$,

$$(\mathscr{S}_{2,r}^{QI} f)(x) = [6f(0) - 8f(1) + 3f(2)] U_{2,-2}(x) + [3f(0) - 3f(1) + f(2)] U_{2,-1}(x)$$

$$+ \sum_{k=0}^{r} f(k) U_2(x-k) + f(r+1) U_{2,r+1}(x)$$

$$+ [3f(r+1) - 3f(r) + f(r-1)] U_{2,r+2}(x), \quad (10.4.64)$$

where, according to the middle line of (10.4.63),

$$U_2(x) := \tfrac{1}{4} N_2(x+2) + N_2(x+1) - \tfrac{1}{4} N_2(x). \tag{10.4.65}$$

Observe from (10.4.64) and (10.4.65), together with (10.4.63), that

$$(\mathscr{S}_{2,r}^{QI} f)(x) = \sum_{k=0}^{2} f(k) \tilde{U}_{2,k}(x) + \sum_{k=3}^{r-2} f(k) U_2(x-k) + \sum_{k=r-1}^{r+1} f(k) \tilde{U}_{2,k}(x),$$

$$x \in [0, r+1], \quad f \in C[0, r+1], \tag{10.4.66}$$

where

$$\tilde{U}_{2,0}(x) \quad := \tfrac{7}{4}N_2(x+2) + \tfrac{1}{4}N_2(x+1) - \tfrac{1}{4}N_2(x);$$

$$\tilde{U}_{2,1}(x) \quad := -N_2(x+2) + N_2(x+1) + N_2(x) - \tfrac{1}{4}N_2(x-1);$$

$$\tilde{U}_{2,2}(x) \quad := \tfrac{1}{4}N_2(x+2) - \tfrac{1}{4}N_2(x+1) + \tfrac{1}{4}N_2(x) + N_2(x-1) - \tfrac{1}{4}N_2(x-2);$$

$$\tilde{U}_{2,r-1}(x) := \tfrac{1}{4}N_2(x-r+3) + N_2(x-r+2) - \tfrac{1}{4}N_2(x-r+1) + \tfrac{1}{4}N_2(x-r);$$

$$\tilde{U}_{2,r}(x) \quad := \tfrac{1}{4}N_2(x-r+2) + N_2(x-r+1) - N_2(x-r);$$

$$\tilde{U}_{2,r+1}(x) := \tfrac{1}{4}N_2(x-r+1) + \tfrac{7}{4}N_2(x-r),$$

$$(10.4.67)$$

with the spline U_2 defined by (10.4.65), and by recalling also the convention $\sum\limits_{j=j_1}^{j_0} a_j := 0$, if $j_0 < j_1$, according to which the middle sum in (10.4.66), that is, the sum containing the spline U_2, is non-vanishing only for $r \geqslant 5$.

(c) For $m = 3$ and $[a,b] = [0, r+1]$ in (10.2.31), and with the knot sequence chosen as the integers as in (10.2.67), that is,

$$\tau_j = j, \quad j = -3, \ldots, r+4, \tag{10.4.68}$$

the formula (10.4.52) yields (see Exercise 10.45) the coefficient values

$$\left. \begin{array}{l} \alpha_{3,j,j} = 0; \\[4pt] \alpha_{3,j,j+1} = -\tfrac{1}{6}; \ \alpha_{3,j,j+2} = \tfrac{4}{3}; \ \alpha_{3,j,j+3} = -\tfrac{1}{6}, \end{array} \right\} \ j = -3, \ldots, r, \tag{10.4.69}$$

from which, together with (10.4.34) in Theorem 10.4.4, as well as (10.2.68) in Theorem 10.2.7, we obtain (see Exercise 10.45)

$$U_{3,-3}(x) = \text{the zero function;}$$

$$U_{3,-2}(x) = -\tfrac{1}{6}N_3(x+3);$$

$$U_{3,-1}(x) = \tfrac{4}{3}N_3(x+3) - \tfrac{1}{6}N_2(x+2);$$

$$U_{3,k}(x) = -\tfrac{1}{6}N_3(x+3-k) + \tfrac{4}{3}N_3(x+2-k) - \tfrac{1}{6}N_3(x+1-k),$$

$$k = 0, \ldots, r+1;$$

$$U_{3,r+2}(x) = -\tfrac{1}{6}N_3(x-r+1) + \tfrac{4}{3}N_3(x-r);$$

$$U_{3,r+3}(x) = -\tfrac{1}{6}N_3(x-r).$$

$$(10.4.70)$$

By using the formula (10.4.33) in Theorem 10.4.4, together with (10.4.12), (10.4.13), as well as the Lagrange interpolation formula (1.2.5) in Theorem 1.2.2, the definition (1.2.1) of the Lagrange fundamental polynomials, and the first line of (10.4.70), we deduce that (see Exercise 10.45), for any $f \in C[0, r+1]$, and $x \in [0, r+1]$,

$$(\mathscr{S}_{3,r}^{QI} f)(x) = [10f(0) - 20f(1) + 15f(2) - 4f(3)]U_{3,-2}(x)$$

$$+ [4f(0) - 6f(1) + 4f(2) - f(3)]U_{3,-1}(x) + \sum_{k=0}^{r+1} f(k)U_3(x-k)$$

$$+ [4f(r+1) - 6f(r) + 4f(r-1) - f(r-2)]U_{3,r+2}(x)$$

$$+ [10f(r+1) - 20f(r) + 15f(r-1) - 4f(r-2)]U_{3,r+3}(x), \quad (10.4.71)$$

where, according to the fourth line of (10.4.70),

$$U_3(x) := -\tfrac{1}{6}N_3(x+3) + \tfrac{4}{3}N_3(x+2) - \tfrac{1}{6}N_3(x+1). \quad (10.4.72)$$

Observe from (10.4.71), (10.4.72) and (10.4.70) that

$$(\mathscr{S}_{3,r}^{QI} f)(x) = \sum_{k=0}^{3} f(k)\tilde{U}_{3,k}(x) + \sum_{k=4}^{r-3} f(k)U_3(x-k) + \sum_{k=r-2}^{r+1} f(k)\tilde{U}_{3,k}(x),$$

$$x \in [0, r+1], \; f \in C[0, r+1], \quad (10.4.73)$$

where

$$\tilde{U}_{3,0}(x) \quad := \tfrac{7}{2}N_3(x+3) + \tfrac{2}{3}N_3(x+2) - \tfrac{1}{6}N_3(x+1);$$

$$\tilde{U}_{3,1}(x) \quad := -\tfrac{14}{3}N_3(x+3) + \tfrac{5}{6}N_3(x+2) + \tfrac{4}{3}N_3(x+1) - \tfrac{1}{6}N_3(x);$$

$$\tilde{U}_{3,2}(x) \quad := \tfrac{17}{6}N_3(x+3) - \tfrac{2}{3}N_3(x+2) - \tfrac{1}{6}N_3(x+1) + \tfrac{4}{3}N_3(x) - \tfrac{1}{6}N_3(x-1);$$

$$\tilde{U}_{3,3}(x) \quad := -\tfrac{2}{3}N_3(x+3) + \tfrac{1}{6}N_3(x+2) - \tfrac{1}{6}N_3(x) + \tfrac{4}{3}N_3(x-1) - \tfrac{1}{6}N_3(x-2);$$

$$\tilde{U}_{3,r-2}(x) := -\tfrac{1}{6}N_3(x-r+5) + \tfrac{4}{3}N_3(x-r+4) - \tfrac{1}{6}N_3(x-r+3)$$

$$+\tfrac{1}{6}N_3(x-r+1) - \tfrac{2}{3}N_3(x-r); \quad \left.\begin{array}{c} \\ \\ \\ \\ \\ \\ \\ \\ \\ \\ \end{array}\right\} \quad (10.4.74)$$

$$\tilde{U}_{3,r-1}(x) := -\tfrac{1}{6}N_3(x-r+4) + \tfrac{4}{3}N_3(x-r+3) - \tfrac{1}{6}N_3(x-r+2)$$

$$-\tfrac{2}{3}N_3(x-r+1) + \tfrac{17}{6}N_3(x-r);$$

$$\tilde{U}_{3,r}(x) \quad := -\tfrac{1}{6}N_3(x-r+3) + \tfrac{4}{3}N_3(x-r+2) + \tfrac{5}{6}N_3(x-r+1)$$

$$-\tfrac{14}{3}N_3(x-r);$$

$$\tilde{U}_{3,r+1}(x) := -\tfrac{1}{6}N_3(x-r+2) + \tfrac{2}{3}N_3(x-r+1) + \tfrac{7}{2}N_3(x-r),$$

with the spline U_3 defined by (10.4.72). Note that the middle sum in (10.4.73), that is, the sum containing U_3, is non-vanishing only for $r \geqslant 7$. ∎

10.5 Local spline interpolation

The non-local interpolation operator $\mathscr{S}_{m,n}^I$ of Section 10.3 is exact on the whole spline space $\sigma_m([a,b]; \tau_1, \ldots, \tau_{n-m})$, but not local, whereas the local quasi-interpolation operator

$\mathscr{S}_{m,r}^{QI}$ of Section 10.4 is exact on the smaller space π_m, but interpolatory (at the spline knots $\{\tau_0,\ldots,\tau_{r+1}\}$) only for $m=1$, as follows from (10.4.56) in Example 10.4.1(a). In this section, we shall establish an explicitly formulated local spline interpolation operator, with exactness on π_m.

To this end, for positive integers m and n such that

$$n \geqslant m, \tag{10.5.1}$$

we choose $r = nm - 1$ in (10.2.31), (10.2.32), thereby yielding, for an arbitrary bounded interval $[a,b]$, a knot sequence $\{\tau_{-m},\ldots,\tau_{(n+1)m}\}$ satisfying

$$\tau_{-m} < \cdots < \tau_0 := a < \tau_1 < \cdots < \tau_{nm-1} < b =: \tau_{nm} < \cdots < \tau_{(n+1)m}, \tag{10.5.2}$$

and, for positive integers p and q such that

$$p + q = m, \tag{10.5.3}$$

we let $\{x_{-p},\ldots,x_{n+q}\}$ be any sequence satisfying

$$x_{-p} < \cdots < x_0 := a < x_1 < \cdots x_{n-1} < b =: x_n < \cdots < x_{n+q}, \tag{10.5.4}$$

with, moreover,

$$x_j = \tau_{mj}, \quad j = 0,\ldots,n. \tag{10.5.5}$$

We shall construct a spline sequence

$$\{V_k : k = -p,\ldots,n+q\} \subset \sigma_m([a,b]; \tau_1,\ldots,\tau_{nm-1}) \tag{10.5.6}$$

satisfying the properties

(a)

$$V_k(x) = 0, \quad x \in [a,b] \setminus (x_{k-q-1},x_{k+p+1}), \quad k = -p,\ldots,n+q; \tag{10.5.7}$$

(b)

$$V_k(x_j) = \delta_{k-j}, \quad j = 0,\ldots,n; \quad k = -p,\ldots,n+q; \tag{10.5.8}$$

(c)

$$\sum_{k=-p}^{n+q} P(x_k)V_k(x) = P(x), \quad x \in [a,b], \quad P \in \pi_m, \tag{10.5.9}$$

and in terms of which we shall then define the approximation operator

$$\mathscr{S} : C[a,b] \to \sigma_m([a,b]; \tau_1, \ldots, \tau_{nm-1}) \qquad (10.5.10)$$

by

$$(\mathscr{S}f)(x) := \sum_{k=-p}^{n+q} f(x_k)V_k(x), \quad x \in [a,b], \quad f \in C[a,b], \qquad (10.5.11)$$

where, analogously to (10.4.12), (10.4.13), the function values $\{f(x_k) : k = 0, \ldots, n\}$ are extended to $\{f(x_k) : k = -p, \ldots, n+q\}$ according to the polynomial extrapolation method

$$f(x_k) := \begin{cases} (\mathscr{P}_m^I f)(x_k), \, k = -p, \ldots, -1; \\ (\widetilde{\mathscr{P}}_m^I f)(x_k), \, k = n+1, \ldots, n+q, \end{cases} \qquad (10.5.12)$$

with $\mathscr{P}_m^I f$ and $\widetilde{\mathscr{P}}_m^I f$ denoting, as in (5.1.2), the polynomials in π_m interpolating f at, respectively, the point sequences $\{x_0, \ldots, x_m\}$ and $\{x_{n-m}, \ldots, x_n\}$, that is,

with

$$\left. \begin{array}{l} \mathscr{P}_m^I f \in \pi_m \; ; \quad \widetilde{\mathscr{P}}_m^I f \in \pi_m, \\[2mm] (\mathscr{P}_m^I f)(x_k) = f(x_k), \quad k = 0, \ldots, m; \\[2mm] (\widetilde{\mathscr{P}}_m^I f)(x_k) = f(x_k), \quad k = n-m, \ldots, n. \end{array} \right\} \qquad (10.5.13)$$

It is immediately evident from the definition (10.5.11), (10.5.12), together with Theorem 5.1.1(a), that $\mathscr{S}$ is linear. Also, observe from (10.5.11), (10.5.7) and (10.5.3) that $\mathscr{S}$ is a local approximation operator, in the sense that, for any fixed $x \in [a,b]$, the value of $(\mathscr{S}f)(x)$ is independent of the function values $\{f(x) : x \in [a,b] \setminus [x_\kappa, x_\lambda]\}$, where κ and λ are integers such that $\lambda - \kappa \leqslant m+2$.

Moreover, (10.5.11) and (10.5.8) imply that $\mathscr{S}$ is an interpolation operator, with

$$(\mathscr{S}f)(x_j) = f(x_j), \quad j = 0, \ldots, n, \quad f \in C[a,b], \qquad (10.5.14)$$

whereas, according to (10.5.11), (10.5.12) and (10.5.9), as well as Theorem 5.1.1(b), $\mathscr{S}$ is exact on π_m, that is,

$$(\mathscr{S}f)(x) = f(x), \quad x \in [a,b], \quad f \in \pi_m. \qquad (10.5.15)$$

Our first step is to establish the following general result, according to which the interpolatory property (10.5.8) is guaranteed if $\{V_{-p}, \ldots, V_{n+q}\}$ satisfies (10.5.7) and (10.5.9).

Theorem 10.5.1. *For positive integers m, n, p and q as in (10.5.1), (10.5.3), let $\{V_{-p}, \ldots, V_{n+q}\}$ denote a function sequence in $C[a,b]$ such that the properties (10.5.7) and (10.5.9) are satisfied. Then the interpolatory property (10.5.8) is satisfied by the sequence $\{V_{-p}, \ldots, V_{n+q}\}$.*

Proof. Let $j \in \{0, \ldots, n\}$ be fixed. It follows from (10.5.9) and (10.5.7) that

$$\sum_{k=j-p}^{j+q} x_k^\ell V_k(x_j) = x_j^\ell, \quad \ell = 0, \ldots, m, \tag{10.5.16}$$

or equivalently, in matrix-vector notation,

$$\begin{bmatrix} 1 & 1 & \cdots & 1 \\ x_{j-p} & x_{j-p+1} & \cdots & x_{j+q} \\ x_{j-p}^2 & x_{j-p+1}^2 & \cdots & x_{j+q}^2 \\ \vdots & \vdots & & \vdots \\ x_{j-p}^m & x_{j-p+1}^m & \cdots & x_{j+q}^m \end{bmatrix} \begin{bmatrix} V_{j-p}(x_j) \\ V_{j-p+1}(x_j) \\ \vdots \\ V_{j+q}(x_j) \end{bmatrix} = \begin{bmatrix} 1 \\ x_j \\ x_j^2 \\ \vdots \\ x_j^m \end{bmatrix}. \tag{10.5.17}$$

It follows from (10.5.3) that the matrix in (10.5.17) is a square $(m+1) \times (m+1)$ matrix. Analogously to the argument leading from (10.4.23) to (10.4.30) (see Exercise 10.46), we now use the fact that the matrix in (10.5.17) is the transpose of an invertible Vandermonde matrix, and apply the Taylor expansion polynomial identity (10.1.11), with $c = 0$, to deduce that, with $\{L_{m,k} : k = j-p, \ldots, j+q\}$ denoting the Lagrange fundamental polynomials with respect to the interpolation points $\{x_{j-p}, \ldots, x_{j+q}\}$, as obtained from (1.2.1), we have, for any $k \in \{j-p, \ldots, j+q\}$,

$$V_k(x_j) = \sum_{\ell=0}^n \frac{L_{m,k}^{(\ell)}(0)}{\ell!} x_j^\ell = L_{m,k}(x_j) = \delta_{k-j}, \tag{10.5.18}$$

by virtue of (1.2.3) in Theorem 1.2.1(a). It then follows from (10.5.18) and (10.5.7) that the interpolatory condition (10.5.8) is indeed satisfied. ∎

Based on Theorem 10.5.1, we therefore proceed to obtain a spline sequence $\{V_{-p}, \ldots, V_{n+q}\}$ satisfying the properties (10.5.6), (10.5.7) and (10.5.9). To this end, for a sequence

$$\{\lambda_{j,k}\} = \{\lambda_{mj+\rho,k} : k = j-p+1, \ldots, j+q+1; \, j = -1, \ldots, n-1; \, \rho = 0, \ldots, m-1\} \subset \mathbb{R}, \tag{10.5.19}$$

we define

$$V_k(x) := \sum_{j=\max\{m(k-q-1),-m\}}^{\min\{m(k+p)-1,nm-1\}} \lambda_{j,k} N_{m,j}(x), \quad x \in [a,b], \; k = -p, \ldots, n+q, \tag{10.5.20}$$

with the B-spline sequence

$$\{N_{m,-m}, \ldots, N_{m,nm-1}\} \subset \sigma_m(\tau_{-m}, \ldots, \tau_{(n+1)m}) \tag{10.5.21}$$

defined by (10.2.32) in terms of a knot sequence $\{\tau_{-m},\dots,\tau_{(n+1)m}\}$ as in (10.5.2) and
(10.5.5). Observe that, with the spline sequence $\{V_{-p},\dots,V_{n+q}\}$ given by (10.5.20), the
spline approximation operator $\mathscr{S}$ in (10.5.11) satisfies the formulation

$$(\mathscr{S}f)(x) = \sum_{\rho=0}^{m-1}\sum_{j=-1}^{n-1}\left[\sum_{k=j-p+1}^{j+q+1}\lambda_{mj+\rho,k}f(x_k)\right]N_{m,mj+\rho}(x),$$

$$x \in [a,b], \quad f \in C[a,b]. \tag{10.5.22}$$

It follows from (10.5.20), together with the second line of (10.2.33) in Theorem 10.2.2(a),
as well as (10.5.3), that the spline sequence $\{V_{-p},\dots,V_{n+q}\}$ satisfies the conditions (10.5.6)
and (10.5.7).

Next, we deduce from (10.5.11), (10.5.22), together with the identity (10.4.4) in Theorem 10.4.2, with $r = nm - 1$, that the polynomial exactness condition (10.5.9) holds if and
only if the sequence (10.5.19) satisfies

$$\sum_{\rho=0}^{m-1}\sum_{j=-1}^{n-1}\left[\sum_{k=j-p+1}^{j+q+1}x_k^\ell\lambda_{mj+\rho,k} - (-1)^\ell\frac{\ell!}{m!}g_{m,mj+\rho}^{(m-\ell)}(0)\right]N_{m,mj+\rho}(x), x \in [a,b], \ell = 0,\dots,m,$$

which, in view of Theorem 10.2.2(d), with $r = nm - 1$, is equivalent to

$$\sum_{k=j-p+1}^{j+q+1}x_k^\ell\lambda_{mj+\rho,k} = (-1)^\ell\frac{\ell!}{m!}\,g_{m,mj+\rho}^{(m-\ell)}(0),$$

$$\ell = 0,\dots,m; \ j = -1,\dots,n-1; \ \rho = 0,\dots,m-1. \tag{10.5.23}$$

Note from (10.5.3) that, for any fixed $\rho \in \{0,\dots,m-1\}$ and $j \in \{-1,\dots,n-1\}$, the
linear system (10.5.23) consists of $m+1$ equations in the $m+1$ unknowns $\{\lambda_{mj+\rho,k} : k = j-p+1,\dots,j+q+1\}$. Moreover, since (10.5.23) has a similar structure to the linear
system (10.4.22), we may now argue as in the steps leading from (10.4.22) to (10.4.52) (see
Exercise 10.47) to explicitly solve the linear system (10.5.23), and thereby, after recalling
also Theorem 10.5.1, completing the proof of the following result (see also Exercise 10.48).

Theorem 10.5.2. *For positive integers m,n,p and q as in* (10.5.1), (10.5.3), *and sequences*
$\{\tau_{-m},\dots,\tau_{(n+1)m}\},\{x_{-p},\dots,x_{n+q}\}$ *satisfying* (10.5.2), (10.5.4), *let the spline approximation operator* $\mathscr{S}_{m,n}^{LI} : C[a,b] \to \sigma_m([a,b];\tau_1,\dots,\tau_{nm-1})$ *be defined by*

$$(\mathscr{S}_{m,n}^{LI}f)(x) := \sum_{\rho=0}^{m-1}\sum_{j=-1}^{n-1}\left[\sum_{k=j-p+1}^{j+q+1}\lambda_{m,mj+\rho,k}f(x_k)\right]N_{m,mj+\rho}(x),$$

$$x \in [a,b], \quad f \in C[a,b], \tag{10.5.24}$$

where the function values $\{f(x_k) : k = 0,\ldots,n\}$ *are extended to* $\{f(x_k) : k = -p,\ldots,n+q\}$
by means of the polynomial extrapolation method (10.5.12), (10.5.13); *where the coefficient*
sequence (10.5.19) *is given by*

$$\lambda_{m,mj+\rho,j+k} := \frac{1}{m!} \frac{\displaystyle\sum_{\{v_{-p+1},\ldots,v_{q+1}\}\backslash\{v_k\}\in per\{1,\ldots,m\}} \prod_{k\neq\ell=-p+1}^{q+1} (\tau_{mj+\rho+v_\ell} - x_{j+\ell})}{\displaystyle\prod_{k\neq\ell=-p+1}^{q+1} (x_{j+k} - x_{j+\ell})},$$

$$k = -p+1,\ldots,q+1; \quad j = -1,\ldots,n-1; \quad \rho = 0,\ldots,m-1, \qquad (10.5.25)$$

and, moreover, where $\{N_{m,-m},\ldots,N_{m,nm-1}\}$ *is the B-spline sequence, as in* (10.2.32), *with*
respect to the knot sequence $\{\tau_{-m},\ldots,\tau_{nm-1}\}$. *Then* $\mathscr{S}_{m,n}^{LI}$ *is linear, and* $\mathscr{S}_{m,n}^{LI}$ *is a local*
interpolation operator, with

$$(\mathscr{S}_{m,n}^{LI}f)(x_j) = f(x_j), \quad j = 0,\ldots,n, \quad f \in C[a,b], \qquad (10.5.26)$$

and such that the polynomial exactness condition

$$(\mathscr{S}_{m,n}^{LI}P)(x) = P(x), \quad x \in [a,b], \quad P \in \pi_m, \qquad (10.5.27)$$

is satisfied. Also, $\mathscr{S}_{m,n}^{LI}$ *satisfies the formulation*

$$(\mathscr{S}_{m,n}^{LI}f)(x) = \sum_{k=-p}^{n+q} f(x_k)V_{m,k}(x), \quad x \in [a,b], \quad f \in C[a,b], \qquad (10.5.28)$$

where the splines $\{V_{m,-p},\ldots,V_{m,n+q}\}$ *are given by*

$$V_{m,k}(x) := \sum_{j=\max\{m(k-q-1),-m\}}^{\min\{m(k+p)-1,nm-1\}} \lambda_{m,j,k}N_{m,j}(x), \quad x \in [a,b], \quad k = -p,\ldots,n+q, \qquad (10.5.29)$$

and satisfy the properties (10.5.6) - (10.5.9), *with* $V_k = V_{m,k}, k = -p,\ldots,n+q$.

We proceed to show that, in certain neighbourhoods of the endpoints of the interval $[a,b]$,
the interpolation operator $\mathscr{S}_{m,n}^{LI}$ of Theorem 10.5.2 corresponds to the polynomial interpolation operators $\mathscr{P}_m$ and $\widetilde{\mathscr{P}}_m$ in (10.5.13), as follows.

Theorem 10.5.3. *The local spline interpolation operator* $\mathscr{S}_{m,n}^{LI}$ *of Theorem 10.5.2 satisfies*
the formulation

$$(\mathscr{S}_{m,n}^{LI}f)(x) = \begin{cases} (\mathscr{P}_m^I f)(x), & x \in [a,x_p], \\ \displaystyle\sum_{k=0}^{n} f(x_k)V_{m,k}(x), & x \in (x_p,x_{n-q}) \text{ (if } n > m), \\ (\widetilde{\mathscr{P}}_m^I f)(x), & x \in [x_{n-q},b], \end{cases} \quad f \in C[a,b], \qquad (10.5.30)$$

with $\mathscr{P}_m^I$ *and* $\widetilde{\mathscr{P}}_m^I$ *denoting the polynomial interpolation operators as in* (10.5.13)

Proof. Let $f \in C[a,b]$, and suppose first

$$x \in [a,x_p] = [x_0, x_{m-q}],\qquad(10.5.31)$$

from (10.5.4) and (10.5.3), according to which it follows from (10.5.28) and (10.5.7) that

$$(\mathscr{S}_{m,n}^{LI}f)(x) = \sum_{k=-p}^{m} f(x_k)V_{m,k}(x).\qquad(10.5.32)$$

Next, observe from the first line of (10.5.12), as well as the middle line of (10.5.13), that

$$f(x_k) = (\mathscr{P}_m^I f)(x_k), \quad k = -p,\ldots,m.\qquad(10.5.33)$$

By using the notation $\mathscr{P}_m^I f := P_m^I$ as in (5.1.2), we deduce from (10.5.32) and (10.5.33), together with (10.5.27), and $P_m^I \in \pi_m$, that

$$\sum_{k=-p}^{m} f(x_k)V_{m,k}(x) = \sum_{k=-p}^{m} (\mathscr{P}_m^I f)(x_k)V_{m,k}(x) = (\mathscr{S}_{m,n}^{LI}P_m^I)(x) = P_m^I(x) = (\mathscr{P}_m^I f)(x),$$

which, together with (10.5.32), proves the first line of (10.5.30). The middle line of (10.5.30) is an immediate consequence of (10.5.28) and (10.5.7), together with (10.5.3), whereas the proof of the third line of (10.5.30) is similar to the proof of the first line of (10.5.30). ∎

For the case $n = m$, we note from (10.5.30), (10.5.13) and (10.5.3) that $\mathscr{S}_{m,m}^{LI} = \mathscr{P}_m^I = \widetilde{\mathscr{P}}_m^I$, according to which $\mathscr{S}_{m,n}^{LI}$, with $n \geq m$, can be interpreted as a spline extension of the polynomial interpolation operator $\mathscr{P}_m^I$.

We observe furthermore that, since $\mathscr{S}_{m,n}^{LI}f \in \sigma_m([a,b];\tau_1,\ldots,\tau_{nm-1})$ for each $f \in C[a,b]$, it follows from (10.2.29) that

$$\mathscr{S}_{m,n}^{LI}f \in C^{m-1}[a,b], \quad f \in C[a,b],\qquad(10.5.34)$$

according to which the piecewise formulation in (10.5.30) yields, for each $f \in C[a,b]$, an interpolant $\mathscr{S}_{m,n}^{LI}f$ with $m-1$ continuous derivatives also at the breakpoints x_p and x_{n-q} in (a,b).

We proceed to derive, for $n \geq 2m+1$, a convenient expression of the formulation (10.5.30) in Theorem 10.5.3, as follows. Let $\{L_{m,k} : k = 0,\ldots,m\}$ and $\{\widetilde{L}_{m,n-k} : k = 0,\ldots,m\}$ denote the Lagrange fundamental polynomial sequences corresponding to, respectively, the point sequences $\{x_0,\ldots,x_m\}$ and $\{x_n,\ldots,x_{n-m}\}$, that is, from (1.2.1),

$$\left.\begin{array}{ll} L_{m,k}(x) & := \displaystyle\prod_{k\neq\ell=0}^{m} \frac{x-x_\ell}{x_k-x_\ell}, \quad k=0,\ldots,m; \\[4mm] \widetilde{L}_{m,n-k}(x) & := \displaystyle\prod_{k\neq\ell=0}^{m} \frac{x-x_{n-\ell}}{x_{n-k}-x_{n-\ell}}, \quad k=0,\ldots,m. \end{array}\right\}\qquad(10.5.35)$$

It follows from (10.5.30) in Theorem 10.5.3, together with the Lagrange interpolation formula (1.2.5) in Theorem 1.2.2, as well as (5.1.2), that

$$(\mathscr{S}_{m,n}^{LI}f)(x) = \begin{cases} \sum_{k=0}^{m} f(x_k)L_{m,k}(x), & x \in [a,x_p], \\ \sum_{k=0}^{n} f(x_k)V_{m,k}(x), & x \in (x_p,x_{n-q}) \text{ (if } n > m), \\ \sum_{k=0}^{m} f(x_{n-k})\widetilde{L}_{m,n-k}(x), & x \in [x_{n-q},b], \end{cases} \quad f \in C[a,b],$$

(10.5.36)

and by means of which we can now prove the following representation formula for $\mathscr{S}_{m,n}^{LI}$.

Theorem 10.5.4. *For $n \geqslant 2m+1$, the local spline interpolation operator $\mathscr{S}_{m,n}^{LI}$ of Theorem 10.5.2 satisfies the formulation*

$$(\mathscr{S}_{m,n}^{LI}f)(x) = \sum_{k=0}^{n} f(x_k)W_{m,k}(x), \quad x \in [a,b], \quad f \in C[a,b], \tag{10.5.37}$$

where the function sequence $\{W_{m,k} : k = 0,\ldots,n\}$ is given by

$$W_{m,k}(x) := \begin{cases} L_{m,k}(x) \,, x \in [a,x_p], \\ V_{m,k}(x) \,, x \in (x_p,b], \end{cases} k = 0,\ldots,m; \tag{10.5.38}$$

$$W_{m,k}(x) := V_{m,k}(x), \quad k = m+1,\ldots,n-m-1 \ (\text{if } n \geqslant 2m+2); \tag{10.5.39}$$

$$W_{m,n-k}(x) := \begin{cases} \widetilde{L}_{m,n-k}(x), x \in [x_{n-q},b], \\ V_{m,n-k}(x), x \in [a,x_{n-q}), \end{cases} k = 0,\ldots,m, \tag{10.5.40}$$

with the Lagrange fundamental polynomials $\{L_{m,k} : k = 0,\ldots,m\}$, $\{\widetilde{L}_{m,n-k} : k = 0,\ldots,m\}$ given as in (10.5.35), and where, moreover,

$$\{W_{m,k} : k = 0,\ldots,n\} \subset \sigma_m([a,b];\tau_1,\ldots,\tau_{nm-1}). \tag{10.5.41}$$

Proof. First, note that the formulation (10.5.37)–(10.5.40) is an immediate consequence of (10.5.36). It therefore remains to prove (10.5.41).

To this end, for any fixed $k \in \{0,\ldots,m\} \cup \{n-m,\ldots,n\}$, we choose $f \in C[a,b]$ such that

$$f(x_k) = \delta_{k-\ell}, \quad \ell = 0,\ldots,n,$$

for which (10.5.37) then yields

$$(\mathscr{S}_{m,n}^{LI}f)(x) = W_{m,k}(x), \quad x \in [a,b],$$

and thus, since also $\mathscr{S}_{m,n}^{LI} : C[a,b] \to \sigma_m([a,b];\tau_1,\ldots,\tau_{nm-1})$, we deduce that

$$W_{m,k} \in \sigma_m([a,b];\tau_1,\ldots,\tau_{nm-1}), \quad k \in \{0,\ldots,m\} \cup \{n-m,\ldots,n\}, \tag{10.5.42}$$

which, together with (10.5.39) and (10.5.6), completes the proof of (10.5.41).   ∎

Observe in particular from (10.5.42) and (10.2.29) that the piecewise definitions (10.5.38) and (10.5.40) yield functions with $m-1$ continuous derivatives also at the breakpoints x_p and x_{n-q} in (a,b).

The formulation (10.5.37) in Theorem 10.5.4 enables us to show that the exactness property (10.5.27) of $\mathscr{S}_{m,n}^{LI}$ holds with π_m replaced by a larger space, namely the range of the operator $\mathscr{S}_{m,n}^{LI}$, as given in the following result.

Theorem 10.5.5. *For $n \geqslant 2m+1$, the range of the local spline interpolation operator $\mathscr{S}_{m,n}^{LI}$ of Theorem 10.5.2 satisfies*

$$\text{range } (\mathscr{S}_{m,n}^{LI}) = \text{span}\{W_{m,j} : j = 0,\ldots,n\}, \tag{10.5.43}$$

with the sequence $\{W_{m,j} : j = 0,\ldots,n\}$ defined by (10.5.38) - (10.5.40) in Theorem 10.5.4. Moreover, range $(\mathscr{S}_{m,n}^{LI})$ is a subspace of the spline space $\sigma_m([a,b];\tau_1,\ldots,\tau_{nm-1})$, with

$$\pi_m \subset \text{range } (\mathscr{S}_{m,n}^{LI}); \tag{10.5.44}$$

$$\dim \text{range } (\mathscr{S}_{m,n}^{LI}) = n+1, \tag{10.5.45}$$

and the exactness condition

$$(\mathscr{S}_{m,n}^{LI}f)(x) = f(x), \quad x \in [a,b], \quad f \in \text{range } (\mathscr{S}_{m,n}^{LI}). \tag{10.5.46}$$

is satisfied.

Proof. First, observe that the results (10.5.43) and (10.5.44) are immediate consequences of, respectively, (10.5.37) in Theorem 10.5.4, and (10.5.27) in Theorem 10.5.2. Also, it follows from (10.5.43) and (10.5.41) that range $(\mathscr{S}_{m,n}^{LI})$ is a subspace of $\sigma_m([a,b];\tau_1,\ldots,\tau_{nm-1})$.

Next, we note from (10.5.43) that the result (10.5.45) will follow if we can show that $\{W_{m,j} : j = 0,\ldots,n\}$ is a linearly independent set. Hence we let the coefficient sequence $\{c_0,\ldots,c_n\} \subset \mathbb{R}$ be such that

$$\sum_{k=0}^{n} c_k W_{m,k}(x) = 0, \quad x \in [a,b]. \tag{10.5.47}$$

Now observe from (10.5.38)–(10.5.40), together with (10.5.8), as well as (1.2.3) in Theorem 1.2.1, that

$$W_{m,k}(x_\ell) = \delta_{k-\ell}, \quad k,\ell = 0,\ldots,n. \tag{10.5.48}$$

For any fixed $\ell \in \{0,\ldots,n\}$, we now choose $x = x_\ell$ in (10.5.47), and use (10.5.48), to obtain

$$0 = \sum_{k=0}^{n} c_k W_{m,k}(x_\ell) = \sum_{k=0}^{n} c_k \delta_{k-\ell} = c_\ell,$$

and thus $c_0 = \cdots = c_n = 0$, according to which $\{W_{m,j} : j = 0,\ldots,n\}$ is a linearly independent set, and it follows that (10.5.45) holds.

Finally, to prove the exactness condition (10.5.46), let $f \in$ range $(\mathscr{S}_{m,n}^{LI})$, that is, from (10.5.43),

$$f(x) = \sum_{j=0}^{n} \alpha_j W_{m,j}(x), \quad x \in [a,b], \tag{10.5.49}$$

for some coefficient sequence $\{\alpha_0,\ldots,\alpha_n\} \subset \mathbb{R}$. Now substitute (10.5.49) into the formula (10.5.37) of Theorem 10.5.4, to deduce by means of (10.5.48) that, for any $x \in [a,b]$,

$$(\mathscr{S}_{m,n}^{LI} f)(x) = \sum_{k=0}^{n} \left[\sum_{j=0}^{n} \alpha_j W_{m,j}(x_k) \right] W_{m,k}(x)$$

$$= \sum_{k=0}^{n} \left[\sum_{j=0}^{n} \alpha_j \delta_{j-k} \right] W_{m,k}(x) = \sum_{k=0}^{n} \alpha_k W_{m,k}(x) = f(x),$$

from (10.5.49), and thereby completing our proof of (10.5.46). ∎

The following explicit formulations follow from (10.5.38)–(10.5.40), together with (10.5.35), and the formula (10.5.29) in Theorem 10.5.2, as well as (10.5.5), the second line of (10.2.33) in Theorem 10.2.2(a), and (10.5.3).

Theorem 10.5.6. *The splines $\{W_{m,k} : k = 0,\ldots,n\}$, as defined in (10.5.38)–(10.5.40) of Theorem 10.5.4, satisfy the formulations*

$$W_{m,k}(x) = \begin{cases} \prod\limits_{\substack{k \neq \ell = 0}}^{m} \dfrac{x - x_\ell}{x_k - x_\ell}, & x \in [a, x_p], \\ \sum\limits_{j=m(p-1)}^{m(k+p)-1} \lambda_{m,j,k} N_{m,j}(x), & x \in (x_p, x_{p+k+1}), \\ 0, & x \in [x_{p+k+1}, b], \end{cases} \quad k = 0,\ldots,m; \tag{10.5.50}$$

$$W_{m,k}(x) = \begin{cases} \sum\limits_{j=m(k-q-1)}^{m(k+p)-1} \lambda_{m,j,k} N_{m,j}(x), & x \in (x_{k-q-1}, x_{k+p+1}), \\ 0, & x \in [a, x_{k-q-1}] \cup [x_{k+p+1}, b], \end{cases}$$

$$k = m+1,\ldots,n-m-1 \ (\text{if } n \geqslant 2m+2); \tag{10.5.51}$$

$$W_{m,n-k}(x) = \begin{cases} \prod_{\substack{k\neq\ell=0}}^{m} \dfrac{x-x_{n-\ell}}{x_{n-k}-x_{n-\ell}}, & x\in[x_{n-q},b], \\[2mm] \displaystyle\sum_{j=m(n-k-q-1)}^{m(n-q)-1} \lambda_{m,j,n-k}N_{m,j}(x), & x\in(x_{n-q-k-1},x_{n-q}), \\[2mm] 0, & x\in[a,x_{n-q-k-1}], \end{cases} \Bigg\} k=0,\dots,m,$$

$$(10.5.52)$$

where the sequence

$$\{\lambda_{m,j,k}\} = \{\lambda_{m,mj+\rho,k}:k=j-p+1,\dots,j+q+1; \ j=p-1,\dots,n-q-1; \ \rho=0,\dots,m-1\}$$

$$(10.5.53)$$

is defined as in (10.5.25), and where $\{N_{m,m(p-1)},\dots,N_{m,m(n-q)-1}\}$ *is the B-spline sequence, as formulated in (10.2.32), with respect to the knot sequence* $\{\tau_{m(p-1)},\dots,\tau_{m(n-q+1)}\}$, *which is a subsequence of the knot sequence* $\{\tau_{-m},\dots,\tau_{(n+1)m}\}$ *as in (10.5.2).*

In view of (10.5.6) and (10.5.3), a natural choice for the integer pair $\{p,q\}$ is

$$p = \lfloor\tfrac{1}{2}(m+1)\rfloor; \quad q = \lfloor\tfrac{1}{2}m\rfloor, \qquad (10.5.54)$$

thereby, for each fixed $k \in \{-p,\dots,n+q\}$, placing the index k, for which (10.5.8) gives $V_k(x_k) = V_{m,k}(x_k) = 1$, as close as possible to the middle of the index sequence $\{k-q-1,\dots,k+p+1\}$ in (10.5.7).

Example 10.5.1. (a) In Theorem 10.5.2, choose $m = 1, n \geqslant 3$, and, by following (10.5.54), let $p = 1, q = 0$. We then calculate from (10.5.25) and (10.5.5) that, for $j = -1,\dots,n-1$,

$$\lambda_{1,j,j} = 0; \quad \lambda_{1,j,j+1} = 1, \qquad (10.5.55)$$

which, together with (10.5.24), yields the local linear spline interpolation operator

$$(\mathscr{S}_{1,n}^{LI}f)(x) = \sum_{j=0}^{n} f(x_j)N_{1,j-1}(x), \quad x\in[a,b], \quad f\in C[a,b], \qquad (10.5.56)$$

according to which, as expected from (10.4.56), (10.4.57) and (10.5.5), with $m = 1$, we have $\mathscr{S}_{1,n}^{LI} = \mathscr{S}_{1,r}^{QI}$, the local linear spline quasi-interpolation operator in (10.4.54), with $r = n - 1$.

(b) In Theorem 10.5.2, choose $m = 2$, $n \geqslant 5$, and, by following (10.5.54), let $p = q = 1$. By using the formula (10.5.25), as well as (10.5.5), we obtain (see Exercise 10.49) the coefficients

$$\begin{aligned} \lambda_{2,2j,j} &= \frac{1}{2}\frac{(x_{j+1}-\tau_{2j+1})(x_{j+2}-x_{j+1})}{(x_{j+1}-x_j)(x_{j+2}-x_j)}, \\[2mm] \lambda_{2,2j,j+1} &= \frac{1}{2}\frac{(\tau_{2j+1}-x_j)(x_{j+2}-x_{j+1})+(x_{j+1}-x_j)(x_{j+2}-\tau_{2j+1})}{(x_{j+1}-x_j)(x_{j+2}-x_{j+1})}, \\[2mm] \lambda_{2,2j,j+2} &= -\frac{1}{2}\frac{(x_{j+1}-x_j)(x_{j+1}-\tau_{2j+1})}{(x_{j+2}-x_j)(x_{j+2}-x_{j+1})}, \end{aligned} \Bigg\} j=0,\dots,n-2;$$

$$(10.5.57)$$

$$\begin{aligned}
\lambda_{2,2j+1,j} &= -\frac{1}{2}\frac{(\tau_{2j+3}-x_{j+1})(x_{j+2}-x_{j+1})}{(x_{j+1}-x_j)(x_{j+2}-x_j)}, \\
\lambda_{2,2j+1,j+1} &= \frac{1}{2}\frac{(x_{j+1}-x_j)(x_{j+2}-\tau_{2j+3})+(\tau_{2j+3}-x_j)(x_{j+2}-x_{j+1})}{(x_{j+1}-x_j)(x_{j+2}-x_{j+1})}, \\
\lambda_{2,2j+1,j+2} &= \frac{1}{2}\frac{(x_{j+1}-x_j)(\tau_{2j+3}-x_{j+1})}{(x_{j+2}-x_j)(x_{j+2}-x_{j+1})},
\end{aligned}\left.\vphantom{\begin{aligned}1\\1\\1\end{aligned}}\right\} \; j=0,\ldots,n-2,$$

$$(10.5.58)$$

which can now be substituted into the formulas (10.5.50)–(10.5.52) to obtain the sequence $\{W_{m,k}:k=0,\ldots,n\}$, and thereby yielding the local quadratic spline interpolation operator $\mathscr{S}_{2,n}^{LI}$, as formulated in (10.5.37). ∎

We proceed to consider the important special case of Theorem 10.5.2 where both the sequences (10.5.2) and (10.5.4) are equispaced, with, by keeping in mind also the constraint (10.5.5),

$$\tau_j = a + j\left(\frac{b-a}{mn}\right), \quad j=-m,\ldots,(n+1)m; \qquad (10.5.59)$$

$$x_j = a + j\left(\frac{b-a}{n}\right), \quad j=-p,\ldots,n+q. \qquad (10.5.60)$$

The following result then holds.

Theorem 10.5.7. *In Theorem 10.5.2, suppose that the sequences $\{\tau_{-m},\ldots,\tau_{(n+1)m}\}$ and $\{x_{-p},\ldots,x_{n+q}\}$ are equispaced as in (10.5.59), (10.5.60). Then*

$$(\mathscr{S}_{m,n}^{LI}f)(x) = \sum_{k=-p}^{n+q} f(x_k)V_m\left(\frac{mn}{b-a}(x-a)-mk\right), \quad x\in[a,b], \quad f\in C[a,b], \quad (10.5.61)$$

where the spline V_m is given by

$$V_m(x) := \sum_{j=-m(q+1)}^{mp-1} \lambda_{m,j}N_m(x-j), \qquad (10.5.62)$$

with

$$\lambda_{m,mj+\rho} := \frac{(-1)^{q+1-j}\binom{m}{p-1-j}}{(m!)^?} \sum_{\substack{\{v_{-p+1},\ldots,v_{q+1}\}\setminus\{v_{-j}\} \\ \in per\{1,\ldots,m\}}} \prod_{-j\neq\ell=-p+1}^{q+1}\left(\frac{\rho+v_\ell}{m}-\ell\right),$$

$$j=-q-1,\ldots,p-1; \; \rho=0,\ldots,m-1, \qquad (10.5.63)$$

and where N_m is the cardinal B-spline of degree m, as in (10.2.69) of Theorem 10.2.7. Moreover, V_m satisfies the following properties:

(i)

$$\left.\begin{array}{l} V_m \in \sigma_{m,0}(-m(q+1),\dots,m(p+1)), \\ V_m(x) = 0, \quad x \in \mathbb{R} \setminus (-m(q+1), m(p+1)); \end{array}\right\} \tag{10.5.64}$$

with

(ii)

$$V_m(mj) = \delta_j, \quad j \in \mathbb{Z}; \tag{10.5.65}$$

(iii)

$$\sum_{k=-\infty}^{\infty} P(mk)V_m(x-mk) = P(x), \quad x \in \mathbb{R}, \quad P \in \pi_m. \tag{10.5.66}$$

(iv) *For even m, and with p and q chosen as in (10.5.54), that is,*

$$p = q = \tfrac{1}{2}m, \tag{10.5.67}$$

the symmetry condition

$$V_m(-x) = V_m(x), \quad x \in \mathbb{R}, \tag{10.5.68}$$

is satisfied.

Proof. First, by applying (10.2.51) in Theorem 10.2.3, together with (10.5.59), and (3.4.2) in Theorem 3.4.1, as well as the definition (10.2.69) in Theorem 10.2.7, we deduce that, for any $j \in \{-m,\dots,nm-1\}$ and $x \in \mathbb{R}$, we have

$$N_{m,j}(x) = \left[-\frac{(m+1)(b-a)}{mn} \right] \left[-\frac{(mn)^{m+1}}{(m+1)!(b-a)^{m+1}} \sum_{k=0}^{m+1} (-1)^k \binom{m+1}{k} \right.$$
$$\left. \times \left(x - a - (j+k)\left(\frac{b-a}{mn}\right) \right)_+^m \right]$$
$$= \frac{1}{m!} \sum_{k=0}^{m+1} (-1)^k \binom{m+1}{k} \left(\frac{mn}{b-a}(x-a) - j - k \right)_+^m$$
$$= N_m \left(\frac{mn}{b-a}(x-a) - j \right). \tag{10.5.69}$$

Also, by substituting (10.5.59) and (10.5.60) into (10.5.25), and using the fact (see Exercise 10.50(a)) that

$$\prod_{\substack{k \neq \ell = -p+1}}^{q+1} (k - \ell) = (-1)^{q+1-k} \frac{m!}{\binom{m}{p-1+k}}, \quad k = -p+1,\dots,q+1, \tag{10.5.70}$$

as well as the condition (10.5.3), we obtain the formulation

$$\lambda_{m,mj+\rho,j+k} = \frac{(-1)^{q+1-k}\binom{m}{p-1+k}}{(m!)^2} \sum_{\{v_{-p+1},\dots,v_{q+1}\}\setminus\{v_k\}\in \mathrm{per}\{1,\dots,m\}} \prod_{k \neq \ell = -p+1}^{q+1} \left(\frac{\rho + v_\ell}{m} - \ell \right),$$
$$k = -p+1,\dots,q+1; \ j = -1,\dots,n-1; \ \rho = 0,\dots,m-1. \tag{10.5.71}$$

Hence we may use (10.5.71) and (10.5.69) in (10.5.29), to obtain, for $k \in \{-p, \ldots, n+q\}$, and $x \in [a,b]$,

$$V_{m,k}(x) = \frac{(-1)^{q+1-k}}{(m!)^2} \sum_{\rho=0}^{m-1} \sum_{j=\max\{k-q-1,-1\}}^{\min\{k+p-1,n-1\}} (-1)^j \binom{m}{p-1+k-j}$$

$$\times \left[\sum_{\{v_{-p+1},\ldots,v_{q+1}\}\backslash\{v_{k-j}\}\in \text{per}\{1,\ldots,m\}} \prod_{\substack{k-j\neq\ell=-p+1}}^{q+1} \left(\frac{\rho+v_\ell}{m} - \ell \right) \right]$$

$$\times N_m\left(\frac{mn}{b-a}(x-a) - mj - \rho \right)$$

$$= \frac{(-1)^{q+1}}{(m!)^2} \sum_{\rho=0}^{m-1} \sum_{j=\max\{-q-1,-k-1\}}^{\min\{p-1,n-k-1\}} (-1)^j \binom{m}{p-1-j}$$

$$\times \left[\sum_{\{v_{-p+1},\ldots,v_{q+1}\}\backslash\{v_{-j}\}\in \text{per}\{1,\ldots,m\}} \prod_{\substack{-j\neq\ell=-p+1}}^{q+1} \left(\frac{\rho+v_\ell}{m} - \ell \right) \right]$$

$$\times N_m\left(\frac{mn}{b-a}(x-a) - mk - mj - \rho \right). \qquad (10.5.72)$$

Next, we apply (10.5.62), together with the second line of (10.2.33) in Theorem 10.2.2(a), as well as (10.5.63), to deduce that, for any $k \in \{-p, \ldots, n+q\}$ and $x \in [a,b]$,

$$V_m\left(\frac{mn}{b-a}(x-a) - mk \right) = \sum_{j=\max\{-m(q+1),-m(k+1)\}}^{\min\{mp-1,m(n-k)-1\}} \lambda_{m,j} N_m\left(\frac{mn}{b-a}(x-a) - mk - j \right)$$

$$= \frac{(-1)^{q+1}}{(m!)^2} \sum_{\rho=0}^{m-1} \sum_{j=\max\{-q-1,-k-1\}}^{\min\{p-1,n-k-1\}} (-1)^j \binom{m}{p-1-j}$$

$$\times \left[\sum_{\{v_{-p+1},\ldots,v_{q+1}\}\backslash\{v_{-j}\}\in \text{per}\{1,\ldots,m\}} \prod_{\substack{-j\neq\ell=-p+1}}^{q+1} \left(\frac{\rho+v_\ell}{m} - \ell \right) \right]$$

$$\times N_m\left(\frac{mn}{b-a}(x-a) - mk - mj - \rho \right),$$

which, together with (10.5.72), yields

$$V_{m,k}(x) = V_m\left(\frac{mn}{b-a}(x-a) - mk \right), \quad x \in [a,b], \quad k = -p, \ldots, n+q. \qquad (10.5.73)$$

The desired result (10.5.61) now follows from (10.5.28) in Theorem 10.5.2, together with (10.5.73).

It remains to prove the properties (i) - (iv) of the spline V_m.

(i) The property (10.5.64) is an immediate consequence of the definition (10.5.62), together with (10.2.70) in Theorem 10.2.7.

(ii) For any $j \in \{0,\ldots,n\}$ and $k \in \{-p,\ldots,n+q\}$, we may apply (10.5.8) with $V_k = V_{m,k}$, as well as (10.5.60), (10.5.72) and (10.5.73), to deduce that

$$\delta_{j-k} = V_{m,k}(x_j) = V_m(m(j-k)),$$

which, together with the second line of (10.5.64), yields the desired interpolation property (10.5.65).

(iii) Let $P \in \pi_m$, and, for any fixed $x \in \mathbb{R}$, denote by ℓ the (unique) integer for which it holds that $x \in [m\ell, m(\ell+1))$, and in terms of which we now introduce the one-to-one mapping between the intervals $[m\ell, m(\ell+1)]$ and $[a,b]$, as given by

$$\left.\begin{array}{l} \xi = a + \dfrac{b-a}{mn}(x-m\ell), \quad x \in [m\ell, m(\ell+1)]; \\[2mm] x = mn\left(\dfrac{\xi-a}{b-a}\right) + m\ell, \quad \xi \in [a,b]. \end{array}\right\} \qquad (10.5.74)$$

Observe that the polynomial

$$P_\ell(\xi) := P\left(mn\left(\frac{\xi-a}{b-a}\right) + m\ell\right) \qquad (10.5.75)$$

satisfies $P_\ell \in \pi_m$, and, from (10.5.75) and (10.5.60),

$$P_\ell(x_k) = P(m(k+\ell)), \quad k = -p,\ldots,n+q. \qquad (10.5.76)$$

Hence we may apply (10.5.9) with $V_k = V_{m,k}$, together with (10.5.74), (10.5.75), (10.5.76), (10.5.73), and finally the second line of (10.5.64), to obtain

$$P(x) = P_\ell(\xi) = \sum_{k=-p}^{n+q} P_\ell(x_k)V_{m,k}(\xi)$$

$$= \sum_{k=-p}^{n+q} P(m(k+\ell))V_m\left(\frac{mn}{b-a}(\xi-a) - mk\right)$$

$$= \sum_{k=-p+\ell}^{n+q+\ell} P(mk)V_m\left(\frac{mn}{b-a}(\xi-a) + m\ell - mk\right)$$

$$= \sum_{k=-p+\ell}^{n+q+\ell} P(mk)V_m(x-mk) = \sum_{k=-\infty}^{\infty} P(mk)V_m(x-mk),$$

which completes the proof of (10.5.66).

(iv) Suppose $m = 2\mu$ for a positive integer μ, so that (10.5.67) yields $p = q = \mu$. Hence we may apply (10.5.62) and (10.5.63), as well as the first line of (10.2.74) in Theorem 10.2.7, to deduce that, for any $x \in \mathbb{R}$,

$$V_{2\mu}(-x) = \sum_{j=-2\mu(\mu+1)}^{2\mu^2-1} \lambda_{2\mu,j} N_{2\mu}(2\mu+1+x+j)$$

$$= \frac{(-1)^{\mu+1}}{[(2\mu)!]^2} \sum_{j=-\mu-1}^{\mu-1} (-1)^j \binom{2\mu}{\mu-1-j} \sum_{\rho=0}^{2\mu-1} \sum_{\{v_{-\mu+1},\ldots,v_{\mu+1}\}\setminus\{v_{-j}\}\in\text{per}\{1,\ldots,2\mu\}}$$

$$\left[\prod_{-j\neq\ell=-\mu+1}^{\mu+1} \left(\frac{\rho+v_\ell}{2\mu} - \ell \right) \right] N_{2\mu}(2\mu+1+x+2\mu j+\rho)$$

$$= \frac{(-1)^{\mu+1}}{[(2\mu)!]^2} \sum_{j=-\mu-1}^{\mu-1} (-1)^j \binom{2\mu}{\mu+1+j} \sum_{\rho=0}^{2\mu-1} \sum_{\{v_{-\mu+1},\ldots,v_{\mu+1}\}\setminus\{v_{j+2}\}\in\text{per}\{1,\ldots,2\mu\}}$$

$$\left[\prod_{j+2\neq\ell=-\mu+1}^{\mu+1} \left(\frac{\rho+v_\ell}{2\mu} - \ell \right) \right] N_{2\mu}(x-2\mu j-(2\mu-1-\rho))$$

$$= \frac{(-1)^{\mu+1}}{[(2\mu)!]^2} \sum_{j=-\mu-1}^{\mu-1} (-1)^j \binom{2\mu}{\mu-1-j} \sum_{\rho=0}^{2\mu-1} \sum_{\{v_{-\mu+1},\ldots,v_{\mu+1}\}\setminus\{v_{j+2}\}\in\text{per}\{1,\ldots,2\mu\}}$$

$$\left[\prod_{j+2\neq\ell=-\mu+1}^{\mu+1} \left(\frac{2\mu-1-\rho+v_\ell}{2\mu} - \ell \right) \right] N_{2\mu}(x-2\mu j-\rho)$$

$$= \frac{(-1)^{\mu+1}}{[(2\mu)!]^2} \sum_{j=-\mu-1}^{\mu-1} (-1)^j \binom{2\mu}{\mu-1-j} \sum_{\rho=0}^{2\mu-1} \sum_{\{v_{-\mu+1},\ldots,v_{\mu+1}\}\setminus\{v_{-j}\}\in\text{per}\{1,\ldots,2\mu\}}$$

$$\left[\prod_{-j\neq\ell=-\mu+1}^{\mu+1} \left(\frac{-2\mu-1-\rho+v_{2-\ell}}{2\mu} + \ell \right) \right] N_{2\mu}(x-2\mu j-\rho)$$

$$= \frac{(-1)^{\mu+1}}{[(2\mu)!]^2} \sum_{j=-\mu-1}^{\mu-1} (-1)^j \binom{2\mu}{\mu-1-j} \sum_{\rho=0}^{2\mu-1} \sum_{\{v_{-\mu+1},\ldots,v_{\mu+1}\}\setminus\{v_{-j}\}\in\text{per}\{1,\ldots,2\mu\}}$$

$$\left[\prod_{-j\neq\ell=-\mu+1}^{\mu+1} \left(\frac{\rho+(2\mu+1-v_{2-\ell})}{2\mu} - \ell \right) \right] N_{2\mu}(x-2\mu j-\rho)$$

$$= \frac{(-1)^{\mu+1}}{[(2\mu)!]^2} \sum_{j=-\mu-1}^{\mu-1} (-1)^j \binom{2\mu}{\mu-1-j} \sum_{\rho=0}^{2\mu-1} \sum_{\{v_{-\mu+1},\ldots,v_{\mu+1}\}\setminus\{v_{-j}\}\in\text{per}\{1,\ldots,2\mu\}}$$

$$\left[\prod_{-j\neq\ell=-\mu+1}^{\mu+1} \left(\frac{\rho+v_\ell}{2\mu} - \ell \right) \right] N_{2\mu}(x-2\mu j-\rho), \qquad (10.5.77)$$

by virtue of the facts that

$$\{2-\ell : \ell = -\mu+1,\ldots,\mu+1\} = \{-\mu+1,\ldots,\mu+1\},$$

and

$$\{2\mu+1-v_{2-\ell} : v_{2-\ell} \in \{1,\ldots,2\mu\}\} = \{1,\ldots,2\mu\}.$$

The desired result (10.5.68) is now an immediate consequence of (10.5.77), (10.5.62) and (10.5.63). ∎

Example 10.5.2. In Theorem 10.5.7, let $m = 2$, and choose p and q as in (10.5.67), that is, $p = q = 1$. Calculating by means of the formula (10.5.63), we obtain (see Exercise 10.51) the coefficients

$$\{\lambda_{2,-4},\lambda_{2,-3},\lambda_{2,-2},\lambda_{2,-1},\lambda_{2,0},\lambda_{2,1}\} = \{-\tfrac{1}{8},\tfrac{1}{8},1,1,\tfrac{1}{8},-\tfrac{1}{8}\}. \tag{10.5.78}$$

By substituting the coefficient values (10.5.78) into the formulation (10.5.62), we obtain the quadratic spline

$$V_2(x) = -\tfrac{1}{8}N_2(x+4)+\tfrac{1}{8}N_2(x+3)+N_2(x+2)+N_2(x+1)+\tfrac{1}{8}N_2(x)-\tfrac{1}{8}N_2(x-1), \tag{10.5.79}$$

which then satisfies, according to (10.5.64)-(10.5.68) in Theorem 10.5.7, the properties

with
$$\left.\begin{array}{l} V_2 \in \sigma_{2,0}(-4,\ldots,4), \\ V_2(x) = 0, \quad x \in \mathbb{R}\setminus(-4,4); \end{array}\right\} \tag{10.5.80}$$

$$V_2(2j) = \delta_j, \quad j \in \mathbb{Z}; \tag{10.5.81}$$

$$\sum_{k=-\infty}^{\infty} P(2k)V_2(x-2k) = P(x), \quad x \in \mathbb{R}, \quad P \in \pi_2; \tag{10.5.82}$$

$$V_2(-x) = V_2(x), \quad x \in \mathbb{R}. \tag{10.5.83}$$

Also, by using (10.5.79), (10.2.80) and (10.5.83), we obtain (see Exercise 10.51) the explicit formulation

$$V_2(x) = \frac{1}{16}\left\{\begin{array}{ll} 16-7x^2, & x \in [0,1); \\ (x-2)(5x-14), & x \in [1,2); \\ (x-2)(3x-10), & x \in [2,3); \\ -(x-4)^2, & x \in [3,4); \\ 0, & x \in [4,\infty); \end{array}\right\} \tag{10.5.84}$$

$$V_2(x) = V_2(-x), \quad x \in (-\infty, 0). \tag{10.5.85}$$

The corresponding local spline interpolation operator $\mathscr{S}^{LI}_{2,n}$ is then given, according to the formulation (10.5.61) in Theorem 10.5.7, by

$$(\mathscr{S}^{LI}_{2,n}f)(x) = \sum_{k=-1}^{n+1} f(x_k) V_2\left(\frac{2n}{b-a}(x-a) - 2k\right), \ x \in [a,b], \ f \in C[a,b], \tag{10.5.86}$$

where the function values $\{f(x_k) : k = 0, \ldots, n\}$ are extended to $\{f(x_k) : k = -1, \ldots, n+1\}$ by means of quadratic polynomial extrapolation, as in (10.5.12), (10.5.13), with $m = 2$.

∎

Next, by combining Theorems 10.5.4, 10.5.6 and 10.5.7, with particular reference also to (10.5.60) and (10.5.73), and by using the fact (see Exercise 10.50(b)) that

$$\prod_{\substack{k \neq \ell = 0}}^{m} (k - \ell) = (-1)^k \frac{m!}{\binom{m}{k}}, \quad k = 0, \ldots, m, \tag{10.5.87}$$

we obtain the following alternative formulation to (10.5.61) in Theorem 10.5.7.

Theorem 10.5.8. *For $n \geqslant 2m+1$, the local spline interpolation operator $\mathscr{S}^{LI}_{m,n}$ in Theorem 10.5.7 satisfies the formulation*

$$(\mathscr{S}^{LI}_{m,n}f)(x) = \sum_{k=0}^{n} f(x_k) W_{m,k}(x), \quad x \in [a,b], \quad f \in C[a,b], \tag{10.5.88}$$

where the spline sequence $\{W_{m,k} : k = 0, \ldots, n\}$ is given by

$$W_{m,k}(x) = \begin{cases} \dfrac{(-1)^k}{m!}\binom{m}{k} \displaystyle\prod_{\substack{k \neq \ell = 0}}^{m} \left(\dfrac{n}{b-a}(x-a) - \ell\right), & x \in [a, x_p], \\[4mm] V_m\left(\dfrac{mn}{b-a}(x-a) - mk\right), & x \in (x_p, b], \end{cases} \right\} k = 0, \ldots, m; \tag{10.5.89}$$

$$W_{m,k}(x) = V_m\left(\frac{mn}{b-a}(x-a) - mk\right), \quad x \in [a,b], \ k = m+1, \ldots, n-m-1 \,(if\, n \geqslant 2m+2); \tag{10.5.90}$$

$$W_{m,n-k}(x) = \begin{cases} \dfrac{(-1)^k}{m!}\binom{m}{k} \displaystyle\prod_{\substack{k \neq \ell = 0}}^{m} \left(\dfrac{n}{b-a}(x-a) - (n-\ell)\right), & x \subset [x_{n-q}, b], \\[4mm] V_m\left(\dfrac{mn}{b-a}(x-a) - m(n-k)\right), & x \in [a, x_{n-q}), \end{cases} \right\} k = 0, \ldots, m, \tag{10.5.91}$$

with V_m denoting the spline defined by (10.5.62), (10.5.63).

10.6 Uniform convergence and error analysis

Our first objective in this section is to establish, analogously to the Weierstrass results for algebraic polynomials and trigonometric polynomials, in, respectively, Theorem 3.3.4 and Theorem 9.2.1, the result that, for a fixed spline degree m, any function
$f \in C[a,b]$ can be approximated with arbitrary accuracy in the maximum norm by a spline
in $\sigma_m([a,b]; \tau_1, \ldots, \tau_r)$, for a sufficiently dense knot sequence $\{\tau_1, \ldots, \tau_r\}$.

To this end, for integers m and n such that $m \geqslant 0$, $n \geqslant 2$, and any bounded interval $[a,b]$,
we let $\tau_n = \{\tau_{n,-m}, \ldots, \tau_{n,n+m}\}$ denote a knot sequence satisfying

$$\tau_{n,-m} < \cdots < \tau_{n,0} := a < \tau_{n,1} < \cdots < \tau_{n,n-1} < b =: \tau_{n,n} < \cdots < \tau_{n,n+m}, \quad (10.6.1)$$

and define the Schoenberg operator $\mathscr{S}_{m,n}^{SC} : C[a,b] \to \sigma_m([a,b]; \tau_{n,1}, \ldots, \tau_{n,n-1})$ by

$$(\mathscr{S}_{m,n}^{SC} f)(x) := \sum_{j=-m}^{n-1} f(t_{n,j}) N_{m,j}(x), \quad x \in [a,b], \quad f \in C[a,b], \quad (10.6.2)$$

where

$$t_{n,j} := \begin{cases} \max\{a, \frac{1}{2}(\tau_{n,j} + \tau_{n,j+m+1})\}, & j = -m, \ldots, -1; \\ \frac{1}{2}(\tau_{n,j} + \tau_{n,j+m+1}), & j = 0, \ldots, n-m-1; \\ \min\{b, \frac{1}{2}(\tau_{n,j} + \tau_{n,j+m+1})\}, & j = n-m, \ldots, n-1, \end{cases} \quad (10.6.3)$$

and with $\{N_{m,-m}, \ldots, N_{m,n-1}\}$ denoting the B-splines with respect to the knot sequence τ_n,
as obtained from the formulation (10.2.32). The following uniform convergence result then
holds.

Theorem 10.6.1. *For a fixed non-negative integer m, let $\{\tau_n = \{\tau_{n,-m}, \ldots, \tau_{n,n+m}\}$, $n = 2,3,\ldots\}$ denote a sequence of knot sequences satisfying the condition (10.6.1), and with, moreover,*

$$||\tau_n||_\infty := \max\{\tau_{n,j+1} - \tau_{n,j} : j = 0, \ldots, n-1\} \to 0, \quad n \to \infty. \quad (10.6.4)$$

Then the Schoenberg operator sequence $\{\mathscr{S}_{m,n}^{SC} : C[a,b] \to \sigma_m([a,b]; \tau_{n,1}, \ldots, \tau_{n,n-1}) : n = 2,3,\ldots\}$, as defined by (10.6.2), (10.6.3), satisfies the uniform convergence result

$$||f - \mathscr{S}_{m,n}^{SC} f||_\infty := \max_{a \leqslant x \leqslant b} |f(x) - (\mathscr{S}_{m,n}^{SC} f)(x)| \to 0, \quad n \to \infty, \quad f \in C[a,b]. \quad (10.6.5)$$

Proof. Let $f \in C[a,b]$, and choose $\varepsilon > 0$. To prove the uniform convergence result (10.6.5),
we shall prove the equivalent statement that there exists a positive integer $N = N(\varepsilon)$ such
that

$$||f - \mathscr{S}_{m,n}^{SC} f||_\infty < \varepsilon, \quad \text{for} \quad n > N. \quad (10.6.6)$$

To this end, let $x \in [a,b]$ and $n \in \{2,3,\ldots\}$ be fixed, and denote by k the (unique) integer in the set $\{0,\ldots,n-1\}$ for which it holds that

$$x \in [\tau_{n,k}, \tau_{n,k+1}). \tag{10.6.7}$$

By using (10.6.2), together with the identity (10.2.63) in Theorem 10.2.6(b), as well as (10.6.7), (10.6.3), and the second line of (10.2.33) in Theorem 10.2.2(a), we obtain

$$f(x) - (\mathscr{S}^{SC}_{m,n}f)(x) = \sum_{j=-m}^{n-1} [f(x) - f(t_{n,j})]N_{m,j}(x) = \sum_{j=k-m}^{k} [f(x) - f(t_{n,j})]N_{m,j}(x). \tag{10.6.8}$$

It follows from (10.6.8), together with (10.2.62) in Theorem 10.2.6(a), that

$$|f(x) - (\mathscr{S}^{SC}_{m,n}f)(x)| \leqslant \sum_{j=k-m}^{k} |f(x) - f(t_{n,j})|N_{m,j}(x). \tag{10.6.9}$$

Now observe from (10.6.7), (10.6.3), and the definition of $||\tau_n||_\infty$ in (10.6.4), that

$$|x - t_{n,j}| \leqslant \frac{1}{2}(m+1)||\tau_n||_\infty, \quad j = k-m,\ldots,k. \tag{10.6.10}$$

Since $f \in C[a,b]$, we know that f is uniformly continuous on $[a,b]$, according to which there exists a positive number $\delta = \delta(\varepsilon)$ such that

$$x,y \in [a,b], \quad \text{with} \quad |x-y| < \delta \Rightarrow |f(x) - f(y)| < \varepsilon. \tag{10.6.11}$$

Next, we deduce from the condition in (10.6.4) that there exists a positive number $N = N(\varepsilon)$ for which it holds that

$$||\tau_n||_\infty < \frac{2\delta}{m+1}, \quad n > N. \tag{10.6.12}$$

It follows from (10.6.9), (10.6.10), (10.6.11) and (10.6.12) that, for $n > N$,

$$|f(x) - (\mathscr{S}^{SC}_{m,n}f)(x)| < \varepsilon \sum_{j=k-m}^{k} N_{m,j}(x) = \varepsilon \sum_{j=-m}^{n-1} N_{m,j}(x) = \varepsilon, \tag{10.6.13}$$

after again recalling the second line of (10.2.33) in Theorem 10.2.2(a), as well as the identity (10.2.63) in Theorem 10.2.6(b), and thus

$$|f(x) - (\mathscr{S}^{SC}_{m,n}f)(x)| < \varepsilon, \quad x \in [\tau_{n,k}, \tau_{n,k+1}), \quad n > N. \tag{10.6.14}$$

By noting that the right hand side of the inequality (10.6.14) is independent of k, we deduce from (10.6.14) that

$$\max_{a \leqslant x \leqslant b} |f(x) - (\mathscr{S}^{SC}_{m,n}f)(x)| < \varepsilon, \quad n > N,$$

which completes our proof of (10.6.6). ∎

The following analogue of the Weierstrass results in Theorem 3.3.4 and Theorem 9.2.1 is now an immediate consequence of Theorem 10.6.1.

Theorem 10.6.2. *Let* $f \in C[a,b]$, *and let* m *denote any fixed non-negative integer. Then, for each* $\varepsilon > 0$, *there exists a knot sequence* $\{\tau_{-m},\ldots,\tau_{r+m+1}\}$ *and a spline* $S \in \sigma_m([a,b];\tau_1,\ldots,\tau_r)$ *such that*

$$||f - S||_\infty < \varepsilon. \tag{10.6.15}$$

A function $f : [a,b] \to \mathbb{R}$ is called a Lipschitz-continuous function on $[a,b]$ if, analogously to (9.6.35), there exists a constant K_f such that

$$|f(x) - f(y)| \leqslant K_f |x - y|, \quad x, y \in [a,b]. \tag{10.6.16}$$

The constant K_f in (10.6.16) is called a Lipschitz constant for f on $[a,b]$, and the class of all Lipschitz-continuous functions on $[a,b]$ will be denoted by $C^{\text{Lip}}[a,b]$. Analogously to (9.6.36) in Theorem 9.6.3, it can be proved that the inclusions

$$C^1[a,b] \subset C^{\text{Lip}}[a,b] \subset C[a,b] \tag{10.6.17}$$

hold.

The following error bounds can now be proved.

Theorem 10.6.3. *For any integers $m \geqslant 0$ and $n \geqslant 2$, the Schoenberg operator $\mathscr{S}_{m,n}^{SC}$ of Theorem 10.6.1 satisfies the following error bounds:*

(a)

$$||f - \mathscr{S}_{m,n}^{SC} f||_\infty \leqslant \tfrac{1}{2}(m+1)||\tau_n||_\infty K_f, \quad f \in C^{\text{Lip}}[a,b], \tag{10.6.18}$$

   with K_f denoting a Lipschitz constant of f on $[a,b]$;

(b)

$$||f - \mathscr{S}_{m,n}^{SC} f||_\infty \leqslant \tfrac{1}{2}(m+1)||\tau_n||_\infty ||f'||_\infty, \quad f \in C^1[a,b]. \tag{10.6.19}$$

Proof. (a) Let $f \in C^{\text{Lip}}[a,b]$, with Lipschitz constant K_f on $[a,b]$. For any fixed $x \in [a,b]$, and with the integer k chosen, as in the proof of Theorem 10.6.1, to satisfy (10.6.7), it follows similarly that (10.6.9) holds, so that, since also (10.6.3) shows that $\{t_{n,j} : j = -m, \ldots, n-1\} \subset [a,b]$, we may apply (10.6.16) to the right hand side of (10.6.9) to obtain

$$|f(x) - (\mathscr{S}_{m,n}^{SC} f)(x)| \leqslant K_f \sum_{j=k-m}^{k} |x - t_{n,j}| N_{m,j}(x). \tag{10.6.20}$$

By using (10.6.7), (10.6.3), and the definition of $||\tau_n||_\infty$ in (10.6.4), we deduce that

$$|x - t_{n,j}| \leqslant \tfrac{1}{2}(m+1)||\tau_n||_\infty, \quad j = k-m, \ldots, k. \tag{10.6.21}$$

It follows from (10.6.20), (10.6.21), together with the steps leading to (10.6.13), that

$$|f(x) - (\mathscr{S}_{m,n}^{SC} f)(x)| \leqslant \tfrac{1}{2}(m+1)||\tau_n||_\infty K_f. \tag{10.6.22}$$

Since the right hand side of (10.6.22) is independent of k, the inequality (10.6.18) immediately follows.

(b) If $f \in C^1[a,b]$, an application of the mean value theorem as in (9.6.37) shows that $f \in C^{\mathrm{Lip}}[a,b]$, with Lipschitz constant $K_f = ||f'||_\infty$ on $[a,b]$, according to which (10.6.19) follows from (10.6.18). ∎

Observe from (10.6.18) and (10.6.19) in Theorem 10.6.3 that the corresponding convergence rates are governed by the rate at which $||\tau_n||_\infty$ converges to zero in (10.6.4).

We proceed to investigate the issues of uniform convergence, and corresponding convergence rate, with respect to the local spline interpolation operator $\mathscr{S}^{LI}_{m,n}$ of Section 10.5. For integers $m \geqslant 1$ and $n \geqslant 2m+1$, and any bounded interval $[a,b]$, we let $\tau_n = \{\tau_{n,-m}, \ldots, \tau_{n,n+m}\}$ be a knot sequence as in (10.6.1), and denote by $\mathbf{x}_n = \{x_{n,0}, \ldots, x_{n,n}\}$ an interpolation point sequence such that

$$a =: x_{n,0} < \cdots < x_{n,n} := b, \tag{10.6.23}$$

and with, as in (10.5.5),

$$x_{n,j} = \tau_{n,mj}, \quad j = 0, \ldots, n. \tag{10.6.24}$$

According to the representation formula (10.5.37) in Theorem 10.5.4, we then have

$$(\mathscr{S}^{LI}_{m,n} f)(x) = \sum_{j=0}^{n} f(x_{n,j}) W_{m,j}(x), \quad x \in [a,b], \quad f \in C[a,b], \tag{10.6.25}$$

where the splines $\{W_{m,j} : j = 0, \ldots, n\}$ are given by (10.5.50)–(10.5.52), as well as (10.5.25), with $\{x_j : j = 0, \ldots, n\}$ replaced by $\{x_{n,j} : j = 0, \ldots, n\}$. For the sequence $\mathbf{x}_n = \{x_{n,0}, \ldots, x_{n,n}\}$ as in (10.6.23), (10.6.24), we now define the maximum ratio

$$R_n := \max\left\{ \frac{|x_{n,j} - x_{n,j-1}|}{|x_{n,k} - x_{n,k-1}|} : \ j,k = 1, \ldots, n; \ \ |j-k| = 1 \right\}, \tag{10.6.26}$$

according to which $R_n \geqslant 1$, with $R_n = 1$ if and only if the points $\{x_{n,0}, \ldots, x_{n,n}\}$ are equispaced.

We shall rely on the following uniform bound.

Theorem 10.6.4. *For positive integers m and n, with $n \geqslant 2m+1$, the splines $\{W_{m,j} : j = 0, \ldots, n\}$, as given by (10.5.50) - (10.5.52), together with (10.5.25), satisfy the uniform bound*

$$||W_{m,j}||_\infty := \max_{a \leqslant x \leqslant b} |W_{m,j}(x)| \leqslant \left[\sum_{i=1}^{m} R_n^i \right]^m, \quad j = 0, \ldots, n, \tag{10.6.27}$$

with R_n denoting the maximum ratio as defined in (10.6.26).

Proof. First, we shall show that the coefficients in (10.5.25) satisfy the uniform bound

$$|\lambda_{m,j,k}| \le \left[\sum_{i=1}^{m} R_n^i \right]^m, \quad k = 0,\ldots,n; \quad j = 0,\ldots,nm-1. \tag{10.6.28}$$

To this end, we first observe from (10.6.1), (10.6.23) and (10.6.24) that, for any $j \in \{0,\ldots,n-1\}$, $\rho \in \{0,\ldots,m-1\}$, $v \in \{1,\ldots,m\}$, and with the positive integers p and q chosen to satisfy (10.5.3), it holds that

$$|x_{n,j+\ell} - \tau_{n,mj+\rho+v}| \le \begin{cases} |x_{n,j+\ell} - x_{n,j+2}|, & \ell = -p+1,\ldots,0; \\ \max\{|x_{n,j+1} - x_{n,j}|, |x_{n,j+1} - x_{n,j+2}|\}, & \ell = 1; \\ |x_{n,j+\ell} - x_{n,j}|, & \ell = 2,\ldots,q+1. \end{cases} \tag{10.6.29}$$

In order to bound the right hand side of (10.6.29) in terms of the maximum ratio R_n, we shall use the fact that, for any integers $\mu, v \in \{0,\ldots,n\}$, we have

$$|x_{n,\mu} - x_{n,v}| \le \left[\sum_{i=0}^{|\mu-v|-1} R_n^i \right] \begin{cases} |x_{n,\mu} - x_{n,\mu-1}|, & \text{if } \mu > v; \\ |x_{n,\mu} - x_{n,\mu+1}|, & \text{if } \mu < v, \end{cases} \tag{10.6.30}$$

which follows immediately from the definition (10.6.26) if $|\mu - v| = 1$, whereas, if $|\mu - v| \ge 2$, the definition (10.6.26) yields, for $\mu > v$,

$$|x_{n,\mu} - x_{n,v}| = x_{n,\mu} - x_{n,v} = (x_{n,\mu} - x_{n,v+2}) + (x_{n,v+2} - x_{n,v+1}) + (x_{n,v+1} - x_{n,v})$$

$$\le (x_{n,\mu} - x_{n,v+2}) + (1 + R_n)(x_{n,v+2} - x_{n,v+1})$$

$$\le \cdots$$

$$\le (1 + R_n + \cdots + R_n^{\mu-v-1})(x_{n,\mu} - x_{n,\mu-1}),$$

and, for $\mu < v$,

$$|x_{n,\mu} - x_{n,v}| = x_{n,v} - x_{n,\mu} = (x_{n,v} - x_{n,v-1}) + (x_{n,v-1} - x_{n,v-2}) + (x_{n,v-2} - x_{n,\mu})$$

$$\le (1 + R_n)(x_{n,v-1} - x_{n,v-2}) + (x_{n,v-2} - x_{n,\mu})$$

$$\le \cdots$$

$$\le (1 + R_n + \cdots + R_n^{v-\mu-1})(x_{n,\mu+1} - x_{n,\mu}),$$

and thereby proving (10.6.30) for $|\mu - v| \geqslant 2$. By applying the bounds (10.6.30), with $\mu = j + \ell$, to the right hand side of (10.6.29), we deduce the bounds

$$
|x_{n,j+\ell} - \tau_{n,mj+\rho+v}| \leqslant
\begin{cases}
\left[\displaystyle\sum_{i=0}^{-\ell+1} R_n^i\right] |x_{n,j+\ell} - x_{n,j+\ell+1}|, & \ell = -p+1,\ldots,0; \\[2ex]
\max\{|x_{n,j+1} - x_{n,j}|, |x_{n,j+1} - x_{n,j+2}|\}, & \ell = 1; \\[2ex]
\left[\displaystyle\sum_{i=0}^{\ell-1} R_n^i\right] |x_{n,j+\ell} - x_{n,j+\ell-1}|, & \ell = 2,\ldots,q+1.
\end{cases}
$$

$$(10.6.31)$$

Next, we observe from (10.6.23) that, for any $k \in \{-p+1,\ldots,q+1\}$, we have

$$
|x_{n,j+\ell} - x_{n,j+k}| \geqslant
\begin{cases}
|x_{n,j+\ell} - x_{n,j+\ell+1}|, & \text{if } k > \ell; \\[1ex]
|x_{n,j+\ell} - x_{n,j+\ell-1}|, & \text{if } k < \ell.
\end{cases}
$$

$$(10.6.32)$$

Moreover, the definition (10.6.26) gives

$$
\left.
\begin{aligned}
|x_{n,j+\ell} - x_{n,j+\ell+1}| &\geqslant \frac{1}{R_n}|x_{n,j+\ell} - x_{n,j+\ell-1}|; \\[1ex]
|x_{n,j+\ell} - x_{n,j+\ell-1}| &\geqslant \frac{1}{R_n}|x_{n,j+\ell} - x_{n,j+\ell+1}|.
\end{aligned}
\right\}
$$

$$(10.6.33)$$

Now observe from (10.5.25) that

$$
|\lambda_{m,mj+\rho,k}| \leqslant \frac{1}{m!} \sum_{\{v_{-p+1},\ldots,v_{q+1}\}\backslash\{v_k\}\in\text{per}\{1,\ldots,m\}} \prod_{\substack{k\neq\ell=-p+1}}^{q+1} \frac{|x_{n,j+\ell} - \tau_{n,mj+\rho+v_\ell}|}{|x_{n,j+\ell} - x_{n,j+k}|}. \quad (10.6.34)
$$

By applying the bounds (10.6.31), (10.6.32) and (10.6.33) in (10.6.34), and using the fact that, in (10.6.34), the number of terms in the product equals $p + q = m$, from (10.5.3), whereas the number of terms in the sum equals $m!$, we deduce that

$$
|\lambda_{m,j,k}| \leqslant \left[\sum_{i=1}^{\max\{p+1,q+1\}} R_n^i\right]^m, \quad j = 0,\ldots,nm-1; \quad k = 0,\ldots,n. \quad (10.6.35)
$$

Since, moreover, p and q are positive integers satisfying (10.5.3), we have

$$
\max\{p+1, q+1\} \leqslant m, \quad (10.6.36)
$$

which, together with (10.6.35), yields the desired bound (10.6.28).

Our next step is to show that

$$
\left.
\begin{aligned}
\prod_{k\neq\ell=0}^{m} \frac{|x - x_{n,\ell}|}{|x_{n,k} - x_{n,\ell}|} &\leqslant \left[\sum_{i=1}^{m} R_n^i\right]^m, & x \in [a, x_{n,p}], \\[2ex]
\prod_{k\neq\ell=0}^{m} \frac{|x - x_{n,n-\ell}|}{|x_{n,n-k} - x_{n,n-\ell}|} &\leqslant \left[\sum_{i=1}^{m} R_n^i\right]^m, & x \in [x_{n,n-q}, b],
\end{aligned}
\right\} \ k=0,\ldots,m. \quad (10.6.37)
$$

Let $k \in \{0,\ldots,m\}$ and $\ell \in \{0,\ldots,m\} \setminus \{k\}$ be fixed. To prove the first line of (10.6.37), we fix $x \in [a, x_{n,p}] = [x_{n,0}, x_{n,p}]$, and observe that then

$$|x - x_{n,\ell}| \leqslant \max\{|x_{n,\ell} - x_{n,0}|, |x_{n,\ell} - x_{n,p}|\}. \tag{10.6.38}$$

By applying the bounds (10.6.30), we obtain

$$|x_{n,\ell} - x_{n,0}| \leqslant \left[\sum_{i=0}^{\ell-1} R_n^i\right] |x_{n,\ell} - x_{n,\ell-1}|; \tag{10.6.39}$$

$$|x_{n,\ell} - x_{n,p}| \leqslant \begin{cases} \left[\displaystyle\sum_{i=0}^{\ell-p-1} R_n^i\right] |x_{n,\ell} - x_{n,\ell-1}|, & \text{if } \ell > p; \\ \left[\displaystyle\sum_{i=0}^{p-\ell-1} R_n^i\right] |x_{n,\ell} - x_{n,\ell+1}|, & \text{if } \ell < p. \end{cases} \tag{10.6.40}$$

Moreover, analogously to (10.6.32) and (10.6.33),

$$|x_{n,\ell} - x_{n,k}| \geqslant \begin{cases} |x_{n,\ell} - x_{n,\ell+1}|, & \text{if } k > \ell; \\ |x_{n,\ell} - x_{n,\ell-1}|, & \text{if } k < \ell; \end{cases} \tag{10.6.41}$$

$$\left.\begin{array}{l} |x_{n,\ell} - x_{n,\ell+1}| \geqslant \dfrac{1}{R_n} |x_{n,\ell} - x_{n,\ell-1}|; \\ |x_{n,\ell} - x_{n,\ell-1}| \geqslant \dfrac{1}{R_n} |x_{n,\ell} - x_{n,\ell+1}|. \end{array}\right\} \tag{10.6.42}$$

By noting also that the number of terms in the product in the first line of (10.6.37) equals m, an application of the bounds (10.6.38)–(10.6.42) yields the first line of (10.6.37). The proof of the second line of (10.6.37) it similar.

Finally, we deduce from the formulas (10.5.50)-(10.5.52), together with the bounds (10.6.28) and (10.6.37), as well as (10.2.62) in Theorem 10.2.6(a), and the identity (10.2.63) in Theorem 10.2.6(b), that the uniform bound (10.6.27) is satisfied. ∎

By using Theorem 10.6.4, we can now prove, analogously to Theorem 10.6.1, the following uniform convergence result for $\mathscr{S}_{m,n}^{LI}$.

Theorem 10.6.5. *For a fixed positive integer m, let $\{\tau_n = \{\tau_{n,-m}, \ldots, \tau_{n,(n+1)m}\} : n = 2m+1, 2m+2, \ldots\}$ denote a sequence of knot sequences satisfying (10.6.1), and suppose $\{\mathbf{x}_n = \{x_{n,0}, \ldots, x_{n,n}\} : n = 2m+1, 2m+2, \ldots\}$ is a sequence of interpolation point sequences satisfying (10.6.23) and (10.6.24), as well as the conditions*

$$\|\mathbf{x}_n\|_{\infty} := \max\{x_{n,j+1} - x_{n,j} : j = 0, \ldots, n-1\} \to 0, \quad n \to \infty, \tag{10.6.43}$$

and

$$R_n \leqslant R, \quad n = 2m+1, 2m+2, \ldots, \tag{10.6.44}$$

for some constant $R \geqslant 1$, where $\{R_n : n = 2m+1, 2m+2,\ldots\}$ is the maximum ratio sequence as defined in (10.6.26). Then the local spline interpolation operator sequence $\{\mathscr{S}_{m,n}^{LI} : C[a,b] \to \sigma_m([a,b]; \tau_{n,1}, \ldots, \tau_{n,nm-1}) : n = 2m+1,\ 2m+2,\ldots\}$ as defined by (10.5.24), (10.5.25) in Theorem 10.5.2, satisfies the uniform convergence result

$$\|f - \mathscr{S}_{m,n}^{LI} f\|_\infty \to 0, \quad n \to \infty, \quad f \in C[a,b]. \tag{10.6.45}$$

Proof. Let $f \in C[a,b]$, and choose $\varepsilon > 0$. To prove (10.6.45), we shall prove the equivalent statement that there exists a positive integer $N = N(\varepsilon)$ such that

$$\|f - \mathscr{S}_{m,n}^{LI} f\|_\infty < \varepsilon, \quad \text{for} \quad n > N. \tag{10.6.46}$$

To this end, let $x \in [a,b]$ and $n \in \{2m+1, 2m+2,\ldots\}$ be fixed, and denote by k the (unique) integer in the set $\{0,\ldots,n-1\}$ for which it holds that

$$x \in [x_{n,k}, x_{n,k+1}). \tag{10.6.47}$$

Let the spline sequence $\{W_{m,j} : j = 0,\ldots,n\}$ be defined as in (10.5.50)–(10.5.52). By choosing the polynomial f in the polynomial exactness condition (10.5.27) to be identically equal to one, we deduce from the representation formula (10.5.37) in Theorem 10.5.4, together with (10.6.47), and the bottom lines of (10.5.50)–(10.5.52), that

$$\sum_{j=0}^{n} W_{m,j}(x) = \sum_{j=\mu_k}^{v_k} W_{m,j}(x) = 1, \tag{10.6.48}$$

where

$$\left.\begin{array}{l} \mu_k := \max\{k - p, 0\}; \\[4pt] v_k := \min\{k + q + 1, n\}. \end{array}\right\} \tag{10.6.49}$$

Note from (10.6.49) and (10.5.3) that

$$0 < v_k - \mu_k \leqslant p + q + 2 = m + 2. \tag{10.6.50}$$

By using (10.5.37) and (10.6.48), as well as the uniform bound (10.6.27) in Theorem 10.6.4, and the condition (10.6.44), we deduce that

$$\begin{aligned} |f(x) - (\mathscr{S}_{m,n}^{LI} f)(x)| &= \left| \sum_{j=\mu_k}^{v_k} [f(x) - f(x_{n,j})] W_{m,j}(x) \right| \\ &\leqslant \sum_{j=\mu_k}^{v_k} |f(x) - f(x_{n,j})| \, |W_{m,j}(x)| \\ &\leqslant \left[\sum_{i=1}^{m} R^i \right]^m \sum_{j=\mu_k}^{v_k} |f(x) - f(x_{n,j})|. \end{aligned} \tag{10.6.51}$$

Next, we use (10.6.47), (10.6.23) and (10.6.49), together with the definition in (10.6.43), as well as (10.6.36), to obtain, for $j = \mu_k, \ldots, \nu_k$,

$$|x - x_{n,j}| \leqslant \max\{x_{n,k+1} - x_{n,\mu_k}, x_{n,\nu_k} - x_{n,k}\}$$

$$\leqslant \max\{p+1, q+1\}||\mathbf{x}_n||_\infty \leqslant m||\mathbf{x}_n||_\infty. \tag{10.6.52}$$

Since $f \in C[a,b]$, we know that f is uniformly continuous on $[a,b]$, according to which there exists a positive number $\delta = \delta(\varepsilon)$ such that

$$x, y \in [a,b], \quad \text{with} \quad |x - y| < \delta \Rightarrow |f(x) - f(y)| < \frac{\varepsilon}{(m+2)\left[\displaystyle\sum_{i=1}^{m} R^i\right]^m}. \tag{10.6.53}$$

Now deduce from the condition in (10.6.43) that there exists a positive integer $N = N(\varepsilon)$ for which it holds that

$$||\mathbf{x}_n||_\infty < \frac{\delta}{m}, \quad n > N. \tag{10.6.54}$$

It then follows from (10.6.51), (10.6.52), (10.6.53), (10.6.54), and (10.6.50), that

$$|f(x) - (\mathscr{S}_{m,n}^{LI} f)(x)| < \varepsilon, \quad n > N. \tag{10.6.55}$$

Since the right hand side of the inequality (10.6.55) is independent of k, we deduce that

$$\max_{a \leqslant x \leqslant b} |f(x) - (\mathscr{S}_{m,n}^{LI} f)(x)| < \varepsilon, \quad n > N,$$

which is equivalent to the desired inequality (10.6.46). ∎

Observe that, subject to the constraints (10.6.1), (10.6.23) and (10.6.24), the condition in (10.6.43) on the sequence $\{\mathbf{x}_n : n = 2m+1, 2m+2, \ldots\}$ is equivalent to the condition in (10.6.4) on the sequence $\{\tau_n : n = 2, 3, \ldots\}$. Also, note from (10.6.24) that the conditions (10.6.43) and (10.6.44) are satisfied for any choice of the knot subsequence $\{\tau_{n,j} : j = 1, \ldots, nm - 1\} \setminus \{\tau_{n,mj} : j = 1, \ldots, n - 1\}$.

Example 10.6.1. In the special case of an equispaced interpolation point sequence $\{x_{n,j} : j = 0, \ldots, n\}$ satisfying (10.6.23), that is,

$$x_{n,j} = a + j\left(\frac{b-a}{n}\right), \quad j = 0, \ldots, n, \tag{10.6.56}$$

it follows from the definition in (10.6.43) that

$$||\mathbf{x}_n||_\infty = \frac{b-a}{n}, \quad n = 2m+1, \quad 2m+2, \ldots, \tag{10.6.57}$$

whereas the definition (10.6.26) yields

$$R_n = 1, \quad n = 2m+1, 2m+2, \ldots. \tag{10.6.58}$$

It follows from (10.6.57) and (10.6.58) that the conditions in (10.6.43) and (10.6.44) are satisfied, with $R = 1$, according to which Theorem 10.6.5 implies the uniform convergence result (10.6.45) for any choice of the knot subsequence $\{\tau_{n,j} : j = 1, \ldots, nm - 1\} \setminus \{\tau_{n,mj} : j = 1, \ldots, m - 1\}$. ∎

By using an analogous argument to the one in the proof of Theorem 10.6.3, and in particular by referring also to (10.6.51) and (10.6.52) in the proof of Theorem 10.6.5, we obtain the following error bounds.

Theorem 10.6.6. *For any integers $m \geqslant 1$ and $n \geqslant 2m+1$, the local spline interpolation operator $\mathscr{S}_{m,n}^{LI}$ of Theorem 10.6.5 satisfies the following error bounds:*

(a)

$$||f - \mathscr{S}_{m,n}^{LI}f||_\infty \leqslant m \left[\sum_{i=1}^{m} R^i\right]^m ||\mathbf{x}_n||_\infty K_f, \quad f \in C^{\text{Lip}}[a,b], \tag{10.6.59}$$

with K_f denoting a Lipschitz constant of f on $[a,b]$;

(b)

$$||f - \mathscr{S}_{m,n}^{LI}f||_\infty \leqslant m \left[\sum_{i=1}^{m} R^i\right]^m ||\mathbf{x}_n||_\infty ||f'||_\infty, \quad f \in C^1[a,b]. \tag{10.6.60}$$

Observe from (10.6.59) and (10.6.60) in Theorem 10.6.6 that, for $f \in C^{\text{Lip}}[a,b]$, or $f \in C^1[a,b]$, the convergence rate in (10.6.45) of Theorem 10.6.5 is governed by the rate at which $||\mathbf{x}_n||_\infty$ converges to zero in (10.6.43).

We proceed to show how the polynomial exactness property (10.5.27) of $\mathscr{S}_{m,n}^{LI}$ can be used to prove that, in (10.6.45), a convergence rate proportional to $(||\mathbf{x}_n||_\infty)^{m+1}$ is obtained for $f \in C^{m+1}[a,b]$.

For any positive integer m, let $f \in C^{m+1}[a,b]$, so that we may apply Taylor's theorem, as given in (9.3.11), with $c = a$, as well as the truncated power definition in (10.1.13), to obtain

$$f(x) = \sum_{j=0}^{m} \frac{f^{(j)}(a)}{j!}(x-a)^j + \frac{1}{m!}\int_a^b (x-t)_+^m f^{(m+1)}(t)dt, \quad x \in [a,b]. \tag{10.6.61}$$

Let $n \in \{2m+1, 2m+2, \ldots\}$ be fixed, and use the linearity of $\mathscr{S}_{m,n}^{LI}$, together with the polynomial exactness property (10.5.27), as well as the representation formula (10.5.37), to deduce from (10.6.61) that, for any $x \in [a,b]$,

$$(\mathscr{S}_{m,n}^{LI}f)(x) = \sum_{j=0}^{m} \frac{f^{(j)}(a)}{j!}(x-a)^j + \frac{1}{m!}\sum_{j=0}^{n}\left[\int_a^b (x_{n,j}-t)_+^m f^{(m+1)}(t)dt\right]W_{m,j}(x)$$

$$= \sum_{j=0}^{m} \frac{f^{(j)}(a)}{j!}(x-a)^j + \frac{1}{m!}\int_a^b \left[\sum_{j=0}^{n}(x_{n,j}-t)_+^m W_{m,j}(x)\right]f^{(m+1)}(t)dt. \tag{10.6.62}$$

It follows from (10.6.61) and (10.6.62) that

$$f(x) - (\mathscr{S}_{m,n}^{LI}f)(x) = \int_a^b K_{m,n}(x,t)f^{(m+1)}(t)dt, \quad x \in [a,b], \tag{10.6.63}$$

where

$$K_{m,n}(x,t) := \frac{1}{m!}\left[(x-t)_+^m - \sum_{j=0}^{n}(x_{n,j}-t)_+^m W_{m,j}(x)\right], \quad x\in[a,b], \quad t\in[a,b]. \quad (10.6.64)$$

The expression (10.6.63) for the error function $f - \mathscr{S}_{m,n}^{LI}$ is a special case of the Peano theorem, with corresponding Peano kernel $K_{m,n}$. It follows immediately from (10.6.63) that

$$|f(x) - (\mathscr{S}_{m,n}^{LI}f)(x)| \leqslant ||f^{(m+1)}||_\infty \int_a^b |K_{m,n}(x,t)|dt, \quad x\in[a,b]. \quad (10.6.65)$$

Let $x\in[a,b]$ be fixed, and, after recalling also (10.6.23), denote by k the (unique) integer for which it holds that

$$x\in[x_{n,k},x_{n,k+1}). \quad (10.6.66)$$

Now observe from the representation formula (10.5.37) of $\mathscr{S}_{m,n}^{LI}$, together with (10.6.66) and the bottom lines of (10.5.50)–(10.5.52), that the polynomial exactness property (10.5.27) of $\mathscr{S}_{m,n}^{LI}$ implies

$$\sum_{j=\mu_k}^{v_k}(x_{n,j}-t)^m W_{m,j}(x) = \sum_{j=0}^{n}(x_{n,j}-t)^m W_{m,j}(x) = (x-t)^m, \quad t\in[a,b], \quad (10.6.67)$$

with, as in (10.6.49),

$$\left.\begin{array}{l} \mu_k := \max\{k-p,0\}; \\ v_k := \min\{k+q+1,n\}, \end{array}\right\} \quad (10.6.68)$$

and where also

$$\sum_{j=0}^{n}(x_{n,j}-t)_+^m W_{m,j}(x) = \sum_{j=\mu_k}^{v_k}(x_{n,j}-t)_+^m W_{m,j}(x), \quad t\in[a,b]. \quad (10.6.69)$$

By using (10.6.64), (10.6.69), (10.6.23), (10.6.66), (10.1.13), (10.6.67), (10.6.27) in Theorem 10.6.4, the definition in (10.6.43), and finally (10.6.68), we obtain

$$m! \int_a^b |K_m(x,t)|dt = \sum_{\ell=0}^{n-1} \int_{x_{n,\ell}}^{x_{n,\ell+1}} \left| (x-t)_+^m - \sum_{j=\mu_k}^{v_k} (x_{n,j}-t)_+^m W_{m,j}(x) \right| dt$$

$$= \sum_{\ell=0}^{k-1} \int_{x_{n,\ell}}^{x_{n,\ell+1}} \left| (x-t)^m - \sum_{j=\ell+1}^{v_k} (x_{n,j}-t)^m W_{m,j}(x) \right| dt$$

$$+ \int_{x_{n,k}}^{x} \left| (x-t)^m - \sum_{j=k+1}^{v_k} (x_{n,j}-t)^m W_{m,j}(x) \right| dt$$

$$+ \int_{x}^{x_{n,k+1}} \left| \sum_{j=k+1}^{v_k} (x_{n,j}-t)^m W_{m,j}(x) \right| dt + \sum_{\ell=k+1}^{n-1} \int_{x_{n,\ell}}^{x_{n,\ell+1}} \left| \sum_{j=\ell+1}^{v_k} (x_{n,j}-t)^m W_{m,j}(x) \right| dt$$

$$= \sum_{\ell=\mu_k}^{k-1} \int_{x_{n,\ell}}^{x_{n,\ell+1}} \left| \sum_{j=\mu_k}^{\ell} (x_{n,j}-t)^m W_{m,j}(x) \right| dt + \int_{x_{n,k}}^{x} \left| \sum_{j=\mu_k}^{k} (x_{n,j}-t)^m W_{m,j}(x) \right| dt$$

$$+ \int_{x}^{x_{n,k+1}} \left| \sum_{j=k+1}^{v_k} (x_{n,j}-t)^m W_{m,j}(x) \right| dt + \sum_{\ell=k+1}^{v_k} \int_{x_{n,\ell}}^{x_{n,\ell+1}} \left| \sum_{j=\ell+1}^{v_k} (x_{n,j}-t)^m W_{m,j}(x) \right| dt$$

$$\leqslant \left[\sum_{i=1}^{m} R_n^i \right]^m \left[\sum_{\ell=\mu_k}^{k-1} \sum_{j=\mu_k}^{\ell} \int_{x_{n,\ell}}^{x_{n,\ell+1}} (t-x_{n,j})^m dt + \sum_{j=\mu_k}^{k} \int_{x_{n,k}}^{x} (t-x_{n,j})^m dt \right.$$

$$\left. + \sum_{j=k+1}^{v_k} \int_{x}^{x_{n,k+1}} (x_{n,j}-t)^m dt + \sum_{\ell=k+1}^{v_k} \sum_{j=\ell+1}^{v_k} \int_{x_{n,\ell}}^{x_{n,\ell+1}} (x_{n,j}-t)^m dt \right]$$

$$= \left[\sum_{i=1}^{m} R_n^i \right]^m \left[\sum_{j=\mu_k}^{k-1} \sum_{\ell=j}^{k-1} \int_{x_{n,\ell}}^{x_{n,\ell+1}} (t-x_{n,j})^m dt + \sum_{j=\mu_k}^{k} \int_{x_{n,k}}^{x} (t-x_{n,j})^m dt \right.$$

$$\left. + \sum_{j=k+1}^{v_k} \int_{x}^{x_{n,k+1}} (x_{n,j}-t)^m dt + \sum_{j=k+1}^{v_k} \sum_{\ell=k+1}^{j-1} \int_{x_{n,\ell}}^{x_{n,\ell+1}} (x_{n,j}-t)^m dt \right]$$

$$= \left[\sum_{i=1}^{m} R_n^i \right]^m \left[\sum_{j=\mu_k}^{k-1} \int_{x_{n,j}}^{x_{n,k}} (t-x_{n,j})^m dt + \sum_{j=\mu_k}^{k} \frac{(x-x_{n,j})^{m+1} - (x_{n,k}-x_{n,j})^{m+1}}{m+1} \right.$$

$$\left. + \sum_{j=k+1}^{v_k} \frac{(x_{n,j}-x)^{m+1} - (x_{n,j}-x_{n,k+1})^{m+1}}{m+1} + \sum_{j=k+1}^{v_k} \int_{x_{n,k+1}}^{x_{n,j}} (x_{n,j}-t)^m dt \right]$$

$$< \frac{1}{m+1} \left[\sum_{i=1}^{m} R_n^i \right]^m \left[\sum_{j=\mu_k}^{k-1} (x_{n,k}-x_{n,j})^{m+1} + \sum_{j=\mu_k}^{k} (x_{n,k+1}-x_{n,j})^{m+1} \right.$$

$$+ \sum_{j=k+1}^{v_k} (x_{n,j} - x_{n,k})^{m+1} + \sum_{j=k+2}^{v_k} (x_{n,j} - x_{n,k+1})^{m+1} \Bigg]$$

$$\leqslant \frac{1}{m+1} \left[\sum_{i=1}^{m} R_n^i \right]^m (||\mathbf{x}_n||_\infty)^{m+1} \left[\sum_{j=\mu_k}^{k-1} (k-j)^{m+1} + \sum_{j=\mu_k}^{k} (k+1-j)^{m+1} \right.$$

$$\left. + \sum_{j=k+1}^{v_k} (j-k)^{m+1} + \sum_{j=k+2}^{v_k} (j-k-1)^{m+1} \right]$$

$$< \frac{2}{m+1} \left[\sum_{i=1}^{m} R_m^i \right]^m (||\mathbf{x}_n||_\infty)^{m+1} \left[\sum_{j=1}^{(k-\mu_k)+1} j^{m+1} + \sum_{j=1}^{v_k-k} j^{m+1} \right]$$

$$\leqslant \frac{2}{m+1} \left[\sum_{i=1}^{m} R_n^i \right]^m (||\mathbf{x}_n||_\infty)^{m+1} \left[\sum_{j=1}^{p+1} j^{m+1} + \sum_{j=1}^{q+1} j^{m+1} \right]. \qquad (10.6.70)$$

Since the right hand side of (10.6.70) is independent of k, we may now combine (10.6.65) and (10.6.70) to deduce the following convergence rate result.

Theorem 10.6.7. *For any integers $m \geqslant 1$ and $n \geqslant 2m+1$, the local spline interpolation operator $\mathscr{S}_{m,n}^{LI}$ of Theorem 10.6.5 satisfies the error bound*

$$||f - \mathscr{S}_{m,n}^{LI} f||_\infty \leqslant \frac{2 \left[\sum_{i=1}^{m} R^i \right]^m}{(m+1)!} \left[\sum_{j=1}^{p+1} j^{m+1} + \sum_{j=1}^{q+1} j^{m+1} \right] (||\mathbf{x}_n||_\infty)^{m+1} ||f^{(m+1)}||_\infty,$$

$$f \in C^{m+1}[a,b]. \quad (10.6.71)$$

As before, we observe that the error bound (10.6.71) is independent of the knot subsequence $\{\tau_{n,j} : j = 1,\ldots,nm-1\} \setminus \{\tau_{n,mj} : j = 1,\ldots,n-1\}$.

Example 10.6.2. For the case of equispaced interpolation points $\{x_{n,j} : j = 0,\ldots,n\}$ as in (10.6.56) in Example 10.6.1, we deduce from (10.6.71), (10.6.57) and (10.6.58) that

$$||f - \mathscr{S}_{m,n}^{LI} f||_\infty \leqslant \frac{2m^m}{(m+1)!} \left[\sum_{j=1}^{p+1} j^{m+1} + \sum_{j=1}^{q+1} j^{m+1} \right] \left(\frac{b-a}{n} \right)^{m+1} ||f^{(m+1)}||_\infty,$$

$$f \in C^{m+1}[a,b]. \quad (10.6.72)$$

In particular, for the quadratic spline case $m = 2$, and with, as in Example 10.5.1(b), the integers p and q chosen as in (10.5.54), that is, $p = q = 1$, it follows from (10.6.72) that

$$||f - \mathscr{S}_{2,n}^{LI} f||_\infty \leqslant 24 \left(\frac{b-a}{n} \right)^3 ||f'''||_\infty, \quad f \in C^3[a,b]. \qquad (10.6.73)$$

∎

10.7 Spline quadrature and the Gregory rule

According to (8.6.9) in Section 8.6, the composite Newton-Cotes quadrature rule $\mathcal{Q}_{\nu,n}^{NC}$ is obtained by approximating the integrand with a continuous, but non-smooth, piece-wise polynomial interpolant. In this section, we investigate the interpolatory quadrature rule obtained from approximating the integrand f by the local spline interpolant $\mathcal{S}_{m,n}^{LI} f \in C^{m-1}[a,b]$ of Theorem 10.5.8 for even m, as based on equispaced knots and interpolation points, and where the integers p and q are given as in (10.5.54). For positive integers $\mu \in \mathbb{N}$ and $n \geqslant 4\mu + 1$, we therefore let

$$m = 2\mu, \tag{10.7.1}$$

and, from (10.5.54) and (10.7.1), choose

$$p = q = \mu. \tag{10.7.2}$$

Also, for any bounded interval $[a,b]$, we define, as in (10.5.59), (10.5.60), the equi-spaced sequences

$$\tau_{n,j} := a + j\left(\frac{b-a}{2\mu n}\right), \quad j = 0,\ldots,2\mu n; \tag{10.7.3}$$

$$x_{n,j} := a + j\left(\frac{b-a}{n}\right), \quad j = 0,\ldots,n. \tag{10.7.4}$$

According to (10.5.88)–(10.5.91) in Theorem 10.5.8, the local spline interpolation operator $\mathcal{S}_{2\mu,n}^{LI} : C[a,b] \to \sigma_{2\mu}([a,b]; \tau_{n,1}, \ldots, \tau_{n,2\mu n - 1})$ then satisfies, for $n \geqslant 4\mu + 1$, the formulation

$$(\mathcal{S}_{2\mu,n}^{LI} f)(x) = \sum_{j=0}^{n} f(x_{n,j}) W_{2\mu,j}(x), \ x \in [a,b], \ f \in C[a,b], \tag{10.7.5}$$

where the splines $\{W_{2\mu,j} : j = 0,\ldots,n\}$ are given by

$$W_{2\mu,j}(x) = \begin{cases} \dfrac{(-1)^j}{(2\mu)!}\dbinom{2\mu}{j} \displaystyle\prod_{\substack{\ell=0 \\ j\neq\ell}}^{2\mu} \left(\dfrac{n}{b-a}(x-a)-\ell\right), & x \in [a,x_{n,\mu}], \\ V_{2\mu}\left(\dfrac{2\mu n}{b-a}(x-a)-2\mu j\right), & x \in (x_{n,\mu},b], \end{cases} \quad\left.\vphantom{\begin{cases}a\\b\end{cases}}\right\} j = 0,\ldots,2\mu; \tag{10.7.6}$$

$$W_{2\mu,j}(x) = V_{2\mu}\left(\frac{2\mu n}{b-a}(x-a)-2\mu j\right), \quad x \in [a,b],$$

$$j = 2\mu + 1, \ldots, n - 2\mu - 1 \ (\text{if } n \geqslant 4\mu + 2); \tag{10.7.7}$$

$$W_{2\mu,n-j}(x) = \begin{cases} \dfrac{(-1)^j}{(2\mu)!}\dbinom{2\mu}{j} \displaystyle\prod_{\substack{\ell=0 \\ j\neq\ell}}^{2\mu} \left(\dfrac{n}{b-a}(x-a)-(n-\ell)\right), & x \in [x_{n,n-\mu},b], \\ V_{2\mu}\left(\dfrac{2\mu n}{b-a}(x-a)-2\mu(n-j)\right), & x \in [a,x_{n,n-\mu}), \end{cases} \quad\left.\vphantom{\begin{cases}a\\b\end{cases}}\right\}$$

$$j = 0,\ldots,2\mu, \tag{10.7.8}$$

where, as in (10.5.62), (10.5.63), the spline $V_{2\mu}$ is given by

$$V_{2\mu}(x) = \sum_{j=-2\mu(\mu+1)}^{2\mu^2-1} \lambda_{2\mu,j} N_{2\mu}(x-j), \qquad (10.7.9)$$

with

$$\lambda_{2\mu,2\mu j+\rho} = \frac{(-1)^{\mu+1-j}\binom{2\mu}{\mu-1-j}}{[(2\mu)!]^2} \sum_{\substack{\{v_{-\mu+1},\ldots,v_{\mu+1}\}\setminus\{v_{-j}\}\\ \in \mathrm{per}\{1,\ldots,2\mu\}}} \prod_{-j\neq\ell=-\mu+1}^{\mu+1}\left(\frac{\rho+v_\ell}{2\mu}-\ell\right),$$

$$j=-\mu-1,\ldots,\mu-1; \ \rho=0,\ldots,2\mu-1, \qquad (10.7.10)$$

and where $N_{2\mu}$ denotes the cardinal B-spline of degree 2μ, as defined in (10.2.69) of Theorem 10.2.7, with $m=2\mu$.

The spline interpolatory quadrature rule $\mathscr{Q}_{2\mu,n}^{LI}$ for the numerical approximation of the integral

$$\int_a^b f(x)dx, \quad f \in C[a,b], \qquad (10.7.11)$$

is now defined, for $n \geqslant 4\mu+1$, by

$$\mathscr{Q}_{2\mu,n}^{LI}[f] := \int_a^b (\mathscr{S}_{2\mu,n}^{LI}f)(x)dx, \quad f \in C[a,b], \qquad (10.7.12)$$

with the local spline interpolation operator $\mathscr{S}_{2\mu,n}^{LI}: C[a,b] \to \sigma_{2\mu}([a,b];\tau_{n,1},\ldots,\tau_{n,2\mu n-1})$ given as in (10.7.5)–(10.7.10).

Observe from (10.7.12) and (10.7.5) that, for any $f \in C[a,b]$,

$$\mathscr{Q}_{2\mu,n}^{LI}[f] = \int_a^b \left[\sum_{j=0}^n f(x_{n,j})W_{2\mu,j}(x)\right] dx = \sum_{j=0}^n \left[\int_a^b W_{2\mu,j}(x)dx\right] f(x_{n,j}),$$

and thus

$$\mathscr{Q}_{2\mu,n}^{LI}[f] = \sum_{j=0}^n w_{2\mu,j} f(x_{n,j}), \quad f \in C[a,b], \qquad (10.7.13)$$

where the weights $\{w_{2\mu,j}: j=0,\ldots,n\}$ are given by

$$w_{2\mu,j} := \int_a^b W_{2\mu,j}(x)dx, \quad j=0,\ldots,n, \qquad (10.7.14)$$

with the splines $\{W_{2\mu,j}: j=0,\ldots,n\}$ defined as in (10.7.6)–(10.7.10).

We proceed to show that the quadrature rule $\mathscr{Q}_{2\mu,n}^{LI}$ is in fact a trapezoidal rule with endpoint corrections, as made precise below.

Theorem 10.7.1. *For any positive integers μ and n, with $n \geqslant 4\mu + 1$, the spline interpolatory quadrature rule $\mathcal{Q}_{2\mu,n}^{LI}$, as defined by (10.7.12), is a trapezoidal rule with endpoint corrections, in the sense that*

$$\mathcal{Q}_{2\mu,n}^{LI}[f] = \mathcal{Q}_n^{TR}[f] + \frac{b-a}{n} \sum_{j=0}^{2\mu} \gamma_{2\mu,j}[f(x_{n,j}) + f(x_{n,n-j})], \quad f \in C[a,b], \qquad (10.7.15)$$

with $\mathcal{Q}_n^{TR}$ denoting the trapezoidal rule, as given in (8.6.22), (8.6.21), and where

$$\gamma_{2\mu,j} := \begin{cases} \dfrac{1}{(2\mu)!} \displaystyle\int_0^\mu \prod_{\ell=1}^{2\mu} (t-\ell)\,dt + \int_\mu^{\mu+1} V_{2\mu}(t)\,dt - \dfrac{1}{2}, & j = 0, \\[6mm] \dfrac{(-1)^j}{(2\mu)!} \dbinom{2\mu}{j} \displaystyle\int_0^\mu \prod_{j \neq \ell = 0}^{2\mu} (t-\ell)\,dt + \int_\mu^{\mu+1+j} V_{2\mu}(t-2\mu j)\,dt - 1, & j = 1,\ldots,2\mu, \end{cases}$$

$$(10.7.16)$$

with $V_{2\mu}$ denoting the spline defined by (10.7.9), (10.7.10). Moreover, $\mathcal{Q}_{2\mu,n}^{LI}$ satisfies the polynomial exactness condition

$$\mathcal{Q}_{2\mu,n}^{LI}[P] = \int_a^b P(x)\,dx, \quad P \in \pi_{2\mu}. \qquad (10.7.17)$$

Proof. Suppose $f \in C[a,b]$, and, for $n \geqslant 4\mu + 2$, let the integer $j \in \{2\mu + 1, \ldots, n - 2\mu - 1\}$ be fixed. It follows from (10.7.14) and (10.7.7), as well as the second line of (10.5.64) in Theorem 10.5.7, together with (10.7.1) and (10.7.2), that

$$\begin{aligned} w_{2\mu,j} &= \int_a^b V_{2\mu}\left(\frac{2\mu n}{b-a}(x-a) - 2\mu j\right) dx \\ &= \frac{b-a}{n} \int_0^n V_{2\mu}(2\mu t - 2\mu j)\,dt \\ &= \frac{b-a}{n} \int_\mu^{n-\mu} V_{2\mu}(2\mu t - 2\mu j)\,dt \\ &= \frac{b-a}{n} \sum_{k=\mu}^{n-\mu-1} \int_k^{k+1} V_{2\mu}(2\mu t - 2\mu j)\,dt \\ &= \frac{b-a}{n} \sum_{k=\mu}^{n-\mu-1} \int_0^1 V_{2\mu}(2\mu t + 2\mu(k-j))\,dt \\ &= \frac{b-a}{n} \int_0^1 \sum_{k=\mu}^{n} V_{2\mu}(2\mu(t-j) + 2\mu k)\,dt \\ &= \frac{b-a}{n} \int_0^1 \sum_{k=-\infty}^{\infty} V_{2\mu}(2\mu(t-j) + 2\mu k)\,dt \\ &= \frac{b-a}{n} \int_0^1 \sum_{k=-\infty}^{\infty} V_{2\mu}(2\mu(t-j) - 2\mu k)\,dt = \frac{b-a}{n} \int_0^1 dt = \frac{b-a}{n}, \end{aligned}$$

that is,

$$w_{2\mu,j} = \frac{b-a}{n}, \quad j = 2\mu+1,\ldots,n-2\mu-1, \tag{10.7.18}$$

after having used also the fact, as obtained from (10.5.66), together with (10.7.1), that

$$\sum_{k=-\infty}^{\infty} V_{2\mu}(x-2\mu k) = 1, \quad x \in \mathbb{R}. \tag{10.7.19}$$

Next, for any $n \geqslant 4\mu+1$, we prove the symmetry result

$$W_{2\mu,n-j}(a+b-x) = W_{2\mu,j}(x), \quad x \in [a,b], \quad j = 0,\ldots,2\mu. \tag{10.7.20}$$

To prove (10.7.20), let $j \in \{0,\ldots,2\mu\}$ be fixed, and suppose first $x \in [a,x_{n,\mu}]$, so that, from (10.7.4), $a+b-x \in [x_{n,n-\mu},b]$, and thus, from the first line of (10.7.8),

$$W_{2\mu,n-j}(a+b-x) = \frac{(-1)^j}{(2\mu)!}\binom{2\mu}{j} \prod_{\substack{\ell=0\\j\neq\ell}}^{2\mu} \left(\frac{n}{b-a}(b-x)-(n-\ell)\right)$$

$$= \frac{(-1)^j}{(2\mu)!}\binom{2\mu}{j} \prod_{\substack{\ell=0\\j\neq\ell}}^{2\mu} \left(\frac{n}{b-a}(a-x)+\ell\right)$$

$$= \frac{(-1)^j}{(2\mu)!}\binom{2\mu}{j} \prod_{\substack{\ell=0\\j\neq\ell}}^{2\mu} \left(\frac{n}{b-a}(x-a)-\ell\right) = W_{2\mu,j}(x),$$

by virtue of the first line of (10.7.6), and thereby establishing (10.7.20) for $x \in [a,x_{n,\mu}]$. Next, for $x \in (x_{n,\mu},b]$, so that (10.7.4) yields $a+b-x \in [a,x_{n,n-\mu})$, we may apply the second line of (10.7.8), together with the symmetry property (10.5.68) of $V_{2\mu}$ in (iv) of Theorem 10.5.7, to deduce that

$$W_{2\mu,n-j}(a+b-x) = V_{2\mu}\left(\frac{2\mu n}{b-a}(b-x)-2\mu(n-j)\right)$$

$$= V_{2\mu}\left(\frac{2\mu n}{b-a}(a-x)+2\mu j\right) = V_{2\mu}\left(\frac{2\mu n}{b-a}(x-a)-2\mu j\right) = W_{2\mu,j}(x),$$

from the second line of (10.7.6), according to which (10.7.20) also holds for $x \in (x_{n,\mu},b]$, and thereby completing our proof of (10.7.20). By using (10.7.14) and (10.7.20), we deduce that, for any $j \in \{0,\ldots,2\mu\}$,

$$w_{2\mu,n-j} = \int_a^b W_{2\mu,n-j}(x)dx = \int_a^b W_{2\mu,n-j}(a+b-x)dx = \int_a^b W_{2\mu,j}(x)dx = w_{2\mu,j},$$

that is,

$$w_{2\mu,n-j} = w_{2\mu,j}, \quad j = 0,\ldots,2\mu. \tag{10.7.21}$$

Now use (10.7.14) and (10.7.6), together with (10.7.4), as well as the second line of (10.5.64) in Theorem 10.5.7, with (10.7.1) and (10.7.2), to obtain, for $j \in \{0, \ldots, 2\mu\}$,

$$w_{2\mu,j} = \frac{(-1)^j}{(2\mu)!} \binom{2\mu}{j} \int_a^{x_{n,\mu}} \prod_{\substack{\ell=0 \\ j\neq\ell}}^{2\mu} \left(\frac{n}{b-a}(x-a) - \ell \right) dx$$

$$+ \int_{x_{n,\mu}}^b V_{2\mu} \left(\frac{2\mu n}{b-a}(x-a) - 2\mu j \right) dx$$

$$= \frac{b-a}{n} \left[\frac{(-1)^j}{(2\mu)!} \binom{2\mu}{j} \int_0^\mu \prod_{\substack{\ell=0 \\ j\neq\ell}}^{2\mu} (t-\ell)\,dt + \int_\mu^{\mu+1+j} V_{2\mu}(t - 2\mu j)\,dt \right],$$

which, together with (10.7.13), (10.7.18) and (10.7.21), as well as the formulation in (8.6.22), (8.6.21) of the trapezoidal rule $\mathscr{Q}_n^{TR}$, then yields the desired result (10.7.15), (10.7.16). Finally, observe that the polynomial exactness condition (10.7.17) is an immediate consequence of the definition (10.7.12), together with (10.5.27) in Theorem 10.5.2, for $m = 2\mu$. ∎

In general, for integers $m \geqslant 0$ and $n \geqslant m$, and a given coefficient sequence $\{\gamma_0, \ldots, \gamma_m\} \subset \mathbb{R}$ which is independent of n, the quadrature rule

$$\mathscr{Q}_{m,n}^{CT}[f] := \mathscr{Q}_n^{TR}[f] + \frac{b-a}{n} \sum_{j=0}^m \gamma_j [f(x_{n,j}) + f(x_{n,n-j})], \quad f \in C[a,b], \tag{10.7.22}$$

with the points $\{x_{n,j} : j = 0, \ldots, n\}$ as in (10.7.4), and where $\mathscr{Q}_n^{TR}$ denotes the trapezoidal rule in (8.6.22), (8.6.21), is called a trapezoidal rule with endpoint corrections of order m, and for which we proceed to prove, by means of the Euler-Maclaurin formula, the following existence and uniqueness result if m is an even positive integer.

Theorem 10.7.2. *For integers $\mu \geqslant 0$ and $n \geqslant 2\mu$, there exists precisely one sequence $\{\gamma_0, \ldots, \gamma_{2\mu}\}$, which is independent of n, such that $\mathscr{Q}_{2\mu,n}^{CT}$, the corresponding trapezoidal rule with endpoint corrections of order 2μ, as defined by (10.7.22), satisfies the polynomial exactness condition*

$$\mathscr{Q}_{2\mu,n}^{CT}[P] = \int_a^b P(x)dx, \quad P \in \pi_{2\mu}. \tag{10.7.23}$$

Proof. First, we apply the Euler-Maclaurin formula (9.3.12) in Theorem 9.3.3, with $m = \mu$, to obtain the polynomial integral identity

$$\int_a^b P(x)dx = \mathscr{Q}_n^{TR}[P] - \sum_{j=1}^\mu \frac{B_{2j}}{(2j)!} \left[P^{(2j-1)}(b) - P^{(2j-1)}(a) \right] \left(\frac{b-a}{n} \right)^{2j},$$

$$P \in \pi_{2\mu}, \tag{10.7.24}$$

with $\{B_{2j} : j = 1, 2, \ldots\}$ denoting the Bernoulli numbers with even indexes, as defined recursively in (9.3.13). It follows from (10.7.24), together with the definition (10.7.22), that $\{\gamma_0, \ldots, \gamma_{2\mu}\} \subset \mathbb{R}$ is a sequence such that the polynomial exactness condition (10.7.23) is satisfied if and only if $\{\gamma_0, \ldots, \gamma_{2\mu}\}$ satisfies the condition

$$\sum_{j=0}^{2\mu} \gamma_j \left[P(x_{n,j}) + P(x_{n,n-j}) \right] = - \sum_{k=1}^{\mu} \frac{B_{2k}}{(2k)!} \left[P^{(2k-1)}(b) - P^{(2k-1)}(a) \right] \left(\frac{b-a}{n} \right)^{2k-1},$$

$$P \in \pi_{2\mu}. \quad (10.7.25)$$

Next, we apply the Taylor expansion polynomial identity (10.1.11), together with (10.7.4), to obtain, for any fixed $j \in \{0, \ldots, 2\mu\}$,

$$P(x_{n,j}) = \sum_{k=0}^{2\mu} \frac{P^{(k)}(a)}{k!} \left[j \left(\frac{b-a}{n} \right) \right]^k$$

$$= \sum_{k=0}^{\mu} \frac{P^{(2k)}(a)}{(2k)!} j^{2k} \left(\frac{b-a}{n} \right)^{2k} + \sum_{k=1}^{\mu} \frac{P^{(2k-1)}(a)}{(2k-1)!} j^{2k-1} \left(\frac{b-a}{n} \right)^{2k-1}, \quad (10.7.26)$$

and, similarly,

$$P(x_{n,n-j}) = P \left(b - j \left(\frac{b-a}{n} \right) \right)$$

$$= \sum_{k=0}^{2\mu} \frac{P^{(k)}(b)}{k!} \left[-j \left(\frac{b-a}{n} \right) \right]^k$$

$$= \sum_{k=0}^{\mu} \frac{P^{(2k)}(b)}{(2k)!} j^{2k} \left(\frac{b-a}{n} \right)^{2k} - \sum_{k=1}^{\mu} \frac{P^{(2k-1)}(b)}{(2k-1)!} j^{2k-1} \left(\frac{b-a}{n} \right)^{2k-1}. \quad (10.7.27)$$

It follows from (10.7.25), (10.7.26) and (10.7.27) that, for any sequence $\{\gamma_0, \ldots, \gamma_{2\mu}\} \in \mathbb{R}$,

$$\sum_{j=0}^{2\mu} \gamma_j \left[P(x_{n,j}) + P(x_{n,n-j}) \right]$$

$$= \sum_{k=0}^{\mu} \left[P^{(2k)}(a) + P^{(2k)}(b) \right] \left[\sum_{j=0}^{2\mu} \frac{j^{2k}}{(2k)!} \gamma_j \right] \left(\frac{b-a}{n} \right)^{2k}$$

$$- \sum_{k=1}^{\mu} \left[P^{(2k-1)}(b) - P^{(2k-1)}(a) \right] \left[\sum_{j=0}^{2\mu} \frac{j^{2k-1}}{(2k-1)!} \gamma_j \right] \left(\frac{b-a}{n} \right)^{2k-1},$$

according to which the condition (10.7.25) has the equivalent formulation

$$\sum_{k=0}^{\mu} \left[P^{(2k)}(a) + P^{(2k)}(b) \right] \left[\sum_{j=0}^{2\mu} \frac{j^{2k}}{(2k)!} \gamma_j \right] \left(\frac{b-a}{n} \right)^{2k} - \sum_{k=1}^{\mu} \left[P^{(2k-1)}(b) - P^{(2k-1)}(a) \right]$$

$$\times \left[\sum_{j=0}^{2\mu} \frac{j^{2k-1}}{(2k-1)!} \gamma_j - \frac{B_{2k}}{(2k)!} \right] \left(\frac{b-a}{n} \right)^{2k-1} = 0, \quad P \in \pi_{2\mu}. \quad (10.7.28)$$

We claim that $\{\gamma_0, \ldots, \gamma_{2\mu}\} \subset \mathbb{R}$ is a sequence that is independent of n, and such that the condition (10.7.28) is satisfied, if and only if $\{\gamma_0, \ldots, \gamma_{2\mu}\}$ is a solution of the $(2\mu + 1) \times (2\mu + 1)$ linear system

$$\sum_{j=0}^{2\mu} j^k \gamma_j = \begin{cases} \dfrac{B_{k+1}}{k+1}, & k = 1, 3, \ldots, 2\mu - 1; \\ 0, & k = 0, 2, \ldots, 2\mu. \end{cases} \tag{10.7.29}$$

To prove this statement, we observe first that if the sequence $\{\gamma_0, \ldots, \gamma_{2\mu}\} \subseteq \mathbb{R}$ satisfies (10.7.29), then (10.7.28) holds. Suppose next that $\{\gamma_0, \ldots, \gamma_{2\mu}\}$ is a sequence independent of n, and such that the condition (10.7.28) is satisfied. Let $P_\mu \in \pi_{2\mu}$ be defined by

$$P_\mu(x) := (x - a)^{2\mu}, \tag{10.7.30}$$

for which we have

$$\alpha_{\mu,k} := P_\mu^{(2k)}(a) + P_\mu^{(2k)}(b) = \begin{cases} (2k)! \dbinom{2\mu}{2k} (b - a)^{2\mu - 2k}, & k = 0, \ldots, \mu - 1; \\ 2(2\mu)!, & k = \mu, \end{cases} \tag{10.7.31}$$

and

$$\beta_{\mu,k} := P_\mu^{(2k-1)}(b) - P_\mu^{(2k-1)}(a) = (2k - 1)! \binom{2\mu}{2k - 1} (b - a)^{2\mu - 2k + 1}, \quad k = 1, \ldots, \mu. \tag{10.7.32}$$

Observe from (10.7.31) and (10.7.32) that the sequences $\{\alpha_{\mu,k} : k = 0, \ldots, \mu\}$ and $\{\beta_{\mu,k} : k = 1, \ldots, \mu\}$ are independent of n, with also

$$\left. \begin{array}{l} \alpha_{\mu,k} \neq 0, \quad k = 0, \ldots, \mu; \\ \beta_{\mu,k} \neq 0, \quad k = 1, \ldots, \mu. \end{array} \right\} \tag{10.7.33}$$

Let

$$\left. \begin{array}{ll} A_{\mu,k} := \alpha_{\mu,k} \displaystyle\sum_{j=0}^{2\mu} \dfrac{j^{2k}}{(2k)!} \gamma_j, & k = 0, \ldots, \mu; \\ B_{\mu,k} := \beta_{\mu,k} \displaystyle\sum_{j=0}^{2\mu} \left[\dfrac{j^{2k-1}}{(2k-1)!} \gamma_j - \dfrac{B_{2k}}{(2k)!} \right], & k = 1, \ldots, \mu, \end{array} \right\} \tag{10.7.34}$$

according to which, since the sequences $\{\alpha_{\mu,k} : k = 0, \ldots, \mu\}$, $\{\beta_{\mu,k} : k = 1, \ldots, \mu\}$ and $\{\gamma_0, \ldots, \gamma_{2\mu}\}$ are independent of n, it follows that the sequences $\{A_{\mu,k} : k = 0, \ldots, \mu\}$ and $\{B_{\mu,k} : k = 1, \ldots, \mu\}$ are independent of n.

By choosing $P = P_\mu \in \pi_{2\mu}$ in (10.7.28), we deduce from (10.7.31), (10.7.32) and (10.7.34) that

$$\sum_{k=0}^{\mu} A_{\mu,k} \left(\frac{b-a}{n} \right)^{2k} - \sum_{k=1}^{\mu} B_{\mu,k} \left(\frac{b-a}{n} \right)^{2k-1} = 0, \quad n = 2\mu, 2\mu + 1, \ldots,$$

and thus

$$n^{2\mu}\left[\sum_{k=0}^{\mu}A_{\mu,k}\left(\frac{b-a}{n}\right)^{2k}-\sum_{k=1}^{\mu}B_{\mu,k}\left(\frac{b-a}{n}\right)^{2k-1}\right]=0,\quad n=2\mu,2\mu+1,\dots,$$

or equivalently,

$$\sum_{k=0}^{\mu}A_{\mu,k}(b-a)^{2k}n^{2\mu-2k}-\sum_{k=1}^{\mu}B_{\mu,k}(b-a)^{2k-1}n^{2\mu-2k+1}=0,\ n=2\mu,2\mu+1,\dots,$$

and thus, since the zero polynomial is the only polynomial with infinitely many distinct zeros, it holds that

$$\left.\begin{array}{ll}A_{\mu,k}(b-a)^{2k}=0,&k=0,\dots,\mu;\\[2mm]B_{\mu,k}(b-a)^{2k-1}=0,&k=1,\dots,\mu.\end{array}\right\}\qquad(10.7.35)$$

By recalling also (10.7.34) and (10.7.33), we deduce from (10.7.35) that the sequence $\{\gamma_0,\dots,\gamma_{2\mu}\}$ satisfies the conditions

$$\left.\begin{array}{ll}\displaystyle\sum_{j=0}^{2\mu}j^{2k}\gamma_j\ =0,&k=0,\dots,\mu;\\[4mm]\displaystyle\sum_{j=0}^{2\mu}j^{2k-1}\gamma_j=\dfrac{B_{2k}}{2k},&k=1,\dots,\mu,\end{array}\right\}$$

which is equivalent to (10.7.29), and thereby completing our proof of the equivalence of (10.7.28) and (10.7.29).

Now observe that the $(2\mu+1)\times(2\mu+1)$ coefficient matrix A of the linear system (10.7.29) is given by

$$A:=\begin{bmatrix}1&1&1&\cdots&1\\0&1&2&\cdots&2\mu\\0&1^2&2^2&\cdots&(2\mu)^2\\\vdots&&&&\\0&1^{2\mu}&2^{2\mu}&\cdots&(2\mu)^{2\mu}\end{bmatrix}.\qquad(10.7.36)$$

Note from (10.7.36) and (1.1.7) that the transpose A^T of A is a Vandermonde matrix. We may therefore apply Theorem 1.1.2 to deduce that A^T is an invertible matrix. Hence $A=(A^T)^T$ is an invertible matrix, and it follows that there exists a unique solution $\{\gamma_0,\dots,\gamma_{2\mu}\}$ of the linear system (10.7.29), which completes our proof. ∎

The (unique) trapezoidal rule $\mathscr{Q}_{2\mu,n}^{CT}$ with endpoint corrections of order 2μ, and satisfying the polynomial exactness condition (10.7.23), as established in Theorem 10.7.2, is called the Gregory rule of (even) order 2μ, and will be denoted here by the symbol $\mathscr{Q}_{2\mu,n}^{GR}$. By observing from (10.7.16) in Theorem 10.7.1 that the sequence $\{\gamma_{2\mu,j} : j = 0,\dots,2\mu\}$ is independent of n, we immediately deduce the following result from Theorems 10.7.1 and 10.7.2, together with the definition (10.7.12).

Theorem 10.7.3. *For positive integers μ and $n \geqslant 4\mu + 1$, the Gregory rule $\mathscr{Q}_{2\mu,n}^{GR}$ of order 2μ, as established in Theorem 10.7.2, is an interpolatory quadrature rule, with*

$$\mathscr{Q}_{2\mu,n}^{GR}[f] = \int_a^b (\mathscr{S}_{2\mu,n}^{LI} f)(x)dx =: \mathscr{Q}_{2\mu,n}^{LI}[f], \quad f \in C[a,b], \tag{10.7.37}$$

where $\mathscr{Q}_{2\mu,n}^{LI}$ denotes the spline interpolatory quadrature rule defined in (10.7.12), that is, $\mathscr{Q}_{2\mu,n}^{GR}$ is the trapezoidal rule with endpoint corrections of order 2μ, as given by

$$\mathscr{Q}_{2\mu,n}^{GR}[f] = \mathscr{Q}_n^{TR}[f] + \frac{b-a}{n} \sum_{j=0}^{2\mu} \gamma_{2\mu,j}[f(x_{n,j}) + f(x_{n,n-j})], \quad f \in C[a,b], \tag{10.7.38}$$

with the sequence $\{\gamma_{2\mu,j} : j = 0,\dots,2\mu\}$ defined by (10.7.16), and where the points $\{x_{n,j} : j = 0,\dots,n\}$ are given by (10.7.4).

According to (10.7.37) in Theorem 10.7.3, we have shown that, for $n \geqslant 4\mu + 1$, the Gregory rule $\mathscr{Q}_{2\mu,n}^{GR}$ is an interpolatory quadrature rule, as obtained, for any integrand $f \in C[a,b]$, by integrating the local spline interpolant $\mathscr{S}_{2\mu,n}^{LI} f$ of Theorem 10.5.7. Indeed, the coefficients $\{\gamma_{2\mu,j} : j = 0,\dots,2\mu\}$ in (10.7.37) may, for any given positive integers μ and $n \geqslant 4\mu + 1$, be computed by means of the formulations in (10.7.16). We proceed to show how the formula (8.4.38) in Theorem 8.4.1 can be used to establish a more efficient computational method for $\mathscr{Q}_{2\mu,n}^{GR}$, for $n \geqslant 2\mu$. In particular, we shall express the sequence $\{\gamma_{2\mu,j} : j = 0,\dots,2\mu\}$ explicitly in terms of the Laplace coefficients $\{\Lambda_2,\Lambda_3,\dots,\Lambda_{2\mu+1}\}$, as defined in (8.4.37). We shall rely on the following integral polynomial identity.

Theorem 10.7.4. *For any positive integers μ and $n \geqslant 2\mu$, it holds that*

$$\int_a^b P(x)dx = \mathscr{Q}_n^{TR}[P] + \frac{b-a}{n} \sum_{j=0}^{2\mu} \widetilde{\gamma}_{2\mu,j}[P(x_{n,j}) + P(x_{n,n-j})], \quad P \in \pi_{2\mu}, \tag{10.7.39}$$

where

$$\widetilde{\gamma}_{2\mu,j} := \begin{cases} -\sum_{k=1}^{2\mu} \Lambda_{k+1}, & j = 0; \\ (-1)^{j-1} \sum_{k=j}^{2\mu} \binom{k}{j} \Lambda_{k+1}, & j = 1,\dots,2\mu, \end{cases} \tag{10.7.40}$$

with $\mathscr{Q}_n^{TR}$ denoting the trapezoidal rule as in (8.6.22), (8.6.21), where $\{\Lambda_2,\Lambda_3,\dots,\Lambda_{2\mu+1}\}$ are the Laplace coefficients defined in (8.4.37), and with the points $\{x_{n,j} : j = 0,\dots,n\}$ given as in (10.7.4).

Proof. Let $P \in \pi_{2\mu}$. Since $n \geqslant 2\mu$, we have $\pi_{2\mu} \subset \pi_n$, and thus, from the identity (1.2.7) in Theorem 1.2.3, we have

$$P(x) = \sum_{j=0}^{n} P(x_{n,j}) L_{n,j}(x), \qquad (10.7.41)$$

with $\{L_{n,j} : j = 0,\ldots,n\}$ denoting the Lagrange fundamental polynomials, as given in (1.2.1). Hence we may apply the formula (8.4.38) in Theorem 8.4.1, together with (8.4.7), to deduce from (10.7.41) that

$$\int_a^b P(x)dx = \sum_{j=0}^{n} P(x_{n,j}) \left[\int_a^b L_{n,j}(x)dx \right]$$

$$= \sum_{j=0}^{n} P(x_{n,j}) w_{n,j}$$

$$= \frac{b-a}{n} \left[\sum_{j=0}^{n} P(x_{n,j}) \left\{ 1 - (-1)^j \sum_{k=0}^{n} \Lambda_{k+1} \left(\binom{k}{j} + (-1)^n \binom{k}{n-j} \right) \right\} \right]$$

$$= \frac{b-a}{n} \left[\sum_{j=0}^{n} P(x_{n,j}) - \sum_{j=0}^{n} (-1)^j P(x_{n,j}) \sum_{k=j}^{n} \binom{k}{j} \Lambda_{k+1} \right.$$

$$\left. - \sum_{j=0}^{n} (-1)^j P(x_{n,n-j}) \sum_{k=j}^{n} \binom{k}{j} \Lambda_{k+1} \right]$$

$$= \frac{b-a}{n} \left[\sum_{j=0}^{n} P(x_{n,j}) - \sum_{k=0}^{n} \Lambda_{k+1} \sum_{j=0}^{k} (-1)^j \binom{k}{j} \{P(x_{n,j}) + P(x_{n,n-j})\} \right].$$

$$(10.7.42)$$

Suppose $n \geqslant 2\mu + 1$, let $k \in \{2\mu + 1,\ldots,n\}$ be fixed, and define the polynomial

$$\widetilde{P}(x) := P(x) + P(a+b-x), \qquad (10.7.43)$$

so that, since $P \in \pi_{2\mu}$, we have $\widetilde{P} \in \pi_{2\mu}$. By applying (3.4.2) in Theorem 3.4.1, and using (10.7.4), we obtain the divided difference

$$\widetilde{P}[x_{n,0},\ldots,x_{n,k}] = \frac{(-1)^k}{k!} \left(\frac{n}{b-a} \right)^k \sum_{j=0}^{k} (-1)^j \binom{k}{j} \widetilde{P}(x_{n,j}),$$

and thus, by using also (10.7.43) and (10.7.4), we deduce that

$$\sum_{j=0}^{k} (-1)^j \binom{k}{j} \{P(x_{n,j}) + P(x_{n,n-j})\} = (-1)^k k! \left(\frac{b-a}{n} \right)^k \widetilde{P}[x_{n,0},\ldots,x_{n,k}] = 0, \quad (10.7.44)$$

from (2.1.10) in Theorem 2.1.2, together with $k \geqslant 2\mu + 1$, and the fact that $\widetilde{P} \in \pi_{2\mu}$. Hence we may use (10.7.44) in (10.7.42) to obtain, for any integer $n \geqslant 2\mu$,

$$\int_a^b P(x)dx = \frac{b-a}{n}\left[\sum_{j=0}^n P(x_{n,j}) - \sum_{k=0}^{2\mu} \Lambda_{k+1} \sum_{j=0}^k (-1)^j \binom{k}{j}\{P(x_{n,j}) + P(x_{n,n-j})\}\right]$$

$$= \frac{b-a}{n}\left[\sum_{j=0}^n P(x_{n,j}) - \sum_{j=0}^{2\mu} (-1)^j \left\{\sum_{k=j}^{2\mu}\binom{k}{j}\Lambda_{k+1}\right\}\{P(x_{n,j}) + P(x_{n,n-j})\}\right].$$

$$(10.7.45)$$

Since (8.4.54) gives $\Lambda_1 = \frac{1}{2}$, it then follows from (10.7.45), together with (8.6.22), (8.6.21), that (10.7.39), (10.7.40) does indeed hold. ∎

It follows from (10.7.39) in Theorem 10.7.4 that, for any positive integers μ and $n \geqslant 2\mu$, the quadrature rule $\widetilde{\mathscr{D}}_{2\mu,n}$ defined by

$$\widetilde{\mathscr{D}}_{2\mu,n}[f] := \mathscr{Q}_n^{TR}[f] + \frac{b-a}{n}\sum_{j=0}^{2\mu} \widetilde{\gamma}_{\mu,j}[f(x_{n,j}) + f(x_{n,n-j})], \quad f \in C[a,b], \quad (10.7.46)$$

with the sequence $\{\widetilde{\gamma}_{\mu,j} : j = 0,\ldots,2\mu\}$ defined by (10.7.40), satisfies the polynomial exactness condition

$$\widetilde{\mathscr{D}}_{2\mu,n}[P] = \int_a^b P(x)dx, \quad P \in \pi_{2\mu}. \quad (10.7.47)$$

By observing also from (10.7.40) that the coefficient sequence $\{\widetilde{\gamma}_{\mu,j} : j = 0,\ldots,2\mu\}$ is independent of n, we may deduce from the uniqueness statement in Theorem 10.7.2, together with Theorem 10.7.3, that

$$\widetilde{\mathscr{D}}_{2\mu,n} = \mathscr{Q}_{2\mu,n}^{GR}, \quad (10.7.48)$$

and thus, from (10.7.46), (10.7.40) and (10.7.38), we have now established the following explicit formulation.

Theorem 10.7.5. *For any positive integers μ and $n \geqslant 2\mu$, the Gregory rule $\mathscr{Q}_{2\mu,n}^{GR}$, as established in Theorem 10.7.2, satisfies the formulation*

$$\mathscr{Q}_{2\mu,n}^{GR}[f] = \mathscr{Q}_n^{TR}[f] + \frac{b-a}{n}\sum_{j=0}^{2\mu} \gamma_{\mu,j}[f(x_{n,j}) + f(x_{n,n-j})], \quad f \in C[a,b], \quad (10.7.49)$$

where

$$\gamma_{\mu,j} := \begin{cases} -\sum_{k=1}^{2\mu} \Lambda_{k+1}, & j = 0; \\ (-1)^{j-1}\sum_{k=j}^{2\mu}\binom{k}{j}\Lambda_{k+1}, & j = 1,\ldots,2\mu, \end{cases} \quad (10.7.50)$$

with $\{\Lambda_2,\Lambda_3,\ldots,\Lambda_{2\mu+1}\}$ denoting the Laplace coefficients given in (8.4.37).

By using the formula (10.7.50), together with the Laplace coefficient values in (8.4.54), we calculate (see Exercise 10.68), for $\mu = 1, 2, 3$, the coefficient values for $\{\gamma_{2\mu} : j = 0, \ldots, 2\mu\}$ as given in Table 10.7.1.

Table 10.7.1   The coefficients $\{\gamma_{2\mu,j} : j = 0, \ldots, 2\mu\}$ in the formulation (10.7.49) of the Gregory rule $\mathscr{Q}_{2\mu,n}^{GR}$, for $\mu = 1, 2, 3$.

μ	$\{\gamma_{2\mu,j}\}$
1	$\{-\frac{1}{8}, \frac{1}{6}, -\frac{1}{24}\}$
2	$\{-\frac{49}{288}, \frac{77}{240}, -\frac{7}{30}, \frac{73}{720}, -\frac{3}{160}\}$
3	$\{-\frac{3383}{17280}, \frac{6961}{15120}, -\frac{66109}{120960}, \frac{33}{70}, -\frac{31523}{120960}, \frac{1247}{15120}, -\frac{275}{24192}\}$

For the choice $\mu = 1$ in Theorem 10.7.5, the corresponding Gregory rule $\mathscr{Q}_{2,n}^{GR}$ of order 2 is known as the Lacroix rule $\mathscr{Q}_n^{LA}$, that is,

$$\mathscr{Q}_n^{LA} := \mathscr{Q}_{2,n}^{GR}, \quad n = 2, 3, \ldots. \tag{10.7.51}$$

With the notation (8.6.21), that is,

$$f_{n,j} := f(x_{n,j}), \quad j = 0, \ldots, n, \tag{10.7.52}$$

it follows from (10.7.51), (10.7.49), (10.7.50), Table 10.7.1 and (8.6.22) that

$$\mathscr{Q}_n^{LA}[f] = \frac{b-a}{n} \left[\frac{3}{8} f_{n,0} + \frac{7}{6} f_{n,1} + \frac{23}{24} f_{n,2} + f_{n,3} + f_{n,4} + \cdots + f_{n,n-3} + \frac{23}{24} f_{n,n-2} \right.$$

$$\left. + \frac{7}{6} f_{n,n-1} + \frac{3}{8} f_{n,n} \right],$$

$$f \in C[a,b], \quad n = 6, 7 \ldots. \tag{10.7.53}$$

Example 10.7.1. For the numerical approximation of the integral $\int_0^2 f(x)dx$, it follows from (10.7.53), together with (10.7.52) and (10.7.4), that

$$\mathscr{Q}_{12}^{LA}[f] = \frac{1}{6} \left[\frac{3}{8} f(0) + \frac{7}{6} f(\tfrac{1}{6}) + \frac{23}{24} f(\tfrac{1}{3}) + f(\tfrac{1}{2}) + f(\tfrac{2}{3}) + f(\tfrac{5}{6}) + f(1) + f(\tfrac{7}{6}) + f(\tfrac{4}{3}) \right.$$

$$\left. + f(\tfrac{3}{2}) + \frac{23}{24} f(\tfrac{5}{3}) + \frac{7}{6} f(\tfrac{11}{6}) + \frac{3}{8} f(2) \right],$$

which yields, for the case $f(x) = e^x$, the quadrature error

$$\left| \int_0^2 f(x)dx - \mathscr{Q}_{12}^{LA}[f] \right| \approx 1.089 \times 10^{-4}.$$

Observe from Theorem 10.7.5 that, apart from the condition $n \geqslant 2\mu$, the Gregory rule $\mathscr{Q}^{GR}_{2\mu,n}$ does not require the integer n to satisfy any further constraints, as is the case for the composite Newton-Cotes quadrature rule $\mathscr{Q}^{NC}_{\nu,n}$ in Theorem 8.6.1, where it is required that n satisfies the divisibility condition (8.6.3).   ∎

We proceed to analyze the Gregory rule quadrature error

$$\mathscr{E}^{GR}_{2\mu,n}[f] := \int_a^b f(x)dx - \mathscr{Q}^{GR}_{2\mu,n}[f], \quad f \in C[a,b]. \tag{10.7.54}$$

To this end, we first observe from (10.7.54), together with (10.7.37) in Theorem 10.7.3, that, for any $f \in C[a,b]$, we have

$$|\mathscr{E}^{GR}_{2\mu,n}[f]| = \left| \int_a^b f(x)dx - \int_a^b (\mathscr{S}^{LI}_{2\mu,n}f)(x)dx \right|$$

$$= \left| \int_a^b [f(x) - (\mathscr{S}^{LI}_{2\mu,n}f)(x)]dx \right| \leqslant (b-a) \max_{a \leqslant x \leqslant b} |f(x) - (\mathscr{S}^{LI}_{2\mu,n}f)(x)|,$$

that is,

$$|\mathscr{E}^{GR}_{2\mu,n}[f]| \leqslant (b-a)\|f - \mathscr{S}^{LI}_{2\mu,n}\|_\infty, \quad f \in C[a,b]. \tag{10.7.55}$$

By using also the fact that, according to (10.7.4), the results (10.6.57) and (10.6.58) are satisfied, the following convergence result and quadrature error estimates are immediate consequences of (10.6.45) in Theorem 10.6.5, as well as (10.6.59) and (10.6.60) in Theorem 10.6.6, and (10.6.71) in Theorem 10.6.7, with, from (10.7.1) and (10.7.2), $m = 2\mu$ and $p = q = \mu$, together with (10.7.55) and (10.7.54).

Theorem 10.7.6. *The Gregory rule* $\mathscr{Q}^{GR}_{2\mu,n}$ *of Theorem 10.7.5 satisfies:*

(a) *The convergence result*

$$\left| \int_a^b f(x)dx - \mathscr{Q}^{GR}_{2\mu,n}[f] \right| \to 0, \quad n \to \infty, \quad f \in C[a,b]. \tag{10.7.56}$$

(b) *The quadrature error estimates*

(i)

$$\left| \int_a^b f(x)dx - \mathscr{Q}^{GR}_{2\mu,n}[f] \right| \leqslant (b-a)2\mu(2\mu)^{2\mu} \left(\frac{b-a}{n} \right) K_f,$$

$$f \in C^{\text{Lip}}[a,b], \quad n = 4\mu+1, 4\mu+2, \ldots, \tag{10.7.57}$$

with K_f *denoting a Lipschitz constant on* $[a,b]$ *of* f;

(ii)

$$\left| \int_a^b f(x)dx - \mathscr{Q}^{GR}_{2\mu,n}[f] \right| \leqslant (b-a)2\mu(2\mu)^{2\mu} \left(\frac{b-a}{n} \right) \|f'\|_\infty,$$

$$f \in C^1[a,b], \quad n = 4\mu+1, 4\mu+2, \ldots; \tag{10.7.58}$$

(iii)

$$\left| \int_a^b f(x)dx - \mathscr{Q}_{2\mu,n}^{GR}[f] \right| \leqslant (b-a)\frac{4(2\mu)^{2\mu}}{(2\mu+1)!}\left[\sum_{j=1}^{\mu+1} j^{2\mu+1}\right]\left(\frac{b-a}{n}\right)^{2\mu+1}||f^{(2\mu+1)}||_\infty,$$

$$f \in C^{2\mu+1}[a,b], \quad n = 4\mu+1, 4\mu+2, \ldots. \qquad (10.7.59)$$

Example 10.7.2. For the Lacroix rule $\mathscr{Q}_n^{LA}$, as defined in (10.7.51), it follows from (10.7.59) in Theorem 10.7.6 that

$$\left| \int_a^b f(x)dx - \mathscr{Q}_n^{LA}[f] \right| \leqslant 24(b-a)\left(\frac{b-a}{n}\right)^3||f'''||_\infty,$$

$$f \in C^3[a,b], \quad n = 5, 6, \ldots. \qquad (10.7.60)$$

∎

Finally, we show how the result of Theorem 8.5.6 can be used to establish, in Theorem 10.7.7 below, and analogously to (8.6.30) in the composite Newton-Cotes case, an expression in terms of the Laplace coefficients $\{\Lambda_1, \Lambda_2, \ldots, \Lambda_{2\mu+1}\}$ for the quadrature error $\mathscr{E}_{2\mu,n}^{GR}$, for integrands $f \in C^{2\mu+2}[a,b]$, and which will then immediately yield the degree of exactness of $\mathscr{Q}_{2\mu,n}^{GR}$, as well as an optimal quadrature error estimate for $\mathscr{E}_{2\mu,n}^{GR}$.

Theorem 10.7.7. *For positive integers μ and $n \geqslant 2\mu$, let $f \in C^{2\mu+2}[a,b]$, and denote by $\mathscr{Q}_{2\mu,n}^{GR}$ the Gregory rule of order 2μ, as given by (10.7.49), (10.7.50) in Theorem 10.7.5. Then the corresponding quadrature error $\mathscr{E}_{2\mu,n}^{GR}[f]$, as given by (10.7.54), satisfies*

$$\mathscr{E}_{2\mu,n}^{GR}[f] = -\left(\frac{b-a}{n}\right)^{2\mu+3}\left[(n-2\mu-1)\Lambda_{2\mu+2} + 2\Lambda_{2\mu+3}\right]f^{(2\mu+2)}(\xi),$$

$$f \in C^{2\mu+2}[a,b], \qquad (10.7.61)$$

for some point $\xi \in [a,b]$, and where the Laplace coefficients $\Lambda_{2\mu+2}$ and $\Lambda_{2\mu+3}$ are defined as in (8.4.37).

Proof. Let $f \in C^{2\mu+2}[a,b]$. Suppose first $n \geqslant 2\mu+1$, and let $j \in \{0, \ldots, n-2\mu-1\}$ and $x \in [x_{n,j}, x_{n,j+1}]$ be fixed. By applying the Newton interpolation formula (1.3.17) in Theorem 1.3.3, together with (2.1.3) and the error expression (2.1.5) in Theorem 2.1.1, as well as the definitions (1.3.11), (2.1.6) and (10.7.4), we obtain

$$f(x) = f(x_{n,j}) + \sum_{k=1}^{2\mu+1} f[x_{n,j}, \ldots, x_{n,j+k}]\prod_{\ell=0}^{k-1}(x - x_{n,j+\ell})$$

$$+ f[x, x_{n,j}, \ldots, x_{n,j+2\mu+1}]\prod_{\ell=0}^{2\mu+1}(x - x_{n,j+\ell}). \qquad (10.7.62)$$

Now use the definitions (8.4.34) and (8.4.37), as well as the recursion formula (1.3.21) in
Theorem 1.3.4, and the value $\Lambda_1 = \frac{1}{2}$ from (8.4.54), to obtain

$$
\int_{x_{n,j}}^{x_{n,j+1}} \left[f(x_{n,j}) + \sum_{k=1}^{2\mu+1} f[x_{n,j},\ldots,x_{n,j+k}] \prod_{\ell=0}^{k-1} (x - x_{n,j+\ell}) \right] dx
$$

$$
= \frac{b-a}{n} \left[f(x_{n,j}) + \sum_{k=1}^{2\mu+1} f[x_{n,j},\ldots,x_{n,j+k}] \left(\frac{b-a}{n}\right)^k \int_0^1 \prod_{\ell=0}^{k-1} (t - \ell)dt \right]
$$

$$
= \frac{b-a}{n} \left[f(x_{n,j}) - \sum_{k=1}^{2\mu+1} (-1)^k k! f[x_{n,j},\ldots,x_{n,j+k}] \left(\frac{b-a}{n}\right)^k \Lambda_k \right]
$$

$$
= \frac{b-a}{n} \left[f(x_{n,j}) + \frac{1}{2}\{f(x_{n,j+1}) - f(x_{n,j})\} - \sum_{k=2}^{2\mu+1} (-1)^k (k-1)! \Lambda_k \left(\frac{b-a}{n}\right)^{k-1} \right.
$$

$$
\left. \times \{f[x_{n,j+1},\ldots,x_{n,j+k}] - f[x_{n,j},\ldots,x_{n,j+k-1}]\} \right]
$$

$$
= \frac{b-a}{n} \left[\frac{1}{2}f(x_{n,j}) + \frac{1}{2}f(x_{n,j+1}) + \sum_{k=1}^{2\mu} (-1)^k k! \Lambda_{k+1} \left(\frac{b-a}{n}\right)^k \right.
$$

$$
\left. \times (f[x_{n,j+1},\ldots,x_{n,j+k+1}] - f[x_{n,j},\ldots,x_{n,j+k}]) \right]. \qquad (10.7.63)
$$

Next, since the polynomial $\prod_{\ell=0}^{2\mu+1} (x - x_{n,j+\ell})$ does not change sign in the interval
$(x_{n,j}, x_{n,j+1})$, we may apply Theorem 1.4.4, together with the mean value theorem for integrals, as formulated in (8.5.16), as well as (2.1.10) in Theorem 2.1.2, and the definitions
(8.4.34), (8.4.37) and (10.7.4), to deduce the existence of points $\eta_j \in (x_{n,j}, x_{n,j+1})$ and
$\xi_j \in [x_{n,j}, x_{n,j+2\mu+1}]$ such that

$$
\int_{x_{n,j}}^{x_{n,j+1}} f[x, x_{n,j},\ldots,x_{n,j+2\mu+1}] \prod_{\ell=0}^{2\mu+1} (x - x_{n,j+\ell})dx
$$

$$
= f[\eta_j, x_{n,j},\ldots,x_{n,j+2\mu+1}] \int_{x_{n,j}}^{x_{n,j+1}} \prod_{\ell=0}^{2\mu+1} (x - x_{n,j+\ell})dx
$$

$$
= \frac{f^{(2\mu+2)}(\xi_j)}{(2\mu+2)!} \left(\frac{b-a}{n}\right)^{2\mu+3} \int_0^1 \prod_{\ell=0}^{2\mu+1} (t - \ell)dt
$$

$$
= -\left(\frac{b-a}{n}\right)^{2\mu+3} \Lambda_{2\mu+2} f^{(2\mu+2)}(\xi_j). \qquad (10.7.64)
$$

It follows from (10.7.62), (10.7.63) and (10.7.64), together with the formula (3.4.2) in
Theorem 3.4.1, as well as an application of the intermediate value theorem as in the steps
leading to (8.6.29), that, for some point $\widetilde{\xi} \in [a,b]$,

$$\int_a^{x_{n,n-2\mu}} f(x)dx = \sum_{j=0}^{n-2\mu-1} \int_{x_{n,j}}^{x_{n,j+1}} f(x)dx$$

$$= \frac{b-a}{n}\left[\frac{1}{2}\sum_{j=0}^{n-2\mu-1}\{f(x_{n,j})+f(x_{n,j+1})\}+\sum_{k=1}^{2\mu}(-1)^k k!\Lambda_{k+1}\left(\frac{b-a}{n}\right)^k\right.$$

$$\times\{f[x_{n,n-2\mu},\ldots,x_{n,n-2\mu+k}]-f[x_{n,0},\ldots,x_{n,k}]\}-\left(\frac{b-a}{n}\right)^{2\mu+2}\Lambda_{2\mu+2}\sum_{j=0}^{n-2\mu-1}f^{(2\mu+2)}(\xi_j)\Big]$$

$$= \frac{b-a}{n}\left[\frac{1}{2}f(x_{n,0})+\sum_{j=1}^{n-2\mu-1}f(x_{n,j})+\frac{1}{2}f(x_{n,n-2\mu})+\sum_{k=1}^{2\mu}\Lambda_{k+1}\left\{\sum_{\ell=0}^k(-1)^\ell\binom{k}{\ell}\right.\right.$$

$$\left.\times(f(x_{n,n-2\mu+\ell})-f(x_{n,\ell}))\right\}-\left(\frac{b-a}{n}\right)^{2\mu+2}(n-2\mu)\Lambda_{2\mu+2}f^{(2\mu+2)}(\widetilde{\xi})\Big]$$

$$= \frac{b-a}{n}\left[\frac{1}{2}f(x_{n,0})+\sum_{j=1}^{n-2\mu-1}f(x_{n,j})+\frac{1}{2}f(x_{n,n-2\mu})+\left(\sum_{k=1}^{2\mu}\Lambda_{k+1}\right)\right.$$

$$\times\{f(x_{n,n-2\mu})-f(x_{n,0})\}+\sum_{k=1}^{2\mu}\Lambda_{k+1}\left\{\sum_{\ell=1}^k(-1)^\ell\binom{k}{\ell}(f(x_{n,n-2\mu+\ell})-f(x_{n,\ell}))\right\}$$

$$\left.-\left(\frac{b-a}{n}\right)^{2\mu+2}(n-2\mu)\Lambda_{2\mu+2}f^{(2\mu+2)}(\widetilde{\xi})\right]$$

$$= \frac{b-a}{n}\left[\frac{1}{2}f(x_{n,0})+\sum_{j=1}^{n-2\mu-1}f(x_{n,j})+\frac{1}{2}f(x_{n,n-2\mu})+\left(\sum_{k=1}^{2\mu}\Lambda_{k+1}\right)\right.$$

$$\left.\times\{f(x_{n,n-2\mu})-f(x_{n,0})\}+\sum_{\ell=1}^{2\mu}(-1)^\ell\left(\sum_{k=\ell}^{2\mu}\binom{k}{\ell}\Lambda_{k+1}\right)\{f(x_{n,n-2\mu+\ell})-f(x_{n,\ell})\}\right]$$

$$-\left(\frac{b-a}{n}\right)^{2\mu+3}(n-2\mu)\Lambda_{2\mu+2}f^{(2\mu+2)}(\widetilde{\xi}). \qquad (10.7.65)$$

Let $\mathscr{Q}_{2\mu}^{NC}$ denote the Newton-Cotes quadrature rule, as defined in (8.4.2), with respect to the
interval $[x_{n,n-2\mu},x_{n,n}] = [x_{n,n-2\mu},b]$, and the points $\{x_{n,n-2\mu},\ldots,x_{n,n}\}$ as in (10.7.4). But
then (8.4.6) yields

$$\mathscr{Q}_{2\mu}^{NC}[f] = \sum_{j=0}^{2\mu} w_{2\mu,j}f(x_{n,n-2\mu+j}), \qquad (10.7.66)$$

with the Newton-Cotes weights $\{w_{2\mu,j}: j=0,\ldots,2\mu\}$ given explicitly by (8.4.38) in Theorem 8.4.1. Hence we may use (8.4.38) in (10.7.66), as well as the definition of the trape-

zoidal rule $\mathscr{Q}_{2\mu}^{TR}$ in (8.6.22), (8.6.21), to deduce that, with the definition

$$\sigma_{2\mu,j} := 1 - (-1)^j \sum_{k=0}^{2\mu} \Lambda_{k+1} \left\{ \binom{k}{j} + \binom{k}{2\mu-j} \right\}, \qquad j = 0,\ldots,2\mu, \qquad (10.7.67)$$

we have

$$\mathscr{Q}_{2\mu}^{NC}[f] = \mathscr{Q}_{2\mu}^{TR}[f] + \frac{b-a}{n} \sum_{j=0}^{2\mu} \gamma_{2\mu,j}^*[f(x_{n,n-2\mu+j}) + f(x_{n,n-j})], \qquad (10.7.68)$$

where

$$\gamma_{2\mu,j}^* := \begin{cases} \sigma_{2\mu,j} - \frac{1}{2}, & j \in \{0,2\mu\}; \\ \sigma_{2\mu,j} - 1, & j = 1,\ldots,2\mu-1. \end{cases} \qquad (10.7.69)$$

By noting from (10.7.69) and (10.7.67) that the sequence $\{\gamma_{2\mu,j}^* : j = 0,\ldots,2\mu\}$ is independent of n, and noting from the first line of (8.5.60) in Theorem 8.5.5(b), together with $\pi_{2\mu} \subset \pi_{2\mu+1}$, that

$$\mathscr{Q}_{2\mu}^{NC}[P] = \int_{x_{n,n-2\mu}}^b P(x)dx, \qquad P \in \pi_{2\mu},$$

we may now deduce from (10.7.68) and the uniqueness statement in Theorem 10.7.2 that

$$\mathscr{Q}_{2\mu}^{NC}[f] = \mathscr{Q}_{2\mu,2\mu}^{GR}[f], \qquad (10.7.70)$$

with $\mathscr{Q}_{2\mu,2\mu}^{GR}$ denoting the Gregory rule of order 2μ with respect to the interval $[x_{n,n-2\mu}x_{n,n}] = [x_{n,n-2\mu},b]$. It follows from (10.7.70) and the formulation (10.7.49) in Theorem 10.7.5, with $n = 2\mu$, that

$$\mathscr{Q}_{2\mu}^{NC}[f] = \mathscr{Q}_{2\mu}^{TR}[f] + \frac{b-a}{n} \sum_{j=0}^{2\mu} \gamma_{2\mu,j}[f(x_{n,n-2\mu+j}) + f(x_{n,n-j})], \qquad (10.7.71)$$

with the sequence $\{\gamma_{2\mu,j} : j = 0,\ldots,2\mu\}$ given as in (10.7.50).

By applying (10.7.71), together with (8.5.3) and the quadrature error expression in the first line of (8.5.47) in Theorem 8.5.4, as well as the definition (8.6.22), (8.6.21) of the trapezoidal rule $\mathscr{Q}_{2\mu}^{TR}$, we deduce that, for some point $\xi^* \in [x_{n,n-2\mu},b]$,

$$\int_{x_{n,n-2\mu}}^b f(x)dx$$

$$= \mathscr{Q}_{2\mu}^{NC}[f] - \left(\frac{b-a}{n}\right)^{2\mu+3} (2\Lambda_{2\mu+3} - \Lambda_{2\mu+2}) f^{(2\mu+2)}(\xi^*)$$

$$- \frac{b-a}{n} \left[\frac{1}{2}f(x_{n,n-2\mu}) + \sum_{j=n-2\mu+1}^{n-1} f(x_{n,j}) + \frac{1}{2}f(x_{n,n}) - \left(\sum_{k=1}^{2\mu} \Lambda_{k+1} \right) \right.$$

$$\times \{f(x_{n,n-2\mu}) + f(x_{n,n})\} + \sum_{j=1}^{2\mu} (-1)^{j-1} \left(\sum_{k=j}^{2\mu} \binom{k}{j} \Lambda_{k+1} \right) \{f(x_{n,n-2\mu+j}) + f(x_{n,n-j})\} \right]$$

$$- \left(\frac{b-a}{n}\right)^{2\mu+3} (2\Lambda_{2\mu+3} - \Lambda_{2\mu+2}) f^{(2\mu+2)}(\xi^*). \qquad (10.7.72)$$

By adding (10.7.65) and (10.7.72), recalling the definition (8.6.22), (8.6.21) of the trapezoidal rule $\mathscr{Q}_n^{TR}$, and using the formulation (10.7.49), (10.7.50), we obtain (see Exercise
10.69)

$$
\int_a^b f(x)dx = \mathscr{Q}_n^{TR}[f] + \frac{b-a}{n}\left[\left(-\sum_{k=1}^{2\mu}\Lambda_{k+1}\right)\{f(x_{n,0})+f(x_{n,n})\}\right.
$$

$$
\left. + \sum_{j=1}^{2\mu}(-1)^{j-1}\left(\sum_{k=j}^{2\mu}\binom{k}{j}\Lambda_{k+1}\right)\{f(x_{n,j})+f(x_{n,n-j})\}\right] - \left(\frac{b-a}{n}\right)^{2\mu+3}
$$

$$
\times\left[(n-2\mu)\Lambda_{2\mu+2}f^{(2\mu+2)}(\widetilde{\xi}) + (2\Lambda_{2\mu+3}-\Lambda_{2\mu+2})f^{(2\mu+2)}(\xi^*)\right]
$$

$$
= \mathscr{Q}_{2\mu,n}^{GR}[f] - \left(\frac{b-a}{n}\right)^{2\mu+3}\{(n-2\mu-1)\Lambda_{2\mu+2}+2\Lambda_{2\mu+3}\}f^{(2\mu+2)}(\xi),
$$

$$\tag{10.7.73}$$

for some point $\xi \in [a,b]$, after having noted from (8.5.59) in Theorem 8.5.5(a) that
$2\Lambda_{2\mu+3} - \Lambda_{2\mu+2} > 0$ and $\Lambda_{2\mu+2} > 0$, so that $n \geqslant 2\mu + 1$ implies $(n-2\mu)\Lambda_{2\mu+2} > 0$,
and applying the intermediate value theorem as in the steps leading to (8.6.29), with
$g = f^{(2\mu+2)}, m = 1, \{\xi_0,\xi_1\} = \{\widetilde{\xi},\xi^*\}$, and $\alpha_0 = (n-2\mu)\Lambda_{2\mu+2}; \alpha_1 = 2\Lambda_{2\mu+3} - \Lambda_{2\mu+2}$.
The desired quadrature error expression (10.7.61) now follows immediately from (10.7.73)
and (10.7.54).

Finally note that, if $n = 2\mu$, the result (10.7.61) is an immediate consequence of (10.7.70),
together with the error expression in the first line of (8.5.47) in Theorem 8.5.4. ∎

The error expression (10.7.61) in Theorem 10.7.7 can now be used to find the degree of
exactness of the Gregory rule $\mathscr{Q}_{2\mu,n}^{GR}$, as follows.

First, note from (10.7.61) that, for any $n \geqslant 2\mu + 1$,

$$
\mathscr{E}_{2\mu,n}^{GR}[P] = 0, \quad P \in \pi_{2\mu+1}, \tag{10.7.74}
$$

whereas

$$
P(x) = x^{2\mu+2} \Rightarrow \mathscr{E}_{2\mu,n}^{GR}[P] = -\left(\frac{b-a}{n}\right)^{2\mu+3}(2\mu+2)![(n-2\mu-1)\Lambda_{2\mu+2}+2\Lambda_{2\mu+3}] \neq 0,
$$

$$\tag{10.7.75}$$

by virtue of the second line (8.5.59) in Theorem 8.5.5(a).

It follows from (10.7.74) and (10.7.75), together with (10.7.70) and the first line of (8.5.60)
in Theorem 8.5.5(b), for the case $n = 2\mu$, that the following result holds.

Theorem 10.7.8. *For positive integers μ and $n \geqslant 2\mu$, the Gregory rule $\mathscr{Q}_{2\mu,n}^{GR}$, as given by*
(10.7.49), (10.7.50), has degree of exactness $2\mu + 1$.

We proceed to show how the error expression (10.7.61) in Theorem 10.7.7 can be used to obtain a quadrature error bound of the form

$$\left| \int_a^b f(x)dx - \mathscr{Q}_{2\mu,n}^{GR}[f] \right| \leqslant (b-a)K_{2\mu} \left(\frac{b-a}{n} \right)^{2\mu+2} ||f^{(2\mu+2)}||_\infty,$$

$$f \in C^{2\mu+2}[a,b], \qquad (10.7.76)$$

with $K_{2\mu}$ denoting a constant that is independent of n.

We shall rely on the following result for Laplace coefficients.

Theorem 10.7.9. *The Laplace coefficient sequence* $\{\Lambda_1, \Lambda_2, \ldots\}$, *as defined in* (8.4.37), *is strictly decreasing, with, more precisely,*

$$\Lambda_{j+1} < \frac{j}{j+1}\Lambda_j < \Lambda_j, \quad j = 1, 2, \ldots. \qquad (10.7.77)$$

Proof. Let $j \in \mathbb{N}$ be fixed. By using the definitions (8.4.37) and (8.4.34), and, since the polynomial $\prod_{k=0}^{j-1}(k-t)$ does not change sign on the interval $[0,1]$, by applying the mean value theorem for integrals, as formulated in (8.5.16), we deduce the existence of a point $\xi \in (0,1)$ such that

$$\Lambda_{j+1} = -\frac{1}{(j+1)!} \int_0^1 \left[\prod_{k=0}^{j-1}(k-t) \right] (j-t)dt$$

$$= -\frac{j-\xi}{j+1} \left[\frac{1}{j!} \int_0^1 \prod_{k=0}^{j-1}(k-t)dt \right]$$

$$= -\frac{j-\xi}{j+1} \left[\frac{(-1)^j}{j!} \int_0^1 \prod_{k=0}^{j-1}(t-k)dt \right] = \frac{j-\xi}{j+1}\Lambda_j < \frac{j}{j+1}\Lambda_j < \Lambda_j,$$

since $\Lambda_j > 0$, by virtue of (8.4.54) and the second line of (8.5.59) in Theorem 8.5.5(a), and thereby completing our proof of (10.7.77). ∎

Observe from (10.7.77) in Theorem 10.7.9, together with

$$\Lambda_j > 0, \quad j = 1, 2, \ldots, \qquad (10.7.78)$$

as follows from (8.4.54) and the second line of (8.5.59) in Theorem 8.5.5(a), that, for any positive integer μ, we have

$$(2\mu+1)\Lambda_{2\mu+2} - 2\Lambda_{2\mu+3} > 2(\Lambda_{2\mu+2} - \Lambda_{2\mu+3}) > 0. \qquad (10.7.79)$$

It follows from the error expression (10.7.61) in Theorem 10.7.7, as well as (10.7.79) and (10.7.78), that, for any $f \in C^{2\mu+2}[a,b]$, we have

$$\left| \mathscr{E}_{2\mu,n}^{GR}[f] \right| \leqslant (b-a) \left[\Lambda_{2\mu+2} - \frac{(2\mu+1)\Lambda_{2\mu+2} - 2\Lambda_{2\mu+3}}{n} \right] \left(\frac{b-a}{n} \right)^{2\mu+2} ||f^{(2\mu+2)}||_\infty$$

$$< (b-a)\Lambda_{2\mu+2} \left(\frac{b-a}{n} \right)^{2\mu+2} ||f^{(2\mu+2)}||_\infty,$$

which, together with the definition (10.7.54), yields the following quadrature error bound
of the form (10.7.76).

Theorem 10.7.10. *For positive integers μ and $n \geqslant 2\mu$, the Gregory rule $\mathcal{Q}_{2\mu,n}^{GR}$, as given by (10.7.49), (10.7.50), satisfies the error bound*

$$\left| \int_a^b f(x)dx - \mathcal{Q}_{2\mu,n}^{GR}[f] \right| \leqslant (b-a)\Lambda_{2\mu+2} \left(\frac{b-a}{n} \right)^{2\mu+2} ||f^{(2\mu+2)}||_\infty,$$

$$f \in C^{2\mu+2}[a,b]. \qquad (10.7.80)$$

For the Lacroix rule $\mathcal{Q}_n^{LA}$, as defined for $n \geqslant 2$ by (10.7.51), and given for $n \geqslant 6$ by (10.7.53), we see from Theorem 10.7.8 that $\mathcal{Q}_n^{LA}$ has degree of exactness $= 3$, whereas it follows from (10.7.80) in Theorem 10.7.10, together with the value $\Lambda_4 = \frac{19}{720}$ from (8.4.54), that

$$\left| \int_a^b f(x)dx - \mathcal{Q}_n^{LA}[f] \right| \leqslant (b-a)\frac{19}{720} \left(\frac{b-a}{n} \right)^4 ||f^{(4)}||_\infty, \quad f \in C^4[a,b]. \qquad (10.7.81)$$

Comparing the optimal error bound (10.7.81) for the Lacroix rule $\mathcal{Q}_n^{LA}$ with the analogous error bound (8.6.34) for the Simpson rule $\mathcal{Q}_n^{SI}$, we see that the error constant $(= \frac{1}{180})$ for $\mathcal{Q}_n^{SI}$ is smaller than its counterpart $(= \frac{19}{720})$ for $\mathcal{Q}_n^{LA}$. Note however that $\mathcal{Q}_n^{LA}$ is defined for each $n \geqslant 2$, whereas $\mathcal{Q}_n^{SI}$ is defined only for $n = 2, 4, 6, \ldots$.

10.8 Exercises

Exercise 10.1 Verify the smoothness property (10.1.14) of the truncated power function $(\cdot)_+^m$.

Exercise 10.2 For the knot sequence $\{\tau_1, \tau_2, \tau_3\} = \{0, 1, 2\}$, and with the polynomials

$$P_0(x) := 0; \quad P_1(x) = x^2; \quad P_2(x) := -x^2 + 4x - 2; \quad P_3(x) := 2$$

in the representation (10.1.3) of the piecewise polynomial S, prove that S satisfies the continuity condition $S \in C^1(\mathbb{R})$, to deduce that S is a spline in $\sigma_2(0, 1, 2)$.

Exercise 10.3 For the spline S of Exercise 10.2, find the coefficients $\{c_0, c_1, c_2\}$ and $\{d_0, d_1, d_2\}$, the existence and uniqueness of which are guaranteed by the second statement in Theorem 10.1.1, for which it holds that

$$S(x) = \sum_{j=0}^2 c_j x^j + \sum_{j=0}^2 d_j (x-j)_+^2, \quad x \in \mathbb{R}.$$

[*Hint*: Apply the method used to establish the result (10.1.20) in the proof of Theorem 10.1.1.]

Exercise 10.4 As another continuation of Exercise 10.2, by differentiating the polynomials $\{P_0, \ldots, P_3\}$ in Exercise 10.2, prove that $S' \in \sigma_1(0, 1, 2)$, thereby verifying Theorem 10.1.2 for $m = 2, r = 3, \{\tau_1, \tau_2, \tau_3\} = \{0, 1, 2\}$ and with S given as in Exercise 10.2.

Exercise 10.5 According to the case $j = 0$ of Theorem 10.2.2(a), it holds that

$$
N_{m,0}(x) = \begin{cases} 0, & x \in (-\infty, \tau_0); \\ P_{m,k}(x), & x \in [\tau_k, \tau_{k+1}), \quad k = 0, \ldots, m; \\ 0, & x \in [\tau_{m+1}, \infty), \end{cases}
$$

where

$$
P_{m,k} \in \pi_m, \quad k = 0, \ldots, m.
$$

By applying the B-spline formulation (10.2.32), calculate the polynomials $\{P_{m,k} : k = 0, \ldots, m\}$ for each of the following cases:

(a) $m = 1$; $\{\tau_0, \tau_1, \tau_2\} = \{0, 1, 3\}$;

(b) $m = 2$; $\{\tau_0, \tau_1, \tau_2, \tau_3\} = \{0, 1, 3, 4\}$;

(c) $m = 2$; $\{\tau_0, \tau_1, \tau_2, \tau_3\} = \{0, 1, 2, 4\}$;

(d) $m = 3$; $\{\tau_0, \tau_1, \tau_2, \tau_3, \tau_4\} = \{0, 1, 2, 4, 5\}$;

(e) $m = 3$; $\{\tau_0, \tau_1, \tau_2, \tau_3, \tau_4\} = \{0, 1, 2, 3, 5\}$.

Exercise 10.6 Verify the formulas (10.2.48) and (10.2.49).

Exercise 10.7 For $m = 1, j = 0, \{\tau_0, \tau_1, \tau_2\} = \{0, 2, 3\}$, and any fixed $x \in \mathbb{R}$, calculate the right hand side of (10.2.51) by means of the recursive formulation (1.3.21) in Theorem 1.3.4 for divided differences, and verify that the right hand side of the B-spline formulation (10.2.32) is thus obtained for these choices of m, j and $\{\tau_0, \tau_1, \tau_2\}$, as guaranteed by (10.2.51) in Theorem 10.2.3.

Exercise 10.8 For the quadratic B-spline $N_{2,0}$ of Exercise 10.5(b), apply the recursive method based on Theorem 10.2.5 to compute the values $N_{2,0}(1), N_{2,0}(2)$ and $N_{2,0}(3)$.

Exercise 10.9 For the cubic B-spline $N_{3,0}$ of Exercise 10.5(d), and as an extension of Example 10.2.2, apply the recursive method based on Theorem 10.2.5 to compute the values $N_{3,0}(1), N_{3,0}(2)$ and $N_{3,0}(4)$.

Exercise 10.10 Verify the explicit cardinal B-spline formulations (10.2.79), (10.2.80) and (10.2.81) in Example 10.2.3.

Exercise 10.11 Extend Example 10.2.3 by calculating the explicit piecewise polynomial formulation of the cardinal B-spline N_4.

Exercise 10.12 By using the definition (10.1.13), prove that, for a bounded interval $[\alpha, \beta] \subset \mathbb{R}$, and any non-negative integer k, it holds that

$$
\int_\alpha^\beta x_+^k = \frac{\beta_+^{k+1} - \alpha_+^{k+1}}{k+1}.
$$

[*Hint:* Consider separately the three cases $0 \leqslant \alpha < \beta$; $\alpha < 0 \leqslant \beta$; $\alpha < \beta < 0$.]

Exercise 10.13 Prove that the cardinal B-spline N_m, as given by (10.2.69) in Theorem 10.2.7, satisfies the recursive formulation

$$\left.\begin{array}{l} N_0(x) = \begin{cases} 1, & x \in [0,1), \\ 0, & x \in \mathbb{R} \setminus [0,1); \end{cases} \\[3mm] N_m(x) = \displaystyle\int_0^1 N_{m-1}(x-t)dt = \int_{x-1}^x N_{m-1}(t)dt, \quad m = 1,2,\ldots, \end{array}\right\}$$

and for any $x \in \mathbb{R}$.

[*Hint:* Apply Exercise 10.12.]

Exercise 10.14 Use Exercise 10.13 to prove that the cardinal B-spline N_m satisfies the differentiation formula

$$N_m'(x) = N_{m-1}(x) - N_{m-1}(x-1), \quad m \geqslant 2.$$

[Hint: Apply the fundamental theorem of calculus.]

Exercise 10.15 As a continuation of Exercise 10.14, prove inductively the differentiation formula

$$N_m^{(k)}(x) = \sum_{j=0}^k (-1)^j \binom{k}{j} N_{m-k}(x-j), \quad k = 1,\ldots,m-1,$$

where $m \geqslant 2$, and for all $x \in \mathbb{R}$.

Exercise 10.16 Apply Exercise 10.13 to prove recursively that the cardinal B-spline N_m has unit integral, that is,

$$\int_{-\infty}^\infty N_m(x)dx = 1, \quad m = 0,1,\ldots.$$

Exercise 10.17 Let

$$S_m(x) := N_m\left(\frac{x}{2}\right),$$

with N_m denoting the cardinal B-spline of Theorem 10.2.7. By applying the formulation (10.2.69), show that $S_m \in \sigma_{m,0}(0,2,4,\ldots,2m+2)$.

Exercise 10.18 As a continuation of Exercise 10.17, apply Theorem 10.1.3 to show that $S_m \in \sigma_{m,0}(0,1,2,\ldots 2m+2)$, and then deduce from Theorem 10.2.2(d) and Theorem 10.2.7 that there exists a unique coefficient sequence $\{p_{m,j} : j = -m,\ldots,2m+1\} \subset \mathbb{R}$ such that

$$S_m(x) = \sum_{j=-m}^{2m+1} p_{m,j} N_m(x-j), \quad x \in [0,2m+2].$$

Exercise 10.19 As a continuation of Exercises 10.17 and 10.18, show that the spline $\widetilde{S}_m$ defined by

$$\widetilde{S}_m(x) := \begin{cases} S_m(x) - \displaystyle\sum_{j=-m}^0 p_{m,j} N_m(x-j), & x \in (-\infty,1]; \\[3mm] 0, & x \in (1,\infty), \end{cases}$$

satisfies $\widetilde{S}_m \in \sigma_{m,0}(-m,\ldots,0)$.

[*Hint:* Appy (10.2.69) and (10.2.70) in Theorem 10.2.7.]

Exercise 10.20 As a continuation of Exercise 10.19, apply Theorem 10.2.1(b) to deduce that $\widetilde{S}_m$ is the zero spline, and then use this result, together with Exercise 10.17, to obtain

$$\sum_{j=-m}^{-1} p_{m,j} N_m(x-j) = 0, \quad x \in (-m,0].$$

Exercise 10.21 As a continuation of Exercise 10.20, prove that

$$p_{m,j} = 0, \quad j = -m,\ldots,-1.$$

[*Hint:* Apply the second line of (10.2.70) in Theorem 10.2.7(a), and consider successively the intervals $(-m,-m+1], (-m+1,-m+2], \ldots, (-1,0]$, in each case also recalling (10.2.71) in Theorem 10.2.7(b).]

Exercise 10.22 Use an argument analogous to the one in Exercises 10.19 - 10.21 to prove that, in Exercise 10.18, it holds that

$$p_{m,j} = 0, \quad j = m+2,\ldots,2m+1.$$

[*Hint:* Show first, analogously to Exercise 10.19, that the spline S^* defined by

$$S^*(x) := \begin{cases} S_m(x) - \displaystyle\sum_{j=m+1}^{2m+1} p_{m,j} N_m(x-j), & x \in [m+1,\infty); \\ 0, & x \in (-\infty, m+1), \end{cases}$$

satisfies $S^* \in \sigma_{m,0}(2m+2,\ldots,3m+2)$.]

Exercise 10.23 Deduce from Exercises 10.17, 10.18, 10.21 and 10.22, and by applying also the second line of (10.2.70) in Theorem 10.2.7(a), that there exists a bi-infinite sequence $\{p_{m,j} : j \in \mathbb{Z}\} \subset \mathbb{R}$, with

$$p_{m,j} = 0, \ j \notin \{0,\ldots,m+1\}, \tag{$\bullet$}$$

such that

$$N_m(x) = \sum_{j=-\infty}^{\infty} p_{m,j} N_m(2x-j) = \sum_{j=0}^{m+1} p_{m,j} N_m(2x-j), \quad x \in \mathbb{R}, \tag{$*$}$$

according to which the cardinal B-spline N_m is a refinable function, in the sense that N_m has the self-similarity property of being expressible as a linear combination of integer shifts of its own contraction by the factor 2, as is of fundamental importance with respect to the use of N_m as basis (or scaling) function in subdivision and wavelet algorithms.

Exercise 10.24 In order to explicitly calculate the coefficients $\{p_{m,j} : j = 0,\ldots,m+1\}$ in equation $(*)$ of Exercise 10.23, first apply $(\bullet)$ and $(*)$, together with the recursive formulation in Exercise 10.13, to deduce that, for any $x \in \mathbb{R}$, it holds that

$$\sum_{j=0}^{m+2} p_{m+1,j} N_{m+1}(x-j) = N_{m+1}\left(\frac{x}{2}\right)$$

$$= \sum_{j=0}^{m+1} p_{m,j}\left[\int_0^{\frac{1}{2}} N_m(x-2t-j)dt + \int_{\frac{1}{2}}^1 N_m(x-2t-j)dt\right],$$

and then use this result, together with $(\bullet)$, to establish the identity

$$\sum_{j=0}^{m+2}\left[p_{m+1,j} - \frac{1}{2}(p_{m,j}+p_{m,j-1})\right] N_{m+1}(x-j) = 0, \quad x \in \mathbb{R}. \tag{**}$$

Exercise 10.25 As a continuation of Exercise 10.24, consider the identity $(**)$ successively on the intervals $[0,1),[1,2),\ldots,[m,m+1)$, and apply the second line of (10.2.70) in Theorem 10.2.7(b), together with (10.2.71) in Theorem 10.2.7(b), as well as $(\bullet)$, to deduce the Pascal triangle-like recursive formula

$$p_{m+1,j} = \frac{1}{2}(p_{m,j}+p_{m,j-1}), \quad j \in \mathbb{Z}.$$

Exercise 10.26 By using the explicit formulation in Exercise 10.13 of the cardinal B-spline N_0, show that the equation $(*)$ in Exercise 10.23 is satisfied by the coefficient sequence $\{p_{0,j} : j \in \mathbb{Z}\}$ given by

$$p_{0,0} = p_{0,1} = 1; \qquad p_{0,j} = 0, \; j \notin \{0,1\},$$

that is,

$$N_0(x) = N_0(2x) + N_0(2x-1), x \in \mathbb{R}.$$

Exercise 10.27 Apply the results of Exercises 10.25 and 10.26 to prove inductively that the coefficient sequence $\{p_{m,j} : j \in \mathbb{Z}\}$ in equation $(*)$ of Exercise 10.23 is given explicitly by the formula

$$p_{m,j} = \frac{1}{2^m}\binom{m+1}{j}, \quad j \in \mathbb{Z},$$

and thus

$$N_m(x) = \sum_{j=-\infty}^{\infty} \frac{1}{2^m}\binom{m+1}{j} N_m(2x-j) = \sum_{j=0}^{m+1} \frac{1}{2^m}\binom{m+1}{j} N_m(2x-j), \quad x \in \mathbb{R},$$

after having kept in mind also the second line of the binomial coefficient definition (3.2.1).

Exercise 10.28 According to Exercise 10.27, the linear cardinal B-spline N_1 satisfies the identity

$$N_1(x) = \frac{1}{2}N_1(2x) + N_1(2x-1) + \frac{1}{2}N_1(2x-2), \quad x \in \mathbb{R}.$$

Verify this identity by means of the explicit formulation (10.2.79) of N_1.

[*Hint:* Consider successively the intervals $(-\infty, 0), [0, \frac{1}{2}), [\frac{1}{2}, 1), [1, \frac{3}{2}), [\frac{3}{2}, 2), [2, \infty).]$

Exercise 10.29 Explain why the following statement is true: Together, the results of Theorem 10.2.2(a) and Theorem 10.2.6(a) are consistent with the result of Theorem 10.3.1.

Exercise 10.30 Apply the Schoenberg-Whitney theorem, as formulated in Theorem 10.3.2, to prove that, for any $f \in C[0,2]$, there exists precisely one spline $S_{3,4}^I \in \sigma_3([0,2]; 1)$, where the corresponding extended knot sequence is given by $\tau_j = j, j = -3, \ldots, 5$, such that the interpolation conditions

$$(S_{3,4}^I)\left(\frac{j}{2}\right) = f\left(\frac{j}{2}\right), \quad j = 0, \ldots, 4,$$

are satisfied.

Exercise 10.31 As a continuation of Exercise 10.30, apply a method based on matrix inversion as in Example 10.3.1 to obtain, for any $f \in C[0,2]$, and analogously to (10.3.50), an explicit formulation for $(\mathscr{S}_{3,4}^I f)(x), x \in [0,2]$, with $\mathscr{S}_{3,4}^I$ denoting the spline interpolation operator defined as in (10.3.38).

Exercise 10.32 Let the function $S : [0,2] \to \mathbb{R}$ be defined by

$$S(x) := \begin{cases} x^2 - x^3, & x \in [0,1]; \\ -2x^2 + 3x - 1, & x \in (1,2]. \end{cases}$$

Prove that $S \in \sigma_3([0,2]; 1)$, and deduce from Theorem 10.2.2(d), together with (10.2.68) in Theorem 10.2.7, that there exists a unique coefficient sequence $\{c_j : j = -3, \ldots, 1\}$ such that

$$S(x) = \sum_{j=-3}^{1} c_j N_3(x - j), \quad x \in [0,2].$$

Exercise 10.33 By applying the exactness condition (10.3.39) in Theorem 10.3.3(b), and using Exercise 10.31, calculate the coefficients $\{c_j : j = -3, \ldots, 1\}$ of Exercise 10.32.

Exercise 10.34 For $m = 2$ and $m = 3$, and any fixed $j \in \{-m, \ldots, r+m+1\}$, by using (10.2.68) in Theorem 10.2.7, as well the explicit formulations (10.2.80), (10.2.81), verify the result of Theorem 10.3.4 on the zeros of the derivatives $N_{m,j}^{(k)}, k = 1, \ldots, m-1$.

Exercise 10.35 By applying Theorems 10.2.7 and 10.4.2, prove that, for any fixed $r \in \mathbb{N}$, the cardinal B-spline N_m satisfies the identity

$$x^\ell = \frac{\ell!}{m!} \sum_{j=-m}^{r} \tilde{g}_m^{(m-\ell)}(j) N_m(x - j), \quad x \in [0, r+1], \quad \ell = 0, \ldots, m,$$

where the polynomial $\tilde{g}_m \in \pi_m$ is defined by

$$\tilde{g}_m(t) := \prod_{k=1}^{m} (t + k).$$

Exercise 10.36 Prove that the polynomial $\widetilde{g}_m$ of Exercise 10.35 satisfies the formulation

$$\widetilde{g}_m(t) = t^m + \sum_{k=0}^{m-1} \alpha_{m,k} t^k,$$

for a coefficient sequence $\{\alpha_{m,k} : k = 0, \ldots, m-1\} \subset \mathbb{R}$ such that

$$\alpha_{m,m-1} = \frac{m(m+1)}{2}; \qquad \alpha_{m,m-2} = \frac{(m+1)m(m-1)(3m+2)}{24}.$$

[*Hint:* Recall the formulas

$$\sum_{k=1}^{n} k = \frac{n(n+1)}{2}; \quad \sum_{k=1}^{n} k^2 = \frac{n(n+1)(2n+1)}{6}; \quad \sum_{k=1}^{n} k^3 = \left[\frac{n(n+1)}{2} \right]^2 .]$$

Exercise 10.37 Apply the result of Exercise 10.36 in Exercise 10.35 to deduce the identities

$$\left.\begin{array}{l} x = \sum_{j=-m}^{r} \left(j + \dfrac{m+1}{2} \right) N_m(x-j), \qquad\qquad \text{for } m \geqslant 1, \\[4mm] x^2 = \sum_{j=-m}^{r} \left[j^2 + (m+1)j + \dfrac{(m+1)(3m+2)}{12} \right] N_m(x-j), \text{ for } m \geqslant 2, \end{array}\right\} x \in [0, r+1],$$

for any fixed $r \in \mathbb{N}$.

Exercise 10.38 By using Exercise 10.37, derive the formulas

$$\sum_{j=1}^{m} j N_m(j) = \frac{m+1}{2}; \qquad \sum_{j=1}^{m} j^2 N_m(j) = \frac{(m+1)(3m+4)}{12}.$$

[*Hint:* Set $x = 0$ in Exercise 10.37, and apply (10.2.72) in Theorem 10.2.7(c).]

Exercise 10.39 For each of the cases $m = 1, m = 2$, and $m = 3$, calculate the sum

$$\sum_{j=1}^{m} j^\ell N_m(j),$$

for $\ell = 1$ if $m = 1$, and for $\ell = 1, 2$ if $m \geqslant 2$, by applying the formula (10.2.75) in Theorem 10.2.7(f), and then use these results to verify the formulas in Exercise 10.38 for $m = 1, m = 2$, and $m = 3$.

Exercise 10.40 Calculate the matrix in the right hand side of (10.4.5) for $n = 3$, by using the Lagrange fundamental polynomials $\{L_{3,j} : j = 0, \ldots, 3\}$ as obtained in Exercise 1.2. Confirm that the matrix thus obtained is precisely the inverse matrix V_3^{-1} calculated in Exercise 1.1, and thereby verifying Theorem 10.4.3 for this special case.

Exercise 10.41 For each of the cases $n = 1$ and $n = 2$, and with polynomials P and Q as in (10.4.41) of Theorem 10.4.5, calculate separately both sides of the equation (10.4.42), thereby verifying Theorem 10.4.5 for $n = 1$ and $n = 2$.

Exercise 10.42 In Example 10.4.1(b), verify the formulas (10.4.58), (10.4.59), and (10.4.63) - (10.4.67).

Exercise 10.43 Let $\mathscr{S}_{2,1}^{QI} : C[0,2] \to \sigma_2([0,2];1)$ denote the quasi-interpolation spline approximation operator of Example 10.4.1(b), with respect to the extended integer knot sequence $\{\tau_j = j : j = -2,\ldots,4\}$, as in (10.4.62), with $r = 1$. By applying the formulas in (10.4.58), obtain an expression as in (10.4.31) of Theorem 10.4.4 for $(\mathscr{S}_{2,1}^{QI}f)(x), x \in [0,2]$, for any $f \in C[0,2]$.

Exercise 10.44 According to Theorem 10.4.4, the approximation operator $\mathscr{S}_{2,1}^{QI}$ of Exercise 10.43 satisfies the polynomial exactness property (10.4.1), with $m = 2$, that is,

$$(\mathscr{S}_{2,1}^{QI}f)(x) = f(x), \quad f \in [0,2], \quad f \in \pi_2.$$

By choosing, respectively, $f(x) = x$ and $f(x) = x^2$ in this equation, and applying Exercise 10.43, obtain coefficient sequences $\{\alpha_{1,j} : j = -2,\ldots,1\}$ and $\{\alpha_{2,j} : j = -2,\ldots,1\}$ such that

$$\left.\begin{array}{l} x = \displaystyle\sum_{j=-2}^{1} \alpha_{1,j}N_2(x-j), \\ x^2 = \displaystyle\sum_{j=-2}^{1} \alpha_{2,j}N_2(x-j), \end{array}\right\} x \in [0,2],$$

and then verify that these results correspond precisely to the case $m = 2, r = 1$ of Exercise 10.37.

Exercise 10.45 In Example 10.4.1(c), verify the coefficient values (10.4.69), and the formulas (10.4.70) - (10.4.74).

Exercise 10.46 In the proof of Theorem 10.5.1, provide the details in the derivation of (10.5.18) from (10.5.17).

[*Hint:* Use the same argument as the one leading from (10.4.23) to (10.4.30).]

Exercise 10.47 Show that the linear systems (10.5.23) are uniquely solved by (10.5.25) in Theorem 10.5.2.

[*Hint:* Use the same argument as the one leading from (10.4.22) to (10.4.52).]

Exercise 10.48 In Theorem 10.5.2, verify the equivalence of the two formulations (10.5.24) and (10.5.28), (10.5.29) of $(\mathscr{S}_{m,n}^{LI}f)(x), x \in [a,b]$, for any $f \in C[a,b]$.

Exercise 10.49 In Example 10.5.1(b), verify the formulas (10.5.57), (10.5.58).

Exercise 10.50 Provide the details in the derivations of (a) the formulas (10.5.70) and (10.5.71) in the proof of Theorem 10.5.7; (b) the formula (10.5.87).

Exercise 10.51 In Example 10.5.2, verify the coefficient values (10.5.78), as well as the explicit formulation (10.5.84).

Exercise 10.52 Apply Theorem 10.5.7 with $m = 3$, and where p and q are chosen as in (10.5.54), to evaluate the corresponding coefficient sequence $\{\lambda_{3,-6},\ldots,\lambda_{3,5}\}$, and to obtain an explicit piecewise polynomial formulation for $V_3(x), x \in [-6,9]$, analogous to the one given for $V_2(x), x \in [0,4)$, in (10.5.84) of Example 10.5.2. Also, write down the analogue of (10.5.79) - (10.5.82) and (10.5.86) for V_3 and $\mathscr{S}_{3,n}^{LI}$.

Exercise 10.53 For $[a,b] = [0,n]$, and the integer knot case

$$\tau_{n,j} = j, \quad j = -m,\ldots,n+m,$$

of the Schoenberg operator $\mathscr{S}_{m,n}^{SC}$ defined in (10.6.2), (10.6.3), and with also $m \geqslant 2n+1$, apply (10.2.68) and (10.2.72) in Theorem 10.2.7, as well as the first identity in Exercise 10.37, to show that $\mathscr{S}_{m,n}^{SC}$ preserves linear polynomials in the interior of $[0,n]$, in the sense that

$$(\mathscr{S}_{m,n}^{SC}f)(x) = f(x), \qquad x \in [m,n-m], \qquad f \in \pi_1.$$

Exercise 10.54 For any integer $n \geqslant 2$, let $\mathscr{S}_{2,n}^{SC} : C[0,2] \to \sigma_2([0,2]; \tau_{n,1},\ldots,\tau_{n,n-1})$ denote the Schoenberg operator with respect to the uniformly spaced knot sequence

$$\tau_{n,j} := \frac{2j}{n}, \quad j = -2,\ldots,n+2.$$

For the function

$$f(x) = \frac{1}{x+1}, \quad x \in [0,2],$$

calculate explicit polynomial formulations of the spline $\mathscr{S}_{2,2}^{SC}f \in \sigma_2([0,2];1)$ on each of the intervals $[0,1)$ and $[0,2]$.

[*Hint:* Use (10.6.2), (10.6.3), and the explicit piecewise polynomial formulation (10.2.80) in Example 10.2.3 of the quadratic cardinal B-spline N_2.]

Exercise 10.55 As a continuation of Exercise 10.54, find the smallest value of n for which the error bound (10.6.19) in Theorem 10.6.3(b) guarantees that

$$\|f - \mathscr{S}_{2,n}^{SC}f\|_\infty < \frac{1}{10}.$$

Exercise 10.56 Prove that the spline sequence $\{W_{m,k} : k = 0,\ldots,n\}$ of Theorem 10.5.4 satisfies the condition

$$\sum_{k=0}^{n} f(x_k)W_{m,k}(x) = f(x), \quad x \in [a,b], \quad f \in \pi_m,$$

and deduce that

$$\sum_{k=0}^{n} W_{m,k}(x) = 1, \quad x \in [a,b].$$

Exercise 10.57 Deduce by means of (10.5.37) in Theorem 10.5.4, together with (10.5.50) - (10.5.52) in Theorem 10.5.6, that, for $n \geqslant 2m+1$,

$$(\mathscr{S}_{m,n}^{LI}f)(x) = \sum_{k=\max\{0,j-p\}}^{\min\{n,j+q+1\}} f(x_{n,k})W_{m,k}(x), \quad x \in [x_{n,j},x_{n,j+1}), \quad j = 0,\ldots,n-1,$$

where, as implied by (10.6.24) and (10.5.5), we have written $x_{n,j}$ for x_j.

Exercise 10.58 As a continuation of Exercise 10.57, prove that, for $n \geqslant 2m+1$, the operator $\mathscr{S}_{m,n}^{LI}$ is bounded with respect to the maximum norm on $[a,b]$, with Lebesgue constant bounded by

$$\|\mathscr{S}_{m,n}^{LI}\|_{\infty} \leqslant (m+2)\left[\sum_{i=1}^{m} R_n^i\right]^m, \qquad (*)$$

where R_n is the maximum ratio defined in (10.6.26), and satisfying the formulation

$$\|\mathscr{S}_{m,n}^{LI}\|_{\infty} = \max_{0\leqslant j\leqslant n-1} \ \max_{x_{n,j}\leqslant x\leqslant x_{n,j+1}} \sum_{k=\max\{0,j-p\}}^{\min\{n,j+q+1\}} |W_{m,k}(x)|, \qquad (**)$$

with the splines $\{W_{m,k} : k = 0,\dots,n\}$ given as in (10.5.50) - (10.5.52) of Theorem 10.5.6.

Exercise 10.59 Apply the bound $(*)$ in Exercise 10.58 to show that, for $n \geqslant 2m+1$, the Lebesgue constant of the local spline operator $\mathscr{S}_{m,n}^{LI}$ of Theorem 10.5.7, as based on uniformly distributed knots, is bounded independently of n by

$$\|\mathscr{S}_{m,n}^{LI}\|_{\infty} \leqslant (m+2)m^m.$$

Exercise 10.60 Show that, for $n \geqslant 5$, the Lebesgue constant of the quadratic local spline interpolation operator $\mathscr{S}_{2,n}^{LI}$ of Theorem 10.5.7, with $m = 2$, has the value

$$\|\mathscr{S}_{2,n}^{LI}\|_{\infty} = \frac{5}{4},$$

according to which the Lebesgue constant estimate in Exercise 10.59 is a rather crude one for $m = 2$.

[*Hint:* Apply the formula $(**)$ in Exercise 10.58, as well as the formulations (10.5.89)–(10.5.91) in Theorem 10.5.8, together with the explicit piecewise polynomial formulation (10.5.84), and the symmetry property (10.5.85), of the spline V_2, as obtained in Example 10.5.2. Also, apply the fact, as established in Exercise 10.56, that the sequence $\{W_{m,k}(x) : k = 0,\dots,n\}$ sums to one for all $x \in [a,b]$.]

Exercise 10.61 Show that, in Theorem 10.6.5, the interpolation point sequence

$$x_{n,j} = \frac{1}{2}(b-a)\cos\left(\frac{n-j}{n}\pi\right) + \frac{1}{2}(a+b), \quad j = 0,\dots,n,$$

satisfies the condition (10.6.43), with

$$\|\mathbf{x}_n\|_{\infty} \leqslant \frac{\pi}{2}\left(\frac{b-a}{n}\right), \quad n = 2m+1, \ 2m+2,\dots,$$

as well as the condition (10.6.44), with $R = 3$. Observe from (8.3.2) that

$$x_{n,j} = x_{n,j}^{CC}, \quad j = 0,\dots,n,$$

the interpolation points for the Clenshaw-Curtis quadrature rule $\mathscr{Q}_n^{CC}$, and which are concentrated more densely near the endpoints of the interval $[a,b]$.

[*Hint:* Use the mean value theorem to deduce the bound on $\|\mathbf{x}_n\|_\infty$; then, from the definition (10.6.26), by applying also the trigonometric identity

$$\cos A - \cos B = 2\sin\left(\frac{B+A}{2}\right)\sin\left(\frac{B-A}{2}\right),$$

show that

$$R_n = \max_{1\leqslant j\leqslant n-1}\left\{\frac{\sin\left(\frac{2j+1}{2n}\pi\right)}{\sin\left(\frac{2j-1}{n}\pi\right)},\ \frac{\sin\left(\frac{2j-1}{2n}\pi\right)}{\sin\left(\frac{2j+1}{2n}\pi\right)}\right\} = \frac{\sin\left(\frac{3}{2n}\pi\right)}{\sin\left(\frac{1}{2n}\pi\right)},$$

before obtaining the results

$$\frac{d}{dx}\left[\frac{\sin\left(\frac{3\pi}{2x}\right)}{\sin\left(\frac{\pi}{2x}\right)}\right] > 0, \quad x \geqslant 2; \qquad \lim_{x\to\infty}\frac{\sin\left(\frac{3\pi}{2x}\right)}{\sin\left(\frac{\pi}{2x}\right)} = 3,$$

to establish that (10.6.44) holds with $R = 3$.]

Exercise 10.62 As a continuation of Exercise 10.61, show that the error bound (10.6.71) of Theorem 10.6.7 yields

$$\|f - \mathscr{S}_{m,n}^{LI}f\|_\infty \leqslant \frac{2}{(m+1)!}\left[\frac{3}{2}(3^m - 1)\right]^m\left[\sum_{j=1}^{p+1}j^{m+1} + \sum_{j=1}^{q+1}j^{m+1}\right]$$

$$\times\left(\frac{\pi}{2}\right)^{m+1}\left(\frac{b-a}{n}\right)^{m+1}\|f^{(m+1)}\|_\infty, \quad f \in C^{m+1}[a,b].$$

Exercise 10.63 As a continuation of Exercises 10.61 and 10.62, show that the case $m = 2$, $[a,b] = [0,2]$, and where the integers p and q are chosen as in (10.5.54), yields the error bound

$$\|f - \mathscr{S}_{2,n}^{LI}f\|_\infty \leqslant \frac{864\pi^3}{n^3}\|f'''\|_\infty, \quad f \in C^3[0,2].$$

Exercise 10.64 As a continuation of Exercise 10.63, for the function

$$f(x) = \ln(x+2), \quad x \in [0,2],$$

find the smallest value of n for which the error bound in Exercise 10.63 guarantees that

$$\|f - \mathscr{S}_{2,n}^{LI}f\|_\infty < \frac{1}{10}.$$

Exercise 10.65 For a quadratic case $m = 2$ of Theorem 10.5.8, apply the explicit formulations (10.5.89) - (10.5.91), as well as (10.5.84) and (10.5.85), to calculate the constant

$$M := \max_{0\leqslant j\leqslant n}\max_{a\leqslant x\leqslant b}|W_{2,j}(x)|,$$

and where M is independent of a,b and n, thereby showing also that the estimate $M \leqslant 4$, as obtained from (10.6.27) in Theorem 10.6.4, is a rather crude one.

Exercise 10.66 The error estimates (10.6.59) and (10.6.60) in Theorem 10.6.6, as well as (10.6.71) in Theorem 10.6.7, were proved by using, in one of the steps, the uniform bound (10.6.27) of Theorem 10.6.4, as can be seen, for example, in the derivation of (10.6.70), which then yielded Theorem 10.6.7. For the quadratic interpolation operator $\mathscr{S}_{2,n}^{LI}$ based on uniformly spaced knots as in Theorem 10.5.7, and with $p = q = 1$ as in (10.5.54) with $m = 2$, argue as in the proofs of Theorems 10.6.6 and 10.6.7, but with the uniform bound (10.6.27) replaced by the precise value of M, as obtained in Exercise 10.65, to deduce the improved error bounds

$$\|f - \mathscr{S}_{2,n}^{LI}f\|_\infty \leqslant 2M\left(\frac{b-a}{n}\right)K_f, \quad f \in C^{\mathrm{Lip}}[a,b],$$

with K_f denoting a Lipschitz constant of f on $[a,b]$;

$$\|f - \mathscr{S}_{2,n}^{LI}f\|_\infty \leqslant 2M\left(\frac{b-a}{n}\right)\|f'\|_\infty, \quad f \in C^1[a,b];$$

$$\|f - \mathscr{S}_{2,n}^{LI}f\|_\infty \leqslant 6M\left(\frac{b-a}{n}\right)^3\|f'''\|_\infty, \quad f \in C^3[a,b].$$

Exercise 10.67 As a continuation of Exercise 10.66, for the function

$$f(x) = \ln(x+2), \quad x \in [0,2],$$

find the smallest value of n for which, according to the error bounds in Exercise 10.66, with $[a,b] = [0,2]$, it is guaranteed that

$$\|f - \mathscr{S}_{2,n}^{LI}f\|_\infty < \frac{1}{10}.$$

Exercise 10.68 Verify the coefficient values in Table 10.7.1.

Exercise 10.69 In the proof of Theorem 10.7.7, provide the details of the derivation of (10.7.73) from (10.7.65) and (10.7.72).

Exercise 10.70 By using (8.4.54), verify the inequalities (10.7.77) in Theorem 10.7.9 for $j = 1,\dots,9$.

Exercise 10.71 According to Theorem 10.7.8, with $[a,b] = [-1,1]$ and $\mu = 1$, together with the definition (10.7.51), the degree of exactness of the corresponding Lacroix rule $\mathscr{Q}_n^{LA}$ is equal to 3, that is,

$$\mathscr{E}_n^{LA}[f] := \int_{-1}^1 f(x)dx - \mathscr{Q}_n^{LA}[f] = 0, \quad f \in \pi_3,$$

for $n = 2,3,\dots$. Verify this fact for $n = 6$, by explicitly calculating $\mathscr{E}_6^{LA}[f]$ for, respectively,

$$\text{(a) } f(x) = \sum_{j=0}^{3} \alpha_j x^j; \qquad f(x) = x^4,$$

where, in (a), $\{\alpha_0, \dots, \alpha_3\}$ denotes an arbitrary coefficient sequence in $\mathbb{R}$.

Exercise 10.72 As a continuation of Exercise 10.71, for each of the respective integrands

$$\text{(a) } f(x) = \ln(x+2); \qquad \text{(b) } f(x) = x^4,$$

and recalling the precise value obtained in Exercise 8.2 for the integral $\int_{-1}^{1} \ln(x+2)dx$,

verify that the corresponding quadrature error $\mathscr{E}_6^{LA}[f]$ satisfies the upper bound (10.7.81).

Exercise 10.73 Find the constant K for which it holds that

$$\mathscr{E}_{4,n}^{GR}[f] := \left| \int_{-1}^{1} f(x)dx - \mathscr{Q}_{4,n}^{GR}[f] \right| \leqslant \frac{K}{n^6} \|f^{(6)}\|_\infty, \quad f \in C^6[-1,1].$$

[*Hint:* Apply the Gregory quadrature estimate (10.7.80) in Theorem 10.7.10.]

Exercise 10.74 As a continuation of Exercise 10.73, for the integrand f as in Exercise 10.72(a), verify that the corresponding quadrature error $\mathscr{E}_{4,10}^{GR}[f]$ satisfies the upper bound established in Exercise 10.73.

Exercise 10.75 As a further extension of Exercises 8.25 and 9.10, apply the Lacroix rule error estimate in (10.7.81), as well as the estimate for $\mathscr{E}_{4,n}^{GR}[f]$ derived in Exercise 10.73, to calculate the values of $\mathscr{Q}_n^{LA}[f]$ and $\mathscr{Q}_{4,n}^{GR}[f]$, with

$$f(x) = e^{-x^2}, \quad x \in [0,1],$$

where n is the smallest value for which it holds, according to these estimates, that:

$$\left| \int_{-1}^{1} f(x)dx - \mathscr{Q}_n^{LA}[f] \right| < \frac{1}{100}; \qquad \left| \int_{-1}^{1} f(x)dx - \mathscr{Q}_{4,n}^{GR}[f] \right| < \frac{1}{100}.$$

Exercise 10.76 Apply the identity in Exercise 10.27 for the cardinal B-spline N_m to prove that the integral moment sequence

$$\mu_{m,j} := \int_{-\infty}^{\infty} x^j N_m(x)dx = \int_0^{m+1} x^j N_m(x)dx, \quad j = 0, 1, \dots,$$

satisfies the identity

$$\mu_{m,j} = \frac{1}{2^{m+j+1}} \sum_{k=0}^{j} \binom{j}{k} \left[\sum_{\ell=0}^{m+1} \binom{m+1}{\ell} \ell^{j-k} \right] \mu_{m,k}, \quad j = 0, 1, \dots.$$

Exercise 10.77 As a continuation of Exercise 10.76, and by applying Exercise 10.16, show that the integral moment sequence $\{\mu_{m,j} : j = 0, 1, \dots\}$ satisfies the recursive formulation

$$\mu_{m,0} = 1;$$

$$\mu_{m,j} = \frac{1}{2^{m+1}(2^j - 1)} \sum_{k=0}^{j-1} \binom{j}{k} \left[\sum_{\ell=0}^{m+1} \binom{m+1}{\ell} \ell^{j-k} \right] \mu_{m,k}, \quad j = 1, 2, \dots.$$

[*Hint:* Observe that the binomial theorem yields

$$\sum_{\ell=0}^{m+1} \binom{m+1}{\ell} = \sum_{\ell=0}^{m+1} \binom{m+1}{\ell} 1^{m+1-\ell} 1^{\ell} = (1+1)^{m+1} = 2^{m+1}.\bigg]$$

Exercise 10.78 As a continuation of Exercise 10.77, calculate, for each of the cases $m = 1, m = 2$ and $m = 3$, the integral moments $\{\mu_{m,j} : j = 1, 2, 3\}$.

Exercise 10.79 After noting from Exercise 10.16 and (10.2.62) in Theorem 10.2.6(a) that the weight function $w = N_m$ satisfies the conditions (4.2.8) and (4.2.9), apply the three-term recursion formulation (7.4.1) in Theorem 7.4.1, together with the integral moments calculated in Exercise 10.78, to find, for each of the cases $m = 1, m = 2$ and $m = 3$, the orthogonal polynomials $\{P_{m,j}^{\perp} : j = 0, 1, 2\}$ for which it holds that

$$\int_0^{m+1} N_m(x) P_{m,j}^{\perp}(x) P_{m,k}^{\perp}(x) dx = 0, \quad j \neq k, \quad j, k = 0, 1, 2.$$

Exercise 10.80 For each of the cases $m = 1, m = 2$ and $m = 3$, apply Theorem 8.2.2, together with Exercise 10.79, to design the Gauss rule $\mathcal{Q}_n^G$ for the numerical approximation of the integral

$$\int_0^{m+1} N_m(x) f(x) dx, \quad f \in C[0, m],$$

such that the polynomial exactness condition

$$\int_0^{m+1} N_m(x) f(x) dx = \mathcal{Q}_n^G[f], \quad f \in \pi_3,$$

is satisfied. Also, give the corresponding quadrature error estimates (8.2.28) and (8.2.29) in Theorem 8.2.4(b).

Index

401

Printed in the United States
By Bookmasters

Printed in the United States
By Bookmasters